Computational Electromagnetics for RF and Microwave Engineering

This hands-on introduction to computational electromagnetics (CEM) links theoretical coverage of the three key methods – the finite difference time domain (FDTD) method, the method of moments (MoM) and the finite element method (FEM) – to open source MATLAB codes (freely available online) in 1D, 2D, and 3D, together with many practical hints and tips gleaned from the author's 25 years of experience in the field. There is also extensive coverage of leading commercial CEM software, including many application examples. Updated and extensively revised, this second edition includes a new chapter on 1D FEM analysis, and extended 3D treatments of the FDTD, MoM, and FEM, with entirely new 3D MATLAB codes. Coverage of higher-order finite elements in 1D, 2D, and 3D is also provided, with supporting code, in addition to a detailed 1D example of the FDTD method from an FEM perspective. With running examples throughout the book and end-of-chapter problems to aid understanding, this is ideal for professional engineers and senior undergraduate/graduate students who need to master CEM and avoid common pitfalls in writing code and using existing software.

David B. Davidson holds the Square Kilometer Array Research Chair in the Department of Electrical and Electronic Engineering at Stellenbosch University, South Africa, from where he received his Ph.D. in 1991. He is a Senior Member of the IEEE, edits the EM Programmers Notebook in the *IEEE Antennas and Propagation Magazine* and is the recipient of a number of research awards, including the South African National Research Foundation President's Award and the Rector's Award for Excellent Research from Stellenbosch University. His main research interest is CEM; he has published extensively on this topic and has contributed to commercial software.

Computational Electromagnetics for RF and Microwave Engineering

Second Edition

DAVID B. DAVIDSON
Department of Electrical and Electronic Engineering
University of Stellenbosch, South Africa

CAMBRIDGE
UNIVERSITY PRESS

University Printing House, Cambridge CB2 8BS, United Kingdom

One Liberty Plaza, 20th Floor, New York, NY 10006, USA

477 Williamstown Road, Port Melbourne, VIC 3207, Australia

314-321, 3rd Floor, Plot 3, Splendor Forum, Jasola District Centre, New Delhi - 110025, India

79 Anson Road, #06-04/06, Singapore 079906

Cambridge University Press is part of the University of Cambridge.

It furthers the University's mission by disseminating knowledge in the pursuit of
education, learning and research at the highest international levels of excellence.

www.cambridge.org
Information on this title: www.cambridge.org/9780521518918

First published 2005
Second edition 2011

A catalogue record for this publication is available from the British Library

Library of Congress Cataloging in Publication data
Davidson, David B.
Computational electromagnetics for RF and microwave engineering / David B. Davidson. – 2nd ed.
p. cm.
Includes bibliographical references and index.
ISBN 978-0-521-51891-8 (hardback)
1. Electromagnetic fields – Mathematical models. 2. Radio – Mathematics. 3. Microwave devices.
4. Electronic apparatus and appliances. I. Title.
TK454.4.E5D38 2010
621′.384 – dc22 2010032992

ISBN 978-0-521-51891-8 Hardback

Additional resources for this publication at www.cambridge.org/9780521518918

To Amor, Bruce and Ethan

Contents

Preface to the second edition

Six years after the first edition was prepared, it was clear that a revised edition was in order. Continued advances in computational electromagnetics, new capabilities in commercial codes, the continual increase in computational resources, challenging new problems and a new generation of research students and engineers required new material.

Since the first edition appeared, several trends can be noted in the field. Firstly, in terms of commercial companies, there has been a significant shake-out in the market. Whilst not pretending to offer encyclopedic coverage of the large number of commercial codes available, the three codes whose application is discussed in this book, viz. CST, FEKO and HFSS, have further established themselves as amongst the market leaders in their regions of application during this period. These codes have evolved continuously, and this evolution is reflected in places in this revised edition. Secondly, whilst no fundamentally new techniques have been introduced (either in the field in general or in commercial codes in particular), a large number of additional features, improvements and enhancements have continued to extend the utility of these packages. Thirdly, after more than two decades of continual increase in CPU clock speeds in personal computers (which now dominate computational engineering), the last few years have seen clock speeds not only stagnate, but in some cases actually decrease. However, Moore's law has continued to hold sway, but in terms of multi-core and multi-processor systems. Exploiting parallelism has become essential to benefit from new hardware. Related to this has been a re-ignition of interest in unconventional computational platforms, in particular the general-purpose use of graphical processing units. Fourthly, although multi-physics code capability has increased, it has not been a dominant theme of the 2000s; most radio-frequency and microwave applications appear to remain traditional single-discipline problems. Finally, public domain packages remain quite widely used.

Readers familiar with the first edition will note a number of changes, but should find that their favourite material is largely unchanged. By request and suggestion of many users of the first edition, almost no material has been entirely removed, although of course errors have been corrected, unclear arguments rephrased and out-of-date information updated. The most obvious change has been an entirely new chapter, Chapter 9, introducing the finite element method (FEM) in one dimension; this was in response to teaching experience with the first edition, where the more challenging two-dimensional introduction to the FEM was in marked contrast to the one-dimensional introductions to the finite difference time domain (FDTD) and the method of moments (MoM). This new chapter uses the same one-dimensional transmission line problem as in the one-dimensional FDTD introduction, providing a "golden thread" highlighting the similarities and differences between these methods. Another new feature is the greatly expanded coverage of basic three-dimensional implementations of all the methods; accompanying the text, there are now MATLAB codes implementing simple FDTD, MoM and FEM problems in three dimensions. Theoretical material has also been added. For the FDTD, the basics of finite differencing in general is discussed in more detail; the Courant stability condition is now derived; and the three-dimensional FDTD algorithm is now discussed in detail. For the MoM, the problem of scattering from infinitely long cylinders is now addressed and an MoM solution worked through; and the by-now classic EFIE MPIE formulation using triangular RWG elements for scattering from three-dimensional structures is worked in detail, with supporting MATLAB code. For the FEM, the coverage of higher-order vector elements, now routinely used, has been much extended, again with new MATLAB codes; a new section on waveguide dispersion analysis has been added; and the FDTD–FETD analogy is now worked in detail in one-dimension. The section on high-performance computing has been completely rewritten, and benchmarks on contemporary systems added. The use of extrapolation to improve computed results is now introduced in Chapter 1, and an application shown in Chapter 10.

A number of end-of-chapter problems are now included, and assignments in the form of computer coding tasks have been moved from an appendix into the main text. These are primarily for classroom use, although some of the problems are used to further illuminate theoretical issues and may also be of interest for self-study. Solutions to selected problems are available to instructors via the publisher's website. Not all chapters have material suitable for problems. Instructors should note that especially the more advanced coding assignments can be very time-consuming if students are coding entirely from scratch; making some existing material available to students can greatly reduce the time required. This is especially true of the assignments in Chapters 7 and 11.

No specific assignments are given in Chapters 5 and 8. Both chapters consist largely of material which lends itself to assigning as tasks, as well as simple variants on the designs presented, but this is heavily dependent on what software is available to students. If time is pressing, a good alternative is to make available an existing model and ask students to modify it for a different geometry, frequency range, etc. One should also not underestimate the learning curve associated with using a new package for the first time when assigning such tasks.

A number of MATLAB codes complement this text, and are freely available online (as is some other supporting material). An outline of the codes is given in Appendix F. It is hoped that these open-source codes will prove useful for readers needing to develop their own codes. The decision to code in MATLAB was accepted without comment in reviews of the first edition, and it is notable that a number of other texts on the topic either use MATLAB, or have converted to it in the most recent edition. It is probably the most popular prototyping platform in engineering at the time of writing.

Preface to the first edition

On graduating twenty years back, in 1984, my first job was as a research engineer working on computational electromagnetics (CEM) at the National Institute for Aeronautical Systems Technology (as it was then called) of the Council for Scientific and Industrial Research (CSIR) in Pretoria, South Africa. It was an exciting time to be working in this field. Although a number of methods had already been successfully introduced, including the three which will be discussed in detail in this book, major advances were being made in all of these methods, and the power of desktop computers was growing in leaps and bounds. No commercial programs (or codes, as they are generally called) were then available for RF problems, but some US government-sponsored codes, in particular the NEC-2 code, were becoming available for general use.

The 1980s saw the final decade of the Cold War, which in some areas (such as Southern Africa) was far from cold. New military technologies, in particular stealth, were driving CEM to address progressively more electromagnetically complex problems. However, when the Cold War ended, far from CEM work coming to a halt, new commercial markets, such as the rapidly developing market in mobile telephony and personal communication systems, and the proliferation of electronic systems in motor vehicles, continued to drive the technology forward at breakneck speed throughout the 1990s. This was also due to the widespread availability of cheap and progressively more powerful personal computers as a crucial enabling technology.

CEM has now reached a modicum of maturity, with a number of powerful methods available, able to solve problems of real engineering interest at radio frequencies, and with a number of commercial codes available. This has brought a significant change in the profile of CEM practitioners, which has not been fully appreciated in the community at the time of writing. In addition to the traditional group of CEM users – largely academics, post-graduate students and research engineers at large corporations or research establishments – an entirely new generation of users has arisen. Their interest is typically in using an existing commercial packet to solve a particular problem as rapidly as possible. They may well not have any post-graduate exposure to CEM methods, and questions which may appear elementary to CEM researchers (such as which technique is most appropriate for the problem at hand) are actually far from obvious to the beginner in the field; furthermore, marketing can "hype" a particular implementation/technique to the point where it appears omnipotent. Commercial codes aside, even academic papers are not free of such bias.

This book aims to serve the interest of both "traditional" CEM users, primarily academics, researchers and research students, and also this new non-specialist user community in industry. The book aims to fill the gap between traditional undergraduate textbooks, which generally have at most a very cursory discussion of numerical methods; antenna texts, which concentrate only on the analysis of antennas using the methods; and the specialist books on each method which are frequently formidable reading for students, or unnecessarily detailed for engineers whose primary interest is in *using* the powerful CEM codes now available. In this book, the computational methods will generally be introduced using simple one-dimensional or two-dimensional examples, so that the core of the method can be appreciated without being overwhelmed by the problems of handling complex three-dimensional geometries. Following this, the extensions required to deal with the real three-dimensional world of RF engineering are outlined, so that one gains an appreciation for the operation of complex codes. Such is the complexity of general-purpose three-dimensional CEM codes that realistic applications cannot be undertaken with anything a post-graduate student can realistically be expected to develop during a typical course, and product cycles are too short in industry to make the development of general-purpose three-dimensional codes feasible, given that off-the-shelf codes are now available.

Research students will find some features not often described in other books in this field, such as how to go about debugging and verifying a CEM code. Industrial users should find the discussions of the strengths and weaknesses of each method, as well as frequent modelling hints, comprehensive discussions of typical modelling errors, and the necessity of careful evaluation and verification of results, of great interest and utility. In short, the book discusses not only the *science* of CEM modelling, which can be gleaned from (much) reading, but also the *art* of developing and verifying reliable codes and computing reliable data, which is a skill generally derived from (sometimes bitter!) experience.

This book concentrates on the "big three" techniques in CEM – the finite difference time domain (FDTD) method, the method of moments (MoM) and the finite element method (FEM). It was decided to focus on these three methods, since they are the most widely used in the field and all have been implemented in successful commercial codes; some other methods are very briefly discussed so that readers are at least aware of them, but this book makes no pretense of addressing these other methods in any detail. Furthermore, the discussion in this book is focussed exclusively on applications in RF engineering. Methods such as the FEM have been used with great success for magnetostatic problems, such as motor design, but this will not be discussed here at all. A feature not often found in other books at this level is a discussion of stratified media, using the Sommerfeld potentials. Although a theoretically advanced topic, the widespread use of integrated antennas, especially microstrip, has made an appreciation of at least the basics of this approach very important. Finally, the book does not pretend to be a comprehensive text on electromagnetic theory, high-frequency circuit theory, or antenna theory and design. There are a number of superb books addressing these topics and this book is designed to complement, not compete, with them. Frequent references are made to suitable books.

Readers will also note that the level of the material becomes increasingly sophisticated as the book progresses. This is by design. The FDTD is the only method where one can realistically hope to develop useful code oneself in a reasonable timeframe, so the discussion of this method is rather more "nuts and bolts" than for the MoM or FEM. CEM methods can also be approached as essentially an exercise in applied mathematics; although interesting theoretical insights can be thus gained, it is the author's experience that engineers do not readily take to this approach, certainly not for their initial introduction to the methods, so the introductory discussions of at least the FDTD and MoM draw mainly on engineering physics, rather than applied mathematics. Some of the more theoretical approaches to CEM are introduced towards the end of the book, in the chapters on the MoM and FEM. (Perhaps because of the enormous amount of work on the FEM in applied mechanics, this is probably the method with the most well-developed mathematical background.) These include some elementary concepts from functional analysis, with the associated concepts of inner products and weighted residuals, as well as a brief mention of differential forms. A difficult decision was how much of the great volume of recent advances to reflect in the book. Topics such as the fast multipole method have revitalized the MoM in particular, and cannot be ignored, but the treatment of this and some other "research frontier" material is of necessity cursory.

A highly problematic issue was the selection of which commercial CEM codes to use to illustrate complex real-world implementations. One factor influencing this was the availability of a no-cost limited feature version of the software, as in the case of the MoM code FEKO; however, the FDTD and FEM codes discussed are unfortunately *not* available in such a format. The discussion tries to highlight generic features which a code should offer, and how users can exploit these. User-manual style descriptions of how to use particular codes have been avoided as far as possible, so that discussions of one particular code should extend to other commercial codes implementing the same method, at least to a degree. At the time of writing, FEKO supported a type of scripting language, which has been used in places to automate the generation of complex geometries for MoM analysis; the constructs (`FOR` loops, `IF-THEN-ELSE` conditionals) are felt to be sufficiently generic to be useful in other codes supporting similar features.

Where appropriate, references are provided for further reading. In general, only those readily available in the English language archival open literature have been listed. On one or two occasions, internal reports have been included. The engineering community is divided on the use of such references; authors in the USA in particular often reference such reports in journal papers, which often prove frustratingly difficult to locate, sometimes being limited to US distribution only. In consequence, this has only been done when there is no other published version of the material. A similar problem can be encountered with theses; here, however, some significant recent research has necessitated limited reference to recent dissertations, since these results are yet to appear in the archival literature.

The book draws primarily on the literature of Western science. Much work was done on computational electromagnetics in especially the former Soviet Union, but unfortunately little has been translated, and what has been is very difficult reading for electronic engineers trained in the Western tradition; it also tends to be at a much higher theoretical level than the main thrust of this book.

This book is an outgrowth of notes developed over a fifteen year period for a postgraduate course taught by the author at the University of Stellenbosch, South Africa, as well as a short course for industry taught by the author and several colleagues in 1999. Extensive integration of the material was undertaken during the author's sabbatical visit as a Guest Professor at the Delft University of Technology during 2003, where the course was also taught. Chapter 2 is adapted and extended from notes originally prepared by James T. Aberle at Arizona State University, Tempe, AZ, USA, and he is credited accordingly, but the rest of the authorship is that of DBD.

Acknowledgements

Stimulating careers are frequently the result of interactions with interesting people, and I would like to acknowledge a number of exceptional engineering scientists who have either mentored me, worked with me or studied under me. My late father, an electronic engineer, spent much of his career working in the microwave and telecommunications industry in the UK and South Africa and sparked my early interest in electronic engineering; he started his career during the Second World War, working on some of the first radar sets deployed in South Africa (and later North Africa and Italy). John Cloete, Wynand Louw, Derek McNamara and Jan Malherbe gave inspiring undergraduate and post-graduate courses at the University of Pretoria from 1981–1983, which originally fired my interest in this specific field. John and Derek continued as research supervisors for my M.Sc. and Ph.D. research on the MoM from 1986–1991. Dirk Baker gave me my first job at the CSIR in 1984; he is an outstanding antenna engineer and his scepticism of computed results was an invaluable baptism of fire. John Cloete offered me the opportunity to join the University of Stellenbosch in 1988 and we have continued to interact most fruitfully throughout my career.

Rick Ziolkowski taught me the power of the FDTD method during my post-doctoral stay at the University of Arizona in 1993. (Rick made significant contributions to the method and its applications, especially in complex material modelling.) Ron Ferrari and Ricky Metaxas kindly hosted me at Cambridge University during a sabbatical visit in 1997, where I had the opportunity to enrich greatly my knowledge of the FEM during frequent discussions with them and their students. Jim Aberle (Arizona State University) brought novel ideas to the teaching of the FDTD as well as spectral domain MoM methods, during a short course we taught in 1999; his ideas are reflected in places in this book. Leo Ligthart and Alex Yarovoy hosted me during my 2003 sabbatical at Delft University of Technology, during which time I initiated the actual writing of this book; their enthusiasm was very supportive.

Of my research students: in particular, the work of a number of my doctoral students is reflected in places in this volume, especially Frans Meyer – who went on to co-found Electromagnetic Software and Systems (Pty) Ltd., turning research ideas in CEM into commercially successful products – Marianne Bingle, Matthys Botha, Pierre Steyn and Riana Geschke, and I would like to acknowledge their dedication to research excellence here. Frans and Matthys' work in particular is described in some detail in the final chapter. I would also like to thank Matthys for his proofreading and detailed comments on, and suggestions for, the final two chapters (of the first edition), which were most useful. Very useful interactions with a number of engineers (some of them previously my graduate students) at Electromagnetic Software and Systems are also reflected in this book, including Ulrich Jakobus (the original author of FEKO), Johann van Tonder, Isak Theron, Gronum Smith, Danie le Roux and Sam Clarke. Many years of continuing discussions on electromagnetics with my colleagues at the University of Stellenbosch, in particular John Cloete, Petrie Meyer, Howard Reader and Keith Palmer have also influenced the development of this book, as have those with colleagues in electronic engineering in general, in particular Dave Weber and the late David Frost.

I would also like to thank the (South African) National Research Foundation and its predecessor, the Foundation for Research Development, for many years of research funding, in particular grant-holder bursaries, equipment funding and sabbatical support.

Electromagnetic Software and Systems and Computer Simulation Technology kindly provided evaluation copies of FEKO and CST MICROWAVE STUDIO™ respectively. The former also provided the image on which the cover art was based. My thanks to Vanessa Weber for the graphic design she produced from this for the cover.

The love and forebearance of my wife Amor, who was bearing our first child Bruce during much of the period when this book was in preparation, was essential.

Finally, the support of the Cambridge University Press team is much appreciated.

Acknowledgements for the second edition

The second edition was largely prepared whilst I was on a research sabbatical from Stellenbosch University. I would like to gratefully acknowledge the Oppenheimer Memorial Trust for generous funding for travel during this sabbatical, in particular a visit to the University of Manchester. The work of a new generation of my post-graduate students is reflected in this book, in particular Evan Lezar, Danie Ludick, Renier Marchand, Drs Neilen Marais and Julian Swartz, and Andre Young, as well as a productive collaboration with Dr Matthys Botha, who returned to Stellenbosch University for several years (after a post-doctoral appointment at the University of Illinois at Urbana-Champaign) before moving on to industry. As with the first edition, CST and EMSS are again thanked for evaluation copies of MWS and FEKO respectively, as is the National Research Foundation for continued research funding, in particular via the Focus Areas Program. I would also like to acknowledge the Centre for High Performance Computing (CHPC) in Cape Town and its staff, in particular Kevin Coville and Dr Jeff Chen; access to these facilities via a Flagship Project enabled the contemporary performance benchmarking study reported in Chapter 6. The continuing support of Dr Phil Meyler, my editor at the Press, is much appreciated, as is that of the editorial team as a whole, in particular Miss Sarah Matthews, editor (engineering).

Finally, I would like to most gratefully acknowledge the love and support of my wife Amor and sons Bruce and Ethan, to whom this revised edition is dedicated.

To the reader

This book is designed to serve as an introduction to computational electromagnetics for radio-frequency applications. It assumes the reader has completed typical undergraduate courses in electromagnetic field theory, and has some basic knowledge of antenna design and microwave systems.

For readers in a hurry, who already know which of the techniques discussed they would like to learn more about, it is possible to go directly to the relevant chapters, but it would nonetheless be useful first to read the introductory chapter. For those in a hurry, but who need first to find out which method (or methods) to use, this chapter is essential reading.

For readers who intend working through most of the book, it would be best to work through it in the sequence presented, although the chapters on the Sommerfeld formulation and practical applications thereof could be omitted without interrupting the sequence of presentation. A more detailed outline of the book may be found in Section 1.12; this will also assist readers to locate rapidly the parts of the book of interest to them.

At the end of each chapter, a list of references linked to the chapter topic is presented, for further reading and study.

Notation

Throughout this book, the following notation is used. Spatial vectors are indicated as $\boldsymbol{E}$ (in this case, the electric field). Vectors in the linear algebra sense are indicated as $\{x\}$, and matrices as $[A]$. The individual elements of a vector or matrix are of course indicated as x_i or A_{ij} respectively. Otherwise, the notation is as generally encountered in engineering books on this topic. A summary is presented below.

The time convention used for phasor quantities is $e^{j\omega t}$, hence, an e^{-jkr} plane wave propagates in the direction of increasing r. (Note that physics books often adopt the $e^{-i\omega t}$ convention, in which case the sign also changes in the plane wave exponential factor.)

$\nabla\times$	the curl operation
$\nabla\cdot$	the divergence operation
$\times$	the vector cross-product of two vectors
$\boldsymbol{E}$	the (field) vector E
ϵ_0	the permittivity of free space ($\approx 8.854 \times 10^{-12}$ F/m)
ϵ_r	relative permittivity of a dielectric material (dimensionless)
μ_0	the permeability of free space ($4\pi \times 10^{-7}$ H/m)
μ_r	relative permeability of a magnetic material (dimensionless)
c	the speed of light in free space ($\approx 2.9979 \times 10^8$ m/s)
λ	wavelength [m/s] *or* real part of spectral variable k_ρ (the meaning will be clear from the context)
λ_i	simplex coordinate i
$\mathcal{O}(M^n)$	of the order of M^n, formally, $\mathcal{N} = \mathcal{O}(M^n) \Rightarrow \lim_{M\to\infty} \log \mathcal{N} / \log M = n$
$[A]$	the matrix A
a_{ij}	the ijth element of matrix A
$\{x\}$	the (algebraic) vector x
x_i	the ith element of vector $\{x\}$
$\|\|\{x\}\|\|$	the Euclidean norm of the vector $\{x\}$ of length n, $\|\|\{x\}\|\| \equiv \sqrt{\sum_{i=1}^{n} \|x_i\|^2}$
$\equiv$	is defined as
$\forall$	for all
$\|z\|$	absolute value of z
$\Rightarrow$	implies

1 An overview of computational electromagnetics for RF and microwave applications

Even if we do discover a complete unified theory, it would not mean that we would be able to predict events in general... even if we do find a complete set of basic laws, there will still be in the years ahead the intellectually challenging task of developing better approximation methods, so that we can make *useful predictions of the probable outcomes in complicated and realistic situations*.

From [1, pp. 168–169] (the present author's emphasis)

Computations: no-one believes them, except the person who made them.
Measurements: everyone believes them, except the person who made them...

Attributed to the late Professor B. Munk, Ohio State University

1.1 Introduction

Electromagnetics, the study of electrical and magnetic fields and their interaction, has been one of the core technologies of the twentieth century, and shows every sign of continuing this into the twenty-first. Whilst there are many useful ways of subdividing the field, power frequency versus radio frequency, or alternatively quasi-static versus full-wave, is one of the most insightful here. This book focusses exclusively on radio-frequency, full-wave electromagnetic modelling, as typically encountered in communication systems.

The core of modern electromagnetic engineering is of course Maxwell's equations. Written in modern form,[1] they are:

$$\nabla \times \boldsymbol{E} = -\frac{\partial}{\partial t}\boldsymbol{B} \tag{1.1}$$

$$\nabla \times \boldsymbol{H} = \boldsymbol{J} + \frac{\partial}{\partial t}\boldsymbol{D} \tag{1.2}$$

$$\nabla \cdot \boldsymbol{D} = \rho \tag{1.3}$$

$$\nabla \cdot \boldsymbol{B} = 0 \tag{1.4}$$

with the associated constitutive equations

$$\boldsymbol{B} = \mu \boldsymbol{H} \tag{1.5}$$

$$\boldsymbol{D} = \varepsilon \boldsymbol{E} \tag{1.6}$$

[1] Maxwell did not actually write his equations in this form; vector analysis was a late nineteenth-century development.

The actual *solution* of the Maxwell equations is complex, and for realistic problems, approximations are usually required – as indicated by the introductory quote from Hawking, although he had in mind an altogether more ambitious theory (of everything!). The numerical approximation of Maxwell's equations, the subject of this book, is known as *computational electromagnetics* (CEM).

CEM techniques have been available for close on five decades now. These techniques have gestated, grown and matured to the point where they form an invaluable part of current RF and microwave engineering practice [2]. However, the widespread adoption of computational methods to complement the traditional tools of analysis and measurement has attracted criticism, summarized with more than a grain of truth by the second quote at the beginning of the chapter. Ironically, the availability of powerful, commercial codes may well have made the situation worse, not better, since more and more frequently, codes are being applied by users unfamiliar with the basic formulations underlying the codes, and not infrequently to problems for which the codes were not designed. One of the major aims of this book is to make RF computational electromagnetics comprehensible and accessible to a far wider group of RF engineers than has been the case in the past.

CEM is a multi-disciplinary field. Its core disciplines are electromagnetic theory and numerical methods, but for useful implementations, geometric modelling and visualization, computer science and algorithms all have important roles to play. In this book, the focus falls on the core disciplines.

The applications of CEM are legion, and include antennas, biological EM effects, medical diagnosis and treatment, electronic packing and high-speed circuitry, superconductivity, microwave devices, monolithic microwave integrated circuits, law enforcement, environmental issues, materials, avionics, communications, energy generation and conservation, low observable vehicles (stealth), radars and imaging, surveillance and intelligence gathering. In this book, we focus primarily on applications in antennas, wireless communications, radar, and (passive) microwave devices, although an example will be given of a biological EM effect study.

An historical aside – a brief history of electromagnetics

Interest in static electricity and magnetism, of course, dates back to ancient times. The Ancient Greeks circa 400 BC noted that rubbing amber attracted bits of straw, and the Chinese reportedly found lodestones (natural magnets) circa 2600 BC, first using them for burial purposes, and later for navigation. The modern study of electromagnetic phenomena dates to the late eighteenth century, with the great progress in experimental methods by Alessandro Volta (1745–1827), Hans Christian Oersted (1777–1851) and Michael Faraday (1791–1867) on the one hand, and the more mathematical modelling approach of Charles Augustin de Coloumb (1736–1806) and André-Marie Ampère (1775–1836) on the other. Amongst these, the following milestones stand out: the development of the battery by Volta provided a continuous source of electricity for the first time; Coloumb's careful measurements of the electric force resulted in the famous inverse square law; Oersted's 1820 discovery showed that (direct) current deflected a magnet; Ampère developed mathematical laws describing this and the force between current carrying wires; and finally, Faraday's crucial contribution in 1831 showed that a changing magnetic field sets up an alternating current (i.e. an electric field), and for the first time connected two forces of nature which until then had been thought quite independent.

James Clark Maxwell (1831–1879), the most brilliant physicist of the nineteenth century,[a] combined the work of his predecessors in elegant theoretical fashion and postulated that changing electric fields should generate magnetic fields; he then showed that this implied *wave motion*. Hermann Ludwig-Ferdinand Helmholtz (1821–1894) was one of the first to recognize the significance of Maxwell's predictions in this regard; in 1888, his student Heinrich Rudolph Hertz (1857–1894) showed experimentally that electromagnetic fields indeed propagate, and at the speed of light. Oliver Heaviside (1850–1925) also made contributions in this regard, although his work is not widely recognized nowadays [3]. In what we would now describe as the first commercial spin-off of this work, Guglielmo Marconi (1874–1937) was the first to profit financially from the emerging field of wireless.

Electromagnetics was also to have a profound influence on the outstanding physicist of the twentieth century, Albert Einstein (1879–1955). Perhaps less well known than some of his results – certainly amongst the general public – Einstein showed that the magnetic field is the relativistic correction of the electric field, confirming the unified field theoretic nature of Maxwell's electromagnetic theory.

The above is the conventional Western history of electromagnetics. Contributions to the theory of light, intimately connected to electromagnetics, were made by many over an extremely long period of time, including contributions from Arabic scholars. An exceptionally erudite historical perspective may be found in [4].

[a] Maxwell not only unified electricity and magnetism in 1864, he also developed the kinetic theory of gases, before his life was cut tragically short by illness.

1.2 Full-wave CEM techniques

Full-wave CEM methods approximate the Maxwell equations numerically, without any initial physical approximations being made. These are also sometimes called *low-frequency* methods, to distinguish them from asymptotic *high-frequency* methods, but this can be confusing for several reasons.[2] The full-wave techniques which will be studied in this book are the finite difference time domain (FDTD) method; the method of moments (MoM); and the finite element method (FEM). Whilst there are other methods available, these are the most widely used, and all have been implemented in powerful

[2] Firstly, the high-frequency radio band is specifically the spectrum from 3–30 MHz; secondly, the meaning of low and high are entirely relative, and the same methods may be, and are, useful from power frequencies up to the visible spectrum and beyond; and finally, "high-frequency" as a general term in electronic engineering is widely used to distinguish from "power frequency," with the latter usually using quasi-static approaches.

computer codes. These techniques are frequently classified further by whether they are based on integral or differential equations, and by whether they operate in the time or frequency domain. We will discuss this in the context of each method subsequently.

Sometimes, the expressions "static" or "quasi-static" will be used. The former applies obviously to the situation where one is dealing with either steady-state charges (and the associated electric fields) or currents (and the associated magnetic fields). The latter applies to situations where the time rate of change is low enough that the fields still satisfy the static equations to a very good approximation – or put differently, the $(\partial/\partial t)\boldsymbol{B}$ term in Eq. (1.1) is negligible (in which case one obtains electroquasistatics) or similarly for the $(\partial/\partial t)\boldsymbol{D}$ term in Eq. (1.2) (which yields magnetoquasistatics). A very detailed discussion of these approximations and their use may be found in [5]. However, we will not pursue this far in this book, which deals almost entirely with full-wave methods.

There is another class of numerical method for solving the Maxwell equations, generally called the *asymptotic* techniques. These methods require fundamental approximations in the Maxwell equations, the validity of which increases asymptotically with frequency. Examples are physical optics (PO), geometrical optics (GO) and the uniform theory of diffraction (UTD). This is a field of study in its own right. For suitable problems, these methods are very powerful, but the underlying approximations of the physics limits their use for general problems. Furthermore, unlike the full-wave methods, where Moore's law and the resulting increase in computer speed and memory continually extend the limits of applicability, the asymptotic methods have fundamental limits. Hence, in this book, only full-wave methods are considered. However, a hybridization with an asymptotic technique will be discussed as an example of an advanced application.

The full-wave techniques are potentially very accurate. Central to *all* these methods is the idea of *discretizing* some unknown electromagnetic property, typically the surface current for the MoM, and the $\boldsymbol{E}$ field for the FEM and FDTD method. (For the latter, the $\boldsymbol{H}$ field is also discretized.) This process of discretization is also known as *meshing*. It entails subdividing the geometry into a (large) number of small *elements*. These may be one-dimensional segments, two-dimensional surface "patches" (often triangles), three-dimensional tetrahedral elements or a regular three-dimensional "staggered" grid, depending on the problem at hand and the method used. Within each element, a simple functional dependence is assumed for the spatial variation of the unknown – for instance, a linear approximation – but the *amplitude* (and possibly phase) of the unknown is determined by application of the method to the patchwork of elements which approximates the original geometry. This functional dependence is also known as the basis (or expansion) function.[3]

Generally, the accuracy of the methods is related to the discretization (i.e. mesh size). The finer is the mesh, the better is the accuracy of the methods.[4] The largest mesh size (alternatively, the finest geometrical resolution) is limited by the available

[3] With the FDTD method as usually introduced, the fields are sampled at points; it is however possible to define basis functions for the FDTD, a topic we discuss briefly in Chapter 12.

[4] This is not invariably true: limitations imposed by approximations in the formulations may place some lower bound on element size. A classic example is a thin-wire MoM formulation, where using too many segments may violate the underlying thin-wire assumptions. This is discussed in detail in Section 4.3.

Table 1.1 Strengths and weaknesses of CEM methods as widely implemented for open region problems

Formulation	Equation type	Domain	Radiation condition	PEC only	Homogeneous penetrable	Inhomogeneous penetrable
MoM	Integral	Frequency	Yes	⌣	⌣	⌢
FEM	Differential	Frequency	No	⌢	⌣	⌣
FDTD	Differential	Time	No	⌢	⌣	⌣

Key: ⌣ good; ⌢ not optimal.

computational resources. In other fields such as structural mechanics, the mesh fineness is usually determined by the requirement to resolve the *structural geometry* adequately; in radio-frequency electromagnetics, the requirement on the mesh is usually to sample the *phase* adequately. For many years, the CEM community has worked with a rule of thumb of ten segments per wavelength. This was originally derived for wire antenna problems, where the mesh is one-dimensional; for surfaces, this guideline becomes 100 segments per square wavelength (and a similar extension for volumetric meshes to 1000 per cubic wavelength). Much work on better elements has been done to reduce this requirement – it will readily be appreciated that as the dimensionality of the problem goes up, so this becomes progressively more crucial. It should also be noted that when very accurate field data are required – for example, when computing antenna input impedance – a *finer* mesh may be required, at least locally around the feed point of the antenna. Furthermore, this guideline ignores the problem of dispersion in differential equation based solvers, which effectively requires *denser* meshes for electromagnetically larger problems.

Although the full-wave methods share the basic idea of discretization, and indeed have been viewed within a very general framework as simply different implementations of one overarching theoretical formulation, in practice, the methods have quite different challenges for theoreticians, code developers and users, as well as different optimal areas of application, and as such, they will be considered separately in this overview chapter. In Chapter 12, some of the underlying mathematical connections between the methods will emerge.

In the rest of this overview chapter, the MoM, FEM and FDTD method will be reviewed *qualitatively*, emphasizing basic principles such as the underlying formulation (integral/differential equation based, frequency or time domain) and areas of application (perfectly or highly conducting materials versus homogeneous or inhomogeneous penetrable structures; microwave devices versus radiation or scattering analysis). This review is especially designed for readers who have a particular problem to solve, but are not sure which is the best method to use. Details of each method will be found in the subsequent chapters of the book. Key references only are given; a far more extensive list of references will be found at the end of each chapter.

By way of introduction, some of the most important characteristics of the MoM, FEM and FDTD method are presented in Tables 1.1 and 1.2. Table 1.1 provides a comparison of the methods for open region (radiation and scattering) problems. It is important to

Table 1.2 Strengths and weaknesses of CEM methods for guided wave problems

Formulation	Equation type	Domain	Wideband	PEC only	Homogeneous penetrable	Inhomogeneous penetrable
MoM	Integral	Frequency	$\sim$	$\smile$	$\smile$	$\frown$
FEM	Differential	Frequency	$\sim$	$\smile$	$\smile$	$\smile$
FDTD	Differential	Time	$\smile$	$\smile$	$\smile$	$\smile$

Key: $\smile$ good; $\sim$ satisfactory, but not necessarily the best; $\frown$ not optimal.

note that what is presented in this table are the key characteristics of the method *as widely implemented and understood in the CEM community*. As will be seen in the description of each method in the following sections, a number of simplifications have been made in this table: the MoM, for instance, can be seen in a more general sense as including the FEM, although this is not normal usage; and to give another example, the FEM can also operate in the time domain, but there are no commercial implementations of this at present. For the MoM, homogeneous penetrable materials (dielectrics, for instance) can either be modelled using equivalent surface currents or, if the problem consists of layered materials, using a Sommerfeld formulation. This has not been noted in the table, since it depends on the details of the problem. Table 1.2 provides a similar comparison of the methods for guided wave problems.[5] Again, the details of the precise implementation have not been commented on.

This can be further illustrated by studying one of the most significant differences between the methods (as usually deployed), namely meshing requirements. In Figs. 1.1, 1.2 and 1.3, the meshes required by typical MoM, FDTD and FEM codes to handle a problem involving scattering from a homogeneous sphere are compared. MoM codes often use triangular meshes to approximate surfaces, an approach which provides accurate modelling of general geometries. The reduction in dimension afforded by the MoM (here, from a 3D volume to a 2D surface) is clear – the price of this is that every element on the surface now interacts with every other. FDTD codes use a regular "brick," or cuboidal, mesh. This approach is core to the speed of the method, but clearly requires a fine mesh (as here) to model curved geometries with reasonable fidelity. The interaction between elements in an FDTD mesh is local, and the volume need not be materially homogeneous. FEM codes generally use unstructured tetrahedral meshes; similar to triangular surface meshes, tetrahedral meshes offer accurate modelling of fine geometrical details. Again, similiar to the FDTD, the interaction between elements is local, and the volume can be inhomogeneous. (Unlike the FDTD, the irregular, or unstructured, mesh means that geometrically local elements may not be local in the matrix describing the problem[6].) The FEM model as shown in Fig. 1.3 also includes an outer spherical shell,

[5] It is tempting to use the term "closed problems" here, but a number of important guiding structures, such as microstrip, are partially open. It is assumed in this table that FEM and FDTD codes have an appropriate method of terminating this region. Since the energy decays rapidly away from the guiding structure, and this radiation is a secondary effect in most applications, the open boundary is usually less problematic here than in the case of the radiation and scattering problems.

[6] The FDTD is actually a matrix-free method.

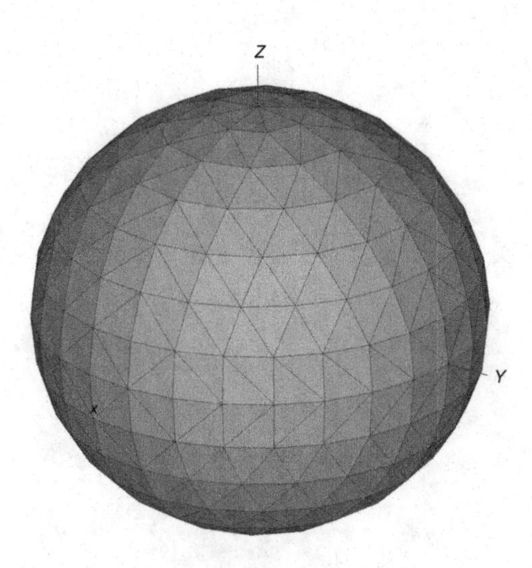

Figure 1.1 A triangular surface mesh of a sphere, as used by typical MoM code.

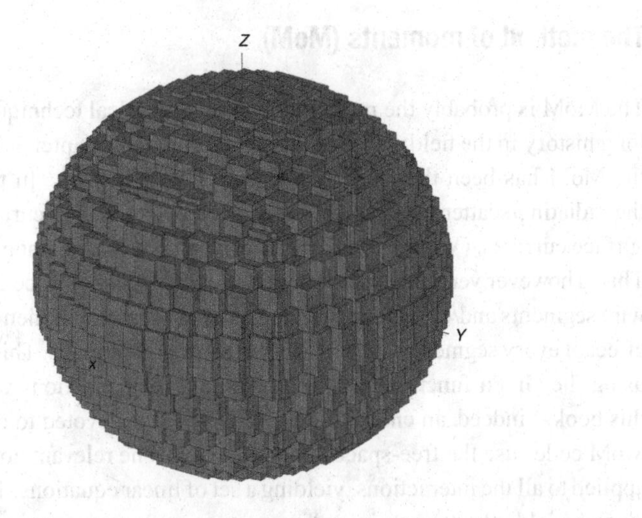

Figure 1.2 A cuboidal volumetric mesh of a sphere, as used by typical FDTD code.

which would usually be free space, providing a "buffer" between the scatterer and whatever mesh closure scheme is applied on the outer boundary to approximate an infinitely large mesh (also known as the "radiation condition"); the FDTD also needs this, but it is not shown in Fig. 1.2. The MoM formulation, which incorporates the radiation condition at formulation level, does not need this. However, and importantly, the MoM surface formulation can only be applied to *homogeneous* scatterers; when dealing with an inhomogeneous scatterer, the MoM also requires a volumetric mesh such as that in Figs. 1.2 or 1.3.

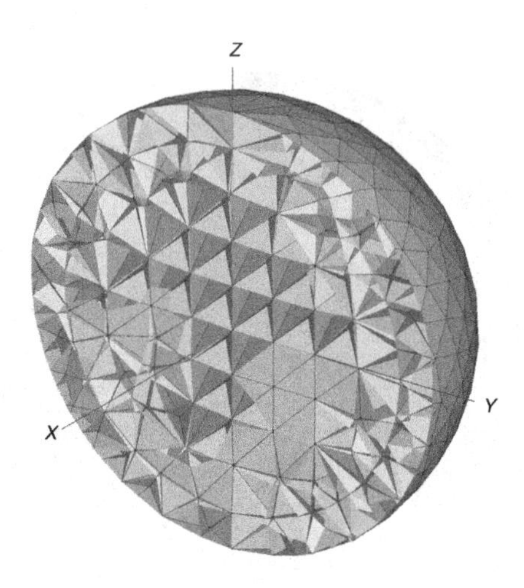

Figure 1.3 A tetrahedral volumetric mesh of a sphere and spherical shell, as used by typical FEM code.

1.3 The method of moments (MoM)

The MoM is probably the most widely used numerical technique in RF CEM, and has a long history in the field; some of this is presented in Chapter 4. For antenna engineering, the MoM has been the most widely used CEM method.[7] In the method of moments, the radiating/scattering structure is replaced by equivalent currents. These are normally surface currents. (Volumetric currents can be used for inhomogeneous dielectric bodies. This is however very expensive computationally.) This surface current is discretized into wire segments and/or surface patches. A matrix equation is then derived, representing the effect of every segment/patch on every other segment/patch. This interaction is computed using the Green function for the problem. (Green functions will be discussed later in this book – indeed, an entire chapter, Chapter 7, is devoted to one such function.) Most MoM codes use the free-space Green function. The relevant boundary condition is then applied to all the interactions, yielding a set of linear equations. The solution of this linear system yields the (approximate) current on each segment/patch. The resulting matrix which must be factored (or used in an iterative solution scheme) is fully populated, with complex valued entries. Typical matrix dimensions range from some hundreds for small antenna problems to several thousand – the upper limit is imposed by computational limitations, either limited memory and/or excessive runtime.

Traditionally, the MoM has been applied in the frequency domain, i.e. single frequency, or monochromatic, sinusoidal excitation, with an $e^{j\omega t}$ convention assumed. The working

[7] The name "method of moments" is peculiar to the CEM community. Perhaps the most descriptive alternative name is the "method of weighted residuals." The term "boundary element method" is frequently used synonymously with MoM, and for surface formulations this is correct, but there are some moment method formulations which use volume, not boundary, elements. We discuss this further in Chapter 4.

variables (unknowns) are thus complex valued, with a magnitude and phase, as for any phasor analysis. Time domain integral equation (TDIE) formulations have been used on occasions, but stability and other issues have proven difficult, and TDIE codes are rare.

The use of the MoM for antenna analysis was given a major boost by the US government's de facto decision during the late 1980s to release the numerical electromagnetic code – method of moments (widely known as NEC-2) into the public domain. NEC-2 is a powerful, general-purpose antenna modelling program, but with no graphical abilities whatsoever and very limited meshing abilities. NEC-2 is discussed in Chapter 5. A later version, NEC-4, added some specialized functionality. At present, there are some excellent commercial codes which offer all the functionality of NEC-2, but with proper graphical user tools and frequently greatly enhanced abilities; examples are FEKO (which will be used quite extensively in this book), SuperNEC, Ensemble and IE3D. (Only SuperNEC is a direct descendant of NEC, the others are independent implementations.) There are also some semi-commercial packages such as GEMACS which are limited to US Department of Defense contractors, and hence not generally available for commercial use world-wide.

The strong points of the MoM (as usually applied) are the following:

- Efficient treatment of perfectly or highly conducting surfaces. Only the surface is meshed; no "air region" around the antenna need be meshed. For wire antennas, the treatment is even more efficient, since only a one-dimensional discretization of the wire is undertaken.
- The MoM automatically incorporates the "radiation condition" – i.e. the correct behavior of the field far from the source (proportional to $1/r$ in free space). This is very important when dealing with radiation or scattering problems.
- The working variable is the current density, from which many important antenna parameters (impedance, gain, radiation patterns, etc.) may be derived, some directly and some via straightforward numerical integration.
- Via the Sommerfeld potentials, efficient formulations may be derived for stratified (layered) media. Important examples are printed antennas, components and feed networks (e.g. microstrip technology) and antenna-above-real-earth calculations.
- The availability of NEC-2 in the public domain – this powerful code has served as the basis for much MoM-based antenna design, and due to the open source nature, has lent itself to all manner of numerical experimentation and improvement.

The weak points of the MoM may be summarized as follows:

- The MoM does not handle electromagnetically penetrable materials as well as differential equation formulations. This is especially true if the material is inhomogeneous. (If the materials are homogeneous, a reasonably efficient fictitious, equivalent surface current formulation may be used, but inhomogeneous materials require fictitious equivalent volumetric currents, and become very expensive computationally.)
- The MoM does not scale gracefully with frequency – for typical applications requiring a surface mesh, the scaling is $\mathcal{O}((kd)^6)$ where kd is the electromagnetic size of the

structure.[8] (This assumes a cubic structure, for simplicity.) Note that this is implies an $\mathcal{O}(f^6)$ scaling – doubling the frequency can result in a runtime 64 times as long! We will see that this is a major problem with *all* the full-wave computational methods, although the details do vary slightly from method to method. For an MoM volumetric mesh, required by an inhomogeneous structure, the scaling is $\mathcal{O}((kd)^9)$; this is so large that such methods are usually very limited in application.

- Some MoM formulations, in particular those based on the magnetic field integral equation (MFIE), require the surface to be *closed*. This is frequently impractical.

In conclusion, the MoM is the preferred method for frequency domain radiation and scattering problems involving perfectly or highly conducting wires and/or surfaces. If the problem involves inhomogeneous dielectric materials, it is unlikely to be the best formulation, but if hybridized with the FEM a very efficient formulation can result.

1.4 The finite difference time domain (FDTD) method

The finite difference time domain (FDTD) method is of a similar vintage to the MoM and FEM in electromagnetics, dating back to the 1960s. Like the FEM, it is partial differential equation based, and one does *not* need a Green function. *Unlike* the FEM, the FDTD method does not use variational functionals or weighted residuals – it directly approximates the differential operators in the Maxwell curl equations, on a grid staggered in time and space. $\boldsymbol{E}$ and $\boldsymbol{H}$ fields are computed on a regular grid, with a marching-on-in-time discretization of time, with field components being offset by $\Delta s/2$ relative to each other and the $\boldsymbol{E}$ and $\boldsymbol{H}$ fields evaluated $\Delta t/2$ apart in time, where Δs and Δt are the spatial and temporal discretizations respectively. This permits a scheme which uses first-order numerical differentiation to provide second-order accuracy. It is also the only widely used CEM scheme to operate in the time domain. (Time domain MoM and FEM formulations have been used, but usually for a rather specialized application. Frequency domain finite difference formulations are also available, but again have never become very popular for general problems.)

Some history of the FDTD method may be found in Chapter 2. For various reasons, the method languished in relative obscurity throughout most of the 1960s and 1970s, but sprang to prominence in the 1980s. There were both technological driving factors behind this – on the one hand, increasing interest in the modelling of inhomogeneous materials, in particular for the assessment of human exposure to RF fields, and on the other, the development of low-observable "stealth" technology – and enabling technology in the shape of the enormous growth in computer power – in particular, memory, for which the FDTD method has a voracious appetite in three dimensions. The development by Berenger of the perfectly matched layer in 1994 solved the previously problematic issue

[8] The notation $\mathcal{O}(x)^p$ means *of the (asymptotic) order of* and indicates to the highest power (p) present in the variable (x); note that it says nothing of the constants. This can be important, since CEM analysis is quite often undertaken in the "pre-asymptotic" region, where lower powers in x may dominate especially runtime.

of mesh termination, and removed the last hurdle to the widespread adoption of the method. In the new millennium, with desktop PCs with hundreds of megabytes available at relatively low cost, the FDTD method has firmly established itself as one of the most popular methods in CEM, both in industry and academia. The apparent simplicity of the basic implementation also means that it is very popular with graduate students in the university research community, where "do-it-yourself" FDTD codes are commonly encountered.

Critics have dismissed the method as a "brute force" technique and, certainly, compared to the mathematical elegance and subtleties of a Sommerfeld integral formulation, the basic method appears to make limited demands on higher mathematics. Most engineers trying to solve tough problems are of course more impressed by how well a code works, rather than by how elegant the formulation is, and the FDTD method has been enormously successful in many diverse applications. Nonetheless, extensions of the FDTD method have required subtle thinking, as have stability proofs, so the "brute force" epithet is undeserved.

The FDTD method is an "explicit" finite difference approach, i.e. no matrix equation is set up and solved.[9] The term comes from the update equations, where the field values at the next timestep are given entirely in terms of the field at this and the previous timesteps. (Implicit finite difference approaches, where the field values at the next timestep at a point in space also involve the values at adjacent points at the next timestep, generate a sparse matrix, which must be solved at each timestep.) This has the great advantage of keeping the required operations very simple – essentially just a stencil involving differencing neighbors in time and space – but does mean that the method is not unconditionally stable. This means that there is an upper limit on the timestep, and it turns out to be rather less than the Nyquist sampling criteria would imply, which is the price one pays for the simplicity of the explicit approach. In three dimensions, the stability criterion (widely known as the Courant limit) is $\Delta t < \Delta s/(\sqrt{3}c)$, where Δs is the grid dimension and c is the speed of light in the mesh.

There are several very good texts on the FDTD method. Kunz and Luebbers' was the first [6], appearing in 1993, but Taflove's 1995 volume [7] (revised twice, [8] and [9], now co-authored with Hagness) and the 1998 companion [10] are the standard reference for the FDTD in CEM. The book offers encyclopediac coverage of the method. (Kunz and Luebbers were unfortunate to publish their book just before the revolutionary perfectly matched layer (PML) was invented by Berenger in 1994, although the book still contains useful material, not least a working FDTD code. This code has served as the basis for a number of academic codes.) The recent volume by Bondeson *et al.* offers a concise introduction to the method, whilst retaining mathematical rigour, and can also be strongly recommended [11]. It is particularly strong on basic aspects of finite differencing, especially when applied to complex exponentials – which underlies both dispersion and stability analyses. There is also a good introductory treatment of the method in the antenna text [12].

[9] In Chapter 12, it is shown that one can alternatively view the FDTD as derived from a diagonalized matrix equation.

The time domain formulation of the FDTD method is both an attractive feature and a drawback. Using wideband sources, the FDTD method can compute a wideband response in one run, whereas frequency domain methods must obviously recompute the system response for each frequency point.[10] However, the majority of RF devices operate over quite a narrow frequency band, and this may be less of an advantage than one might expect. In particular, for high-Q devices, a very large number of time-steps may be required to obtain sufficient frequency resolution.[11] Many systems exhibit *dispersive* properties; examples are waveguides and most real dielectric and magnetic materials. In the frequency domain, this is simple to handle by the obvious expedient of simply changing the material/device properties with frequency, but in the time domain this is more challenging since a convolution is implied. Many techniques have been proposed and implemented to address this issue, but do complicate the method somewhat [9, 10].

Although there are some FDTD codes available on the internet, they are really "toy" codes by comparison with NEC-2, for instance. Commercial versions are available, including CST MICROWAVE STUDIO® and REMCON's XFDTD; the former actually uses the finite integration technique but this is very closely related to the FDTD. It is perhaps surprising that more contenders have not emerged, but this is in no small part because a useful commercial code has to incorporate not only a decent user interface, but also a number of extensions to the standard FDTD to make it generally useful.

The strong points of the FDTD method are the following:

- Exceptionally simple implementation for a full-wave solver – at least an order of magnitude less work than either an MoM or FEM implementation for a basic FDTD implementation. (One should be warned however that there are a number of subtleties which can take a while to appreciate, even with an apparently simple problem. Also, many practical problems require more than just a basic implementation, and the simplicity of the method is often compromised by these extra factors.) It is the only method which one can realistically implement oneself in a reasonable timeframe, although even then only for quite specific problems.
- Very straightforward treatment of material inhomogeneities (as for the FEM).[12]
- Fairly accurate geometrical modelling ability (but not as versatile as the FEM in this regard, due to the "stair-stepping" effect of the regular mesh – see comments below on non-orthogonal grids). (Commercial codes frequently include extensions to the method to improve this, so it is not necessarily a problem.)

[10] Continuing research for frequency domain codes on model based parameter estimation (MBPE) aims to reduce dramatically the number of frequency points required, and good results have been obtained; some commercial frequency domain codes already incorporate this.

[11] Again, work similar to the MBPE, using system approximation techniques, can assist here.

[12] A point worth making here is that for typical RF applications, the dielectric properties of materials are usually the most significant, and relative permittivities at RF and microwave frequencies rarely exceed single figures. For low-frequency magnetoquasistatic problems, magnetic properties are often the most significant, with relative permeabilities which can be very large indeed. In this case, the matter is not quite as simple when accurate modelling is desired, and both the FDTD method and the FEM can exhibit problems. This, however, is not the focus of this book.

- Since the method is a time domain one, wideband data are potentially available from one run.
- Scaling behavior in frequency similar to the other methods. Naively, the scaling is $\mathcal{O}(f^4)$, but more realistically, taking the necessity of controlling accumulated dispersion across a structure kd in size, it is $\mathcal{O}((kd)^6)$, with the same $N \propto (kd)^{1.5}$ assumption to control dispersion as for the FEM, which we will discuss shortly. (This is discussed in detail in Section 3.4.3.) Note that as for the FEM, this is not affected by the material composition of the structure. For wideband systems, this is very attractive, since the other methods have an implied f_n multiplicative term (not shown in the preceding sections), where f_n is the number of frequency points required.
- The PML has made implementing very good absorbing boundary conditions as mesh termination relatively straightforward.

The main drawbacks are the following:

- Inflexible meshing – much work has been done on non-orthogonal FDTD grids, but the method then loses much of its appealing simplicity.
- Some uncertainty about the precise position of boundaries – usually an uncertainly of about $\Delta s/2$. This is due to the offset nature of the $\boldsymbol{E}$ and $\boldsymbol{H}$ field grids.[13]
- Dispersive materials require considerable effort to implement correctly – but it is possible and good results have been obtained.
- As with the FEM, the FDTD method is not as efficient as the MoM when modelling structures consisting entirely of perfectly or highly conducting radiators/scatterers.
- Although considerable theoretical work has been done on higher-order FDTD approaches, none appears to have been successfully implemented in a general-purpose FDTD code. The problem is intimately linked to that mentioned above regarding the ambiguity of the boundary positions.

In conclusion, the FDTD method is the preferred method for wideband systems. Even in its standard Yee form, it is also a strong contender for *any* electromagnetic radiation or scattering problem for which quick answers are needed, great accuracy is not the primary concern, and quite large runtimes and memory usage are acceptable. Furthermore, by using a sufficiently fine mesh, and in particular using various extensions to the standard FDTD method, very accurate results may be obtained; the potential accuracy of the FDTD method should not be underestimated.

1.5 The finite element method (FEM)

The finite element method (FEM) has been widely used in structural mechanics and thermodynamics; its first application in the modern form dates to the 1950s, although its

[13] Work has been done on improving this, typically using some averaging of properties in the FDTD cells on the boundaries, but this can impact on the second-order accuracy of the method.

mathematical roots are older, and the first application in electromagnetics was undertaken in the late 1960s. Chapter 9 gives some more historical background on the method.

As with its main competitor, the FDTD method, the FEM handles inhomogeneous materials and complex geometries with aplomb; these become problems in mesh generation rather than in electromagnetic theory. The FEM may be derived from two viewpoints: one uses variational analysis, the other weighted residuals. Both start with the partial differential equation (PDE) form of Maxwell's equations. The former finds a variational functional whose minimum[14] corresponds with the solution of the PDE, subject to certain boundary conditions. The latter also starts with the PDE form of Maxwell's equations, and then introduces a "weighted" residual (error); using Green's theorem, one of the differentials in the PDE is "shifted" to the weighting functions.[15] For most applications, these procedures result in identical equations. In both cases, the unknown field is discretized using a finite element mesh; typically, triangular elements are used for surface meshes and tetrahedrons for volumetric meshes, although many other types of elements are available. Triangles and tetrahedrons have certain attractive properties best summarized as "simplicial" – these are the simplest geometrical forms with which two-dimensional and three-dimensional regions respectively can be meshed.

Finite element analysis (FEA) can handle two different types of problem, viz. eigenanalysis (source-free) and deterministic (driven) FEA problems.[16] Problems *without* any internal (or external) field source fall into the category of eigenanalysis problems. A classic example is a cavity resonator. What emerges from the analysis is a set of eigenvalues and associated eigenmodes; these represent the resonant frequencies and associated field distribution within the cavity. For microwave dielectric heating, this information can be used to design feed locations and optimize load positioning.[17] It should be noted that eigenanalysis applications are neither time nor frequency but rather eigenvalue domain solvers; using a simple transformation, it is possible to include operating frequency in a waveguide simulation, to compute dispersion curves.

Deterministic problems analyzed using FEA involve a source; the response of the structure to this excitation is then computed. This represents a very large class of electromagnetic engineering applications of the FEM, including antenna, radar cross-section, microwave circuit and periodic structure analyses.

As with the FDTD method, the FEM does *not* include the radiation condition. For closed regions (e.g. waveguide devices or cavities) this is of no concern. However, for open regions (e.g. radiation or scattering problems), this requires special treatment, and

[14] More precisely, extremal point, since it may also be a maximum or stationary point.

[15] From which comes the name "weak" formulation, sometimes encountered in the literature, since the finite element basis functions need only be once differentiable, whereas the wave equation has second-order derivatives.

[16] The MoM and finite difference methods in general can also be used for eigenanalysis, but are not very commonly encountered. Harrington's original text on the MoM included a chapter on eigenvalue problems, but the MoM has not been as widely used as the FEM for this class of problem. The FDTD method is by definition deterministic, and requires a source.

[17] In this real-world application, there is now a source and the problem is strictly speaking no longer an eigenanalysis one, but the source location can be optimized by knowledge of the resonant field behavior within the cavity, since these fields are what one is attempting to excite with the feed.

this must be incorporated using either an artificial absorbing region within the mesh (the numerical analogy of an anechoic chamber) or using a hybridization with the MoM to terminate the mesh.

Traditionally, the FEM has been formulated in the frequency domain, although time domain formulations have also been used for specialized applications.

There are a number of excellent and up-to-date texts on the FEM, including those by Jin (revised in 2002) [13], Silvester and Ferrari [14] (the third edition appearing in 1996), Volakis *et al.* [15] and Peterson *et al.* [16] (both published in 1998, the last also including comprehensive coverage of the MoM). The collection of papers edited by Silvester and Pelosi [17] is also very useful, although quite a number of significant papers have appeared since its 1994 publication. Another useful source is the 1996 volume edited by Itoh *et al.* [18].

Several companies market commercial finite element products for radio-frequency electromagnetics. Ansoft's HFSS package is widely regarded as the market leader; Ansys have a suitable product, and a fairly recent entry, FEMLAB (now COMSOL Multiphysics), has also attracted users.

The strong points of the FEM are the following:

- Very straightforward treatment of complex geometries and material inhomogeneities.
- Very simple handling of dispersive materials (i.e. materials with frequency-dependent properties).
- Ability to handle eigenproblems as above.
- Slightly better frequency scaling than the MoM in its basic form, viz. $\mathcal{O}((kd)^{5.5})$ – although the requirement to mesh a volume rather than a surface means that the number of unknowns in the problem is usually much larger.[18] For a typical mixed first-order scheme, this is the same computational complexity as the FDTD. Depending on the problem, the number of unknowns can be lower than an FDTD solution (due to the better geometrical modelling capability of a tetrahedral mesh), but this is *per frequency point* for the FEM, whereas the FDTD can generate a wideband solution in one run.
- Straightforward extension to higher-order basis functions, which can substantially reduce the asymptotic computational complexity due to lower dispersion. The FEM lends itself to the use of higher-order basis functions; although the book-keeping within an FEM code is a little complicated by this, the theoretical extensions are now well understood. It is also possible to use conformal elements to even better approximate curved geometries.

[18] The exact scaling behavior depends on how efficiently the sparsity of the finite element matrix can be exploited – and the sparsity pattern is problem dependent. The lowest bound on this is $\mathcal{O}(N)$, N being the number of degrees of freedom (unknowns). For a mixed first-order scheme with second-order accuracy, $N \propto ((kd)^{1.5})^3$, where the exponential indicates that as the problem size grows, so the mesh must become proportionally finer to control mesh dispersion. (The further factor of 3 is from the three dimensions.) The effect of dispersion was often overlooked in earlier analyses of differential equation based solvers. It may be shown [11, Section 8.1] that, for the FEM, the number of iterations required for a typical iterative solver is $\propto f$, so the overall computational estimate is $\mathcal{O}((kd)^{5.5})$.

- "Multi-physics" potential – this means the ability to couple EM solutions with, for instance, mechanical or thermal solutions. Due no doubt to the widespread popularity and maturity of the FEM in other fields of engineering, one is starting to see packages which can compute such coupled solutions. It is probably only significant in high-power applications, where thermal effects can be important – either desired, as in the case of microwave dielectric heating, or undesired, such as with high-power transmitter design.

The weak points of the FEM include the following:

- Inefficient treatment of highly conducting radiators when compared to the MoM (due to the requirement to have some mesh between the radiator and the absorber).[19]
- The FEM meshes can become very complex for large three-dimensional structures – indeed, some workers have reported mesh generation times starting to exceed solution time.
- The FEM is rather more complex to implement than the FDTD method. This impacts in particular in terms of the suitability of the FEM for parallel computing. It also implies that "home-built" FEM codes are quite rare compared to such FDTD codes.
- Efficient preconditioned iterative solvers are required when higher-order elements are used; so important is this in commercial applications that these are usually treated as proprietary information, making "do-it-yourself" implementation even more challenging.

In conclusion, the FEM is the preferred method for microwave device simulation and eigenproblem analysis. Using FEM/MoM hybrids, scattering problems involving electromagnetically penetrable media and specialized antenna problems can be accurately and efficiently solved.

1.6 Other methods

The MoM, FEM and FDTD method are the most popular methods in current use. There are a number of other methods which will be encountered in the literature, and some commercial codes are based on these methods. Here we will briefly outline some of them.

1.6.1 Transmission line matrix (TLM) method

The TLM method is conceptually very similar to the FDTD method. Instead of directly discretizing the Maxwell equations, an equivalent array of short transmission lines is used. Historically, the method was especially appealing to engineers with a strong circuit but weak field background, but for most contemporary CEM practitioners the circuit approximation of the field equations seems rather circuitous for field problems. However,

[19] The FEM/MoM hybrids overcome this problem.

it should be commented that the circuit approach can be more direct than the FDTD approach when one is dealing with high-frequency circuits, which is a major reason for continuing work on the TLM. The method has a dedicated following in some circles, and at least one commercial code, Micro-Stripes, is available.

1.6.2 The method of lines (MoL)

The MoL is a specialized method for primarily waveguiding structures. It uses a semi-analytical solution along a number of lines (in its two-dimensional form) and is especially memory efficient. It is also very accurate. Because it requires the extraction of eigenvalues, it can be computationally expensive. Most MoL applications can be done as well with an FEM formulation. A commercial implementation does not appear to be available at present.

1.6.3 The generalized multipole technique (GMT)

The GMT uses multipoles as the basis functions; these are special function solutions of the Maxwell equations. It is not especially similar to any of the methods that we have discussed thus far. It does require some intelligent user input in terms of placing the multipoles. Good results have been obtained for a variety of problems; a good reference is Hafner's book [19], which is also useful for placing the other methods in perspective. (The book will appear somewhat idiosyncratic though for CEM novices.)

1.7 The CEM modelling process

Before we now proceed to study these methods in more detail in subsequent chapters, it is useful to comment on the modelling process in general. Some astonishing claims have been made about the predictive power of Maxwell's equations [20, p. 4]:

> Most physicists believe that if you lock a graduate student in a room and have him perform an electromagnetic calculation correctly, and if you perform an experiment that does not agree with the graduate student's calculation, then you better check your experiment.

Whilst at its heart this observation is true, in that we believe that for non-quantum interactions, Maxwell's equations provide a complete description of electromagnetic phenomena,[20] for many aspects of CEM modelling, one needs to be *extremely* cautious of such sentiments. The modelling process is about the art of acceptable approximation, and this path is strewn with pitfalls.

Firstly, we are replacing the real world with a mathematical model, or put differently, replacing a *real* field problem with an *approximate* one. Here are some examples of possible problems.

[20] By replacing the field vectors with operators, Maxwell's equations become quantum theoretically correct.

Limitations of the mathematical model

Mathematical models of electromagnetic devices usually have some underlying assumptions. An example is the infinitely large planar ground assumed in a Sommerfeld formulation. Most integrated antennas radiate primarily on broadside, so the finite ground of a real antenna is usually not a problem. However, endfire integrated antennas (a Yagi, for instance, photo-etched on a printed circuit board) radiate most strongly along the ground interface, and the main beam on endfire apparently disappears when a Sommerfeld code is used.

Tolerances

Any engineered system has some measure of tolerances. Some are really of little concern; others impact directly on device performance. An example of the former is surface roughness of average dimensions far less than a wavelength; this usually has little impact on the operation of the system. An example of the latter are uncertainties in dielectric constant or overall device dimensions. For antennas relying on standing-wave operation (most wire antennas, microstrip patch structures, etc.) these translate directly into variations in resonant frequency. (Since such devices are usually quite narrowband, this can be highly problematic.)

Manufacturing deviations

The device that is being simulated may differ subtly from what was designed and analyzed, in a more fundamental way than simply due to device tolerances. An example of a frequency selective surface with such a problem will be discussed in the next section in detail; there, an air inclusion was de-tuning the device, and measurements and computations stubbornly refused to agree until the problem was identified.

Simplifications in the formulation

Currents flowing on thin wires are usually approximated by current filaments (in other words, wire thickness is ignored). This can cause problems when the wire thickness no longer justifies this assumption.

Once we have an acceptable approximate field problem, it will then be solved using an *approximate numerical* solution. Once again, there are many pitfalls in this next step.

Finite discretization

This is usually the biggest single limitation on the accuracy of numerical techniques in electromagnetics. There are typically two different types of error which accompany this: one is *interpolation* error, and is the (in)ability of the basis functions to model the field locally; the other is *mesh dispersion* error (also sometimes called pollution error) which is cumulative error through the mesh.[21] Both can usually be controlled by refining the

[21] The MoM does not suffer from mesh dispersion, only the differential equation based FDTD method and FEM do.

mesh. Unfortunately, the computational cost of especially three-dimensional modelling is such that this is not always practical.

Finite problem space
Neither the FDTD method nor the FEM incorporates the radiation condition, and the mesh needs to be truncated at some point when a radiation or scattering problem is undertaken. Absorbing boundary conditions are widely used for this. After creating an (in)adequately refined mesh, this is probably the second largest source of error in FDTD and FEM computations; in reality, the problems are interwoven, since a poor mesh termination scheme requires a larger solution region, which in turn makes it difficult to ensure that the mesh is fine enough.

Numerical approximations
FEM and MoM codes in particular require numerical integration. This is usually done using quadrature or cubature (multi-dimensional numerical integration), and if not carefully done, may easily result in poor performance, in particular for the MoM which involves the integration of nearly singular or singular fields.

Finite machine precision
The infinite number of real (or complex) numbers are represented on a computer in a very finite fashion. Typically, in single precision, 4 bytes (i.e. 16 bits) is used to represent a real number; this gives some 5 significant digits of accuracy. (Double precision uses 8 bytes, and approximately doubles this.) For problems which are ill conditioned (that is, the solution depends rather drastically on small changes in input data) this may not be adequate. This is usually a less serious problem than the others.

1.8 Verification and validation

The discussion of the previous section leads directly to the issues of the verification and validation of code. The former is defined as ensuring that the code has correctly implemented the formulation, and the latter as checking that the formulation as implemented in the code produces results agreeing with reality. However, for users of codes in practice, the processes are integrated, especially since users cannot change commercial codes. Throughout this book, the necessity of validating and verifying code will be continually emphasized, but it is such an important topic in CEM that it deserves this section on its own. Verification and validation have not recently received the attention they deserve in the computational electromagnetics literature; much of the development has been in related fields of computational science and engineering, such as nuclear engineering, where errors in codes, or erroneous use of codes, may have literally devastating consequences; see, for instance, [21] for a comprehensive review.

There are several methods currently in widespread use.

Comparison with analytically computed solutions
This was the classical approach taken by most of the early researchers in the field. Typically, radiation or scattering solutions involving canonical shapes (usually cylinders in two dimensions, and spheres in three dimensions) were used to compare results with those of the code. The problem with this is that it is a necessary, but not sufficient, condition for a code to be working correctly.

Comparison with approximate solutions
Quite often, approximate solutions of electromagnetic problems are known from simplified models, which have usually been experimentally tested, or may even represent experimental data. Many antennas are a good example of this, with parameters such as gain often a design parameter. Comparison of computed results with these provides some reassurance that the code is in the correct "ballpark," although of course this is not a rigorous process.

Comparison with measurements
In a sense, this is the most satisfying and convincing method. However, it is strewn with difficulties. Unlike CEM, where the basic tools have dropped enormously in price, radio-frequency measurement devices have remained expensive, and accurate measurements of radiation or scattering require an anechoic chamber. Making reliable and repeatable measurements is also both a science and an art, and usually requires considerable experience.

Comparison with other CEM codes
This is a relatively recent innovation, prompted both by the availability of powerful general-purpose codes and the difficulty of obtaining reliable measured data. Once again, caution is required. This is one place where the difference between verification and validation can be significant: to give an example, verifying a thin-wire code by comparison to another thin-wire code will not detect a fundamental problem with the thin-wire assumption. In general, this is most convincingly done by comparing results computed using codes implementing different formulations.

We will see examples of the use of all these techniques throughout this book.

1.8.1 An example: a frequency selective surface

The process of validating computations can sometimes lead to enhanced understanding of the device under test. An example is the following, originally presented in [22, 23] for a device called a frequency selective surface (FSS). There are various applications thereof: in this case, a bandpass radome was required. The structure consists of a slot cut in a metallic sheet, which transmits an incident wave when the frequency is such that the slot is resonant.

When an FSS is fabricated, a dielectric support is generally required, lowering the resonant frequency and complicating the analysis. An FDTD code, originally developed

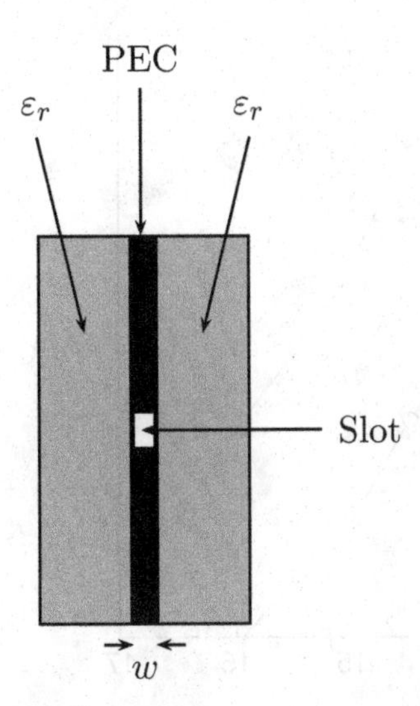

Figure 1.4 A cross-section of a slot forming the FSS in a conductor of *finite* thickness, w, sandwiched between two dielectrics. (Reproduced by permission of IEE from [22].)

by the author, was used to simulate the dielectric support. However, initial results yielded a consistent offset in center frequency between measured and computed data, which was sufficiently large to be problematic. The usual FDTD checks – refining the mesh, and moving the absorbing boundary condition further away – did not solve the problem.

Eventually, the problem was traced to a very subtle manufacturing problem. When manufacturing a dielectric-supported FSS structure, the finite thickness of the metal screen can be surprisingly significant, whether a sandwiched or single-sided support is used; this results effectively in a slot. Although the slot is small in cross-section, the material filling it plays a significant role in the electromagnetic behavior of the device. An example of the slot forming the FSS element in a finite thickness conductor is illustrated in Fig. 1.4. This is a cross-section of a circular ring FSS element, with diameter 5.9055 mm, slot width 0.537 mm and element spacing 10.738 mm.

The FDTD code can accurately predict the effect of different dielectrics, provided the significance of this effect is realized and correctly modelled. Figure 1.5 shows this effect clearly; the resonant frequency is off by around 13% for the FDTD model which (incorrectly) assigned perspex to the slot. Two of the models in Fig. 1.5 used an infinitely thin metal sheet in the FDTD model; in one case perspex of a single FDTD cell thickness was used to model the cavity formed by the slot; in the other case air was assigned to the slot cavity – however, the depth of the actual slot was not entirely correctly simulated. The final model used the correct actual metal thickness and the slot was air filled, also to the correct thickness. The difference in predicted resonance frequency is significant.

Figure 1.6 shows measured and predicted results at normal incidence for a horizontal tri-slot sandwiched between PVC ($\epsilon_r = 2.86$), with petroleum jelly (Vaseline, $\epsilon_r = 2.16$) used to fill the slot which has been carefully modelled with the FDTD mesh. (This particular tri-slot had arm length 3.732 mm, arm width 1.0 mm and inter-element spacing 12.5 mm.) The results demonstrate the accuracy achievable with careful modelling.

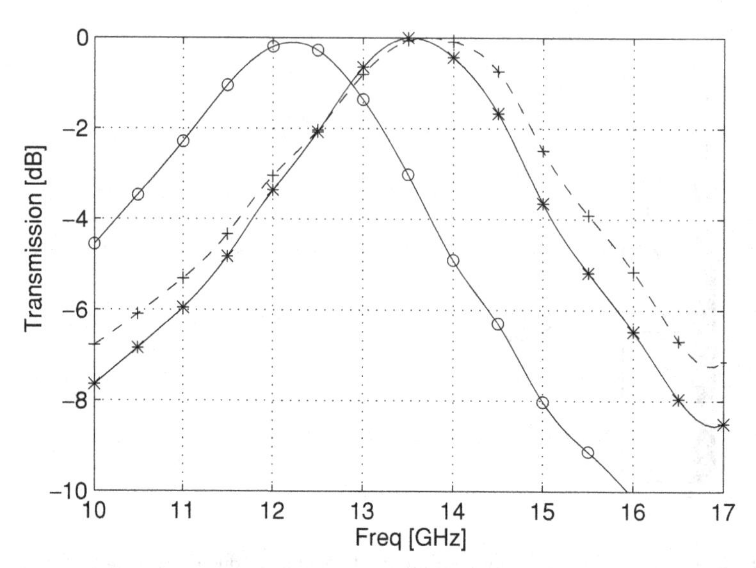

Figure 1.5 Predicted transmission coefficient of an O-ring FSS with one side perspex only. Legend: solid line (o), infinitely thin metal sheet, single-cell perspex in the slot; solid line (x), infinitely thin metal sheet, single-cell air in the slot; dashed line (+), actual 0.268 45 mm thick metal sheet; air in the slot. (Reproduced by permission of IEE from [22].)

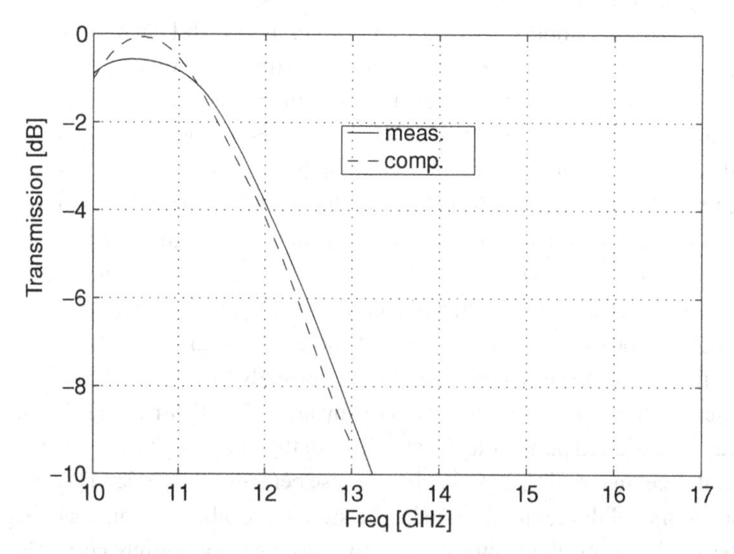

Figure 1.6 Transmission coefficient of PVC sandwiched FSS with petroleum jelly filling in the slot. The solid line is measured data, the broken line is the FDTD simulation. (Reproduced by permission of SAIEE from [23].)

This is an example of a discrepancy between the real and approximate field problems, due in this case to a manufacturing problem. It is especially useful in that it led to improved understanding of the design, and a revised manufacturing process. CEM tools (the FDTD in this case) allowed very quick experimentation to establish that the air inclusion (in this case) was the problem; laboratory experiments with various prototypes would have been very tedious and time consuming indeed.

1.9 Convergence and extrapolation

Most (although not all) CEM formulations converge to a solution as the mesh is refined, and convergence is a topic that will be frequently discussed in this book. The rate of convergence is influenced by a number of factors; if the solution is smooth, then the convergence rate is usually controlled by the polynomial order of approximation. However, if the solution contains singularities, then the rate of convergence may be determined largely by mesh size, and hence be far slower than expected (An example of this may be found in Section 10.2.3.) A useful method which is surprisingly little used in CEM is extrapolation [11, 24].

The basic idea rests on the assumption that the solution I can be written as the desired solution plus an error term which is polynomial in terms of mesh size h:

$$I(h) = I_0 + I_p h^p. \tag{1.7}$$

The desired result is I_0, which is the result extrapolated to zero mesh size. Although the rate of convergence may sometimes be known analytically, non-smooth solutions can impact on this, and it is best to find p from several solutions. The simplest way to do this is to compute a sequence of solutions on a geometric sequence of cell size, that is $h_i/h_{i+1} = h_{i+1}/h_{i+2}$, in which case, subject to the assumption of Eq. (1.7), one can estimate the order of convergence of the solution as [11, Eq. (2.4)]

$$p = \ln\left[\frac{I(h_i) - I(h_{i+1})}{I(h_{i+1}) - I(h_{i+2})}\right] \Bigg/ \ln\left[\frac{h_i}{h_{i+1}}\right] \tag{1.8}$$

Knowing p, one can then apply a polynomial fit in h. MATLAB makes this particularly easy; for instance, if $p = 2$ (typical of low-order solutions) then the MATLAB command

```
pfit = polyfit(h.^2,I,1)
```

gives a first-order fit to the square of mesh size. In the MATLAB convention for this function, the second component of `pfit` is the coefficient for the constant term, i.e. the desired result I_0. (For a more extensive discussion, refer to [11, Chapter 4].) One limitation with extrapolation is that while it can be readily applied to a single, integrated quantity, such as capacitance or radar cross-section, it is more difficult to apply to a radiation pattern, for instance. Another, more serious, limitation is that not all CEM formulations are truly convergent – the thin-wire formulation in Chapter 4 is a case in point. An example of the application of this may be found in Section 10.2.3.

In later parts of this text, errors will frequently be presented on log–log axes – see, for instance, Fig. 9.6 – from which rates of convergence can also be obtained. This analysis is subtly different to that above, in that for such plots, the reference solution (i.e. I_0 is) *known*. In this case, what is being plotted is the error, ε:

$$\varepsilon(h) = I_p h^p, \tag{1.9}$$

and when plotted on log–log axes, this results in a straight line, with slope p:

$$\log \varepsilon = \log I_p + p \log h. \tag{1.10}$$

Equation (1.9) assumes polynomial convergence. Not all CEM algorithms converge thus. Some converge only within a limited range of h (such as the thin-wire MoM formulation mentioned above), whilst others, such as hp adaptive FEM schemes, are designed to converge *exponentially*. In such cases, the above analysis no longer applies.

1.10 Extending the limits of full-wave CEM methods

It should be clear from the preceding sections that no one CEM method should lay claim to being able to address all problems with optimum efficiency. Both the FEM and FDTD method are theoretically capable of addressing any arbitrary problem, but both can be unnecessarily or prohibitively expensive computationally for problems more suited to the MoM, and in practice of course *no* full-wave method still works at asymptotically high frequencies.

The computational cost of the major methods has been briefly reviewed in this chapter. Although the individual methods vary, it is clear that in the case of *all* the full-wave solvers, as the frequency increases, so the mesh must become finer (and thus the number of unknowns become larger. For the differential-equation based solvers, controlling dispersion further results in the mesh size scaling non-linearly in frequency). We have seen that although the methods do not scale identically, *all* of them scale badly with frequency – in basic form, the computational cost of the MoM, FDTD and FEM all scale at between the fifth to sixth power of frequency, and all require large amounts of memory for electromagnetically large problems. Much ingenuity has been devoted to developing new or modified full-wave methods with better scaling behavior – the present class of "fast methods" being an excellent case in point – and to exploiting high-performance computers, in particular parallel computing.

"Fast" methods, including the fast multipole method and the adaptive integral method, aim to reduce the *asymptotic* cost of the methods. Put very simply, the methods replace the traditional direct matrix solution algorithms with iterative solvers, and use methods to approximate the interaction between parts of the mesh which are separated by some reasonable distance (usually at least a few wavelengths). The matrix-vector product – which lies at the heart of iterative solvers – is implemented using a fast technique similar to the FFT, which reduces the cost from $\mathcal{O}(N^2)$ to $\mathcal{O}(N \log N)$ per iteration. Recent work has claimed an asymptotic dependence of $\mathcal{O}(kd)^3$. This appears very attractive indeed, but one should be warned that this is an *asymptotic* calculation and there are some very large constants associated with this, as well as some possibly optimistic assumptions about rapid convergence of the iterative solver. Hence this attractive scaling behavior only manifests itself for electromagnetically very large problems. The convergence of iterative methods is also very problem dependent, so a particular analysis may not yield the expected asymptotic behavior if the solver should converge unexpectedly slowly. For an overview of fast methods, see Chew *et al.*'s review paper [25]. We will discuss fast methods in Chapter 6.

Whilst great advances have been made, the full-wave techniques eventually make impossible demands on even the largest supercomputers, and asymptotic techniques become important. These methods generally use rays as field propagators, and essentially localize electromagnetic interaction, describing the field at a point as the sum of the direct, reflected, and various diffracted rays, all of which originate at points (or sometimes lines) on the structure. With these methods, there is no concept of discretization of an unknown *field* – although the surface may well be approximated by facets.[22] These methods generally rely on the asymptotic nature of some underlying integral or series solution, and the approximation improves with frequency. An excellent overview of this may be found in [26]. Because the asymptotic techniques rely on approximations of the physics from the start, they do not lend themselves as well as the full-wave methods to general-purpose computer programs. However, when *hybridized* with the full-wave methods, some very significant extensions to the frequency range of full-wave codes become possible. Jakobus has made significant contributions using hybrid MoM/PO approaches [27, 28] and much of this work is reflected in the commercial program FEKO. Again, in Chapter 6, hybrid methods will be discussed in more detail.

To paraphrase Hafner [19], CEM is a field which depends not just on "big ideas," but also on getting lots of details right. This chapter has concentrated on the former, with the aim of providing the CEM beginner with some idea of what method is appropriate for what problem. Actually *implementing* a reliable CEM code makes enormous demands on the latter, and requires an on-going process of validation. One should be warned that even the most apparently straightforward method (the FDTD) is not as straightforward to implement as one might expect; development times for even the most specialized CEM codes involve at least months of work, and powerful, general-purpose codes involve many years of effort.

1.11 CEM: the future

Prediction is very difficult, especially about the future.

Niels Bohr

CEM has passed through several phases: the 1960s and 1970s saw primarily work on CEM formulations; the 1980s saw the techniques starting to receive significant acceptance by non-specialists; the 1990s saw the first widely available commercial codes for radio-frequency electromagnetic problems appear on the market; and the 2000s saw

[22] In the case of physical optics (PO), the surface current is indeed discretized, but the amplitude of the current is *assumed* in terms of the known incident field, rather than being computed from a matrix equation enforcing a boundary condition.

their widespread adoption in the market. What does the next decade or so hold in store?

Firstly, it appears that we can look forward to continuing giant strides in computer performance – although we will increasingly have to look to parallel computing to exploit it. Looking back over the last two decades, a typical PC has increased its clock speed from some tens to megahertz to several gigahertz, while memory sizes have grown from under one megabyte to several gigabytes, and disk sizes have increased from ten or twenty megabytes to hundreds of gigabytes. (Workstations have largely disappeared as a separate product line, having been effectively replaced by top-end PCs.) This revolution in affordable computing has revolutionized potential CEM applications for engineers based in industry. However, as of the early 2000s, clock speeds have stagnated, and Moore's Law has manifested itself in terms of multiple processors ("cores").

Whilst computational capacity has increased (doubling at least every two years), algorithmic improvements have been even faster in major codes, perhaps in the order of a doubling in speed every year[23] (although this has been rather more step-wise than improvements in computing). The result of this is that over a decade, the time required to run an algorithm may decrease by as much as $2^{11} = 2048$. This may appear fanciful, but results in Section 6.9.4 substantiate dramatic reductions in runtime for at least one technique in a commercial code, and over a period rather less than a decade. Even if a factor of a couple of thousand in one decade is optimistic, it is clear that a least a reduction in runtime for a particular problem from a week to hours, or hours to minutes, or minutes to seconds, may be expected on average over a decade. (Of course, our response as users is generally to increase either problem size or modelling fidelity, or both, meaning that runtimes tend to remain more constant over time than one might expect!)

CEM theory has also advanced enormously since the first work in the 1960s, and far more RF electromagnetic problems are potentially amenable to a CEM solution. Much work can be expected on hybridizing methods – the benefits of FEM/MoM and MoM/PO hybrids have been noted in this chapter, and will be discussed in more detail later in this book. Another hybrid of which more may be seen is the combination of full-wave modelling techniques with quasi-static analysis, within the same modelling interface. Several codes already support methods which use a quasi-static analysis as a boundary condition for a full-wave solution, but at present, there is in general no feedback from the full-wave solver to the quasi-static one. This is likely to be driven strongly by the signal integrity community.

Intelligently refining meshes automatically is also an important topic, both in the research community and in commercial codes. Closely linked to this are methods for estimating errors in computed solutions; how this can be done will be briefly described in Chapter 12. Another significant trend is the incorporation of automatic optimizers using full-wave CEM tools for the analysis part of each iterative step in the optimization procedure. A number of commercial packages already incorporate such abilities.

[23] This has been dubbed "Moore's Law of Electromagnetics" by Dr Cendes, founder of ANSOFT.

It can also be expected that the user interfaces will continue to improve, making modelling complex three-dimensional devices quicker and easier. Furthermore, it is notable that some commercial packages are starting to offer more than one method within the same graphical user interface. A point that has been made often in this chapter, and will continue to be made frequently in this book, is that one should chose the appropriate method for the problem at hand; working within a consistent user interface, it will be far easier for users to exploit the full power of the CEM techniques available.

In the first edition, a trend which was predicted was the use of increasingly powerful commercial packages, and a decline in the number (or at least use) of CEM "freeware." It was argued that this reflected both the difficulty (and hence expense) of developing general-purpose CEM packages; unless government sponsored (such as NEC), the usual way to recover the cost of developing and maintaining code is licensing. Whilst the former has certainly occurred, public domain EM software has, if anything, increased in availability, perhaps reflecting the ethos of open-source software which emerged during the past decade. Public domain CEM codes have also been able to leverage the availability of public domain meshers from the general computational engineering community to permit modelling of more general structures. Nonetheless, such codes generally have far less well integrated pre-and post-processors than commercial codes; RF and microwave engineers who simply want to model a structure will almost certainly continue to opt for commercial packages, and non-specialist modellers generally seek codes with an established reputation and large following. Hence, despite more public domain codes being available, commercial codes are indeed likely to continue to entrench their dominance.

CEM methods have also emerged in restricted, special-purpose implementations, embedded into other modelling packages, such as optics, or expert systems for antenna design.[24] Sometimes, these are commercial products, othertimes for in-house applications. This has been dubbed the "commoditization" of CEM. These are interesting new departures in CEM, continuing to demand highly skilled practitioners. (Such engineers should find the open-source codes in this second edition of considerable utility in developing new products, in particular the three-dimensional ones.) In related fields in engineering such as computational fluid dynamics, there are some very interesting new business models; there is at least one code[25] which is open-source and freely available, but has an associated commercial company supporting it, whose income derives entirely from offering training courses and extensions to the engine.

The above trends in open-source software notwithstanding, it remains true that, overall, CEM developers should expect an increasing number of non-expert users of CEM tools (in much the same way that FEM analysis is now routinely taught to undergraduate civil and mechanical engineers, and routinely used in industrial design). Codes increasingly need to be robust, incorporating warnings of inappropriate meshing, etc., for users without an extensive post-graduate training in electromagnetics. Electromagnetics remains a challenging discipline, and educating users of CEM tools, as well as

[24] An example is a new product, Antenna Magus, which is in essence an expert system for antenna design.
[25] OpenFOAM.

making the tools more robust, will become increasingly important – it is hoped that this book will contribute to the former.

1.12 A "road map" of this book

This book comprises essentially three parts. The first part, Chapters 2 and 3, deals with the finite difference time domain method, in one, two and three dimensions respectively. Chapter 2 uses a simple transmission line problem to introduce many of the basic ideas of the FDTD method. In the second edition, additional introductory material on the basics of finite differencing has been added, as has a more comprehensive discussion of the important topics of stability and dispersion. Chapter 3 goes on to extend these ideas to two dimensions, and considers a number of the issues raised when handling radiation and scattering in free space, in particular the use of absorbing boundary conditions. In this context, an example is given of a perfectly matched layer. The three-dimensional FDTD method is discussed and the development of a simple eigenvalue code is described. (This three-dimensional coverage is new in the second edition.) Following this, examples of the use of a commercial three-dimensional code are presented. These two chapters form an integrated unit.

The second part, Chapters 4–8, deals with the method of moments. Here, the five chapters largely alternate theoretical development with practical application. Chapters 4 and 5 form a unit, first introducing MoM theory for thin-wire antennas and scattering from infinite cylinders (the latter new in the second edition), and then applying it using both a commercial and a public domain code. Chapter 6, on modelling surfaces (and also volumes) using the MoM, is largely self-contained. In the second edition, a new section, Section 6.3, has been added, with develops a mixed-potential electric field integral equation for electromagnetic scattering by surfaces of arbitrary shape in detail, with supporting code. The material in Chapter 6 on the hybrid MoM/PO, as well as on high-performance computing and fast methods, could be omitted without interrupting the flow of the book on a first reading. (The material on high-performance computing has also been extensively revised and updated in the second edition.) Chapters 7 and 8 form a further unit on the theory and application of the Sommerfeld mixed potential integral equation approach to modelling stratified media (of which microstrip antennas are the most widely encountered application at radio frequencies). The material in Chapter 7 is amongst the most theoretically challenging in the book, and could be omitted or covered only superficially, whilst still allowing time for some of the examples in Chapter 8 to be studied. Similar comments also apply of course to readers in industry whose prime focus is on using the MoM.

The third and final part, Chapters 9–12, is devoted to the finite element method. Chapter 9, new in the second edition, revisits the one-dimensional transmission line problem of Chapter 2, introducing finite elements in one dimension. Chapter 10 considers two-dimensional FEM problems, using the quasi-static analysis of a microstrip transmission line as the vehicle (again, new material in the second edition), before going into two-dimensional vector element FEM theory; it is also used to illustrate the solution of an eigenvalue problem. New in the second edition is a detailed development of a mixed second-order FEM solution for this problem. There is also a new section outlining the FEM formulation for the two-dimensional waveguide dispersion problem. Chapter 11 considers three-dimensional vector problems, with much new material in the second edition, including the development of a code using mixed first- and mixed second-order elements. The material in the last chapter, Chapter 12, is primarily to sensitize readers to more advanced formulations and applications (primarily, but not exclusively, of the FEM). Again, new material has been added in the second edition, most noticeably in terms of a detailed analysis of the FDTD interpreted as an FETD scheme.

A number of open-source MATLAB codes complement this text. An outline thereof is given in Appendix F.

For a course on CEM methods, there is more material here than can be covered in a typical semester course, and instructors can be guided by the above discussion regarding what to omit. End-of-chapter problems expand on theoretical points in the text; the end-of-chapter assignments are largely concerned with developing CEM code. They are intended primarily for use in a formal classroom environment, but would be useful for self-study as well.

Regarding the other appendices: good antenna and electromagnetic texts usually include material on vector calculus, and it is assumed that the reader has at least one, so repeating it here seems superfluous. Instead, the appendices contain material which the author has found useful specifically in CEM, and which is not easy to find in the literature in one place.

A final comment. Electromagnetics, antenna engineering and microwave circuit design are all extremely well-established fields, with excellent textbooks available. This book is designed to complement, not compete, with them. It is a text specifically on the theory and applications of *computational* electromagnetics. It is assumed that the reader has a suitable reference in his or her field of interest, so this book does not define antenna radiation patterns, S-parameters or other well-known and widely used concepts in this field. For readers who would like to add to their libraries, the following can be highly recommended. On electromagnetics in general, a very comprehensive reference is [29]; for antenna engineering, [30], [12] or [31] are all excellent references, as is [32] for microwave circuits and systems. There are many older texts which would also of course be suitable; the above are highlighted since they are all currently in print and have almost all been recently revised.

References

[1] S. Hawking, *A Brief History of Time*. Bantam, 1988.

[2] E. K. Miller, "A selective survey of computational electromagnetics," *IEEE Trans. Antennas Propagat.*, **36**, 1281–1305, September 1988.

[3] P. J. Nahin, *Oliver Heaviside: Sage in Solitude*. New York: IEEE Press, 1988.

[4] R. S. Elliott, *Electromagnetics: History, Theory and Applications*. Piscataway, NJ: IEEE Press, 1993.

[5] H. A. Haus and J. R. Melcher, *Electromagnetic Fields and Energy*. Englewood Cliffs, NJ: Prentice-Hall, 1989.

[6] K. S. Kunz and R. J. Luebbers, *The Finite Difference Time Domain Method for Electromagnetics*. Boca Raton, FL: CRC Press, 1993.

[7] A. Taflove, *Computational Electrodynamics: the Finite Difference Time Domain Method*. Boston, MA: Artech House, 1995.

[8] A. Taflove and S. Hagness, *Computational Electrodynamics: the Finite Difference Time Domain Method*. Boston, MA: Artech House, 2nd edn., 2000.

[9] A. Taflove and S. Hagness, *Computational Electrodynamics: the Finite Difference Time Domain Method*. Boston: Artech House, 3rd edn., 2005.

[10] A. Taflove, *Advances in Computational Electrodynamics: the Finite Difference Time Domain Method*. Boston, MA: Artech House, 1998.

[11] A. Bondeson, T. Rylander and P. Ingelström, *Computational Electromagnetics*. New York, NY: Springer Science, 2005.

[12] W. L. Stutzman and G. A. Thiele, *Antenna Theory and Design*. New York: Wiley, 2nd edn., 1998.

[13] J.-M. Jin, *The Finite Element Method in Electromagnetics*. New York: Wiley, 2nd edn., 2002.

[14] P. P. Silvester and R. L. Ferrari, *Finite Elements for Electrical Engineers*. Cambridge: Cambridge University Press, 3rd edn., 1996.

[15] J. Volakis, A. Chatterjee and L. Kempel, *Finite Element Method for electromagnetics: Antennas, Microwave Cicuits and Scattering Applications*. Oxford & New York: Oxford University Press and IEEE Press, 1998.

[16] A. F. Peterson, S. L. Ray and R. Mittra, *Computational Methods for Electromagnetics*. Oxford & New York: Oxford University Press and IEEE Press, 1998.

[17] P. P. Silvester and G. Pelosi, *Finite Elements for Wave Electromagnetics*. New York: IEEE Press, 1994.

[18] T. Itoh, G. Pelosi and P. P. Silvester, eds., *Finite Element Software for Microwave Engineering*. New York: John Wiley and Sons, 1996.

[19] C. Hafner, *Post-Modern Electromagnetics: Using Intelligent Maxwell Solvers*. New York: Wiley, 1999.

[20] W. C. Chew, J. Jin, E. Michielssen and J. Song, *Fast and Efficient Algorithms in Computational Electroamagentics*. Boston, MA: Artech House, 2001.

[21] W. L. Oberkampf and T. G. Trucano, "Verification and validation benchmarks," *Nuclear Engineering and Design*, **238**(3), 716–743, 2008.

[22] D. B. Davidson, A. G. Smith and J. J. van Tonder, "The analysis, measurement and design of frequency selective surfaces," in *Proceedings of the 10th International Conference on Antennas and Propagation*, vol. 1, pp. 1.156–1.160. Edinburgh: IEE, April 1997.

[23] D. B. Davidson, A. G. Smith and J. J. van Tonder, "FDTD analysis and Gaussian beam measurement of frequency selective surfaces," *Trans. SAIEE*, **88**, 72–81, September 1997.

[24] M. M. Bibby, A. F. Peterson and C. M. Coldwell, "Use of extrapolation to improve accuracy and enhance confidence in numerical results," *IEEE Antennas Propagat. Mag.*, **50**, 150–155, August 2008.

[25] W. C. Chew, J. Jin, C. Lu, E. Michielssen and J. M. Song, "Fast solution methods in electromagnetics," *IEEE Trans. Antennas Propagat.*, **45**, 533–543, March 1997.

[26] P. H. Pathak, "High frequency techniques for antenna analysis," *Proc. IEEE*, **80**, 44–65, January 1992.

[27] U. Jakobus and F. M. Landstorfer, "Improved PO-MM hybrid formulation for scattering from three-dimensional perfectly conducting bodies of abritrary shape," *IEEE Trans. Antennas Propagat.*, **43**, 162–169, February 1995.

[28] U. Jakobus and F. M. Landstorfer, "Improvement of the PO-MM hybrid method by accounting for effects of perfectly conducting wedges," *IEEE Trans. Antennas Propagat.*, **43**, 1123–1129, October 1995.

[29] C. A. Balanis, *Advanced Engineering Electromagnetics*. New York: John Wiley and Sons, 1989.

[30] C. A. Balanis, *Antenna Theory: Analysis and Design*. New York: Wiley, 2nd edn., 1997.

[31] J. D. Kraus and R. J. Marhefka, *Antennas for All Applications*. Boston, MA: McGraw-Hill, 3rd edn., 2002.

[32] D. M. Pozar, *Microwave Engineering*. New York: Wiley, 2nd edn., 1998.

2 The finite difference time domain method: a one-dimensional introduction

David B. Davidson and James T. Aberle

2.1 Introduction

The finite difference time domain method, usually referred to as the FDTD, is a particular implementation of a general class of methods known as finite difference techniques. The FDTD is so widely used in the CEM community that, although finite difference methods cover a wide spectrum of complexity and accuracy, it is the FDTD which is almost always implied in CEM when finite differences are mentioned.

Finite difference methods are numerical methods in which derivatives are directly approximated by finite difference quotients. The general class of such methods is the most intuitive numerical approach, and was the first to be extensively developed by the scientific computing community. To this day, it probably remains the most universally applicable numerical technique and the one most widely used for scientific computation. As just discussed, for dynamic problems in CEM, the most popular is the FDTD. The opening discussion in this chapter will discuss finite differences in general, before moving on to the specifics of the FDTD.

At this point, a general comment about the philosophy underlying the mathematical treatment of the computational algorithms in this book would be in order. Although we endeavor not to be "sloppy" mathematically, the emphasis in this book is in presenting well-known methods for well-known problems in CEM, rather than on the basic mathematical requirements of the methods, as one would expect to find in an applied mathematics text, for instance. An example of the type of issue which we will gloss over, at least initially, is the differentiation of discontinuous functions, which requires the generalized (weak) derivative, properly the field of functional analysis. Fortunately, the physics-based problems we are addressing usually do not evidence the type of pathological behavior which can (rightly so) concern mathematicians, and issues such as existence proofs will generally be treated superficially, if at all, in this book. The reader should be cautioned about applying the methods discussed here in other fields of engineering or applied science without first mastering the underlying theory in more detail. Suitable references for such reading are Arfken and Weber [1, Chapter 9] (on partial differential equations in general) and Press *et al.* [2, Chapter 20] (on finite differencing). Richtmyer and Morton's text on finite difference methods applied to initial value problems remains the classic reference [3]; although out of print, copies may be found on specialist online booksellers.

2.2 An overview of finite differences

2.2.1 Partial differential equations

As commented above, the finite difference method is the workhorse of scientific computing for the solution of partial differential equations (PDEs), of which Maxwell's are of course a prime example and the main focus of this book. Traditionally, PDEs have been classified as *elliptic, parabolic* or *hyperbolic*. This classification derives from considering the following linear operator representation[1] of the PDE (here, in two variables, for simplicity):

$$\mathcal{L} = a\frac{\partial^2}{\partial x^2} + 2b\frac{\partial^2}{\partial x \partial y} + c\frac{\partial^2}{\partial y^2} + d\frac{\partial}{\partial x} + e\frac{\partial}{\partial y} + f \tag{2.1}$$

The discriminant, $D = ac - b^2$, determines the PDE type. For $D > 0$, the PDE is elliptic; for $D = 0$, the PDE is parabolic; and $D < 0$, the PDE is hyperbolic. As an example, the one-dimensional wave equation, with v the wave speed,

$$\left(\frac{1}{v^2}\frac{\partial^2}{\partial^2 t} - \frac{\partial^2}{\partial^2 x}\right) u = 0 \tag{2.2}$$

has $D = -1/v^2 < 0$ and is hyperbolic. It is also often convenient to write this as

$$\frac{\partial^2 u}{\partial^2 t} = v^2 \frac{\partial^2 u}{\partial^2 x} \tag{2.3}$$

Prototypical PDEs are the hyperbolic wave equation above, the parabolic diffusion equation

$$\frac{\partial u}{\partial t} = \frac{\partial}{\partial x}\left(D\frac{\partial u}{\partial x}\right) \tag{2.4}$$

with D the diffusion coefficient, and the elliptic Poisson equation

$$\frac{\partial^2 u}{\partial x^2} + \frac{\partial^2 u}{\partial y^2} = \rho(x, y) \tag{2.5}$$

with $\rho(x, y)$ the given source term. If $\rho(x, y) = 0$, this is the Laplace equation.

Although traditional, for numerical solution of PDEs the classification as *initial value* versus *boundary value* is often more insightful (a point explored in some detail in [2, Chapter 20]). For instance, the Laplace equation has no time dependence, and the solution is determined entirely by the boundary values. However, both the diffusion equation and the wave equation have solutions evolving in time, and initial conditions must be specified. (For both of these, boundary values are also usually required.) The FDTD method is one method of solving the initial value wave problem implied by Maxwell's two curl equations. Some alternate finite difference (FD) schemes in the time domain are explored in the end-of-chapter problems.

[1] Readers wanting more background on linear operators are referred to Chapter 4.

2.2.2 The basic solution procedure

The basic steps of any finite difference method can be summarized as follows:

- Divide the solution region into a grid of nodes.
- Approximate derivatives in the given partial differential equation by finite differences involving the value of the solution at various nodes.
- Solve the finite difference equations for the value of the solution at each node subject to boundary and/or initial conditions. If operating in the time domain, this amounts to finding the values at the next timestep. This process is variously called time marching, time integration, or specifically in the context of the FDTD, "leap-frogging."

The FDTD method, being a time domain approach, is an initial value method (although material boundaries are of course included). Finite difference methods in general can operate as either boundary value or initial value methods.

2.2.3 Approximating derivatives using finite differences

Central to all finite difference methods is the approximation of derivatives with finite differences. From the basic definition of the derivative of a function, various numerical approximations can be proposed. However, these are usually derived from a Taylor series expansion, since this provides a handle on the error. Depending on whether the "next," "previous" or "central" nodes are involved, one obtains forward, backward or central differencing as follows:

Forward difference formula for first derivative

$$\frac{dU(x)}{dx} = \frac{U(x+\Delta x) - U(x)}{\Delta x} - \frac{(\Delta x)}{2}\frac{d^2U}{dx^2} + \mathcal{O}(\Delta x)^2 \tag{2.6}$$

Backward difference formula for first derivative

$$\frac{dU(x)}{dx} = \frac{U(x) - U(x-\Delta x)}{\Delta x} + \frac{(\Delta x)}{2}\frac{d^2U}{dx^2} + \mathcal{O}(\Delta x)^2 \tag{2.7}$$

Central difference formula for first derivative

$$\frac{dU(x)}{dx} = \frac{U(x+\Delta x) - U(x-\Delta x)}{2\Delta x} - \frac{(\Delta x)^2}{6}\frac{d^3U}{dx^3} + \mathcal{O}(\Delta x)^4 \tag{2.8}$$

These expressions are obtained by performing a Taylor series expansion of the function around x. Let us consider the derivation of the central difference formula. The Taylor series expansion about x_0, evaluated at $x_0 + \Delta x$, is

$$U(x_0 + \Delta x) = U(x_0) + \Delta x \left.\frac{\partial U}{\partial x}\right|_{x=x_0} + \frac{(\Delta x)^2}{2}\left.\frac{\partial^2 U}{\partial x^2}\right|_{x=x_0} + \frac{(\Delta x)^3}{6}\left.\frac{\partial^3 U}{\partial x^3}\right|_{x=x_0} + \frac{(\Delta x)^4}{24}\left.\frac{\partial^4 U}{\partial x^4}\right|_{x=\xi} \tag{2.9}$$

ξ is a point located in the interval $(x_0, x_0 + \Delta x)$. This can alternatively be written as

$$U(x_0 + \Delta x) = U(x_0) + \Delta x \left.\frac{\partial U}{\partial x}\right|_{x=x_0} + \frac{(\Delta x)^2}{2}\left.\frac{\partial^2 U}{\partial x^2}\right|_{x=x_0} + \frac{(\Delta x)^3}{6}\left.\frac{\partial^3 U}{\partial x^3}\right|_{x=x_0} + \frac{(\Delta x)^4}{24}\left.\frac{\partial^4 U}{\partial x^4}\right|_{x=x_0} + \mathcal{O}(\Delta x)^5 \tag{2.10}$$

A similar expansion is performed about x_0, evaluated at $x_0 - \Delta x$:

$$U(x_0 - \Delta x) = U(x_0) - \Delta x \left.\frac{\partial U}{\partial x}\right|_{x=x_0} + \frac{(\Delta x)^2}{2}\left.\frac{\partial^2 U}{\partial x^2}\right|_{x=x_0} - \frac{(\Delta x)^3}{6}\left.\frac{\partial^3 U}{\partial x^3}\right|_{x=x_0} + \frac{(\Delta x)^4}{24}\left.\frac{\partial^4 U}{\partial x^4}\right|_{x=x_0} + \mathcal{O}(\Delta x)^5 \tag{2.11}$$

Subtracting the two expressions, grouping terms, dividing by Δx and noting that the remaining terms in Δx cancel, we obtain Eq. (2.8).

A mathematical aside – finite difference approximations of the second derivative

If, instead of differencing Eqs. (2.10) and (2.11) as above, we add them, we obtain a formula for the central difference approximation of the *second* derivative of the function. The result, with a remainder term of second order, is

$$\frac{d^2U(x)}{dx^2} = \frac{U(x + \Delta x) - 2U(x) + U(x - \Delta x)}{(\Delta x)^2} - \frac{(\Delta x)^2}{12}\frac{d^4 U}{dx^4} + \mathcal{O}(\Delta x)^4 \tag{2.12}$$

(Note that the terms in $(\Delta x)^3$ also cancel.) In the FDTD, we will not initially use this formula, but it turns out that the FDTD scheme is also second-order accurate. The reason is that the central difference formula for the second derivative can also be derived by combining the expressions for the forward and backward derivatives to first order, which is what the FDTD effectively does. We will make use of this expression when investigating the numerical dispersion and stability properties of the scheme.

Although Eqs. (2.6), (2.7) and (2.8) appear similar – indeed, the first part of each is identical – the remainder (error) terms are not. For both forward and backward differencing, the error is proportional to the cell length (Δx) (also known as a first-order scheme), but for central differencing, it is proportional to the square thereof, or alternatively, a second-order scheme.[2] Clearly, in the limit $\Delta x \to 0$, the central difference formula will converge more rapidly to the true value of derivative.

This idea of direct discretization of the derivatives underlies the FD method; one should rather view this as a class of methods, since there are a variety of choices which one can make with regard to the specific FD algorithm. Before moving onto the FDTD method, one last general point should be made: FD methods can be either *implicit* or *explicit*. This is particularly relevant when time is one of the variables. An *implicit* method requires the solution of a set of simultaneous equations – a matrix equation – in order to evaluate the unknowns. (The resulting matrix is generally highly *sparse*, i.e. has only a few non-zero entries. Efficient FD solvers exploit this to save both memory and computational time.) From a physics viewpoint, with an explicit method, the "next" value at a point is a function not only of the current and past values at this and the surrounding points, but also the "next" values of some or all of these. In an *explicit* method, each unknown can be obtained directly in terms of given or previously computed values. Physically, the next value is computed entirely from current or past values. Explicit methods do *not* require any matrix solution. However, they usually have some maximum timestep size, which, if exceeded, produces instability, generally known as the *Courant limit*, or the *Courant–Friedrichs–Lewy (CFL) condition*, after their 1928 work.

It should be noted that there are other methods for obtaining numerical derivatives. By using more points, higher-order schemes can be derived. However, the Yee scheme, to be discussed, does not readily accommodate these in general.

Finite differencing is widely used in computational engineering. Differential equations representing different physics bring different challenges. In wave propagation problems, such as those we will be studying, propagation errors – in particular phase – are usually most worrisome, hence our focus on dispersion later in our studies. In fields such as computational fluid dynamics, transport errors – in particular the (non-)conservation of mass – are usually of most concern [2, Section 20.1.3].

2.3 A very brief history of the FDTD

The FDTD is based on a particular FD scheme (Yee's algorithm) that is applied to Maxwell's curl equations in the time domain. It is an explicit marching-in-time procedure that simulates the propagation and interaction of electromagnetic waves in a region of space. At present, the FDTD is probably the most popular numerical method for the solution of RF electrodynamic problems, due to its simplicity and generality.

The algorithm was first proposed by Yee in 1966 [4]. For around a decade, the method attracted little, if any, attention; the computational electromagnetics community was primarily exploring the method of moments during this period. In 1975, Taflove and Brodwin obtained the correct stability criteria, and computed sinusoidal steady-state solutions using the method. In 1977, Holland, Kunz and Lee applied the method to

[2] A reminder: if a function $\sigma(x)$ is said to be $\mathcal{O}(x^n)$, then there exists some constant A such that $\sigma(x) < Ax^n$, $\forall x$.

electromagnetic pulse problems. In 1981, Mur obtained the first numerically stable, second-order accurate absorbing boundary conditions. From then on, the popularity of the method grew in leaps and bounds, as a number of theoretical issues were solved in rapid succession, culminating in Berenger's perfectly matched layer in 1994. The rapid adoption of the method was also due to the explosive growth in especially personal computing; in 1966, realistic applications of the FDTD made what were then outrageous demands on contemporary computers, whereas those of the MoM were decidedly more modest; by the 1990s, Moore's law had ensured that many realistic FDTD simulations could be undertaken on a PC in minutes or at most hours. Hence both theoretical developments and technological progress played crucial roles in the development of the method.

Theoretical work on the FDTD continues to this day, although the main thrust of most work is now in terms of applications. A detailed chronology, with extensive references, may be found in [5, Section 1.5].

2.4 A one-dimensional introduction to the FDTD

2.4.1 A one-dimensional model problem: a lossless transmission line

To introduce the FDTD algorithm, we will consider a lossless transmission line problem. From basic transmission line theory, the reader will be aware that for transverse electromagnetic (TEM) modes, there is a one-to-one correspondence between electric fields and voltage, and magnetic fields and current. Hence in the following, although we use voltage and current, this is fully equivalent to a field description of a TEM transmission line.

> **A reminder – TEM modes**
>
> As noted in the main text, for fields which are entirely transverse electromagnetic in nature, there is a one-to-one correspondence between electric fields and voltage, and magnetic fields and current. The best known example is a coaxial line. If the voltage between the inner (radius a) and outer (radius b) is V, then the radial electric field is $\frac{V}{\ln(b/a)}\frac{1}{r}$, and with current I, the circumferential magnetic field is $I/2\pi r$. The simplest example is the parallel plate waveguide, separation d, at potential difference V, where the electric field is V/d and the surface current density and magnetic field are numerically equal, although orthogonal in space since $\boldsymbol{J}_s = \hat{n} \times \boldsymbol{H}$.
>
> For guiding structures supporting more complex modes, such as transverse electric (TE) or transverse magnetic (TM), a correspondence may still be found but it is no longer unique.

The well-known equivalent circuit of an infinitesimal piece of transmission line is shown in Fig. 2.1. L is the inductance per unit length and C is the capacitance per unit length of the lossless transmission line. On this section of line, the voltage and current

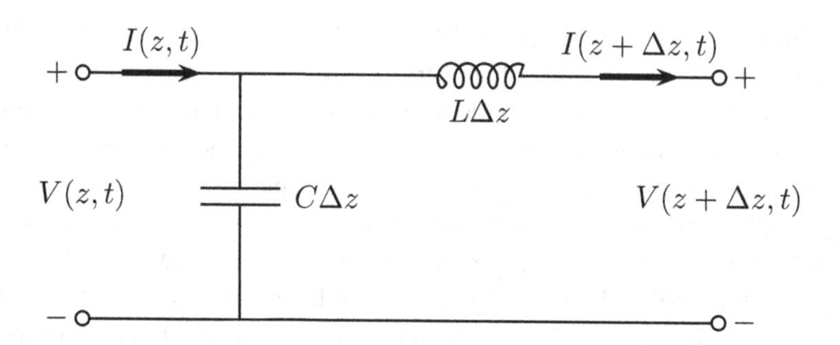

Figure 2.1 Infinitesimal section of a one-dimensional transmission line.

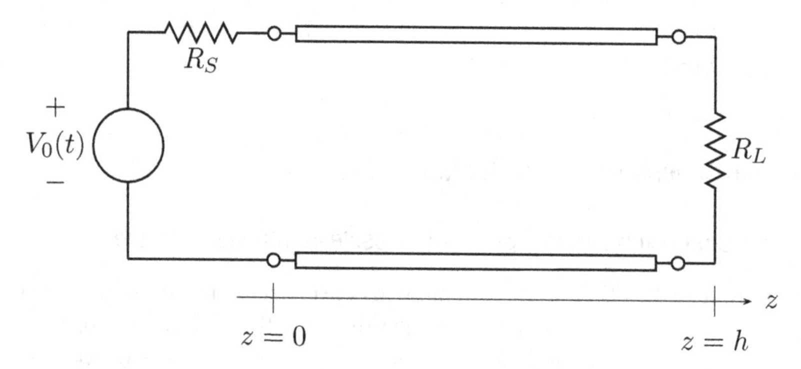

Figure 2.2 Model transmission line problem.

on the line are described by a pair of coupled first-order differential equations, frequently known as the telegraphist's equations:

$$\frac{\partial I(z,t)}{\partial z} = -C\frac{\partial V(z,t)}{\partial t} \tag{2.13}$$

$$\frac{\partial V(z,t)}{\partial z} = -L\frac{\partial I(z,t)}{\partial t} \tag{2.14}$$

As already noted, the transmission line equations are a special case of Maxwell's curl equations in one dimension.

At this stage, we could decouple the differential equations to obtain the wave equation (which is a second-order partial differential equation) for the voltage on the line as a function of position and time. (This is the approach generally taken in introductory electromagnetics texts; the result is the one-dimensional wave equation, in either voltage or current.) However, we will instead work directly with the coupled pair of first-order equations.

Consider the following transmission line circuit problem, illustrated in Fig. 2.2. Assume $L = 1$ H/m, $C = 1$ F/m, $h = 0.25$ m, $R_S = 1\,\Omega$ and $R_L = 2\,\Omega$. (Note that this choice of L and C produces a characteristic impedance of $1\,\Omega$, and velocity of propagation of 1 m/s. Clearly, this is a normalized version of the actual problem; normalized equations such as these are quite frequently used in physics [6, 7].) The following, then, will be our model problem.

The model 1D problem

Determine the phasor representation of the steady-state response $V(z)$ versus z for

$$V_0(t) = \cos(8\pi t), \qquad t > 0 \tag{2.15}$$

The boundary conditions (BCs) at $z = 0$ and $z = h$ are:

$$V(0,t) = V_0(t) - R_S I(0,t) \tag{2.16}$$

$$V(h,t) = R_L I(h,t) \tag{2.17}$$

Take the initial conditions (ICs) to be:

$$V(z,0) = I(z,0) = 0 \tag{2.18}$$

A mathematical aside – classification of this problem

This problem is a deterministic, interior problem controlled by a hyperbolic partial differential equation, with mixed boundary conditions. It is deterministic since there is a source. It is an interior problem since the domain lies inside the boundaries. The vector wave equation is a hyperbolic partial differential equation, and the boundary conditions involve both voltage and current.

The exact solution can of course be readily derived used standard transmission line theory. (This is extremely useful, since it will permit us to test the accuracy of our solution.) Noting that the source is matched, the result is

$$V_{ss}(z) = V^+ \left(e^{-j\beta(z-h)} + \Gamma e^{j\beta(z-h)}\right) \tag{2.19}$$

where

$$V^+ = 1/2 \tag{2.20}$$

$$\Gamma = 1/3 \tag{2.21}$$

$$\beta = 8\pi \text{ rad/m} \tag{2.22}$$

Before moving on to the FDTD solution, it should be noted that the above solution is the *phasor* – i.e. frequency domain – solution of the problem. The excitation is a single-frequency sinusoid, radial frequency $\omega = 8\pi$ rad/s, or $f = 4$ Hz. Since the speed of propagation is 1 m/s, the phase constant/wavenumber is also 8π rad/m, and the wavelength is $1/4$ m. The FDTD is a time domain solver, so we need to bear in mind that we are either going to have to convert the above solution into the time domain, or transform our FDTD solution into the frequency domain. The Fourier transform will of course provide the connection. We will also have to bear in mind that the above is the *steady-state* solution of the problem; there is also the transient part of the solution, which the FDTD solution is also going to include. We will discuss how to deal with these issues subsequently.

2.4.2 FDTD solution of the one-dimensional lossless transmission line problem

The first step in obtaining an FDTD solution is to set up a regular grid in space and time. The points on this grid can be designated as (z_k, t_n), where

$$z_k = (k-1)\Delta z, \qquad k = 1, 2, \ldots, N_z \tag{2.23}$$

$$\Delta z = \frac{h}{N_z - 1}, \qquad N_z \geq 2 \tag{2.24}$$

$$t_n = (n-1)\Delta t, \qquad n = 1, 2, 3, \ldots \tag{2.25}$$

$$\Delta t = \frac{T}{M-1}, \qquad M \geq 2 \tag{2.26}$$

As noted in the introductory remarks of this chapter, additional grid points at half-time and half-space points are now also introduced. These additional points can be designated as $(z_{k+1/2}, t_{n+1/2})$, where

$$z_{k+1/2} = \left(k - \frac{1}{2}\right)\Delta z, \qquad k = 1, 2, \ldots, N_z - 1 \tag{2.27}$$

$$t_{n+1/2} = \left(n - \frac{1}{2}\right)\Delta t, \qquad n = 1, 2, 3, \ldots \tag{2.28}$$

We shall compute $V(z,t)$ at the points (z_k, t_n), and $I(z,t)$ at the points $(z_{k+1/2}, t_{n+1/2})$, i.e. the voltage and currents are computed at offset locations in space *and also in time*. We now have two two-dimensional arrays representing the voltage and current. In each array, a row represents the temporal evolution of the field at a particular point in space, and a column represents the spatial distribution of the field at a particular point in time. (This is very convenient in understanding the method, but we should note now that the FDTD generally stores *only* two or three rows of each array – we will see subsequently how this is possible.)

To assist us in imposing the mixed BCs at $z = 0$ and $z = h$, two additional fictitious columns outside of the boundaries of the problem will be introduced, corresponding to

$$z_{1/2} = -\frac{1}{2}\Delta z \tag{2.29}$$

$$z_{N_z+1/2} = h + \frac{1}{2}\Delta z \tag{2.30}$$

Similarly, to assist in the imposition of the initial conditions at $t = 0$, an additional row will be introduced corresponding to

$$t_{1/2} = -\frac{1}{2}\Delta t \tag{2.31}$$

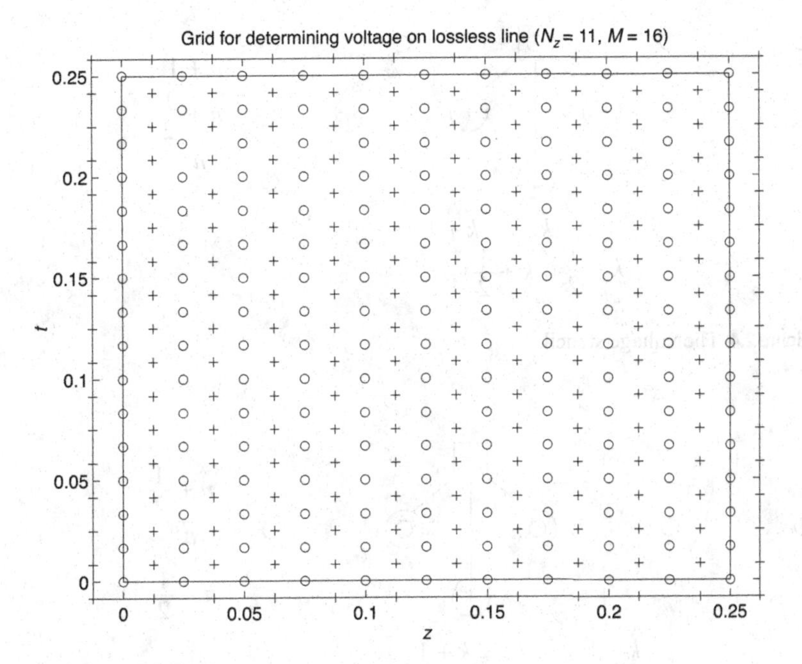

Figure 2.3 The Yee grid.

Figure 2.3 shows these grid points graphically. The "o" indicates points at which $V(z, t)$ is computed, and "+" points at which $I(z, t)$ is computed. (As drawn here, vertical cuts correspond to temporal evolution at a particular point in space, horizontal cuts to spatial distribution at a particular point in time.)

The first transmission line equation, Eq. (2.13), is approximated at z_k and $t_{n+1/2}$ using central differencing in both space and time, i.e.

$$\frac{\partial I\left(z_k, t_{n+1/2}\right)}{\partial z} \approx \frac{I_{k+1/2}^{n+1/2} - I_{k-1/2}^{n+1/2}}{(\Delta z)} \tag{2.32}$$

$$\frac{\partial V\left(z_k, t_{n+1/2}\right)}{\partial t} \approx \frac{V_k^{n+1} - V_k^n}{(\Delta t)} \tag{2.33}$$

Thus, the update equation for V may be obtained as

$$V_k^{n+1} = V_k^n - \frac{\Delta t}{C\Delta z}\left(I_{k+1/2}^{n+1/2} - I_{k-1/2}^{n+1/2}\right) \tag{2.34}$$

This update equation for V may be represented schematically by the "computational molecule" or "stencil" shown in Fig. 2.4. From this, it is clear that the update equation can be used for $k = 2, \ldots, N_z - 1$ and $n \geq 2$. Special update equations must be devised from the initial and boundary conditions to treat $n = 1$ and $k = 1$ and $k = N_z$.

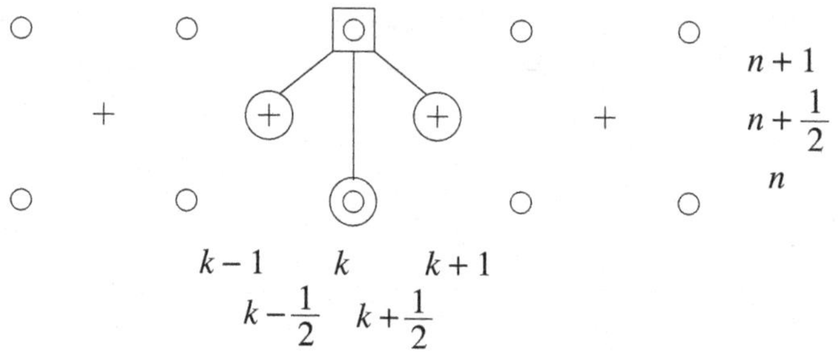

$n+1$
$n+\frac{1}{2}$
n
$k-1$ k $k+1$
$k-\frac{1}{2}$ $k+\frac{1}{2}$

Figure 2.4 The voltage stencil.

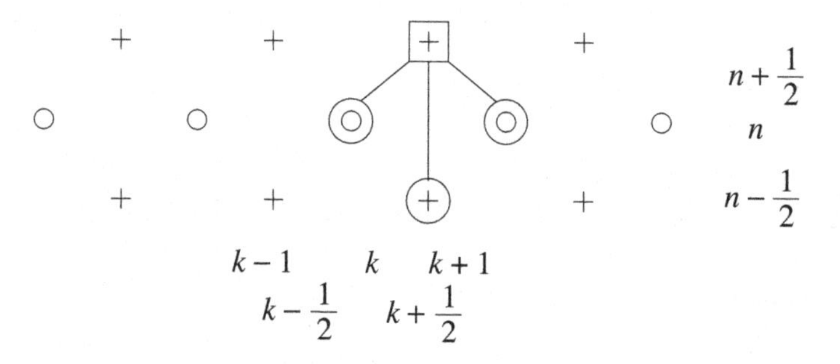

Figure 2.5 The current stencil.

The second transmission line equation, Eq. (2.14), is approximated at $z_{k+1/2}$ and t_n using central differencing in both space and time, i.e.

$$\frac{\partial V\left(z_{k+1/2}, t_n\right)}{\partial z} \approx \frac{V_{k+1}^{n} - V_{k}^{n}}{(\Delta z)} \tag{2.35}$$

$$\frac{\partial I\left(z_{k+1/2}, t_n\right)}{\partial t} \approx \frac{I_{k+1/2}^{n+1/2} - I_{k+1/2}^{n-1/2}}{(\Delta t)} \tag{2.36}$$

Thus, the update equation for I may be obtained as

$$I_{k+1/2}^{n+1/2} = I_{k+1/2}^{n-1/2} - \frac{\Delta t}{L \Delta z}\left(V_{k+1}^{n} - V_{k}^{n}\right) \tag{2.37}$$

The update equation for I may be represented schematically as in Fig. 2.5. Again, it is clear that the update equation can be used for $k = 1, \ldots, N_z - 1$ and $n \geq 2$. Special update equations must be devised from the ICs to treat $n = 1$. However, no special treatment for the boundaries at $z = 0$ or $z = h$ needs to be implemented.

The update equation for V must be modified at $k = 1$ and $k = N_z$ to incorporate the BCs into the solution. Consider the update equation for V at $k = 1$ and $k = N_z$:

$$V_1^{n+1} = V_1^n - \frac{\Delta t}{C\Delta z}\left(I_{3/2}^{n+1/2} - I_{1/2}^{n+1/2}\right) \tag{2.38}$$

$$V_{N_z}^{n+1} = V_{N_z}^n - \frac{\Delta t}{C\Delta z}\left(I_{N_z+1/2}^{n+1/2} - I_{N_z-1/2}^{n+1/2}\right) \tag{2.39}$$

The values of $I_{1/2}^{n+1/2}$ and $I_{N_z+1/2}^{n+1/2}$ must be obtained from the BCs. Consider the BC at $z = 0$:

$$V(0, t_n) + R_S I(0, t_n) = V_0(t_n) \tag{2.40}$$

Using

$$V(0, t_n) = V_1^n \tag{2.41}$$

$$I(0, t_n) \approx \frac{1}{2}\left[I_{1/2}^{n+1/2} + I_{3/2}^{n+1/2}\right] \tag{2.42}$$

the discretized BC gives

$$I_{1/2}^{n+1/2} = -I_{3/2}^{n+1/2} - \frac{2}{R_S}V_1^n + \frac{2}{R_S}V_0(t_n) \tag{2.43}$$

A mathematical aside – semi-implicit approximations

Equation (2.43) is sometimes known as a semi-implicit approximation, since the "next" value at point 1/2 also uses the "next" value at point 3/2. This is also used when conduction currents are included in a full-wave solver. Although widely and successfully used, this approximation can degrade both the stability and accuracy of the solver.

Consider the BC at $z = h$:

$$V(h, t_n) - R_L I(h, t_n) = 0 \tag{2.44}$$

Using

$$V(h, t_n) = V_{N_z}^n \tag{2.45}$$

$$I(h, t_n) \approx \frac{1}{2}\left[I_{N_z-1/2}^{n+1/2} + I_{N_z+1/2}^{n+1/2}\right] \tag{2.46}$$

the discretized BC gives

$$I_{N_z+1/2}^{n+1/2} = -I_{N_z-1/2}^{n+1/2} + \frac{2}{R_L}V_{N_z}^n \tag{2.47}$$

Using the values of $I_{1/2}^{n+1/2}$ and $I_{N_z+1/2}^{n+1/2}$ derived from the BCs, we obtain the update equations for V at $k=1$ and $k=N_z$:

$$V_1^{n+1} = \left(1 - \frac{2\Delta t}{R_S C \Delta z}\right) V_1^n - \frac{2\Delta t}{C\Delta z} I_{3/2}^{n+1/2} + \frac{2\Delta t}{R_S C \Delta z} V_0(t_n) \tag{2.48}$$

$$V_{N_z}^{n+1} = \left(1 - \frac{2\Delta t}{R_L C \Delta z}\right) V_{N_z}^n + \frac{2\Delta t}{C\Delta z} I_{N_z-1/2}^{n+1/2} \tag{2.49}$$

To start the FD scheme, we need to obtain the values of V_k^1 and $I_k^{3/2}$ for $k=1,\ldots,N_z$.

The values of V_k^1 may be obtained from the initial condition $V(z,0)=0$. Hence, $V_k^1=0$ for $k=1,\ldots,N_z$.

The update equation for I for $n=1$ is

$$\begin{aligned} I_{k+1/2}^{3/2} &= I_{k+1/2}^{1/2} - \frac{\Delta t}{L\Delta z}\left(V_{k+1}^1 - V_k^1\right) \\ &= I_{k+1/2}^{1/2} \end{aligned} \tag{2.50}$$

The value of $I_{k+1/2}^{1/2}$ must be obtained from the initial condition. Consider the initial condition:

$$I(z_{k+1/2}, 0) = 0 \tag{2.51}$$

Clearly, I must be zeroed at all points at both timesteps $1/2$ and $3/2$.

Summary: FDTD scheme for the model problem

In summary, the FD scheme for this problem is

$$V_k^1 = 0, \qquad \text{for } k = 1, \ldots, N_z \tag{2.52}$$

$$I_{k+1/2}^{3/2} = 0, \qquad \text{for } k = 1, \ldots, N_z - 1 \tag{2.53}$$

For $n \geq 2$,

$$V_1^n = \left(1 - \frac{2\Delta t}{R_S C \Delta z}\right) V_1^{n-1} - \frac{2\Delta t}{C\Delta z} I_{3/2}^{n-1/2} + \frac{2\Delta t}{R_S C \Delta z} V_0(t_{n-1}) \tag{2.54}$$

$$V_k^n = V_k^{n-1} - \frac{\Delta t}{C\Delta z}\left(I_{k+1/2}^{n-1/2} - I_{k-1/2}^{n-1/2}\right), \qquad \text{for } k = 2, \ldots, N_z - 1 \tag{2.55}$$

$$V_{N_z}^n = \left(1 - \frac{2\Delta t}{R_L C \Delta z}\right) V_{N_z}^{n-1} + \frac{2\Delta t}{C\Delta z} I_{N_z-1/2}^{n-1/2} \tag{2.56}$$

$$I_{k+1/2}^{n+1/2} = I_{k+1/2}^{n-1/2} - \frac{\Delta t}{L\Delta z}\left(V_{k+1}^n - V_k^n\right), \qquad \text{for } k = 1, \ldots, N_z - 1 \tag{2.57}$$

Programming aspects: avoiding half-steps

Half-integer values are inconvenient to program. To avoid them, we can simply make the following changes:

$$n + 1/2 \longrightarrow n \tag{2.58}$$

$$k + 1/2 \longrightarrow k \tag{2.59}$$

However, *this is only a matter of notational convenience, the voltages and currents are still located at the relevant points and times, with half-offsets as appropriate!* This must always be kept in mind. This also extends to both two- and three-dimensional FDTD solvers.

Programming aspects: "in-place" operations

A careful study of the update equations above shows that once all the next values of voltage (i.e. V_k^n) have been obtained, the current values $V_k^{(n-1)}$ are never needed again. (The update equation for current, viz. $I_{k+1/2}^{n+1/2}$, requires only V_k^n.) As such, it is usual practice in an FDTD code to overwrite $V_k^{(n-1)}$ with V_k^n at the end of each timestep. Indeed, it can be done immediately on a point-by-point basis, but it is usually more convenient to do this all in one vector update. Hence only four vectors need be stored, two for voltage and two for current. If, for some reason, one wants the complete time history at a point (or plane or volume) then this is usually stored in a separate array. In signal processing, such operations are known as "in place" operations.

Programming aspects: reducing the "operation count"

It is also possible to reduce the number of operations per timestep (and reduce memory requirements in inhomogeneous problems) if the following change of variables is made:

$$\tilde{V}_k^n = \frac{C\Delta z}{\Delta t} V_k^n \tag{2.60}$$

The algorithm becomes:

$$\tilde{V}_k^1 = 0, \qquad \text{for } k = 1, \ldots, N_z \tag{2.61}$$

$$I_k^1 = 0, \qquad \text{for } k = 1, \ldots, N_z - 1 \tag{2.62}$$

For $n \geq 2$,

$$\tilde{V}_1^n = (1 - \beta_1)\, \tilde{V}_1^{n-1} - 2I_1^{n-1} + \frac{2}{R_S} V_0(t_{n-1}) \tag{2.63}$$

$$\tilde{V}_k^n = \tilde{V}_k^{n-1} - \left(I_k^{n-1} - I_{k-1}^{n-1}\right), \qquad \text{for } k = 2, \ldots, N_z - 1 \tag{2.64}$$

$$\tilde{V}_{N_z}^n = (1 - \beta_2)\, \tilde{V}_{N_z}^{n-1} + 2I_{N_z-1}^{n-1} \tag{2.65}$$

$$I_k^n = I_k^{n-1} - r\left(\tilde{V}_{k+1}^n - \tilde{V}_k^n\right), \qquad \text{for } k = 1, \ldots, N_z - 1 \tag{2.66}$$

$$\beta_1 = \frac{2\Delta t}{R_S C \Delta z} \tag{2.67}$$

$$\beta_2 = \frac{2\Delta t}{R_L C \Delta z} \tag{2.68}$$

$$r = \frac{(\Delta t)^2}{LC(\Delta z)^2} \tag{2.69}$$

Obtaining and evaluating preliminary results

As commented in the opening of this chapter, the FDTD computes results in the *time* domain. The analytical (phasor – steady state) solution is in the *frequency* domain. As

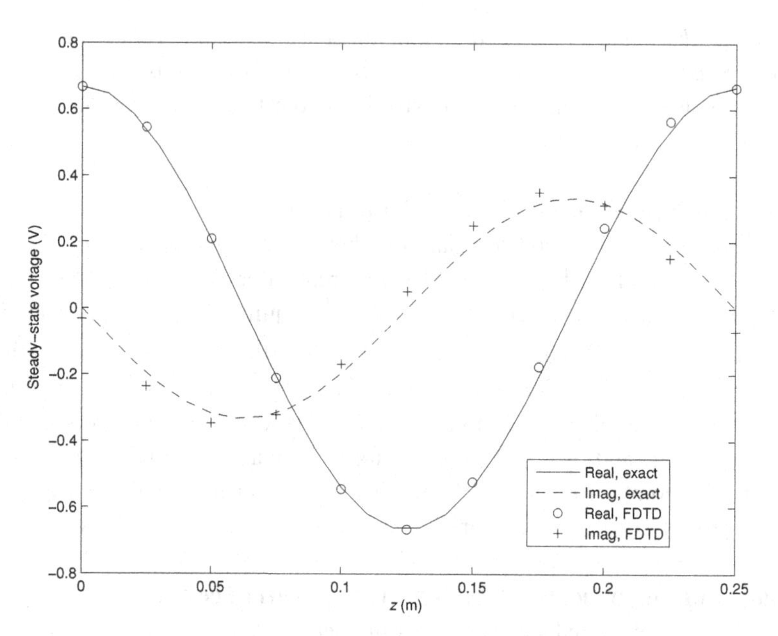

Figure 2.6 Solution for $N_z = 11$, $M = 64$ and $\epsilon = 0.002$.

is well known from circuit theory, the response of a system is the superposition of the transient and steady-state responses. By Fourier transforming the solution at different times, and noting the change in the solution, we can effectively eliminate the transient part of the solution. An estimate of the phasor representation of the steady-state response is obtained at the end of each period (of the sinusoid of 4 Hz frequency, i.e. every 250 ms) by evaluating the Fourier coefficient at frequency $\omega = 8\pi$ rad/s of the time domain data from the FDM solution. In computing the response, time domain data are stored for one period, and then overwritten. Steady state is taken to be achieved when the normalized RMS discrepancy between consecutive estimates is less than some positive error bound, i.e. $D_{\text{rms}} \leq \epsilon > 0$. As a measure of accuracy, we evaluate the normalized RMS error of Yee's algorithm with respect to the exact solution, viz. (E_{rms}). The result of this is given in Fig. 2.6. This particular solution required $N = 6$ for convergence with $E_{\text{rms}} = 0.0432$. Note that since this is a phasor, the result is of course complex, with both real and imaginary parts.

2.4.3 Accuracy, convergence, consistency and stability of the method

For any numerical method, important questions which one must pose include the following:

Accuracy

The degree to which the numerical solution *to the approximate field problem* approximates the exact solution *to the approximate field problem*.

Consistency

A finite difference equation is said to be *consistent* with a PDE provided that the local discretization error tends to zero as the mesh density increases, or alternatively, the mesh increment decreases. This is another statement of *convergence*: the numerical solution should converge to the exact solution as the mesh is refined (for this FDTD problem, this implies $\Delta z \to 0$).

Stability

A process (e.g. a finite difference scheme) is said to be *stable* if and only if errors introduced at any stage in the process remain bounded throughout the entire evolution of the process.

The Lax equivalency theorem – also sometimes called the Lax–Richtmyer theorem – states

> Given a properly posed[3] initial-value problem and a finite-difference approximation to it that satisfies the consistency condition, stability is the necessary and sufficient condition for convergence [3, p. 45].

Stability proofs can be readily obtained; an example of the method introduced by von Neumann will be discussed later in this chapter.

With regard to accuracy, we will first investigate this by numerical experimentation. In short, we are verifying the FDTD scheme. We have discussed this topic in Chapter 1; it is so important that further comments are in order at this point. Right at the start, we must stress that we can only meaningfully talk about accuracy of our numerical model with respect to the field problem which we posed – what we defined as the approximate field problem in Chapter 1. This problem is almost always a simplified version of the real-world problem. For instance, in our transmission line model, we assume no loss; no matter how good our FDTD solution, if the transmission line we are modelling has significant loss, our solution cannot be an accurate simulation of the real problem. Hence, to verify a numerical model, results are often compared to a known analytical solution of the same approximate field problem.

In practice of course, we want to use EM simulators to model the real world, and for this, comparison with measured data is highly desirable or even essential in many cases. Good agreement between measured data and numerically computed results indicates that all the important physics of the problem has been captured in the field description of the problem; the numerical approximation of the field problem is accurate and reliable; and (a point all too often overlooked) that reliable measurements on properly calibrated equipment have been made.

In Fig. 2.7, the effect of decreasing the size of timestep (or as plotted, by equivalently increasing the number of time points per period) is investigated. In Fig 2.8, the effect of decreasing the size of spatial step is investigated (again, the plot shows the equivalent effect of increasing the number of spatial points along the length of the line).

[3] A problem is said to be *properly posed* if: (i) a unique solution exists; (ii) the solution depends continuously on the initial and/or boundary conditions.

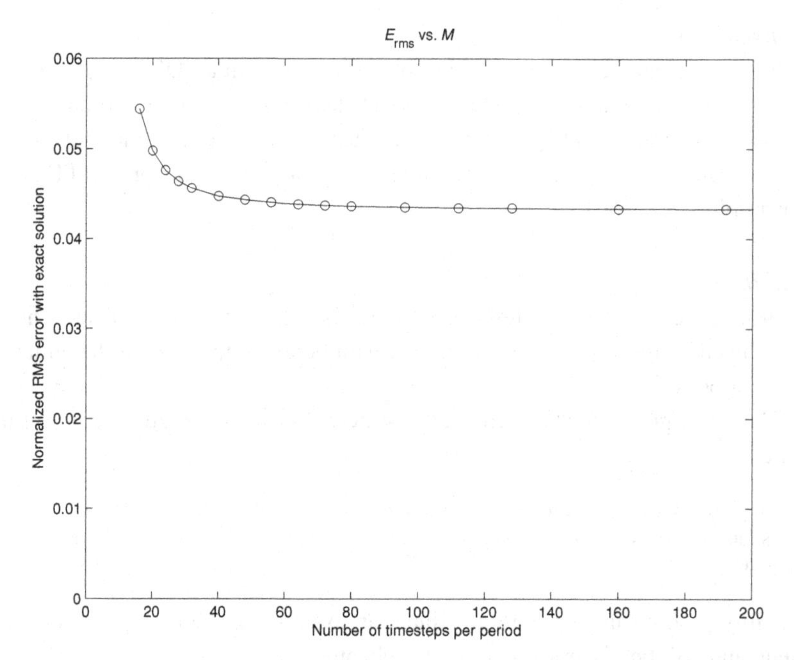

Figure 2.7 Normalized RMS error with exact solution versus number of time points per period (M) with $N_z = 11$. Note that Yee's algorithm is unstable in this case for $M < 16$.

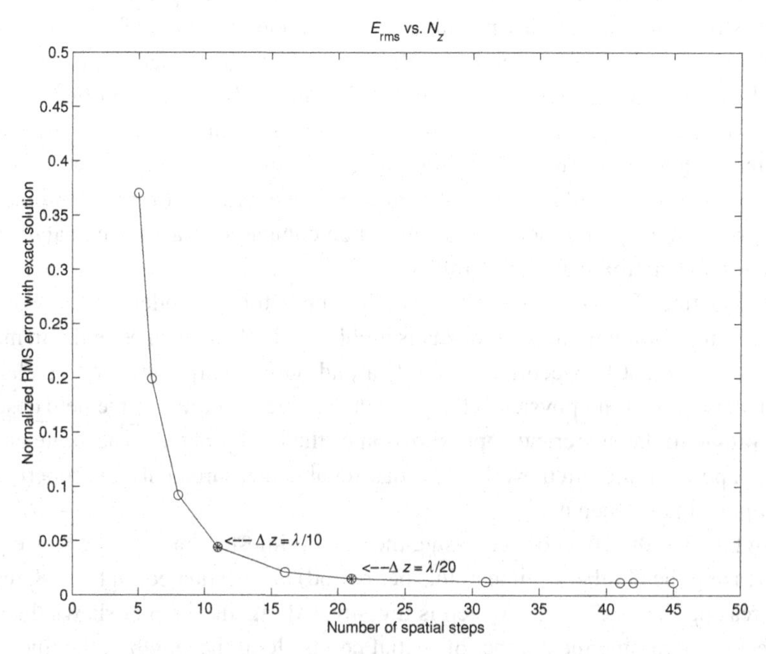

Figure 2.8 Normalized RMS error with exact solution versus number of spatial points (N_z) with $M = 64$. Note that Yee's algorithm is unstable in this case for $N_z > 45$.

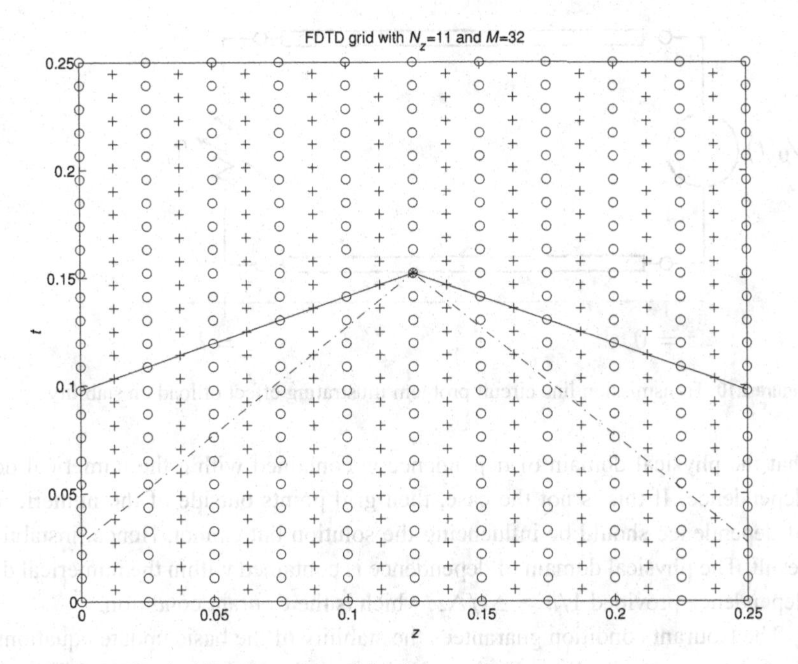

Figure 2.9 Physical interpretation of the Courant limit.

In both cases, one notes that stability imposes limits on the discretization. As already hinted at, with an explicit method such as the FDTD, there will be some maximum timestep size. As Fig. 2.7 implies, this will also be linked to spatial step size. The oft-quoted stability criterion for Yee's algorithm in one, two or three dimensions is

$$\frac{u\Delta t}{\Delta s} \leq \frac{1}{\sqrt{n}} \tag{2.70}$$

where Δs is the length of a side of a uniform cell and n is the number of space dimensions in the problem. For our one-dimensional problem, the above becomes $S = u\Delta t/\Delta z \leq 1$, or

$$\frac{u\Delta t}{\Delta z} \leq 1 \tag{2.71}$$

where $u = 1/\sqrt{LC}$ is the velocity of propagation on the line.

The above stability criterion is also called the *Courant condition*, and it can be derived using von Neumann's method applied to Yee's algorithm. The essential idea is to discretize a known plane wave in the algorithm, and require that its amplitude remain bounded as timestepping progresses; this will be derived later in this chapter.

A physical interpretation of the *Courant condition* may be obtained by considering both the *numerical domain of dependence* and the *physical domain of dependence* for an arbitrary point in the grid. This is illustrated in Fig. 2.9. The region within the solid lines is the *numerical domain of dependence* and the region within the dashed lines is the *physical domain of dependence*. The solid lines have slopes of magnitude $\Delta t/\Delta z$ and the dashed lines have slopes of magnitude $1/u$. Yee's algorithm is stable provided

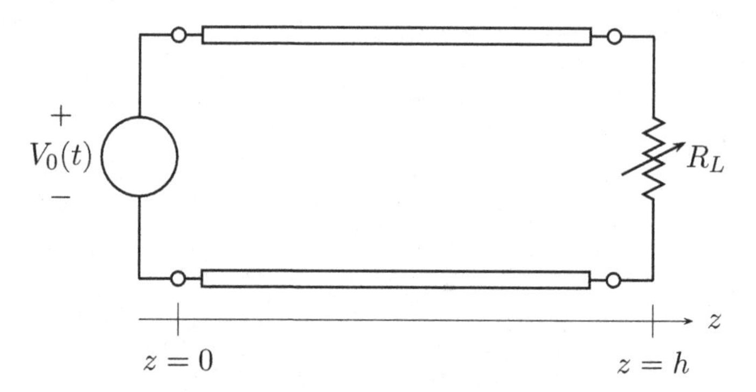

Figure 2.10 Transmission line circuit problem illustrating effect of load on stability.

that the physical domain of dependence is contained within the numerical domain of dependence. If this is not the case, then grid points outside of the numerical domain of dependence should be influencing the solution but cannot. Hence, instability is the result. The physical domain of dependence is contained within the numerical domain of dependence provided $1/u \geq \Delta t/\Delta z$, which is the Courant condition.

The Courant condition guarantees the stability of the basic update equations derived from the transmission line equations. However, it does not guarantee stability of the overall algorithm. Additional stability criteria exist for the update equations at the boundaries. Unfortunately, these are not usually known analytically, and numerical experimentation is often required. In practice, many FDTD simulations use either perfect electrical conductors (PECs) or absorbing boundary conditions on the exterior boundaries; the former simply zero the tangential fields, the latter aim to match the interior wave properties as far as possible. Hence, this is generally not as serious a problem as this example might lead one to believe. Nonetheless, it is a point worth bearing in mind.

We have already seen that, for our particular example, we experience instability for values of S greater than about $1/2$ (S is the fraction of the 1D Courant limit; $S = 1$ implies one is at the limit). Consider the transmission line circuit problem shown in Fig. 2.10. Assume $L = 1$ H/m, $C = 1$ F/m, $h = 0.25$ m and R_L is allowed to vary. Figure 2.11 shows the number of periods required for convergence of the solution and normalized RMS error with the exact solution versus reflection coefficient at the load.[4] Computations were made with $N_z = 11$, $M = 64$ and $\epsilon = 0.002$. The algorithm is found to be unstable for values of R_L equal to or less than about 0.15 Ω, in spite of the very small value of $S = 0.0252$.

Further on the topic of stability, consider the normalized RMS error with the exact solution versus the number of periods used in the calculation for $R_L = 200$ Ω, as shown in Fig. 2.12. Note that, in the context of a 1 Ω system, this load is almost an open circuit. So-called "*late time instabilities*" in Yee's algorithm are rumoured to manifest themselves when dealing with high Q structures – such as this example – that require a large number of timesteps for convergence to the steady state. These instabilities

[4] For $R_L < 1$ Ω, the value of E_{rms} is normalized by dividing by the maximum value of voltage on the line.

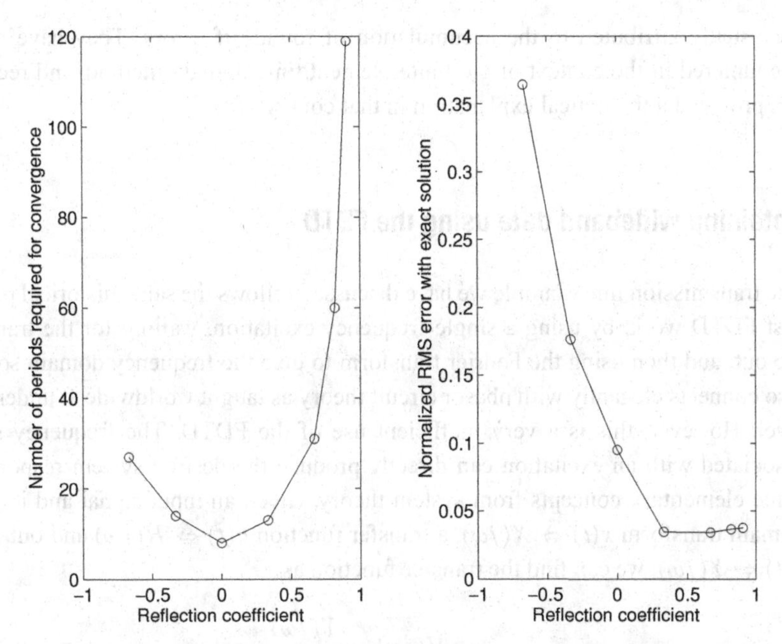

Figure 2.11 Number of periods required for convergence of the solution and normalized RMS error with the exact solution versus reflection coefficient at the load.

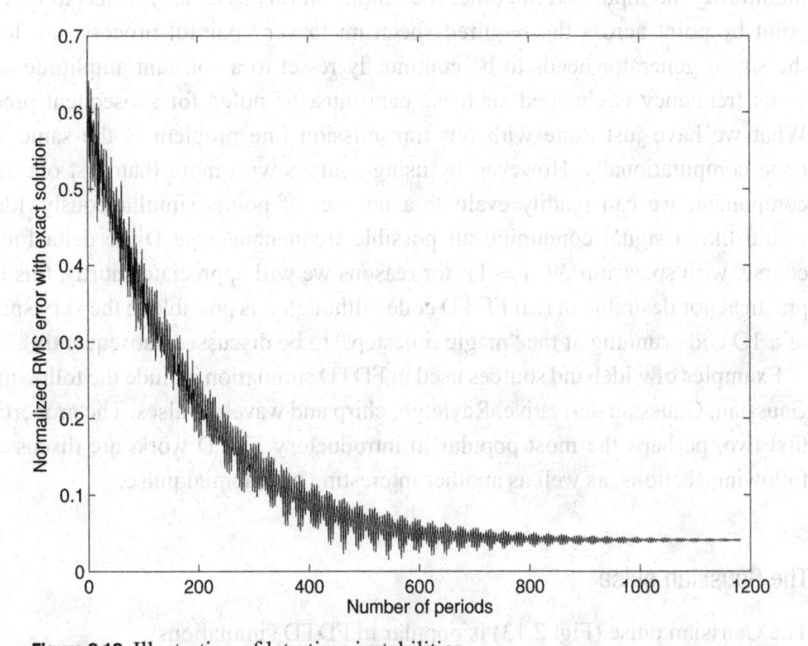

Figure 2.12 Illustration of late-time instabilities.

are usually attributed to the accumulation of round-off errors. They have also been encountered in the context of the finite element time domain method, and recent work has provided a theoretical explanation in that context [8].

2.5 Obtaining wideband data using the FDTD

The transmission line example we have discussed follows the same historical path as the first FDTD work, by using a single-frequency excitation, waiting for the transients to die out, and then using the Fourier transform to give the frequency domain solution. It also connects elegantly with phasor circuit theory as taught worldwide at undergraduate level. However, this is a very inefficient use of the FDTD. The frequency spectrum associated with an excitation can directly produce the desired system response using some elementary concepts from system theory. Given an input signal and its $s(= j\omega)$ domain transform $x(t) \Leftrightarrow X(j\omega)$, a transfer function $h(t) \Leftrightarrow H(j\omega)$ and output signal $y(t) \Leftrightarrow Y(j\omega)$, we can find the transfer function as

$$H(j\omega) = \frac{Y(j\omega)}{X(j\omega)} \tag{2.72}$$

In introductory courses in circuit theory, one may have been asked to measure a transfer function in the laboratory, using a signal generator and an oscilloscope, with one channel monitoring the input and the other the output; in this case, $H(j\omega)$ has to be computed point by point across the required spectrum (a very painful process, not least since the signal generator needs to be continually re-set to a constant amplitude and phase as its frequency is changed, or these data must be noted for subsequent processing). What we have just done with our transmission line problem is the same, although done computationally. However, by using sources with more than just one frequency component, we can readily evaluate a number of points simultaneously. Ideally, we would like a signal containing all possible frequencies (the Dirac delta function, of course, with spectrum $X(s) = 1$); for reasons we will appreciate shortly, this is neither practical nor desirable in real FDTD code (although it is possible in the very special case of a 1D code running at the "magic timestep," to be discussed subsequently).

Examples of wideband sources used in FDTD simulation include the following forms: Gaussian, Gaussian derivative, Rayleigh, chirp and wavelet pulses. The properties of the first two, perhaps the most popular in introductory FDTD work, are discussed in the following sections, as well as another interesting polynomial pulse.

2.5.1 The Gaussian pulse

The Gaussian pulse (Fig. 2.13) is popular in FDTD simulations:

$$v_0(t) = \frac{1}{\sqrt{2\pi}\sigma} e^{-(t-m)^2/2\sigma^2} \tag{2.73}$$

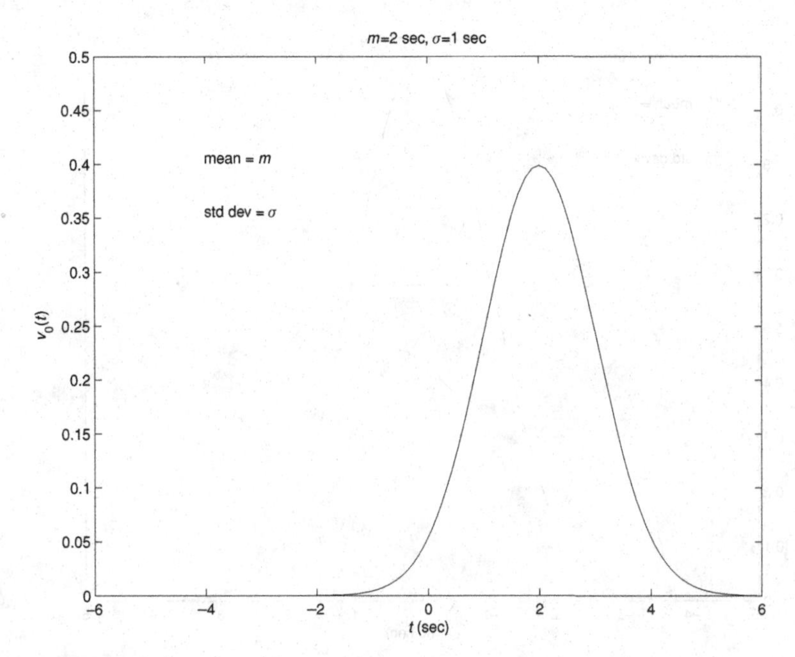

Figure 2.13 A Gaussian pulse.

It has the advantage of having an analytically known spectrum – one of the peculiarities of the Fourier transform is that the spectrum of a Gaussian pulse is also a Gaussian (Fig. 2.14):

$$V_0(\omega) = e^{-j\omega m}\, e^{-\omega^2\sigma^2/2} \tag{2.74}$$

The energy contained in the pulse is also readily obtained:

$$\begin{aligned} E &= \int_{-\infty}^{\infty} v_0^2(t)\, dt = \frac{1}{2\pi}\int_{-\infty}^{\infty} |V_0(\omega)|^2\, d\omega \\ &= \frac{1}{2\sigma\sqrt{\pi}} \end{aligned} \tag{2.75}$$

However, the Gaussian pulse has some significant disadvantages. The most important are:

- it exists for all time, including $t < 0$;
- it has a strong frequency component at $\omega = 0$, i.e. DC.

The former requires that the pulse be windowed at some time (i.e. set to zero) which means there is a slight discontinuity of switch-on. The latter is a more subtle point; it turns out the static (DC) component can cause problems with charge build-up in FDTD grids[5] and it is better to avoid strong DC spectral components in FDTD simulations.

[5] Showing this is beyond the scope of this introductory discussion – for a detailed analysis, refer to [9].

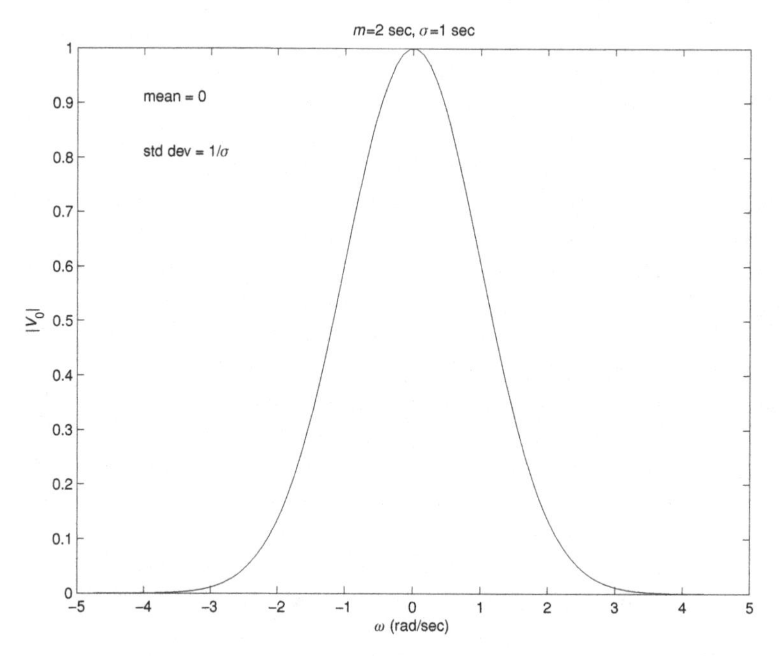

Figure 2.14 Spectrum of a Gaussian pulse.

2.5.2 The Gaussian derivative pulse

A simple variant on the Gaussian, namely its derivative (Fig. 2.15), is also very popular in FDTD simulations, since it removes the DC component. It is defined as follows:

$$v_0(t) = \frac{-1}{\sqrt{2\pi}} \frac{(t-m)}{\sigma^3} e^{-(t-m)^2/2\sigma^2} \tag{2.76}$$

The spectrum of the Gaussian derivative pulse is (Fig. 2.16):

$$V_0(\omega) = j\omega e^{-j\omega m} e^{-\omega^2\sigma^2/2} \tag{2.77}$$

The energy of a Gaussian derivative pulse is also easily computed:

$$\begin{aligned} E &= \int_{-\infty}^{\infty} V_0^2(t)\, dt = \frac{1}{2\pi} \int_{-\infty}^{\infty} |V_0(\omega)|^2\, d\omega \\ &= \frac{1}{4\sigma^3\sqrt{\pi}} \end{aligned} \tag{2.78}$$

2.5.3 A polynomial pulse

A pulse with finite support and interesting properties is the following, of quartic polynomial form:

$$f(t) = \begin{cases} (1-t^2)^4, & \forall |t| \le 1 \\ 0, & \text{otherwise} \end{cases} \tag{2.79}$$

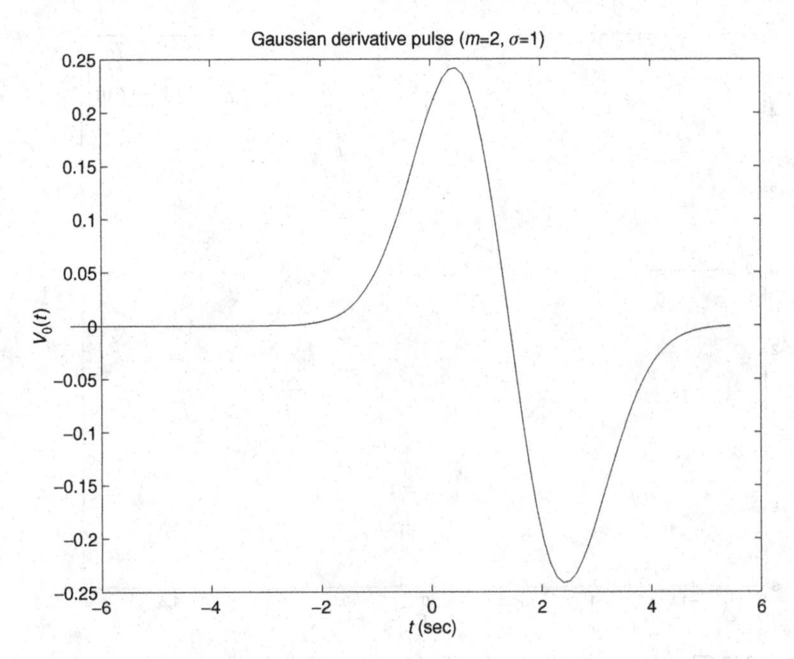

Figure 2.15 A Gaussian derivative pulse.

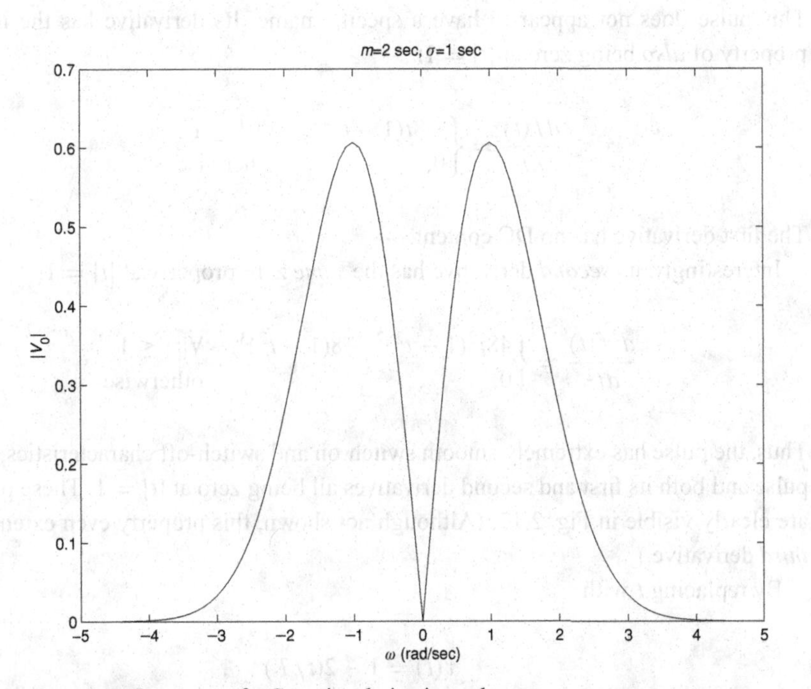

Figure 2.16 Spectrum of a Gaussian derivative pulse.

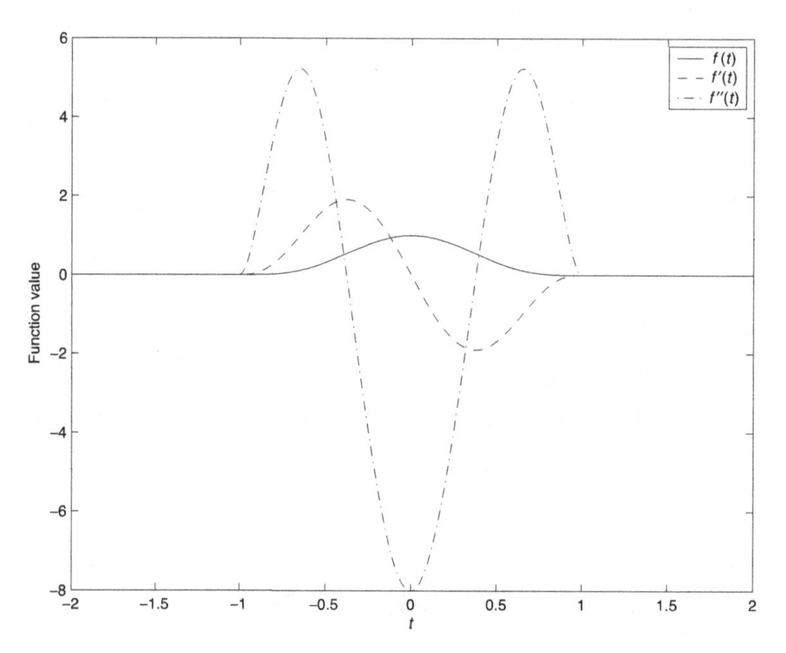

Figure 2.17 The $(1 - t^2)^4$ pulse, and its first and second derivatives.

This pulse does not appear to have a specific name. Its derivative has the important property of *also* being zero at $|t| = 1$:

$$\frac{df(t)}{dt} = \begin{cases} -8t(1-t^2)^3, & \forall |t| \leq 1 \\ 0, & \text{otherwise} \end{cases} \tag{2.80}$$

The first derivative has no DC content.

Interestingly, its *second* derivative has the *same* zero property at $|t| = 1$:

$$\frac{d^2 f(t)}{dt^2} = \begin{cases} 48t^2(1-t^2)^2 - 8(1-t^2)^3, & \forall |t| \leq 1 \\ 0, & \text{otherwise} \end{cases} \tag{2.81}$$

Thus, the pulse has extremely smooth switch-on and switch-off characteristics, with the pulse and both its first and second derivatives all being zero at $|t| = 1$. These properties are clearly visible in Fig. 2.17. (Although not shown, this property even extends to the *third* derivative.)

By replacing t with

$$\tau(t) = 1 - 2(t/T) \tag{2.82}$$

in the above, a pulse is obtained with switch-on time $\tau = 0$ and duration T.

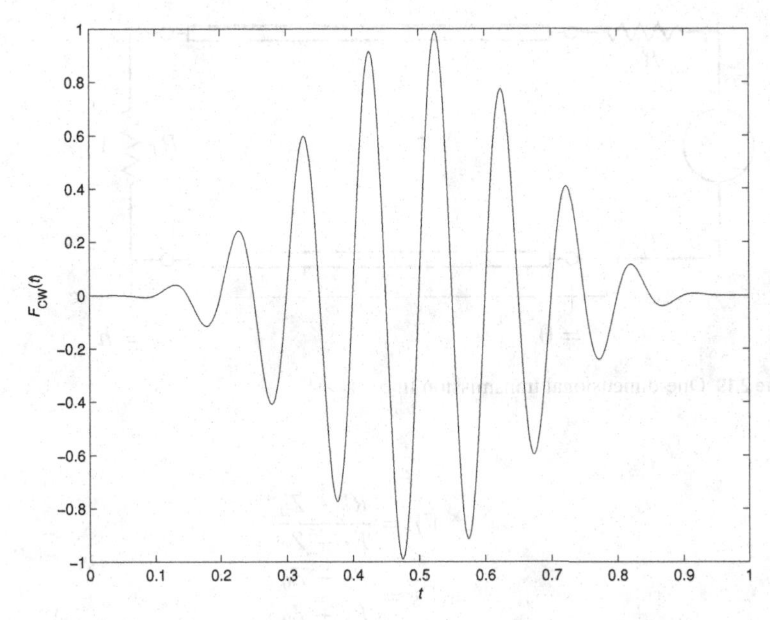

Figure 2.18 A ten-cycle windowed sinusoid, defined over the interval [0, 1].

The Fourier transform of these pulses must be computed numerically.

This pulse is especially suitable for use as a windowed sinusoid (continuous wave):

$$F_{CW} = [1 - \tau(t)^2]^4 \ \sin[m(2\pi t/T)] \tag{2.83}$$

with integer m controlling the number of cycles in the pulse. Clearly, $m = 1$ corresponds to one cycle only, since the windowing function is non-zero only in the interval $t = [0, T]$. An example of a ten cycle windowed sinsusoid is shown in Fig. 2.18.

This specific pulse, and its use as a window, appear to have been introduced in [10], although windowed sinusoids have been quite widely used in FDTD analysis.

2.5.4 The 1D transmission line revisited from a wideband perspective

We will now revisit our model 1D transmission line problem, and pose a slightly different question.

Find the frequency response $V_L(\omega)/V_0(\omega)$ of the transmission line circuit shown in Fig. 2.19 from $0 \leq \omega \leq 16\pi$ rad/s. Assume $L = 1$ H/m, $C = 1$ F/m, $h = 0.25$ m, $R_S = 0.5\,\Omega$ and $R_L = 2\,\Omega$.

Using standard frequency domain transmission line analysis, we can obtain an exact solution for this problem as

$$\frac{V_L(\omega)}{V_0(\omega)} = \frac{Z_0}{Z_0 + R_S} \frac{1 + \Gamma_L}{e^{j\beta h} - \Gamma_L \Gamma_S e^{-j\beta h}} \tag{2.84}$$

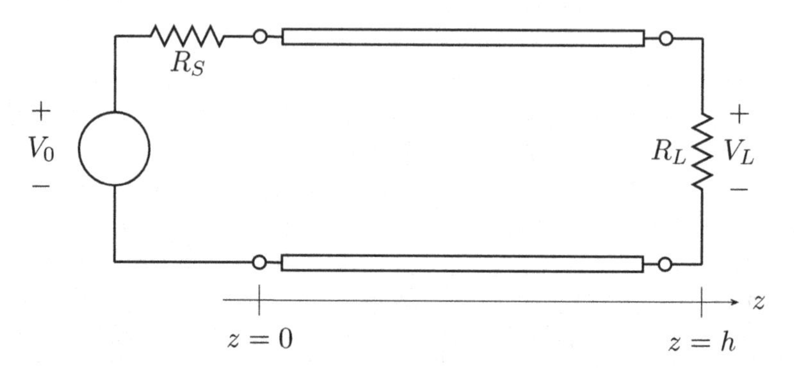

Figure 2.19 One-dimensional transmission line.

where

$$\Gamma_L = \frac{R_L - Z_0}{R_L + Z_0} \tag{2.85}$$

$$\Gamma_S = \frac{R_S - Z_0}{R_S + Z_0} \tag{2.86}$$

$$Z_0 = \sqrt{\frac{L}{C}} \tag{2.87}$$

$$\beta = \omega\sqrt{LC} \tag{2.88}$$

To obtain the transfer function of the circuit using a Gaussian pulse source, we need to do the following:

- Set the bandwidth of the source to be wide enough to cover the frequencies of interest by choosing the standard deviation of the frequency spectrum of the source to be equal to $\omega_{\max}$, the maximum radian frequency of interest, i.e.

$$1/\sigma = \omega_{\max} \tag{2.89}$$

  In the results to be shown, $\omega_{\max} = 16\pi$ was used. This is sufficient to demonstrate the behavior of the transfer function over frequency, as well as the dispersive nature of the FDTD algorithm (more on this subsequently).
- Set the mean value of the time domain source function to be equal to four standard deviations so that the source can safely be assumed to be zero for $t \leq 0$, i.e.

$$m = 4\sigma = 4/\omega_{\max} \tag{2.90}$$

- Choose a space step such that

$$\Delta z \leq \frac{\lambda_{\min}}{10} \tag{2.91}$$

  or, equivalently,

$$\Delta z \leq \frac{\pi}{5\omega_{\max}\sqrt{LC}} \tag{2.92}$$

- Choose a timestep that satisfies both the stability criterion for Yee's algorithm (*Courant condition*) and the required *Nyquist sampling rate* for the highest frequency in the pulse:

$$\Delta t \leq \min\left(\Delta z\sqrt{LC}, \frac{\pi}{4\omega_{\max}}\right) \tag{2.93}$$

where we assumed that the highest frequency in the pulse is $4\omega_{\max}$. (All finite-time sources have a theoretically infinite spectrum; we have to decide some reasonable upper limit on the spectrum. Recall that the Nyquist theorem states that a signal with maximum frequency content f_m must be sampled at at least twice this frequency, i.e. $\Delta t = 1/2f_m$, or $\Delta t = \pi/\omega_m$ in terms of radian frequency. In this case, we chose $\omega_m = 4\omega_{\max}$. Be careful not to confuse $\omega_{\max}$, the maximum frequency of interest, with ω_m, the maximum frequency present in the simulation!) Remember that the Courant condition does not guarantee the stability of the update equations at the boundaries.
- Use the FDTD update equations to let the system evolve during the source "on" time, which can be taken to be $0 \leq t \leq m + 4\sigma$. At the end of this time, compute the total energy of the source as[6]

$$E_{\text{source}} = \int_0^{m+4\sigma} v_0^2(t)\,dt \tag{2.94}$$

Also, compute the Fourier transform of the response and the first estimate of its total energy as

$$E_L^{(1)} = \int_{-\infty}^{\infty} |V_L(\omega)|^2\,d\omega \tag{2.95}$$

- Allow the time evolution of the system to proceed. Periodically interrupt the time evolution to compute the Fourier transform of the response and a new estimate of the total energy of the response. Stop the time evolution of the system when the difference between the Kth and $(K-1)$th estimates of the total energy of the response normalized to the total energy of the source is less than or equal to some positive error bound, i.e.

$$\frac{\left|E_L^{(K)} - E_L^{(K-1)}\right|}{E_{\text{source}}} \leq \epsilon > 0 \tag{2.96}$$

and the total energy of the response is greater than some small fraction of the total energy of the source.

[6] This can be done conveniently in MATLAB using the `trapz` function.

2.5.5 Estimating the Fourier transform

The Fourier transform $X(\omega)$ of a time domain signal $x(t)$, for angular frequencies $\omega = 2\pi f$ is defined as

$$X(\omega) = \int_{-\infty}^{\infty} x(t)\, e^{-j\omega t}\, dt \tag{2.97}$$

and inverse transform

$$x(t) = \frac{1}{2\pi} \int_{-\infty}^{\infty} X(f)\, e^{j\omega t}\, d\omega \tag{2.98}$$

The pair are also often written as

$$X(f) = \int_{-\infty}^{\infty} x(t)\, e^{-j2\pi f t}\, dt \tag{2.99}$$

and inverse transform

$$x(t) = \int_{-\infty}^{\infty} X(f)\, e^{j2\pi f t}\, df \tag{2.100}$$

We will approximate the Fourier transform using the discrete Fourier transform (DFT) defined by

$$X(k) = \sum_{n=1}^{N} x(n)\, e^{-j2\pi(k-1)(n-1)/N}, \qquad 1 \leq k \leq N \tag{2.101}$$

Signal processing experts sometimes view the two as entirely different transforms, and indeed, there are significant differences: the Fourier transform is defined for aperiodic signals, whereas the DFT automatically renders the signals periodic (at the Nyquist frequency); the Fourier transform is continuous, the DFT is discrete. However, we can very usefully approximate the Fourier transform with the DFT if we bear this in mind, ensure that we satisfy the sampling theorem and note that the DFT as defined above is missing the correct normalization. By replacing the infinite limits in Eq. (2.97) with 0 and $T = N\Delta t$, and then approximating the integral as a finite sum with $\Delta t = T/N$, we see that the DFT approximates the Fourier transform, but with a Δt scale factor missing, and also with the signal repeated with period T.

The DFT can be confusing when first used in this context, since the DC component is *not* in the middle as one might expect, but is rather the first component $k = 1$. Some definitions of the DFT include a $1/N$ scaling factor in the forward transform; others include this in the inverse transform. The DFT implementation in MATLAB (`fft`) uses the latter convention. The DFT yields N discrete frequency samples with a spacing $\Delta f = 1/T = 1/N\Delta t$. The number of samples N is usually taken to be a power of 2 (also sometimes known as radix-2) so that efficient algorithms, specifically the FFT, can be used to compute the DFT.[7] For an even number of samples N, the actual frequencies

[7] In Chapter 6, the fast Fourier transform (FFT) is discussed in some detail.

are defined as

$$f_k = (k - 1)\Delta f \tag{2.102}$$

for $k = 1, 2, \ldots, N/2$, and

$$f_k = (k - N - 1)\Delta f \tag{2.103}$$

for $k = N/2 + 1, \ldots, N$ (the negative frequencies). The frequency at $k = N/2 + 1$ (which can equally validly be viewed as a positive frequency) $\mp(N/2)\Delta f$ is also known as the *folding frequency* or the *Nyquist frequency*, and the Fourier transform is symmetric about this.

An aside – gaining confidence with the DFT (and FFT)

Despite undergraduate exposure, the DFT can remain rather mysterious to many students. One way to gain confidence with the DFT is to Fourier transform simple signals, whose transforms are known. Consider a cosine signal of angular frequency 1 rad/s. Its period is 2π s, and its frequency $1/(2\pi)$ Hz. (From elementary courses on signal theory, it will be recalled that its Fourier transform is $\pi\{\delta(\omega + 1) + \delta(\omega - 1)\}$.) Let us take eight samples over one period (remember that we must take more than two to satisfy the sampling theorem!). These should be equally spaced from $t = 0$ to $t = (7/8)2\pi$. (Including the sample at $t = 2\pi$ would be incorrect, since this point has already been included at $t = 0$.) This can be achieved very simply in MATLAB by using the command:

```
t=linspace(0,(7/8)*2*pi,8).
```

Now we create the cosine signal:

```
x=cos(t)
```

and apply the DFT (implemented as the FFT) to this:

```
X=fft(x).
```

The result is the following vector:

```
X=[0 4 0 0 0 0 0 4]
```

Inserting the $\Delta t = T/N$ scale factor, with $T = 2\pi$ and $N = 8$ in this case, which MATLAB omits, this vector is

```
X=pi*[0 1 0 0 0 0 0 1]
```

and we immediately recognize the positive frequency component $X(k = 1)$ at $f_1 = 1/(2\pi)$, and negative frequency component $X(k = 8)$ at $f_8 = -1/(2\pi)$. (Note that the FFT is complex, but by choosing a signal with even time symmetry, only the real parts of the Fourier transform are non-zero.)

Since most of our applications of the Fourier transform will be in computing *ratios* of spectra, the constants are not of great concern, but should be included for completeness.

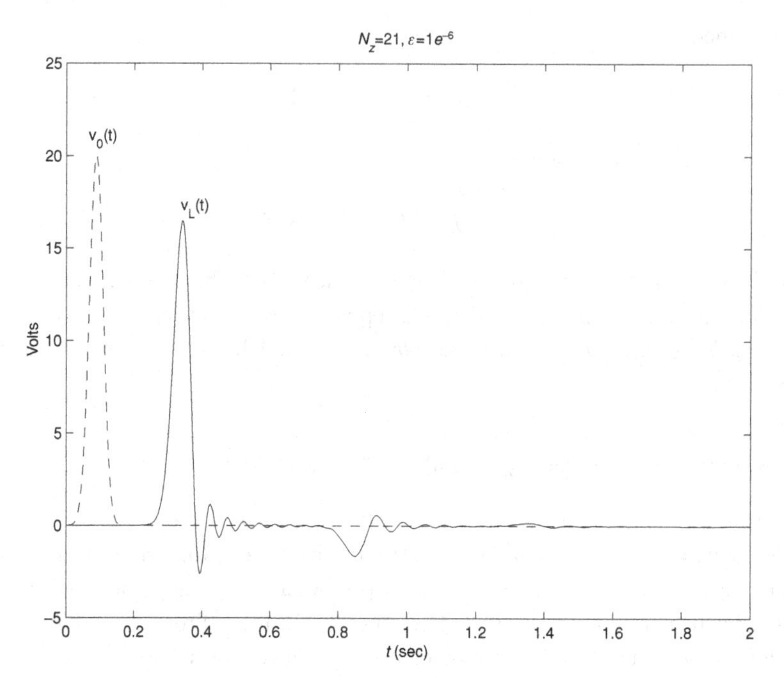

Figure 2.20 Generator and load voltages in the time domain for a Gaussian pulse.

The highest (non-aliased) frequency in the spectrum produced by the FFT is $F_{\max} = 1/2\Delta t$ and the frequency points are spaced by $\Delta f = 1/N\Delta t$. Additional frequency points (i.e. smaller values of Δf) can be obtained by *zero-padding* of the time domain data. The spectrum obtained by zero-padding of the time domain data is equivalent to that obtained by sinc-interpolation of the frequency domain data. (As an aside, we note that zero-padding to improve frequency resolution is a questionable practice, since no additional real data have been added to the system.)

To compare the FDTD solution to the exact solution, define the normalized RMS error with respect to the exact solution as

$$E_{\text{rms}} = \sqrt{\frac{1}{N}\sum_{r}\left|\frac{V_L^{(\text{FDTD})}(\omega_r)}{V_0^{(\text{FDTD})}(\omega_r)} - \frac{V_L^{(\text{exact})}(\omega_r)}{V_0^{(\text{exact})}(\omega_r)}\right|^2} \tag{2.104}$$

2.5.6 Simulation using Gaussian and Gaussian derivative pulses

The FDTD solution with Gaussian pulse excitation required 694 total timesteps for convergence and results in $E_{\text{rms}} = 0.189$. Results are shown in Figs. 2.20, 2.21 and 2.22. The "ringing" in Fig. 2.20 is characteristic in telecommunications theory of a wideband signal on a dispersive channel, and we will see shortly that the FDTD indeed has dispersive properties.

The FDTD solution with Gaussian derivative pulse excitation requires 820 total timesteps for convergence and results in $E_{\text{rms}} = 0.190$. The generator and load voltages in the time domain for a Gaussian derivative pulse are shown in Fig. 2.23.

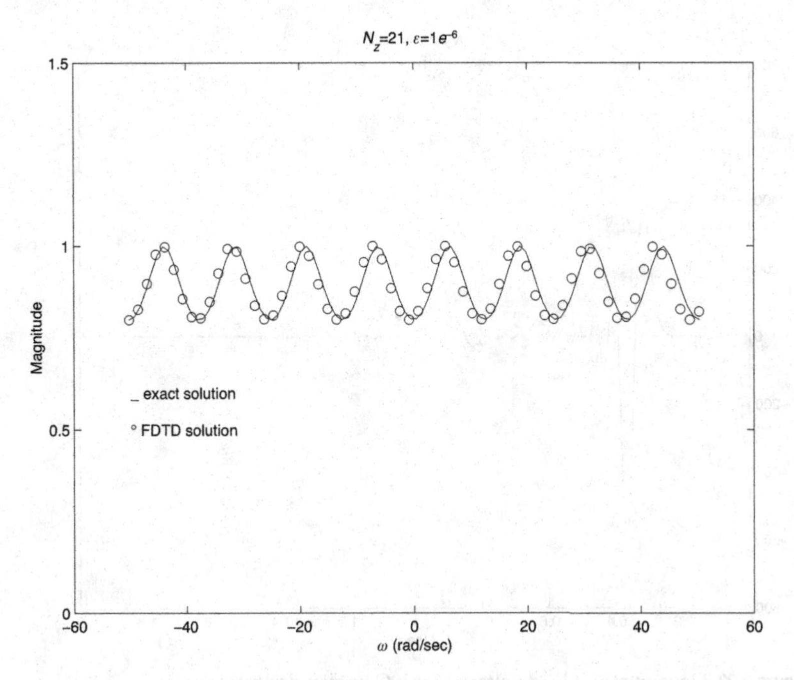

Figure 2.21 Magnitude of the transfer functions for exact and FDTD solutions.

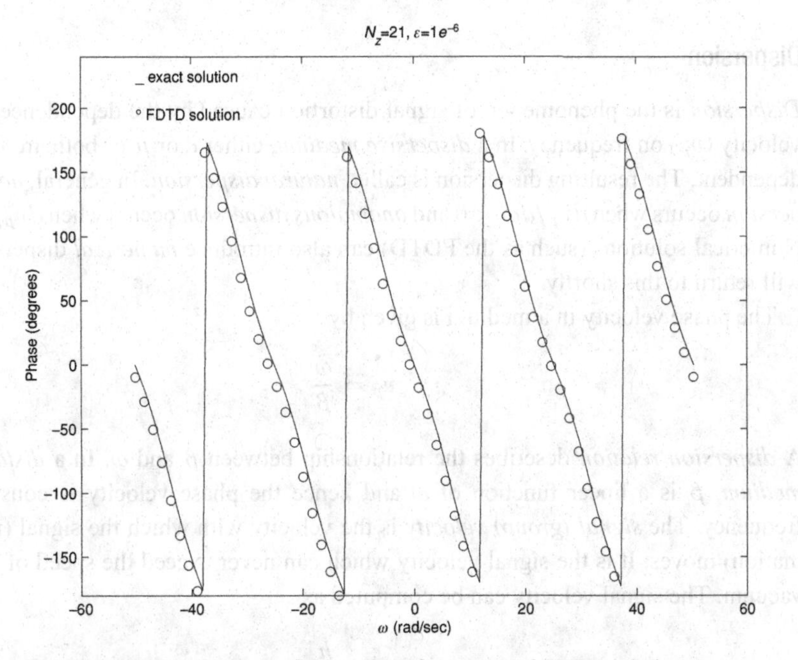

Figure 2.22 Phase of the transfer functions for exact and FDTD solutions.

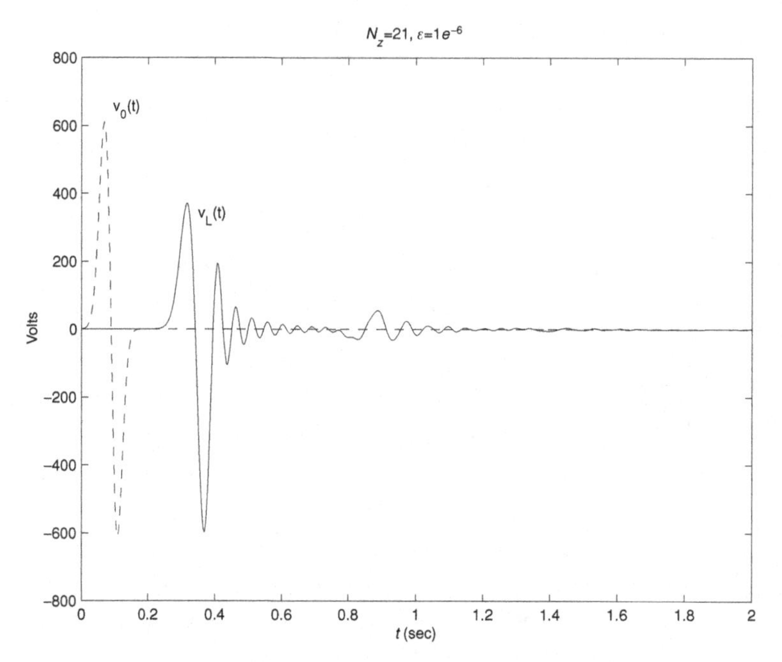

Figure 2.23 Generator and load voltages for a Gaussian derivative pulse.

2.6 Numerical dispersion in FDTD simulations

2.6.1 Dispersion

Dispersion is the phenomenon of signal distortion caused by the dependence of phase velocity (v_p) on frequency. In a *dispersive medium*, either ϵ or μ or both are frequency dependent. The resulting dispersion is called *natural dispersion*. In general, *normal dispersion* occurs when $dv_p/d\omega < 0$ and *anomalous dispersion* occurs when $dv_p/d\omega > 0$. Numerical solutions (such as the FDTD) can also introduce *numerical* dispersion – we will return to this shortly.

The phase velocity in a medium is given by

$$v_p = \frac{\omega}{\beta} \tag{2.105}$$

A *dispersion relation* describes the relationship between β and ω. In a *distortionless medium*, β is a linear function of ω and hence the phase velocity is constant with frequency. The *signal (group) velocity* is the velocity with which the signal (i.e. information) moves. It is the signal velocity which can never exceed the speed of light in a vacuum. The signal velocity can be computed as

$$v_g = \frac{h}{T_d} \tag{2.106}$$

where T_d is the delay time experienced by the signal in travelling over the distance h.

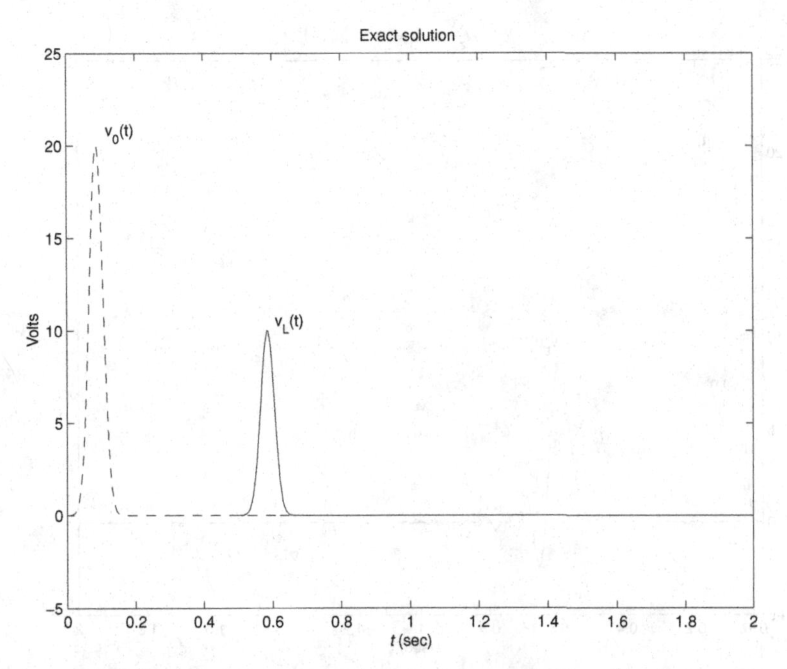

Figure 2.24 Theoretical generator and load voltages in the time domain for a Gaussian pulse.

Assume that the signal at $z = 0$ is given by $v_0(t)$, and that the signal at $z = h$ is given by $v_L(t)$. The delay time in travelling from $z = 0$ to $z = h$ (T_d) is the value of τ which maximizes the cross-correlation between $v_L(t)$ and $v_0(t)$, i.e.

$$\chi_L(\tau) = \int_{\infty}^{\infty} v_L(t)\, v_0(t - \tau)\, dt \tag{2.107}$$

In a distortionless medium, the group velocity is equal to the phase velocity.

A numerical algorithm can introduce *numerical dispersion*, even when waves are propagating in a distortionless medium. Yee's FDTD algorithm causes numerical dispersion. We will illustrate this by comparing theoretical and FDTD results for our simple transmission line circuit shown in Fig. 2.19. Assume $L = 1$ H/m, $C = 1$ F/m, $h = 0.5$ m, $R_S = 1\ \Omega$ and $R_L = 1\ \Omega$. This transmission line is distortionless with $\beta = \omega\sqrt{LC}$. The phase velocity on the line (v_p) is a constant versus frequency and is equal to 1 m/s. The source and load impedances are equal to the characteristic impedance of the line. Hence, there are no reflections at either end of the line. The theoretical generator and load voltages in the time domain for the following Gaussian pulse excitation are shown in Fig. 2.24:

$$v_L(t) = \frac{1}{2} v_0\,(t - 0.5) \tag{2.108}$$

Compare these with the results computed in the time domain using the FDTD shown in Fig. 2.25. The ringing clearly visible on the load voltage is the result of numerical dispersion.

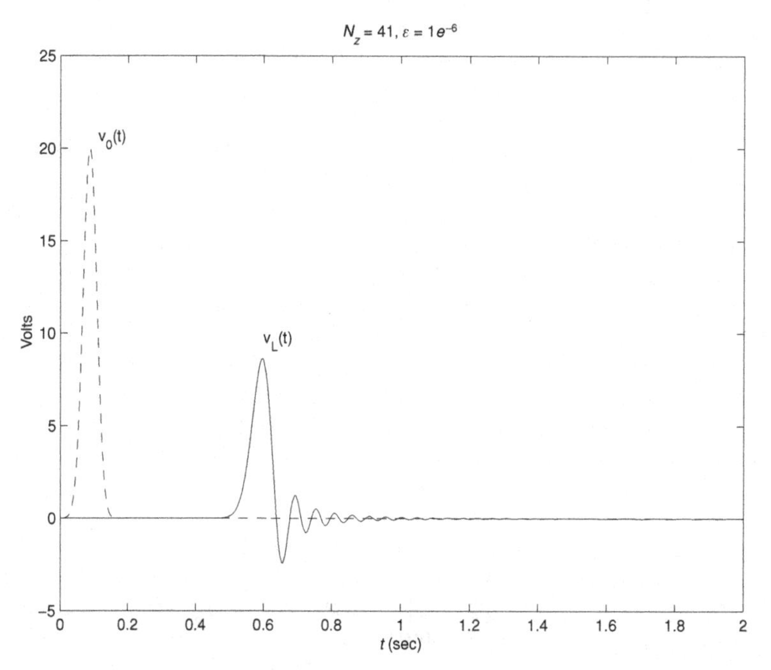

Figure 2.25 FDTD generator and load voltages in the time domain for a Gaussian pulse.

2.6.2 Derivation of the dispersion equation

To obtain the numerical dispersion relation resulting from Yee's algorithm, we assume monochromatic plane-wave trial solutions. Substituting these trial solutions into the update equations and performing some straightforward algebraic manipulations yields the numerical dispersion relation. The procedure is as follows.

Firstly, assume trial solutions of plane-wave form. In continuous space-time, a z-propagating plane wave has the form $e^{j\omega t}e^{-j\beta z}$. In discretized form, and allowing for arbitrary amplitude, this becomes

$$V_k^n = A\, e^{j\omega n\Delta t}\, e^{-j\beta k\Delta z} \tag{2.109}$$

A similar equation can be written for the discretized current:

$$I_k^n = B\, e^{j\omega(n+1/2)\Delta t}\, e^{-j\beta(k+1/2)\Delta z} \tag{2.110}$$

noting the offset between voltage and current.

These are now substituted into the update equations (2.34) and (2.37) to obtain the expression for the *next* timestep:

$$V_k^{n+1} = A\, e^{j\omega n\Delta t}\, e^{-j\beta k\Delta z} - \frac{\Delta t\beta}{c\Delta z}\left(e^{j\omega n\Delta t}\, e^{j\omega\Delta t/2}[e^{-j\beta(k+1/2)\Delta z} - e^{-j\beta(k-1/2)\Delta z}]\right) \tag{2.111}$$

Obviously, the last exponential term can be simplified as a sinusoid.

The *crucial* step in the derivation is to recognize that the discretized plane wave can *also* be written as

$$V_k^{n+1} = A\, e^{j\omega(n+1)\Delta t}\, e^{-j\omega k \Delta z} \tag{2.112}$$

Since these two equations represent the same wave (albeit via the FDTD update and the analytical solution respectively), we can equate them. Thus, equating Eqs. (2.111) and (2.112), noting that for a plane wave the ratio of voltage to current is $Z_0 = \sqrt{L/C}$, and simplifying Eq. (2.111), we obtain the *dispersion equation*:

$$\sin\left(\frac{\omega \Delta t}{2}\right) - \frac{\Delta t}{\sqrt{LC}\Delta z} \sin\left(\frac{\beta \Delta z}{2}\right) = 0 \tag{2.113}$$

In the limit as $\Delta z \to 0$ (and thus, from the Courant limit, $\Delta t \to 0$), the small argument approximation (Taylor series expansion) of the sine function can be applied, and the expression becomes the exact (dispersionless) relation for the transmission line. This is important, because it indicates that *dispersion in an FDTD mesh can be controlled by making the mesh sufficiently fine.* This is a general result, and applies in 2D and 3D (although the dispersion equation is more complex, of course).

It is useful to note that the same result can be obtained from the equivalent second-order central difference scheme, which can be derived by applying Eq. (2.12) to the second-order wave equation which is obtained by eliminating either V or I between Eqs. (2.13) and Eq. (2.14). For instance, if one eliminates I, and assumes that L is constant, the wave equation in V is given by

$$\frac{\partial^2 V(z,t)}{\partial z^2} - \frac{1}{u^2}\frac{\partial^2 V(z,t)}{\partial t^2} = 0 \tag{2.114}$$

and the FDTD-equivalent update scheme, using central differencing, is

$$V_k^{n+1} = 2V_k^n - V_k^{n-1} + \left(\frac{u\Delta t}{\Delta z}\right)^2 \left[V_{k+1}^n - 2V_k^n + V_{k-1}^n\right] \tag{2.115}$$

This result can also be obtained by eliminating I in the FDTD update equations. We will make use of this shortly, when deriving the stability criteria.

2.6.3 Some closing comments on dispersion in FDTD grids

Given ω, L, C, Δt and Δz, the above non-linear equation can be solved numerically for β, allowing us to determine the phase velocity as a function of frequency. This is shown graphically in Fig. 2.26.

The exact group velocity is $v_g^{\text{exact}} = 1$ m/s, and the group velocity resulting from using Yee's algorithm varies over the range of frequencies simulated in our model problem from this value to around 0.984 m/s, a difference exceeding 10%.

As a closing comment on the subject of dispersion, it is interesting to note, from Eq. (2.113), that if the FDTD simulation is run *at the Courant limit*, viz. $\Delta t = \Delta z/c$, with $c = 1/\sqrt{LC}$, the term in front of the second sinusoid becomes unity, hence the sinusoids are equal and hence their arguments, thus $\omega/\beta = \Delta z/\Delta t \equiv v$, in other words, *there is no dispersion*. This is also sometimes known as the "magic" timestep. This

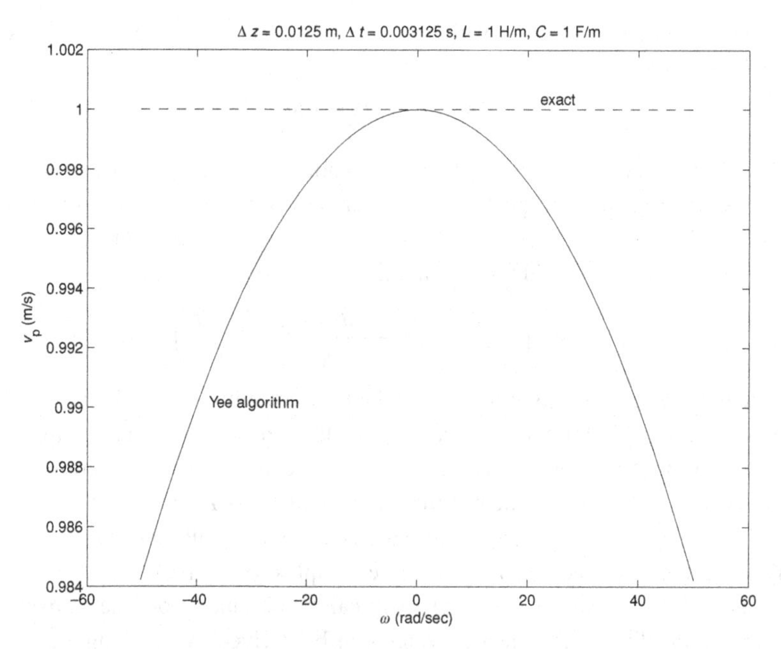

Figure 2.26 Phase velocity as a function of frequency.

implies that an FDTD simulation run at this timestep can (in theory at least) handle Dirac delta functions (of infinitely wide bandwidth). Unfortunately, this does *not* extend to two or three dimensions in general. In 2D and 3D, it turns out that dispersion is minimized (but not eliminated) by operating at the Courant limit. FDTD beginners often run their codes well below the Courant limit, believing that their results will be better with a smaller timestep, but due to numerical dispersion, this is *not* the case. The reason is that the errors due to the spatial and temporal discretizations cancel to some extent, and the cancellation is best at the Courant limit (cancelling completely in 1D; and partially cancelling in 2D or 3D). This is explored in an end-of-chapter problem.

We can summarize this rather counter-intuitive fact as follows: *FDTD codes should be run as close to the Courant limit as possible.* It should also be noted that numerical dispersion is frequency dependent, and worsens rapidly above a certain frequency. As such, when using a wideband source, we should be careful to ensure that we use a source whose spectrum does not have significant frequency content in this region. This is where rules-of-thumb such as ten cells per wavelength criteria used earlier in this chapter arise; we appreciate here that the concept of "wavelength" is rather nebulous in the case of a wideband simulation, and we should rather interpret this as the wavelength corresponding to the maximum frequency of interest – often chosen as the point where the spectrum of the source is $1/e$ of its maximum value (this is -8.6859 dB; -10 dB is also sometimes used). It must be appreciated that these are guidelines rather than exact rules. It should also be appreciated that these rules arose in an era when structures being simulated where at most a wavelength or two in size; for larger structures, as can now be undertaken, a *finer* discretization is required since dispersion accumulates over the length of the simulation. This is explored in more detail in Section 3.4.3 of the next chapter.

2.7 The Courant stability criterion derived by von Neumann analysis

In much the same way in which the dispersion equation was derived, the Courant stability criterion may be derived. It is convenient to work with the equivalent second-order system, Eq. (2.115). We follow the classic approach proposed by von Neumann. Solutions of the following form are considered:

$$V_k^n = \xi^n e^{-j\beta z}|_{z=k\Delta z} \tag{2.116}$$

Depending on the magnitude of ξ, the solution will be growing, decaying or purely oscillating in time (the last for $\xi = 1$). Now, Eq. (2.116) is substituted into Eq. (2.115), yielding

$$\begin{aligned} \xi^{n+1} e^{-j\beta k\Delta z} &= 2\xi^n e^{-j\beta k\Delta z} - \xi^{n-1} e^{-j\beta k\Delta z} \\ &+ \left(\frac{u\Delta t}{\Delta z}\right)^2 [\xi^n e^{-j\beta k\Delta z} e^{-j\beta\Delta z} - 2\xi^n e^{-j\beta k\Delta z} + \xi^n e^{-j\beta k\Delta z} e^{j\beta\Delta z}] \end{aligned} \tag{2.117}$$

Dividing out by $e^{-j\beta k\Delta z}$ and also by ξ^{n-1}, then using the identities $2\cos\theta = e^{j\theta} + e^{-j\theta}$ and $1 - \cos 2\theta = 2\sin^2\theta$, one finally obtains the following equation in ξ:

$$\xi^2 - 2A\xi + 1 = 0 \tag{2.118}$$

with

$$A = 1 - 2S^2 \sin^2 \frac{\beta\Delta z}{2} \tag{2.119}$$

Here,

$$S = \frac{u\Delta t}{\Delta z} \tag{2.120}$$

Equation (2.118) is a quadratic equation in ξ, and using the standard formula for the roots of a quadratic equation, one finds

$$\xi = A \pm \sqrt{A^2 - 1} \tag{2.121}$$

If $|\xi| > 1$, a growing (and hence unstable) solution will occur. This in turn can only happen if $|A| > 1$. From Eq. (2.119), the maximum value of the $\sin^2$ function occurs when $\beta\Delta z = \pi$ (and odd integer multiples thereof). The first occurrence corresponds to $\lambda = 2\Delta z$, the Nyquist limit; put differently, these are the harmonics whose wavelengths are comparable to about twice the mesh spacing [3, p. 13]. At this point, $|A| > 1$ for $S > 1$. Hence, for a stable solution, we require

$$S = \frac{u\Delta t}{\Delta z} \leq 1 \tag{2.122}$$

which is of course the Courant limit in 1D. (For $|A| < 1$, there are two complex conjugate solutions, such that $|\xi|^2 = \left|A \pm j\sqrt{1 - A^2}\right|^2 = 1$, i.e. the roots lie on the unit circle.)

An important point about stability should be made at this juncture. This type of von Neumann analysis precludes *exponentially* growing instabilities (at least in the interior region; comments have already been made about boundary conditions sometimes

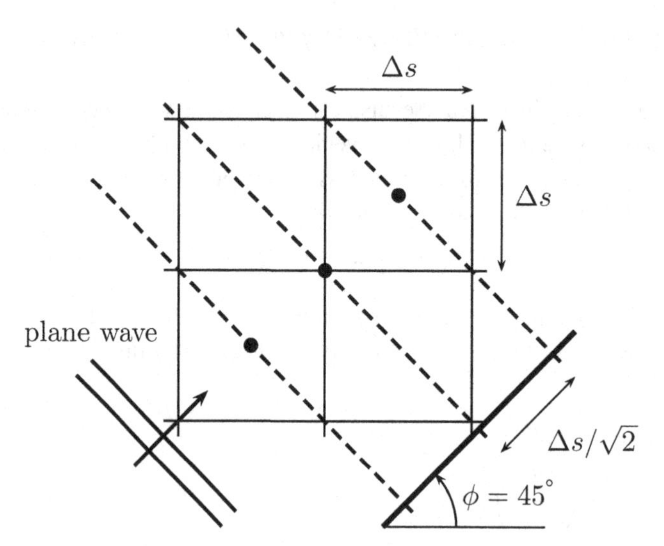

Figure 2.27 A plane wave propagating along the diagonal of a square cell with mesh size Δs in both dimensions, showing the equivalent 1D mesh size $\Delta s/\sqrt{2}$ from which the 2D Courant condition follows. After [12], ©1998 IEEE, reprinted with permission.

resulting in instability). However, there are other forms of instability, such as linearly growing modes, which have been encountered in FDTD analysis when operating at the magic time step [11], and also recently in finite element time domain analysis [8]. Early texts on initial value problems, such as [3], devoted some considerable attention to other forms of stability analysis, but nowadays this appears to have been largely forgotten, and stability is widely (even if incompletely) viewed as synonymous with the Courant limit.

At this point, it is useful to comment that we can immediately derive the 2D and 3D limits, at least on a mesh of uniform discretization. The idea is the following: a uniform plane wave propagating at 45° across a 2D FDTD mesh, with spatial discretization Δs, is equivalent to one propagating on a 1D mesh but with the equivalent 1D mesh size foreshortened as $\Delta_{1D} = \Delta s/\sqrt{2}$; see Fig. 2.27. (Note that the effect is at its most pronounced at $\phi = 45°$ as shown; however, a real FDTD code must be able to handle propagation in all directions, so it is the worst case which determines the limit.)

On a 3D mesh, the foreshortening is $\Delta_{1D} = \Delta s/\sqrt{3}$. From these, the 2D and 3D Courant limits follow immediately for a uniform mesh, as summarized below:

$$\Delta t = \frac{\Delta s}{u} \quad \text{in 1D} \tag{2.123}$$

$$\Delta t = \frac{\Delta s}{u\sqrt{2}} \quad \text{in 2D} \tag{2.124}$$

$$\Delta t = \frac{\Delta s}{u\sqrt{3}} \quad \text{in 3D} \tag{2.125}$$

Note that in a mesh with different materials, the speed of light, u, will vary from material to material. It is the largest u (i.e. smallest Δt) which must be used in this case.

2.8 Conclusion

In this chapter, we have used a very simple one-dimensional transmission line example to introduce the FDTD algorithm. We have seen from first principles how to derive the update equations; this has also given us a handle of the accuracy of the method. Due to its second-order nature, the Yee algorithm is surprisingly accurate. The important issue of stability has been discussed in some detail, including a derivation using the von Neumann method, and we have seen that the Courant stability criterion is a necessary, but not sufficient, condition for stability – the boundary conditions can also cause instabilities, although as we have commented, in most FDTD simulations this is not usually a major cause of concern.

Although the FDTD method can be used in the frequency domain, by simply waiting for the transients to die out – and indeed, our first example did just that – this is an inefficient use of the method, which is capable of generating wideband data in one run. This has been discussed in depth in this chapter.

Finally, the fact that the FDTD method has numerical dispersion has been discussed, as well as the implications. Importantly, and perhaps counter-intuitively, FDTD codes should be run as close to the stability limit as possible to minimize dispersion.

With some very simple substitutions, one can solve one-dimensional TEM field problems using the same theory that we have introduced. However, we prefer now to move into two and three dimensions, and immediately address field problems there. This is the topic of the next chapter.

References

[1] G. B. Arfken and H. J. Weber, *Mathematical Methods for Physicists*. Burlington, MA: Elsevier, 2nd edn., 2005.

[2] W. H. Press, S. A. Teukolsky, W. Vettering and B. R. Flannery, *Numerical Recipes: the Art of Scientific Computing*. Cambridge: Cambridge University Press, 3rd edn., 2007.

[3] R. D. Richtmyer and K. Morton, *Difference Methods for Initial-value Problems*. New York: Wiley, 2nd edn., 1967 (reprint, Malabar, FA: Krieger, 1994).

[4] K. Yee, "Numerical solution of initial boundary value problems involving Maxwell's equation in isotropic media," *IEEE Trans. Antennas Propagat.*, **AP-14**, 302–307, May 1966.

[5] A. Taflove and S. Hagness, *Computational Electrodynamics: the Finite Difference Time Domain Method*. Boston, MA: Artech House, 3rd edn., 2005.

[6] R. P. Feynmann, R. B. Leighton and P. Sands, *The Feynmann Lectures on Physics*, vol. 1. Reading, MA: Addison-Wesley, 1963.

[7] R. P. Feynmann, R. B. Leighton and P. Sands, *The Feynmann Lectures on Physics*, vol. 2. Reading, MA: Addison-Wesley, 1963.

[8] R. A. Chilton and R. Lee, "The discrete origin of FETD-Newmark late time instability, and a correction scheme," *J. Comput. Phys.*, **224**, 1293–1306, June 2007.

[9] C. L. Wagner and J. B. Schneider, "Divergent fields, charge, and capacitance in FDTD simulations," *IEEE Trans. Microwave Theory Tech.*, **46**, 2131–2136, December 1998.

[10] D. Davidson and R. W. Ziolkowski, "Body-of-revolution finite-difference time-domain modelling of space-time focusing by a three-dimensional lens," Special Issue on 3D Scattering, *J. Opt. Soc. Am. A*, **11**, 1471–1490, April 1994.
[11] M. S. Min and C. H. Teng, "The instability of the Yee scheme for the magic time step," *J. Comput. Phys.*, **166**, 418–424, 2001.
[12] A. F. Peterson, S. L. Ray and R. Mittra, *Computational Methods for Electromagnetics*. Oxford & New York: Oxford University Press and IEEE Press, 1998.

Problems and assignments

Problems

P2.1 Apply Eq. (2.12) to Eq. (2.114), to obtain Eq. (2.115). Show that the same result can be obtained by eliminating I in the first-order FDTD equations, Eqs. (2.34) and (2.37).

P2.2 Expand the derivation of the Courant condition sketched in Section 2.7. Starting with Eq. (2.116), follow the steps indicated to derive Eq. (2.122).

P2.3 As outlined in the text, the Yee scheme has a number of very desirable properties. There are many other possible schemes using finite differences, which generally have one or other major deficiency. This problem explores one such scheme, namely the *forward time centered space* (FTCS) scheme, for updating the one-way (advective) wave equation in u (see Section 3.2.5 for more on this):

$$\frac{\partial u}{\partial t} = -v\frac{\partial u}{\partial x} \tag{2.126}$$

were v is the velocity of propagation. The time derivative term is approximated using forward Euler differencing, Eq. (2.6). The spatial derivative term is approximated using central differencing, Eq. (2.8). The resulting, explicit, FTCS update scheme is given by:

$$u_k^{n+1} = u_k^n - v\Delta t\left(\frac{u_{k+1}^n - u_{k-1}^n}{2\Delta x}\right) \tag{2.127}$$

The scheme appears attractive in that there are no half-integer time or space steps. However, using von Neumann stability analysis, show that

$$\xi(\beta) = 1 + j\frac{v\Delta t}{\Delta x}\sin\beta\Delta x \tag{2.128}$$

since $|\xi(\beta)| > 1\ \forall\beta$, this scheme is thus *unconditionally unstable* and hence of no utility at all.[8]

[8] The algorithm has been succinctly described as "a fine example of an algorithm that is easy to derive, takes little storage, and executes quickly. Too bad it doesn't work!" [2, p. 1032].

P2.4 Exploring the above scheme further, a simple correction proposed by Lax stabilizes it. One replaces the term u_i^n in Eq. (2.127) by its average value:

$$u_k^n \to \frac{1}{2}(u_{k+1}^n + u_{k-1}^n) \tag{2.129}$$

so that the update equation becomes

$$u_k^{n+1} = \frac{1}{2}(u_{k+1}^n + u_{k-1}^n) - v\Delta t \left(\frac{u_{k+1}^n - u_{k-1}^n}{2\Delta x} \right) \tag{2.130}$$

For this scheme, show that the amplification factor is

$$\xi = \cos\beta\Delta x + j\frac{v\Delta t}{\Delta x}\sin\beta\Delta x \tag{2.131}$$

and hence that the stability requirement, ($|\xi|^2 \leq 1$), also leads to the 1D Courant limit, viz. Eq. (2.122).

P2.5 By using a Taylor series expansion of the sine functions, show that Eq. (2.113) results in

$$\omega = v\beta[1 + \mathcal{O}(\beta^2\Delta z^2)] \tag{2.132}$$

so the deviation from the correct dispersion relationship, $\omega = v\beta$, is of second order, as expected.

Assignment

A2.1 Write a program to implement the 1D FDTD analysis of a transmission line, as discussed in this chapter. In particular, repeat the results given for the single-frequency source (Fig. 2.6), and also for the wideband source (Figs. 2.20, 2.21 and 2.22). Also investigate the effects of other termination conditions, such as a matched load.

3 The finite difference time domain method in two and three dimensions

3.1 Introduction

In the previous chapter, the basic concepts of the finite difference time domain method were introduced via a one-dimensional example. We will briefly reprise the issues one must attend to when doing an FDTD simulation, as follows:

- An FDTD mesh (or grid) must be created for the problem. (This is trivial in 1D, requires a little thought in 2D, and becomes quite a major problem in 3D.)
- This mesh must be fine enough – i.e. Δs must be no more than perhaps one-tenth of the minimum wavelength (i.e. maximum frequency) of interest (Δs represents the spatial step size; quite often, Δx, Δy and Δz are chosen equal and Δs is used as shorthand for this).
- The time step Δt must satisfy the Courant limit (but be as close to this as possible to minimize dispersion).
- Boundary conditions (the source and load resistors in our 1D example) must be specified.
- An appropriate signal shape (e.g. differentiated Gaussian) with suitable time duration for the desired spectral content must be chosen. Also, in general, its spatial position must be specified. (In the transmission line example, it was fixed as the source voltage generator.)

In this chapter, we will study the FDTD method in two and three dimensions. Firstly, we will develop a 2D simulator for a problem of scattering in free space. Following this, a very important development, the perfectly matched layer absorbing boundary condition, will be discussed and implemented. This is followed by a brief discussion of the extension to three dimensions. We conclude the chapter with a discussion of the use of CST MICROWAVE STUDIO^{TM}, a commercial electromagnetics simulation package which includes an FDTD solver.

3.2 The 2D FDTD algorithm

We will now apply these ideas to a free-space scattering problem in two dimensions. Firstly, we remind the reader that although the real world is obviously three dimensional, many useful problems can be solved when one of the dimensions is much longer than the other two. In this case, we generally assume that the field solution does not *vary* in this dimension – often arbitrarily chosen to be the z-direction, which allows us to simplify the analysis greatly. (A note: assuming that there is no *variation* in z, for instance, does *not* preclude $\hat{z}$-*directed* fields; this point can sometimes cause confusion.) In electromagnetics, this assumption permits us to *decouple* the Maxwell equations into two sets of fields or modes, as they are often called: transverse magnetic and transverse electric.[1] Any field *subject to the assumption of no variation in z* can be written as the sum of these modes:

Transverse magnetic
TM, often written TM_z, modes contain the following field components: $E_z(x, y, t)$, $H_x(x, y, t)$ and $H_y(x, y, t)$.

Transverse electric
TE, often written TE_z, modes contain the following field components: $H_z(x, y, t)$, $E_x(x, y, t)$ and $E_y(x, y, t)$.

At the risk of repetition, there is *no z* variation in any of the above fields.

3.2.1 Electromagnetic scattering problems

When an electromagnetic field encounters a target,[2] currents are excited on it, which in turn re-radiate. This process is called "electromagnetic scattering." Obvious applications are in radar, and also in multi-path analysis for radio-wave propagation. Since the Maxwell equations are linear, the fields are often decomposed into an *incident* field $\boldsymbol{E}^{\text{inc}}$ and a *scattered* field $\boldsymbol{E}^{\text{scat}}$. The overall field, called the *total* field $\boldsymbol{E}^{\text{tot}}$, is then:

$$\boldsymbol{E}^{\text{tot}} = \boldsymbol{E}^{\text{inc}} + \boldsymbol{E}^{\text{scat}} \tag{3.1}$$

By definition, the incident field is the field which would exist if the scatterer were absent. This is very useful; often, this will be a plane wave which can easily be expressed mathematically in closed form. We will see shortly how useful this idea can be when studying scattering.

3.2.2 The TE_z formulation

At this stage, we could solve either (or both) transverse modes; the FDTD process is essentially identical. We will chose the TE_z formulation, because TE_z waves exhibit

[1] Readers who have previously studied waveguide analysis will immediately recognize these concepts.
[2] Because most of the original work was done for radar applications, the military term "target" is frequently used for describing the scatterer in such circumstances.

interesting behavior when scattering off circular targets – *creeping waves* are excited on the structure, i.e. a wave "attaches" itself to the cylinder, goes around the target and then comes back towards the source, potentially in or out of phase with the incident field. (TM_z waves do not do this; the reason is that the boundary conditions are different.)

The TE_z mode set is described by the following parts of Maxwell's equations:

$$\frac{\partial E_x}{\partial t} = \frac{1}{\varepsilon}\left(\frac{\partial H_z}{\partial y} - \sigma E_x\right) \tag{3.2}$$

$$\frac{\partial E_y}{\partial t} = \frac{1}{\varepsilon}\left(-\frac{\partial H_z}{\partial x} - \sigma E_y\right) \tag{3.3}$$

$$\frac{\partial H_z}{\partial t} = \frac{1}{\mu}\left(\frac{\partial E_x}{\partial y} - \frac{\partial E_y}{\partial x}\right) \tag{3.4}$$

We will simplify these further by assuming that the materials are lossless:

$$\frac{\partial E_x}{\partial t} = \frac{1}{\varepsilon}\frac{\partial H_z}{\partial y} \tag{3.5}$$

$$\frac{\partial E_y}{\partial t} = -\frac{1}{\varepsilon}\frac{\partial H_z}{\partial x} \tag{3.6}$$

$$\frac{\partial H_z}{\partial t} = \frac{1}{\mu}\left(\frac{\partial E_x}{\partial y} - \frac{\partial E_y}{\partial x}\right) \tag{3.7}$$

In the transmission line case of the previous chapter it will be recalled that we chose "half-step" increments for the current. We will apply the same idea to developing a 2D FDTD solution of the above equations. We will make the following choices:

$$\begin{aligned} x_i &= (i-1)\Delta x, \qquad i = 1, 2, \ldots, N_x \\ \Delta x &= \frac{X}{N_x - 1}, \qquad N_x \geq 2 \end{aligned} \tag{3.8}$$

$$\begin{aligned} y_j &= (j-1)\Delta y, \qquad j = 1, 2, \ldots, N_y \\ \Delta y &= \frac{Y}{N_y - 1}, \qquad N_y \geq 2 \end{aligned} \tag{3.9}$$

$$\begin{aligned} t_n &= (n-1)\Delta t, \qquad n = 1, 2, 3, \ldots \\ \Delta t &= \frac{T}{M - 1}, \qquad M \geq 2 \end{aligned} \tag{3.10}$$

Here, X and Y are the dimensions of the region we will be gridding (in the x and y directions) and N_x and N_y obviously are the number of cells in each dimension. It is traditional, but certainly not essential, to associate the indices i, j and k in an FDTD code with x, y and z, and m or n with t.

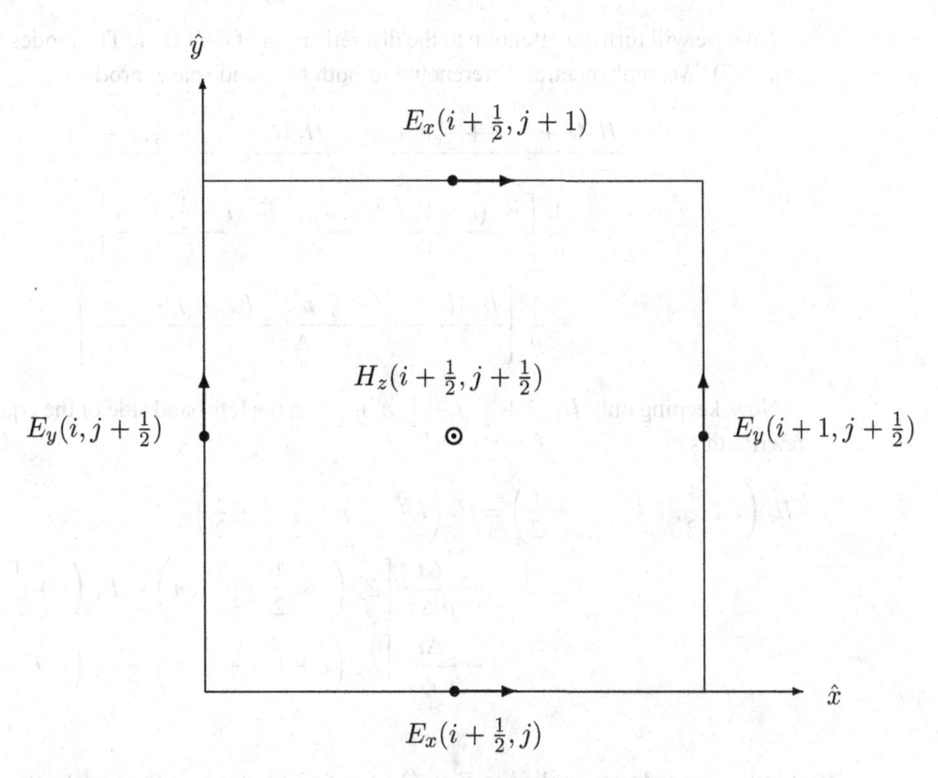

Figure 3.1 The Yee grid for the i, jth cell for the FDTD 2D TE$_z$ mode.

> **Coding hints – the indices i, j and MATLAB**
>
> At this point, it is worth sounding a warning that using these traditional indices can cause very frustrating problems in MATLAB, where i, and j, are usually defined as $\sqrt{-1}$. A useful programming habit to develop is instead to use ii and jj as indices.

A similar array of half-index points will also be defined:

$$x_{i+1/2} = (i - 1/2)\Delta x, \qquad i = 1, 2, \ldots, N_x \tag{3.11}$$

$$y_{j+1/2} = (j - 1/2)\Delta y, \qquad j = 1, 2, \ldots, N_y \tag{3.12}$$

$$t_{n+1/2} = (n - 1/2)\Delta t, \qquad n = 1, 2, 3, \ldots \tag{3.13}$$

Following Yee's choice, we will locate $H_z(i, j, n)$ at $x_{i+1/2}; y_{j+1/2}; t_{n+1/2}$. $E_x(i, j, n)$ will be located at $x_{i+1/2}; y_j; t_n$ and $E_y(i, j, n)$ at $x_i; y_{j+1/2}; t_n$. This choice is far from random; it provides a spatial grid with the magnetic field H_z surrounded in space by the electric fields – $E_x(i, j, k)$ and $E_x(i, j+1, k)$, $E_y(i, j, k)$ and $E_y(i+1, j, k)$ – and offset in time by $\Delta t/2$. The spatial locations are indicated in Fig. 3.1.

Now we will turn our attention to the discretization of the FDTD TE_z modes. Consider Eq. (3.7). We apply central differencing to both time and space, producing:

$$\frac{H_z\left(i+\frac{1}{2},j+\frac{1}{2},n+\frac{1}{2}\right)-H_z\left(i+\frac{1}{2},j+\frac{1}{2},n-\frac{1}{2}\right)}{\Delta t}$$
$$=\frac{1}{\mu}\left[\frac{E_x\left(i+\frac{1}{2},j+1,n\right)-E_x\left(i+\frac{1}{2},j,n\right)}{\Delta y}\right]$$
$$-\frac{1}{\mu}\left[\frac{E_y\left(i+1,j+\frac{1}{2},n\right)-E_y\left(i,j+\frac{1}{2},n\right)}{\Delta x}\right] \tag{3.14}$$

Now, keeping only $H_z\left(i+\frac{1}{2},j+\frac{1}{2},n+\frac{1}{2}\right)$ on the left-hand side of the equation, we rewrite this as

$$H_z\left(i+\frac{1}{2},j+\frac{1}{2},n+\frac{1}{2}\right)=H_z\left(i+\frac{1}{2},j+\frac{1}{2},n-\frac{1}{2}\right)$$
$$+\frac{\Delta t}{\mu\Delta y}\left[E_x\left(i+\frac{1}{2},j+1,n\right)-E_x\left(i+\frac{1}{2},j,n\right)\right]$$
$$-\frac{\Delta t}{\mu\Delta x}\left[E_y\left(i+1,j+\frac{1}{2},n\right)-E_y\left(i,j+\frac{1}{2},n\right)\right] \tag{3.15}$$

Similar procedures, applied to Eqs. (3.5) and (3.6), produce the update equations for the E-field components:

$$E_x\left(i+\frac{1}{2},j,n+1\right)=E_x\left(i+\frac{1}{2},j,n\right)+\frac{\Delta t}{\varepsilon\Delta y}\left[H_z\left(i+\frac{1}{2},j+\frac{1}{2},n+\frac{1}{2}\right)\right.$$
$$\left.-H_z\left(i+\frac{1}{2},j-\frac{1}{2},n+\frac{1}{2}\right)\right] \tag{3.16}$$

$$E_y\left(i,j+\frac{1}{2},n+1\right)=E_y\left(i,j+\frac{1}{2},n\right)-\frac{\Delta t}{\varepsilon\Delta x}\left[H_z\left(i+\frac{1}{2},j+\frac{1}{2},n+\frac{1}{2}\right)\right.$$
$$\left.-H_z\left(i-\frac{1}{2},j+\frac{1}{2},n+\frac{1}{2}\right)\right] \tag{3.17}$$

Just as in the 1D case, the half space and time increments are inconvenient to program, and we will refer simply to i, j, n for E_x, E_y and H_z, but keeping in mind the actual locations. We will also assume $\Delta x=\Delta y=\Delta s$. This allows us the simplify the above to the following:

$$H_z(i,j,n)=H_z(i,j,n-1)+\frac{\Delta t}{\mu\Delta s}\left[E_x(i,j+1,n)\right.$$
$$\left.-E_x(i,j,n)+E_y(i,j,n)-E_y(i+1,j,n)\right] \tag{3.18}$$

$$E_x(i,j,n+1)=E_x(i,j,n)+\frac{\Delta t}{\varepsilon\Delta s}\left[H_z(i,j,n)-H_z(i,j-1,n)\right] \tag{3.19}$$

$$E_y(i,j,n+1)=E_y(i,j,n)-\frac{\Delta t}{\varepsilon\Delta s}\left[H_z(i,j,n)-H_z(i-1,j,n)\right] \tag{3.20}$$

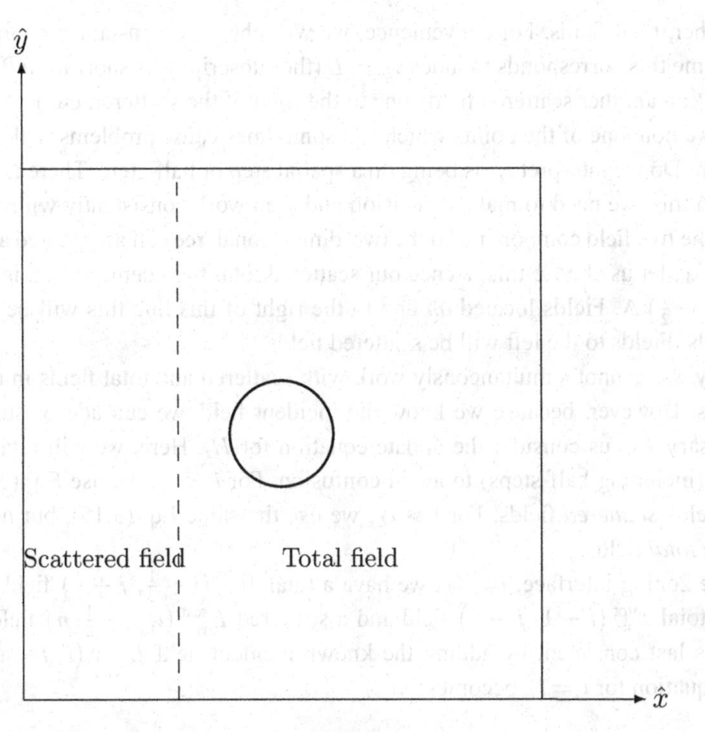

Figure 3.2 The scatterer and surrounding FDTD zones, showing scattered field and total field regions.

Note that when the electric fields are updated, the magnetic field values used are the newly updated ones.

We now have our update equations, and the Courant limit for two dimensions:

$$\Delta t \leq \frac{\Delta s}{\sqrt{2}c} \tag{3.21}$$

where c is the (largest) speed of light in the FDTD region (in non-vacuum regions, the speed of light is of course slowed). We are not quite ready to program, however. There are two things we still need to consider: injecting a source and terminating the mesh.

3.2.3 Including a source: the scattered/total field formulation

If we want to study scattering, we need a method for simulating a plane wave. (Usually, scattering problems assume that whatever source setup the incident field is far removed from the scatterer, and hence the field incident on the target is a uniform plane wave.) The simplest method for doing this is to exploit the concepts of incident, scattered and total fields introduced in Section 3.2.1. Since the Maxwell equations are linear, and we will only work with linear materials here, we can use the FDTD to solve for either the scattered or total fields. (Remember that the incident field is assumed known in this type of formulation.) We will split the computational area into two zones using a (non-physical) line, as in Fig. 3.2: in one region, we will have only scattered fields, and in the other, total fields. For convenience, we will choose a constant x coordinate; we will assume this corresponds to index $i_L = L$ (the subscript L is short for left; we could also position another scattered field zone to the right of the scatterer, etc.).

Now we note one of the points which can sometimes cause problems with the FDTD algorithm. Do we interpret i_L as being on a spatial step or half-step? There is no unique answer to this, we need to make a decision and then work consistently with this. Since three of the five field components in the two-dimensional Yee cell are located at half-step values of x, let us choose this. Hence our scattered/total field demarcation is located at $x_L = \left(L - \frac{1}{2}\right)\Delta$. Fields located *on* and to the right of this line this will be chosen as total fields. Fields to the left will be scattered fields.

Clearly, we cannot simultaneously work with scattered and total fields in the update equations. However, because we know the incident field, we can add or subtract this as necessary. Let us consider the update equation for H_z. Here, we will retain the full notation (including half-steps) to avoid confusion. For $i < i_L$, we use Eq. (3.15), with all the fields *scattered* fields. For $i > i_L$, we use the same Eq. (3.15), but now all the fields are *total* fields.

On the zoning interface, $i = i_L$, we have a total $H_z^{\text{tot}}\left(i + \frac{1}{2}, j + \frac{1}{2}\right)$ field, total E_x^{tot} fields, a total $E_y^{\text{tot}}\left(i + 1, j + \frac{1}{2}\right)$ field and a scattered $E_y^{\text{scat}}\left(i, j + \frac{1}{2}, n\right)$ field. We can make this last consistent by adding the known incident field $E_y^{\text{inc}}y\left(i, j + \frac{1}{2}, n\right)$. The update equation for $i = i_L$ becomes:

$$\begin{aligned} H_z^{\text{tot}}\left(i_L + \frac{1}{2}, j + \frac{1}{2}, n + \frac{1}{2}\right) &= H_z^{\text{tot}}\left(i_L + \frac{1}{2}, j + \frac{1}{2}, n - \frac{1}{2}\right) \\ &+ \frac{\Delta t}{\mu \Delta y}\left[E_x^{\text{tot}}\left(i_L + \frac{1}{2}, j + 1, n\right) - E_x^{\text{tot}}\left(i_L + \frac{1}{2}, j, n\right)\right] \\ &- \frac{\Delta t}{\mu \Delta x}\left[E_y^{\text{tot}}\left(i_L + 1, j + \frac{1}{2}, n\right)\right. \\ &\left. - E_y^{\text{scat}}\left(i_L, j + \frac{1}{2}, n\right) - E_y^{\text{inc}}\left(i_L, j + \frac{1}{2}, n\right)\right] \end{aligned} \tag{3.22}$$

For the E_y component located at $x_L = (L - 1)$, i.e. just to the left of the interface, all the fields in the update equations are scattered, except for the H_z field located at $x_L = \left(L - \frac{1}{2}\right)\Delta_x$. The update equation for this becomes:

$$\begin{aligned} &E_y^{\text{scat}}\left(i_L, j + \frac{1}{2}, n + 1\right) \\ &= E_y^{\text{scat}}\left(i_L, j + \frac{1}{2}, n\right) - \frac{\Delta t}{\varepsilon \Delta x}\left[H_z^{\text{tot}}\left(i_L + \frac{1}{2}, j + \frac{1}{2}, n + \frac{1}{2}\right)\right. \\ &\left. - H_z^{\text{inc}}\left(i_L + \frac{1}{2}, j + \frac{1}{2}, n + \frac{1}{2}\right) - H_z^{\text{scat}}\left(i_L - \frac{1}{2}, j + \frac{1}{2}, n + \frac{1}{2}\right)\right] \end{aligned} \tag{3.23}$$

The update equation for the other component, E_x, involves either only total fields (for $i \geq i_L$) or only scattered fields (for $i < i_L$) and hence can be used without change.

As an example, if the incident field is a plane wave, propagating in the x-direction, in free space, with time history $x(t)$, with a z-polarized magnetic field, the expressions for the incident fields are:

$$E^{\text{inc}} = x\left(t - t_{D_E}\right)\hat{y} \tag{3.24}$$

$$H^{\text{inc}} = \frac{1}{\eta_0} x\left(t - t_{D_H}\right)\hat{z} \tag{3.25}$$

$\eta_0 = \sqrt{\mu_0/\varepsilon_0}$ is the wave impedance of free space. t_D is the delay time from some arbitrary start location. For the problem shown in Fig. 3.2, this could conveniently be taken as $x = 0$. For the magnetic fields located at $x_L = \left(L - \frac{1}{2}\right)\Delta x$, the delay time is $t_{D_H} = \left(L - \frac{1}{2}\right)\Delta x/c$; for the electric fields located at $x_L = (L-1)\Delta x$, the delay time is $t_{D_E} = (L-1)\Delta x/c$. In short, the half-delta difference in spatial position of the fields must be taken into account. Note that these delay times are only valid for the specific case of a field propagating only in the $\hat{x}$-direction. Formulas are easily derived for plane waves propagating in other directions, but the above is sufficient for now.

Considering how simple it was to include the 1D source, one might wonder why this apparently much more complex approach is necessary in 2D. It is possible to include a simple line source in 2D in much the same way as in 1D, by simply specifying the value of the source at a particular point in the mesh. This however radiates *cylindrical*, not plane, waves; hence, this approach is not useful for most scattering problems. However, it is convenient for initial code testing, and also for checking the operation of absorbing boundary conditions. The next idea that springs to mind is simply to drive a *line* of points in the mesh with some source function. The problem with this is more subtle; suffice it to say for now that although this seems like a simple approach, it does not give good results in practice.

3.2.4 Meshing the scatterer

The process of generating a suitable FDTD grid for a problem is often called "meshing." As already indicated, this can be a formidable problem in general. We will be using a very simple test problem – a circular cylinder.[3] This will allow us to make a very simple "mesher." We will place the cylinder, radius a, at a convenient location in the mesh and then simply compute the distance to a point in the mesh; if this distance exceeds a, the point lies outside the cylinder, if it is less than or equal to a, it lies inside or on the surface. Since the E_x and E_y field components are offset in space, we must do this for each component. As a first pass, we will make the cylinder highly conducting, indeed perfectly conducting, so that the (total) fields inside the cylinder are zero. The appropriate boundary condition will be to zero the fields tangential to the cylinder.

[3] "Cylinder" is the general mathematical description of any object generated by translating a two-dimensional cross-section along its normal. For instance, a "cylinder" may be square. (In normal English usage, a cylinder is round.) The full mathematical term for what is commonly called a cylinder is a "right circular cylinder."

The above sounds very straightforward. It is only when coding that a whole number of problematic issues suddenly appear. The first is that we have spoken about "tangential" fields. With a round cylinder, the tangent will only lie in the $\pm\hat{x}$ or $\pm\hat{y}$ directions at four points (top, bottom, right and left in Fig. 3.2). Elsewhere, in all the other FDTD cells which the boundary of the cylinder passes through, we are only going to be able to *approximate* the boundary, and because we use a rectangular grid, the resulting approximation is often called a "stair-step" approximation.

This problem emerged because we are modelling round (or more generally, curvilinear) structures with a rectangular grid. But even if we only model rectangular structures which can be aligned to the FDTD grid, another problem still remains. Refer back to Fig. 3.1. Now, instead of modelling a PEC (perfect electrical conductor) scatterer, let us rather model a cylinder made of some dielectric material, with permittivity ε_R. In the update equations, we need to specify the value for $\varepsilon = \varepsilon_R \varepsilon_0$. Assume we do this for $E^y\left(i+1, j+\frac{1}{2}\right)$. Now, what do we do with the two E^x components located $\Delta x/2$ to the left of this interface? If we set ε_R for them as well, the interface has effectively been "moved" slightly to the left, and now we have the same problem with $E^y\left(i, j+\frac{1}{2}\right)\ldots$ If we do not, the interface is then located somewhere between $\left(i+\frac{1}{2}\right)\Delta x$ and $(i+1)\Delta x$. Again, this is a problem without a simple answer. Due to the half-step offsets in the FDTD Yee grid, there is an uncertainty about the precise position of material interfaces in the basic Yee algorithm. Since it is a maximum of a half-cell, and the cells are usually quite small, it is normally acceptable, but can be problematic. ("Averaging" methods have been used successfully to correct this, and to improve the modelling of curvilinear structures, but we will not consider these at present.)

One final issue still remains to be solved before we develop a 2D FDTD code for scattering off a cylinder: how do we terminate the mesh? The problem is the following: we want to simulate a free-space environment, which means that waves scattered off the target should radiate radially away to infinity, diminishing in strength and eventually disappearing. Clearly, we cannot make an FDTD grid sufficiently large to simulate this. If one has seen an anechoic chamber used for antenna measurements, one will know that antenna designers have a similar problem; they have solved this by coating the walls of the anechoic chamber with an absorbing material. This, effectively, is what we will attempt to do now.

3.2.5 Absorbing boundary conditions

The field of absorbing boundary conditions (ABCs) attracted much research throughout the 1980s and early 1990s. Two methods have historically been pursued: radiation BCs and absorbing BCs. The term ABC is also used more generally for both. The former modifies the FDTD update equations; the latter modifies the material properties in the mesh.

Having really good ABCs, and here is meant ABCs with a reflection coefficient less than -60 or -70 dB, means that it is possible to bring the ABC close to the radiating/scattering structure, "wasting" as few Yee cells as possible meshing up free space. Due to the great interest in the field, one will find a large number of references on the topic. Later in this chapter, we will introduce a revolutionary boundary condition, the perfectly matched layer, but for the time being, we will use a very simple ABC. The idea is the following, for a $-x$ travelling wave on plane $x = 0$. It uses the concept of one-way wave equation, also known as the advective equation, with a wave solution $f(x + ct)$, travelling only in the $-\hat{x}$ direction:

$$\left[\frac{\partial}{\partial x} - \frac{1}{c}\frac{\partial}{\partial t}\right]\phi(x, t) = 0 \tag{3.26}$$

$\phi(x, t)$ represents one of the components of the wave. This leads then to a 1D ABC, as follows. We impose this one-way wave equation on a wave incident on a surface normal to $\hat{x}$:

$$\left.\frac{\partial}{\partial x}\phi(x, t)\right|_{x=0} = \left.\frac{1}{c}\frac{\partial}{\partial t}\phi(x, t)\right|_{x=0} \tag{3.27}$$

Applying *forward differencing* in x and t, one obtains:

$$\phi_1^n - \phi_0^n \approx \frac{\Delta x}{c\Delta t}\left(\phi_0^{n+1} - \phi_0^n\right) \tag{3.28}$$

Finally, rewrite this to give the desired ABC:

$$\phi_0^{n+1} = \phi_0^n\left(1 - \frac{c\Delta t}{\Delta x}\right) + \frac{c\Delta t}{\Delta x}\phi_1^n \tag{3.29}$$

This analysis must be repeated at the boundary $x = x_{\max}$. In this case, the relevant one-way wave equation, with solution in this case $f(x - ct)$, travelling only in the $+\hat{x}$ direction, is

$$\left[\frac{\partial}{\partial x} + \frac{1}{c}\frac{\partial}{\partial t}\right]\phi(x, t) = 0 \tag{3.30}$$

Imposed on a wave incident on a surface normal to $\hat{x}$, the wave is again "absorbed." This leads then to the other 1D ABC:

$$\left.\frac{\partial}{\partial x}\phi(x, t)\right|_{x=x_{\max}} = -\left.\frac{1}{c}\frac{\partial}{\partial t}\phi(x, t)\right|_{x=x_{\max}} \tag{3.31}$$

Applying *backward differencing* in x and forward differencing in t as before, one obtains:

$$\phi_{N_x}^n - \phi_{N_x-1}^n \approx -\frac{\Delta x}{c\Delta t}\left(\phi_{N_x}^{n+1} - \phi_{N_x}^n\right) \tag{3.32}$$

Finally, rewrite this to give the desired ABC:

$$\phi_{N_x}^{n+1} = \phi_{N_x}^n\left(1 - \frac{c\Delta t}{\Delta x}\right) + \frac{c\Delta t}{\Delta x}\phi_{N_x-1}^n \tag{3.33}$$

Interestingly, the equation is identical in form to Eq. (3.29). The extension to $\pm\hat{y}$ propagating waves on the planes $y = 0$ and $y = y_{\max}$ is obvious.

As noted, ϕ was used here; clearly, we need to apply this to the various tangential field components at each boundary. Note that we only need apply it to either $\boldsymbol{E}$ or $\boldsymbol{H}$; once we establish one of the fields "outside" the computational domain, the usual update equations, combined of course with the half-space step offset, establishes the other.

Because this ABC used forward differencing, it is only accurate to first order. (Remember that the Yee scheme has second-order accuracy.) It is "exact" in 1D; in 2D and 3D, for paraxial incidence, reflection coefficients of $\Gamma < -25$ dB may be obtained, but it degrades rapidly off-normal. Mur, in 1981, published a more complete first-order ABC, as well as a second-order one. Details are available in [1, Chapter 6]. These first- and second-order Mur ABCs are still widely used, owing to their simplicity and reasonable effectiveness; however, commercial codes should also offer perfectly matched layers.

We now have all the tools needed to produce a 2D FDTD simulation of electromagnetic scattering from a cylinder in free space – we already have suitable wideband pulses from our 1D work. We will now proceed to develop the simulator.

3.2.6 Developing the simulator

There are a number of issues to consider when turning this algorithm into code. Although we will not be excessively concerned with computational efficiency initially, it is good practice nonetheless to consider some issues. Firstly, division is a much more expensive process in terms of computing time than multiplication. Equation (3.18) contains a term $\Delta t/\mu\Delta s$, and Eqs. (3.19) and (3.20) both contain the term $\Delta t/\varepsilon\Delta s$. Usually, there will only be a few different material regions in an FDTD code. So, it would be better to store these as an array representing material properties, perform the division once before the timestepping starts, and then simply use the relevant value of this array at each stage. One of these is needed per field component:

$$\begin{aligned} H_z(i,j,n) = {} & H_z(i,j,n-1) + D_{Hz}(i,j)\,[E_x(i,j+1,n) \\ & - E_x(i,j,n) + E_y(i,j,n) - E_y(i+1,j,n)] \end{aligned} \tag{3.34}$$

$$E_x(i,j,n+1) = E_x(i,j,n) + C_{Ex}(i,j)\,[H_z(i,j,n) - H_z(i,j-1,n)] \tag{3.35}$$

$$E_y(i,j,n+1) = E_y(i,j,n) - C_{Ey}(i,j)\,[H_z(i,j,n) - H_z(i-1,j,n)] \tag{3.36}$$

with

$$C_{Ex}(i,j) = \frac{\Delta t}{\varepsilon([i-1/2]\Delta s, [j-1]\Delta s)} \tag{3.37}$$

$$C_{Ey}(i,j) = \frac{\Delta t}{\varepsilon([i-1]\Delta s, [j-1/2]\Delta s)} \tag{3.38}$$

$$D_{Hz}(i,j) = \frac{\Delta t}{\mu([i-1/2]\Delta s, [j-1/2]\Delta s)} \tag{3.39}$$

where the (x, y) coordinates at which ε and μ are to be evaluated are explicitly indicated. The previous discussion in Section 3.2.4 regarding the exact position of material interfaces refers again.

Coding hints – programming the update equations efficiently

The obvious way of programming Eqs. (3.34)–(3.36) is to use a double-loop (a DO-loop in FORTRAN, a FOR-loop in many other languages, including MATLAB). However, with MATLAB, this is not a good idea. The problem is that MATLAB is an interpreted language, as opposed to a compiled one, and only runs efficiently when its (highly optimized) vector commands can be used by the interpreter. So, an update such as Eq. (3.36) is best programmed as in Fig. 3.3 – note that the "..." is the MATLAB line continuation character.

```
E_y_n(2:N_x,2:N_y) = E_y_nmin1(2:N_x,2:N_y) ...
    - C(2:N_x,2:N_y).*( H_z_n(2:N_x,2:N_y)  - H_z_n(1:N_x-1,2:N_y) )
```

Figure 3.3 MATLAB code stub for updating E_y.

This looks somewhat cryptic on a first reading: the key operation is `H_z_n(2:N_x,2:N_y) - H_z_n(1:N_x-1,2:N_y)` which effectively shifts the second occurrence of the `H_z_n` array along its first dimension (corresponding to x) and permits the difference to be formed as a vector operation. It is also clear why the indices must run from 2 to N_x, rather than from 1 to N_x (and similarly along the second dimension); otherwise, the operation would refer to non-existing array elements at 0 when shifted. These, the boundary values, must be computed separately. The `.*` operation in MATLAB denotes element-by-element multiplication (also sometimes known as the *outer* product of two matrices).

A point to note when coding is that because the FDTD algorithm is *explicit*, the new values that we compute at a point are *not* affected by the new values at any other points. Hence, we do not need to take particular care at the line corresponding to scattered/total field interface $i = L$. We can update values at this point as usual with a vector operation, and then overwrite them with the correct values. (Obviously, the values of H_z, for instance, must be correct before we start the updates of the electric fields, and vice versa.) Although this involves a small amount of unnecessary computation – in this case, we compute the values along the line separating the scattered/total field twice – the savings in code complexity are so significant that this is almost universal practice in FDTD codes. In the code stub shown in Fig. 3.4, we show the update for the $\boldsymbol{H}$ field, demonstrating this idea. The semicolons at the end of each line prevent the results being written to the command window, which is essential with the large datasets which the FDTD can easily generate. `gaussder` is a function which returns a suitable differentiated Gaussian.

With the 1D FDTD, the algorithm is simple enough that it is relatively easy to program correctly. However, our 2D FDTD simulator is already sufficiently complex that to try to program it in its entirety in one go is likely to lead to great frustration. There are no less than three major, different types of errors that can be made. How to test the code systematically, and locate likely errors, will now be discussed.

```
H_z_n(1:N_x-1,1:N_y-1) = H_z_nmin1(1:N_x-1,1:N_y-1) ...
    + D(1:N_x-1,1:N_y-1).*( E_x_nmin1(1:N_x-1,2:N_y)   - E_x_nmin1(1:N_x-1,1:N_y-1) ...
                          + E_y_nmin1(1:N_x-1,1:N_y-1) - E_y_nmin1(2:N_x,1:N_y-1) ) ;
E_y_nmin1_inc = ones(1,N_y)*gaussder((m-1)*delta_t - (L-1)*delta_s/c,m_offset,sigma) ;
H_z_n(L,1:N_y-1) = H_z_nmin1(L,1:N_y-1) ...
    + D(L,1:N_y-1).*( E_x_nmin1(L,2:N_y)     - E_x_nmin1(L,1:N_y-1) ...
                    + E_y_nmin1(L,1:N_y-1) + E_y_nmin1_inc(1:N_y-1) - E_y_nmin1(L+1,1:N_y-1)) ;
```

Figure 3.4 MATLAB code stub for updating H_z.

Coding hints – frequently made errors in MATLAB

MATLAB is an excellent environment for quickly testing and demonstrating algorithms. However, from the viewpoint of programming, it has a number of "features" which would be seen as deficiencies in most programming languages. The most prominent of these is that it is *not* a strictly typed language – indeed, MATLAB has many properties of a scripting language. This means that variables do not need to be declared before they are used. The advantage is convenience; the drawback is reliability. Firstly, one can accidentally overwrite an existing variable; in particular, `i` and `j` offer suffer this fate. A variant of this is that a subtle spelling error creates a different (and usually undefined) variable. Some other errors frequently made in MATLAB, in particular by programmers used to other languages, include:

Indices in `for` *loops*
The correct format for the `for` loop indices is `for ii=1:N_x`, for example. FORTRAN programmers in particular are inclined to code this as `for ii=1,N_x`, which is incorrect in MATLAB.

Testing equality versus assignment
The correct logical expression to test if `ii` is equal to `jj` is `if ii == jj` (as in C). Again, FORTRAN programmers often code this as `if ii = jj`, which *assigns* the value of `jj` to `ii`.

Both these errors are especially frustrating to locate; MATLAB executes the former incorrectly (or at least incorrectly in terms of the programmer's expectations), and earlier versions also executed the latter (later versions issue a warning).

Implementing the update equations

The easiest mistakes to make here are with the indices. In particular, the repetitiveness of FDTD equations encourages cutting-and-pasting, and one has be *very* careful to correct all the indices (and also field subscripts) when doing this. A simple test which can be used is to note that an FDTD update equation involving (say) the x component of a field on the right-hand side *never* involves a partial derivative (which is of course a difference equation in the code) in x (i.e. the first index). For instance, look at the term in the update for H_z (Fig. 3.4):

```
E_x_nmin1(1:N_x-1,2:N_y)    - E_x_nmin1(1:N_x-1,1:N_y-1)
```

```
H_z_n(1:N_x-1,1:N_y-1) = H_z_nmin1(1:N_x-1,1:N_y-1) ...
    + D(1:N_x-1,1:N_y-1).*( E_x_nmin1(1:N_x-1,2:N_y)    - E_x_nmin1(1:N_x-1,1:N_y-1) ...
                        + E_y_nmin1(1:N_x-1,1:N_y-1) - E_y_nmin1(2:N_x,1:N_y-1) ) ;
H_z_n(N_x/2,N_y/2) = gaussder((m-1)*delta_t,m_offset,sigma);
```

Figure 3.5 MATLAB code stub for updating H_z, using a point (line) source.

```
E_y_n(1,:) = 0;
E_y_n(N_x,:) = 0;
E_x_n(:,1) = 0;
E_x_n(:,N_y) = 0;
```

Figure 3.6 MATLAB code stub for setting PEC boundaries.

Clearly, the following would be incorrect:

```
E_x_nmin1(2:N_x,2:N_y)    - E_x_nmin1(1:N_x-1,2:N_y)
```

It is essential to check the update equations by very carefully reading through each one as programmed.

To check that the update equations are working, a very simple source at one point can be used. Physically, this represents an infinitely long line source. Instead of the full scattered/total field approach shown in Fig. 3.4, the code in Fig. 3.5 injects a source of cylindrical waves in the center of the mesh. Again, note that the source update at $(N_x/2, N_y/2)$ simply overwrites the just updated value. Note also that the E field update equations in this case are simply those of free space. Also combined with this, the outer boundaries at this stage can simply be set as PECs by zeroing the relevant tangential electric field components; see Fig. 3.6 for an example.

Implementing the plane-wave source

Once one has confidence that the update equations are working, one can proceed to test the full scattered/total field formulation, incorporating the plane-wave source. Now, one needs to start thinking about the electromagnetics of the problem. In the one-dimensional case, we simplified matters by using a set of equations with the speed of light set to 1 m/s. Now, we are working with the real world, and $c \approx 3 \times 10^8$ m/s. Since we are primarily interested in radio-frequency (RF) problems, we will select an RF source, with Gaussian derivative shape, with frequency content in the gigahertz range. It turns out to be convenient to select a signal with $\sigma \approx 1 \times 10^{-10}$; this produces a signal with peak spectral amplitude at about 1.5 GHz; reference to Fig. 2.16[4] shows that at around twice the frequency of peak spectral amplitude, the spectrum has decayed to around 30% of the peak value. In the present case, this is 3 GHz; the wavelength in free space is 10 cm (0.1 m) and now we have some guidelines to setting Δs: we should make this around 1/10 of the wavelength at 3 GHz, viz. $\Delta s = 0.01$ m. (Note that we must be careful to work in SI units!) Δt will be set by the Courant limit (a maximum of 23.587 ps, when using the exact value for c).

For testing the code, it is tempting to set N_x and N_y quite small, for instance, 5 or 10. Whilst this is occasionally necessary when something is really wrong and one is having to

[4] That plot was normalized to $\sigma = 1$; the extension is obvious.

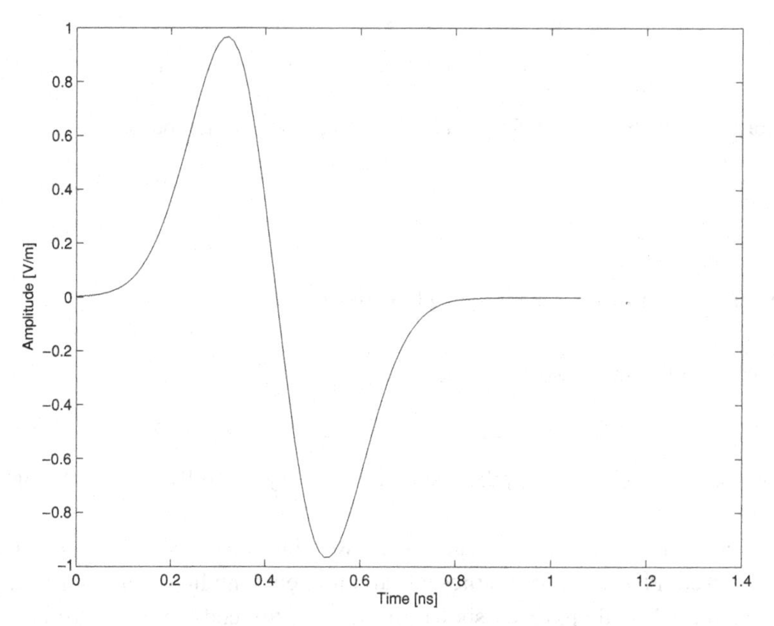

Figure 3.7 Gaussian derivative pulse used for 2D FDTD simulation.

step through the code, it is actually a bad idea in general. The reason is that the absorbing boundary conditions are not included yet, and the temporary PEC boundaries suggested above result of course in the wave reflecting back. With small domains, these reflections mean that it is not possible to observe the field develop and propagate properly. A good test uses $N_x = 200$ and $N_y = 100$ (corresponding physically with $\Delta s = 0.01$ m to an area $0.2 \times 0.1\,\text{m}^2$). The scattered/total field zone is placed at $L = 50$.

The Gaussian derivative pulse defined in Eq. (2.76) was obtained by differentiating Eq. (2.73), and has inconvenient amplitude behavior (being proportional to $1/\sigma^2$). The following pulse has a far more convenient, almost normalized amplitude:

$$v_0(t) = \frac{-4}{\sqrt{2\pi}} \frac{(t-m)}{\sigma} e^{-(t-m)^2/2\sigma^2} \tag{3.40}$$

Its time history for $\sigma = 1 \times 10^{-10}$, and with $m = 4\sigma$, is shown in Fig. 3.7. The peak amplitude is 0.9670, at 0.3322 ns.

Coding hints – a normalized Gaussian derivative pulse

The following equation defines a properly normalized Gaussian derivative pulse:

$$v_0(t) = -\frac{e^{1/2}}{\sigma} (t-m)\, e^{-(t-m)^2/2\sigma^2} \tag{3.41}$$

The normalizing constant $e^{1/2}/\sigma$ provides a unit peak amplitude at $t - m = \pm\sigma$. Since the results in this chapter do not require this, the signal in Eq. (3.40) is used in the following discussion.

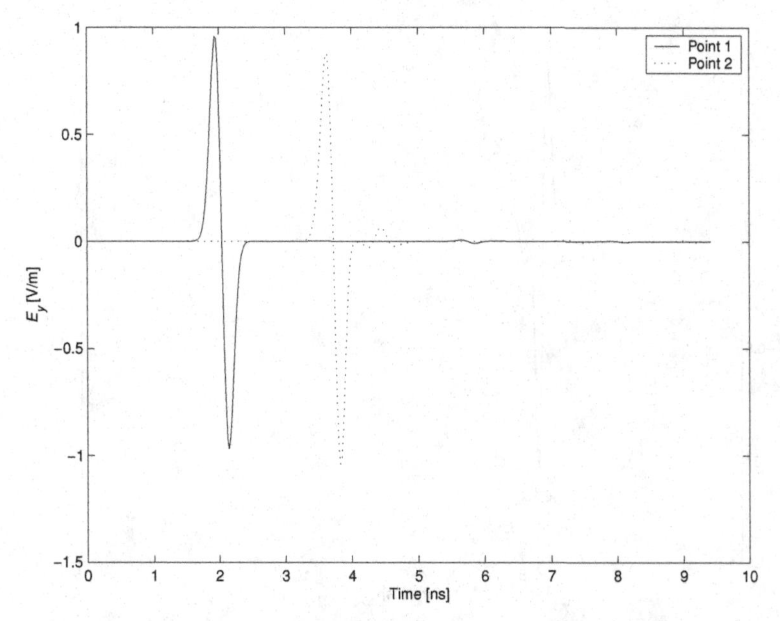

Figure 3.8 Gaussian derivative pulse at a point just to the right of the scattered/total field zone.

This can now be injected into the scattered/total field code. We can monitor the H_z field (scaled by η_0, to give a peak value close to unity) or, equivalently, E_y. We will do this at point 1, with indices $(L+1=51; 50)$, just to the right of the scattered/total field interface, and at point 2, with indices $(101; 50)$. The result, for $M = 400$ timesteps, is shown in Fig. 3.8. The first peak value is at 1.9577 ns, the second, at 3.6559 ns. Now, we establish whether this checks with basic physics. The time difference between the peaks of these pulses is 1.6982 ns. In free space, it should take 1.6678 ns propagating at the speed of light to cover the distance of $50\Delta x = 0.5$ m. This is a difference of around 1.8%. This is very probably due to numerical dispersion. To confirm this, the problem should be rerun, using a finer mesh. If this is done with Δs reduced by half to 0.005 m, the time difference reduces to 1.6746 ns, corresponding to an error of around 0.41%, and confirming that numerical dispersion was indeed the cause of the problem.

The above results demonstrate a working code. If, however, one is not this fortunate, where does one look for the errors? The first thing to do is to ensure that the source really is working correctly. In MATLAB, the source was implemented as a function, in a separate m-file. This allows one to write a short test routine to see what the signal looks like. If this is correct, then the likely errors are in the scattered/total field equations. Be especially careful to ensure that the half-step offsets are correctly taken care of *in both space and time*.

Implementation of the ABC

Now that we have a code with working update equations, and can inject a plane wave into it, the PEC boundaries must be replaced with ABCs. An implementation note in passing: in order to test the ABCs, it is sufficient initially, using the plane-wave source

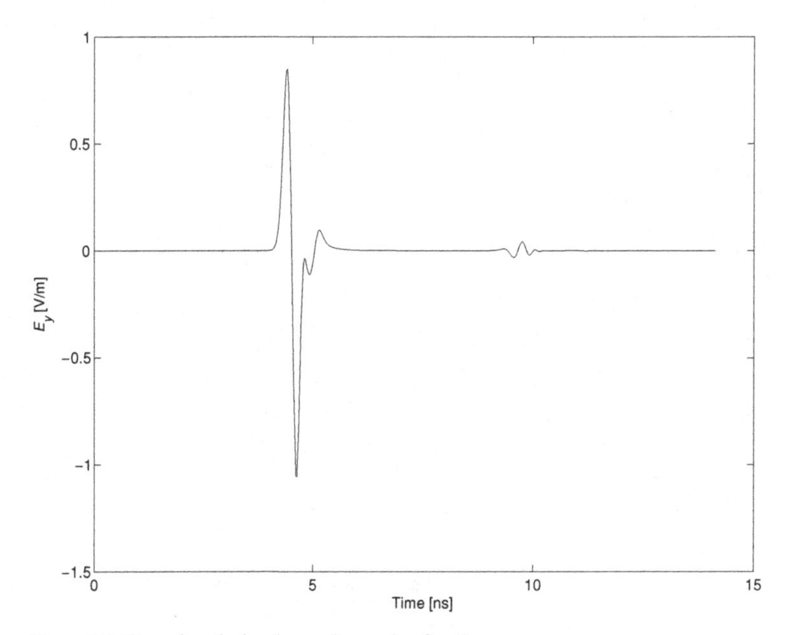

Figure 3.9 Gaussian derivative pulse and reflection.

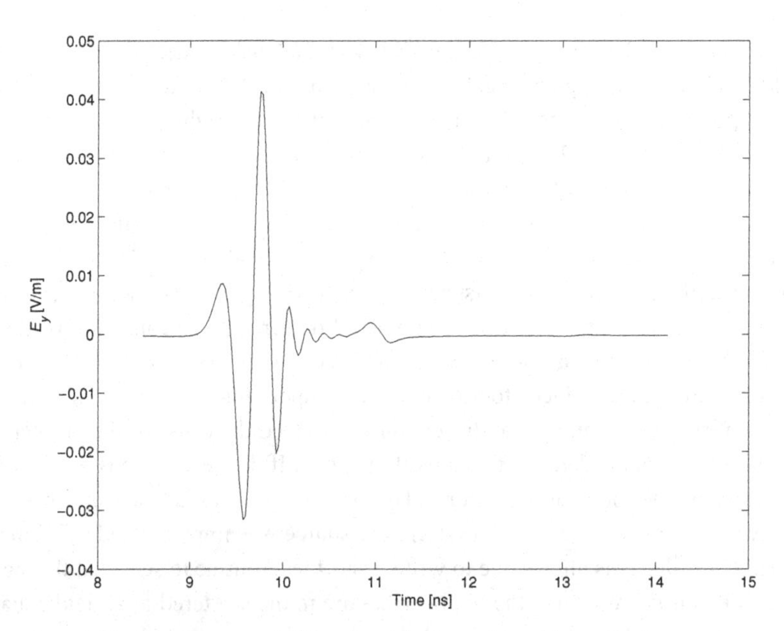

Figure 3.10 An enlargement of the signal in the region of the reflection.

in the previous section, to implement only the ABC at N_x, keeping PEC ABCs at the other boundaries. This permits one to get one set of ABCs working first.

Monitoring the signal at a location mid-way between the zone interface at $i = L$ and the right-hand boundary at $i = N_x$, the signal shown in Fig. 3.9 is recorded for $M = 600$. "Zooming-in" on the reflection, we see Fig. 3.10. The first (negative) peak, with a value of around −0.03 V/m, corresponds to the reflection of the first (positive) peak, which was around 0.8 (see Fig. 3.9), so the reflection coefficient of the ABC is around −30 dB.[5]

3.2.7 FDTD analysis of TE scattering from a PEC cylinder

Now that the basic FDTD code is working, we are in position to study TE_z scattering from a PEC cylinder. Again, we will work in the microwave region. A convenient dimension will be a radius of $a = 0.03$ m. As a first pass, we will choose a rectangular domain, 2 m × 1 m (we will see shortly why we chose this). We will choose $\Delta s = 0.005$ m; this will allow a moderate approximation of the curvature of the cylinder. Even so, this means that the stair-step approximation of the cylinder will be quite crude – across the diameter of the cylinder there are only six cells – and we should bear this in mind when interpreting the results we will generate.

The simplest method of introducing the cylinder into the mesh is by simply zeroing the relevant $C(i, j)$ coefficients, see Eqs. (3.37) and (3.38). This ensures that the relevant electric fields inside and on the surface of the cylinder are zero. It is tempting to do the same with the magnetic fields; this however is incorrect, since it effectively also forces the tangential magnetic fields to zero at the cylinder's "surface," which is not the correct boundary condition.

We want to compute the echo width of the target, usually abbreviated σ_w. It is defined as follows:

$$\sigma_w = \lim_{\ell \to \infty} 2\pi\ell \frac{|E^{\text{scat}}|^2}{|E^{\text{inc}}|^2} \tag{3.42}$$

In an FDTD simulation, some finite limit on ℓ is essential. The conventional 3D criterion for establishing the onset of the far-field, viz. $\ell > 2D^2/\lambda$, where D is the largest dimension of the target, $D = 2a$ in this case, can be used. If we set $\ell \approx 1$ m, the minimum wavelength (and hence maximum frequency) at which this still satisfies the far-field criterion is around 7 mm, or over 40 GHz, so this is more than adequate for our purposes.

We also now appreciate how convenient the scattered/total field formulation is; we can immediately obtain the scattered field by placing our sample point in the scattered field zone. Here, we have the following considerations: we would like to be as far away from the cylinder as possible, but since the reflected signal can be expected to be quite small, we should also be far away enough from the left-hand wall that we can "gate out" unwanted reflections – remember that our ABC is far from perfect. Since we are only going to look at *back-scattered* fields, we can place the scatterer to the right in our grid.

[5] One could compute this more accurately but all that is needed at present is a "ball-park" figure. The correct method for numerically evaluating reflection off ABCs is to run two simulations, one using a reference solution computed on a very large grid, and the other a much smaller grid using the ABC. The reflection is then computed by subtracting the reference solution from the ABC-corrupted solution. We will do this when we evaluate the PML ABC later in this chapter.

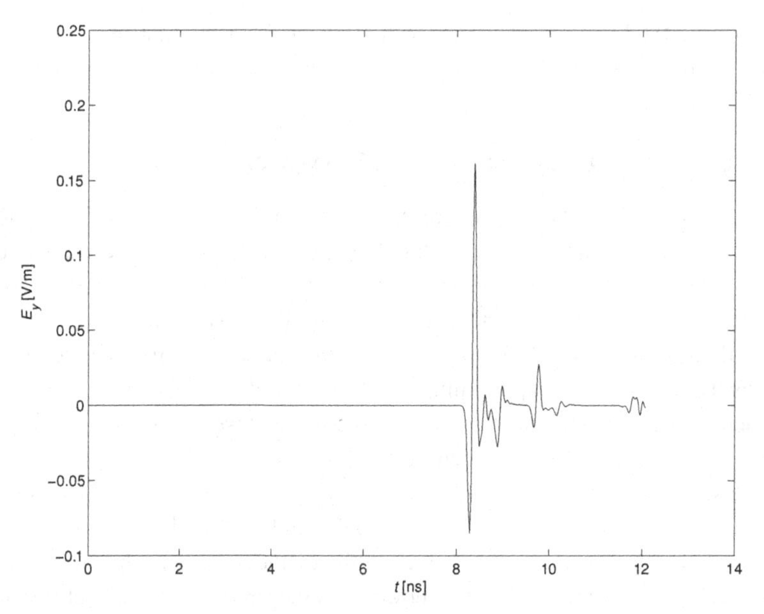

Figure 3.11 Back-scattered signal from the PEC cylinder. Medium mesh, $\Delta s = 0.0025$ m, $\sigma = 5 \times 10^{-11}$.

With these considerations in mind, then, we make the following choices:

- Locate the cylinder at $x = 1.5$ m, $y = 0.5$ m.
- Place the scattered/total field boundary at $x = 1$ m.
- Record the scattered field at $x = 0.5$ m, $y = 0.5$ m (i.e. 1 m away from the target, and 0.5 m away from the closest walls).

We will see shortly (Fig. 3.13) that TE$_z$ back-scattering from a PEC cylinder increases rapidly with frequency up to a first resonance. This occurs when $ka \approx 0.8$, which for our cylinder with $a \approx 0.03$ corresponds to a frequency of just over 1 GHz. We also want to be able to capture the next resonances, so we need a signal with significant frequency content in this region. A differentiated Gaussian pulse with $\sigma = 5.0 \times 10^{-11}$ has a spectrum peaking at around 3.2 GHz, which will be adequate here. A longer pulse would work from the viewpoint of spectral content, but this shorter pulse is convenient for another reason we will see shortly.

Finally, we note that Eq. (3.42) is a frequency domain expression. The Fourier transforms of both the scattered and the incident fields must be computed, and divided pointwise.[6] Note also that this expression, being a power ratio, requires squaring the magnitude of the resultant transforms. (The phase information is irrelevant here.)

The back-scattered signal computed with the FDTD, with grid and problem as set up above, is shown in Fig. 3.11.[7] Although we can go ahead and transform this, we

[6] In MATLAB, the ./ operation.

[7] Readers with the first edition will note that the FDTD results in the following figures show better fidelity than those in the first edition, due to some code improvements and post-processing changes. One of these

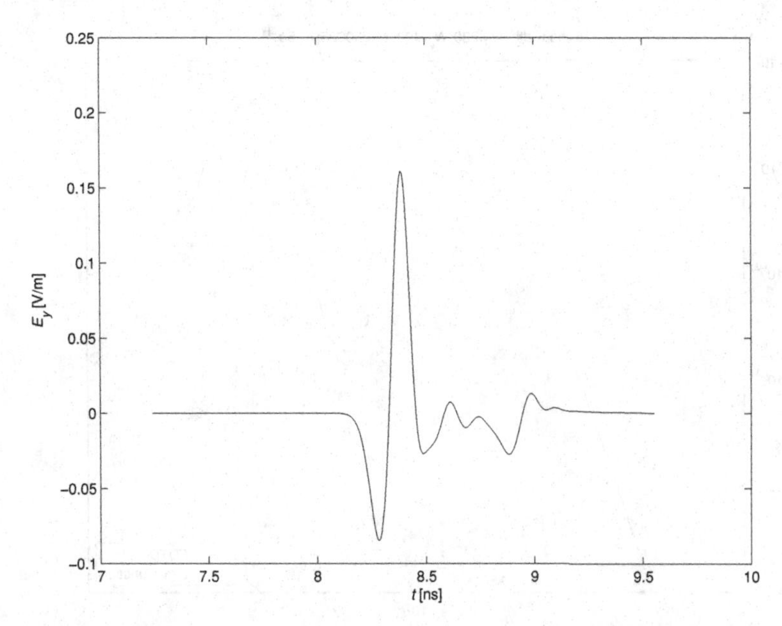

Figure 3.12 Windowed back-scattered signal from the PEC cylinder. Medium mesh, $\Delta s = 0.0025\,\text{m}$, $\sigma = 5 \times 10^{-11}$.

should note that the main signal (the specular reflection and the creeping wave – see next paragraph) lies in the region 8–9.5 ns (this can be confirmed by increasing the computational domain to 4 m × 2 m, for instance). The signal at 12 ns is almost certainly an unwanted reflection of some type; it is usual practice to remove such artifacts by "windowing." Although quite sophisticated windows exist, here it is sufficient simply to zero the signal outside this window. This is shown in Fig. 3.12. Finally, the echo width is plotted in Fig. 3.13, normalized by πa and compared to results computed using an exact eigenfunction solution [2, Figs. 12–34]. (The frequency axis is also normalized; this is usual practice with canonical shapes such as cylinders. k is the free-space wavenumber, and a the cylinder radius.)

The results in Fig. 3.13 show reasonable agreement at the first resonance, but the comparison is quite poor for the next resonances. To improve this, we first need to understand the *physics* of the scattering process. The first peak is simply energy which reflects directly off the cylinder, back in the direction of propagation. (This is the reflection which asymptotic methods, such as geometrical optics, would compute.) The next peak is due to energy which attaches itself to the top (bottom) of the cylinder, and "creeps" around the shadowed side of the cylinder before detaching itself from the bottom (top). Clearly, this signal travels a longer distance than the direct reflection; depending on the cylinder's size, it may reinforce the direct reflection or partially cancel it. This then accounts for the peak at around $ka \approx 2$. The extra distance travelled is

has been to include a scattered/total field formulation on all sides of the cylinder, and not just the on the left-hand side of Fig. 3.2, to be discussed shortly. The "window" has also been tightened. Treatment of the creeping wave has also been improved.

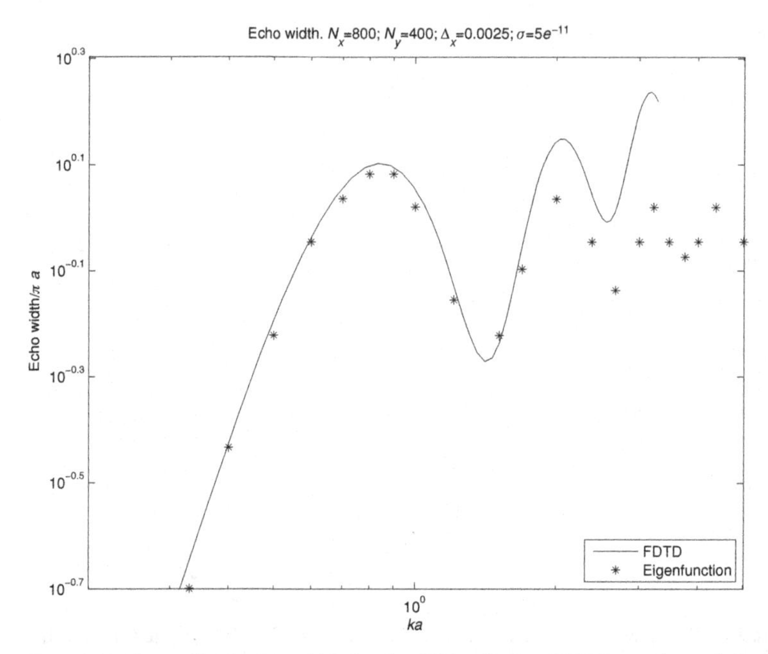

Figure 3.13 Normalized echo width for the PEC cylinder: FDTD results and eigenfunction solution. Medium mesh, $\Delta s = 0.0025$ m, $\sigma = 5 \times 10^{-11}$.

$a + \pi a + a = (2 + \pi)a$; if travelling at the speed of light, this signal would experience a delay of about 514 ps. However, a creeping wave is a slow wave, propagating at somewhat less then c [3, Chapter 11]. If we inspect Fig. 3.12, we can see these two signals; the (negative) peak of the direct reflection is at around 8.4 ns and the (same) peak of the creeping wave is at around 9.05 ns, i.e. around 600 ps apart, consistent with the slow wave behavior.

The problem is that the approximation of the round cylinder with the FDTD stair-step approximation is inadequate at higher frequencies; clearly, we need to refine the mesh. Time domain results using a longer signal with $\sigma = 1 \times 10^{-10}$ comparing a finer mesh (with $\Delta s = 0.00125$) with a coarse mesh ($\Delta s = 0.005$) are shown in Fig. 3.14. Results for both this finer mesh, the medium mesh and a coarser mesh are shown in Fig. 3.15. The eigenfunction solution has been computed from the analytical expression [4, Eq. (11–117), p. 612]. When the time domain results in Fig. 3.14 are compared, one finds that with the coarser mesh, the creeping wave is overestimated, corresponding to the inaccurate results at higher frequencies for the coarser mesh solutions. This is almost certainly as a result of the stair-step approximation of the cylinder; it is apparent that the results in Fig. 3.15 converge to the analytical solution as the mesh is refined.

It is also possible to extend the approach described in Section 3.2.3. Instead of zoning the computational region only as in Fig. 3.2 – with fields on the left of the fictitious interface being scattered fields, and those on the right being total – one can introduce *three* additional such interfaces, one on the right-hand side of the figure, and one each on the upper and lower sides. (The results shown in Figs. 3.14 and 3.15 also use this

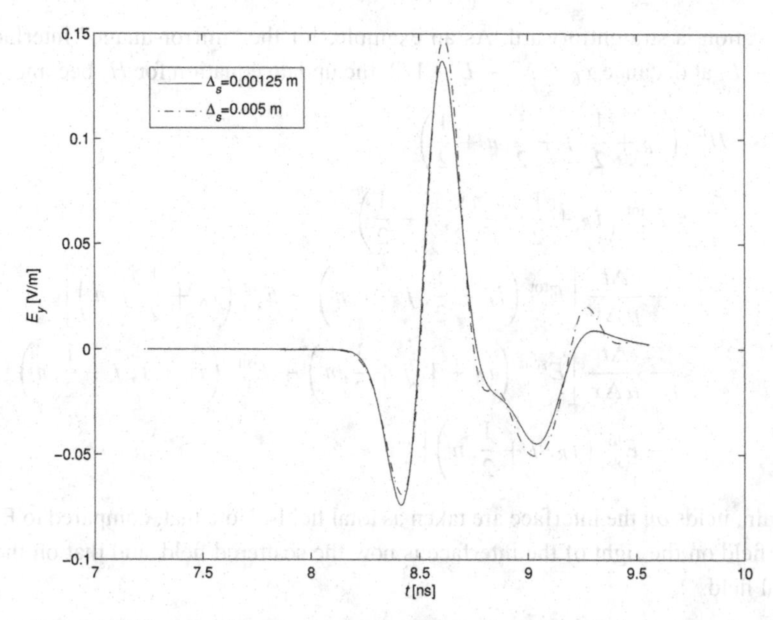

Figure 3.14 Windowed back-scattered signal from the PEC cylinder, comparing the coarse and fine mesh solutions. For both solutions, $\sigma = 1 \times 10^{-10}$.

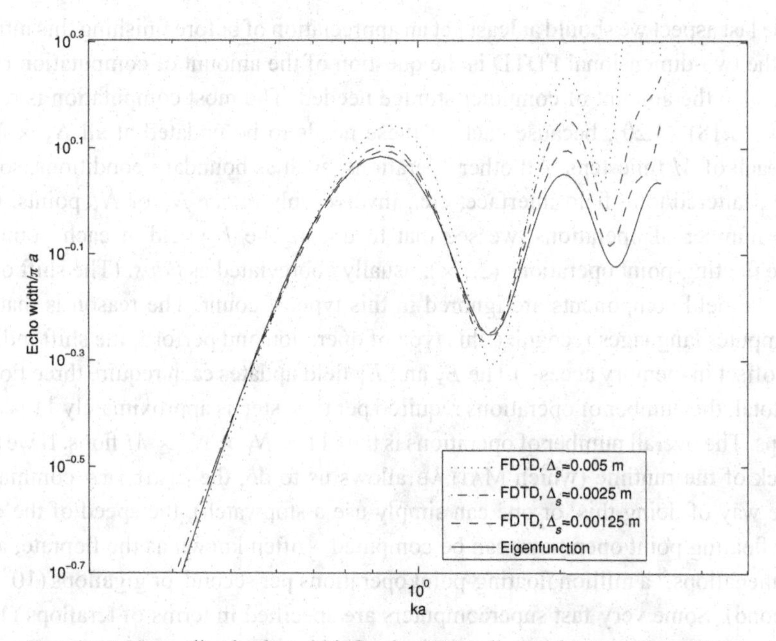

Figure 3.15 Normalized echo width for the PEC cylinder showing three different FDTD results (all with $\sigma = 1 \times 10^{-10}$) compared with the eigenfunction solution.

extension.) Hence, there is now a "buffer" region right around the scatterer, consisting entirely of scattered fields, before one reaches the outer boundaries. This permits the incident field to be introduced cleanly for other angles of incidence of the plane-wave excitation. One needs to derive additional formulas similar to Eqs. (3.22) and (3.23); the derivation is straightforward. As an example, for the "mirror-image" interface in cell $N_x - L$, at distance $x_R = N_x - L - 1/2$, the update equation for H_z becomes:

$$
\begin{aligned}
H_z^{\text{tot}}&\left(i_R+\frac{1}{2}, j+\frac{1}{2}, n+\frac{1}{2}\right) \\
&= H_z^{\text{tot}}\left(i_R+\frac{1}{2}, j+\frac{1}{2}, n-\frac{1}{2}\right) \\
&\quad + \frac{\Delta t}{\mu\Delta y}\left[E_x^{\text{tot}}\left(i_R+\frac{1}{2}, j+1, n\right) - E_x^{\text{tot}}\left(i_R+\frac{1}{2}, j, n\right)\right] \\
&\quad - \frac{\Delta t}{\mu\Delta x}\left[E_y^{\text{scat}}\left(i_R+1, j+\frac{1}{2}, n\right) + E_y^{\text{inc}}\left(i_R+1, j+\frac{1}{2}, n\right)\right. \\
&\quad \left. - E_y^{\text{tot}}\left(i_R, j+\frac{1}{2}, n\right)\right]
\end{aligned}
\tag{3.43}
$$

Again, fields *on* the interface are taken as total fields. Note that, compared to Eq. (3.22), the field on the right of the interface is now the scattered field, and that on the left, the total field.

3.2.8 Computational aspects

One last aspect we should at least get an appreciation of before finishing this introduction to the two-dimensional FDTD is the question of the amount of computation required – and also the amount of computer storage needed. The most computation is required in Eqs. (3.18)–(3.20), because each of these needs to be updated at all $N_x \times N_y$ points at each of M timesteps. All other operations, such as boundary conditions, sources via the scattered/total field interface, etc., involve only *either* N_x or N_y points. Counting the number of operations, we see that to update the H_z field at each point requires five floating-point operations ($\pm$, $\times$), usually abbreviated as *flops*. (The shift operations on the field components are ignored in this type of count. The reason is that efficient computer languages recognize this type of operation and perform the shift indirectly by an offset in memory access.) The E_x and E_y field updates each require three flops. Thus, in total, the number of operations required per time step is approximately $11 \times N_x \times N_y$ flops. The overall number of operations is thus $11 \times N_x \times N_y \times M$ flops. If we also keep track of the runtime (which MATLAB allows us to do, the `cputime` command being one way of doing this, or one can simply use a stopwatch), the speed of the computer for floating point operations can be computed – often known as the floprate, and given as megaflops,[8] a million floating-point operations per second, or gigaflops (10^9 flops per second). Some very fast supercomputers are specified in terms of teraflops (10^{12} flops

[8] The results in this book were originally prepared largely on an IBM A31 notebook computer in 2003. The machine had a Pentium® 4, 1.8 GHz, 512 MB RAM. According to this test, the computer produced around 11.7 megaflops, which is quite slow; the clock speed on its own of course says little about especially floating-point speed. However, it is quite possible that an implementation in FORTRAN or C would be much faster; factors of two orders of magnitude are not unusual when converting code which does not readily vectorize in MATLAB to FORTRAN, etc. In the present context, one would expect a less dramatic speed-up, given the vector nature of the update equations as coded.

per second) or even petaflops (10^{15} flops per second). Now it can be appreciated that halving the mesh size will increase the runtime by a factor of $2^3 = 8$; to put this into practical terms, a run which took perhaps some minutes with one mesh may take an hour or so with a mesh twice as fine.

The analysis just performed leads to a field of computational science known as complexity analysis. What is of interest is the *asymptotic* computational cost of the algorithm. For CEM algorithms, this is usually performed on a square region, dimension d per side, with $\Delta x = \Delta y = \Delta s$ (or in 3D, a cube), thus $N_x = N_y = N$; furthermore, we note that the number of timesteps M is also essentially proportional to N. Hence the runtime is proportional to N^3 for the 2D FDTD algorithm; alternatively, we describe this as an $\mathcal{O}(N)^3$ algorithm.

This analysis in terms of number of unknowns is correct. Since N is inversely proportional to Δs, which in turn is often assumed to be inversely proportional to frequency f (via rules of thumb such as $\Delta s < 1/\lambda_{\min}$), the 2D FDTD algorithm is also often viewed as $\mathcal{O}(f)^3$ or equivalently, noting that $k_{\max}d$ is the size of the region in wavelengths,[9] $\mathcal{O}(k_{\max}d)^3$. This, however, is optimistic. The problem is that the assumption that Δs, and hence N, is directly proportional to $\lambda_{\min}$ is incorrect as the electromagnetic size of the problem increases. The reason is numerical dispersion in the FDTD grid. As an example, a phase error of 5% over a region of one wavelength results in around an 18° cumulative error, probably acceptable; the same percentage error over a region ten wavelengths in length will produce a cumulative error of 180°, clearly *unacceptable.* The dispersion error can be reduced by using a finer mesh. A more realistic assumption is that $N \propto (k_{\max}d)^{1.5}$; hence the 2D FDTD algorithm has an asymptotic complexity of $\mathcal{O}(k_{\max}d)^4$–$\mathcal{O}(k_{\max}d)^{4.5}$, depending on whether the number of time steps is assumed proportional to N or $k_{\max}d$. (One will find both in the literature.)

Regarding storage, the 2D FDTD does not make especially heavy demands on modern computers. The amount of storage required is the following, per cell:

- three field components – times two, for past and present;
- three material constants.

There are $N_x \times N_y$ cells, so the total storage required is $9N_x \times N_y$. In MATLAB, each real number is stored in double precision, requiring 8 bytes (most conventional languages, such as C and FORTRAN, permit the user to choose single or double precision). The storage in bytes is $72N_x \times N_y$. Of course, there are some other variables to store as well, but these are generally the largest. The fine mesh solution of the PEC discussed in the previous section required around 92 Mbytes of storage.

3.3 The PML absorbing boundary condition

3.3.1 An historical perspective

By the early 1990s, the FDTD method had become very popular. However, the problem of terminating the mesh remained problematic. As we have seen, simple ABCs such as

[9] Since $k_{\max} = 2\pi/\lambda_{\min}$.

the first-order one already outlined only provide -20 to -30 dB of absorption, and then only close to normal incidence; whilst there were already better ones available, they were non-trivial to implement, and battled to provide more than -50 dB or so. By comparison, good anechoic chambers were able to provide 70 dB or more of dynamic range. Most of the work on ABCs had concentrated on analytical ABCs, using the properties of the wave operators. However, another type of absorber had also been experimented with – perhaps inspired by the pyramidal absorbers used in anechoic chambers. This was the use of absorbing material at the periphery of the mesh. As we will shortly see, a material with both electric and magnetic loss (carefully chosen in the correct ratios) can provide a perfect match, but only at normal incidence. The advantage of this is that the update equations do not need to be modified. Early efforts had achieved some success, but only worked well near normal incidence.

In 1994, Berenger published a truly seminal paper [5].[10] His idea, like most really good ones, was in essence quite simple. He noted that the problem with artificial absorbers was their inability to operate over a wide range of incidence angles, and proposed that the solution was to increase the degrees of freedom available to provide the match. He proposed a method to do this in two dimensions, by "splitting" one of the field components in two – in the case of the TE_z problem we have investigated, it is H_z which is thus treated, viz. $H_z = H_{zx} + H_{zy}$ – and assigning *different* electric and magnetic loss to each component. Despite the initially worrisome nature of the split field, he showed that the result was what he called a perfectly matched layer (PML) which, in theory at least, absorbed incident waves of *all* polarizations, at *all* frequencies, and at *all* angles of incidence. Furthermore, the wave transmitted into the PML had the *same* wave speed as the incident wave, the *same* characteristic impedance, but *attenuated* (potentially rapidly) in the normal direction. All that was needed to implement the absorber was to modify the FDTD update equations in the PML region to accommodate the split field. Perhaps even more incredibly, "corner regions" of a mesh, which had long caused problems, could be treated by simply overlapping an x-attenuating and a y-attenuating PML.

This almost appeared to good to be true in 1994; within an extremely short time, the entire FDTD community identified the crucial importance of Berenger's work, validated it independently, and quickly extended it to three dimensions. Furthermore, two different approaches were quickly introduced to avoid the split field formulation, whilst retaining the superb performance of the PML. The one approach used "stretched coordinates," and was independently introduced by Chew and Weedon [6] and Rappaport [7]; the other used an anisotropic medium with uniaxial permittivity and permeability tensors, and was introduced by Sacks *et al.* [8]; the latter approach is generally known as the UPML formulation (Uniaxial PML). The stretched coordinate formulation is rather mathematical in nature, but is very useful for other coordinate systems; the UPML, due to

[10] In retrospect, ideas in CEM can often be attributed to several independent inventors, but his invention was unique and certainly deserving of the subsequent accolades. It is interesting that he appears to have published nothing on the FDTD in English language journals prior to this, although he had worked on ABCs before, publishing in French.

its physical plausibility (usually described as Maxwellian), is probably the most popular contemporary approach. Note that even the UPML material is nonetheless fictitious; however, Ziolkowski has investigated the physical realizabilty of such material [9].

In this chapter, we are going to use Berenger's original split field formulation. The reason is that it is both the simplest and also the most efficient approach *in two dimensions*. Using the UPML, for instance, requires introducing the electric and magnetic flux vectors, $\boldsymbol{D}$ and $\boldsymbol{B}$, which doubles the amount of storage required in the UPML region, whereas the split field formulation requires only one extra field component. Additionally, using the UPML requires that we deal with dispersive materials: although this is not too difficult to implement, it is additional complexity we choose to avoid now. It is important to note that this benefit accrues only in two dimensions; in three dimensions, there is little to choose between the formulations from the viewpoint of efficiency, since *all* fields must then be split in the Berenger approach. Furthermore, dispersive materials with the specific form required by the UPML can be quite efficiently handled by the FDTD. Should the reader want to undertake a three-dimensional implementation, a detailed discussion of the UPML approach is available in [1, Chapter 7] and would be the present author's recommendation.

3.3.2 A numerical absorber – pre-Berenger

Before discussing Berenger's contribution, we will review the case of a normally matched numerical absorber. Our presentation is based on Gedney's approach [1, Chapter 7] and we very largely use his notation here. Firstly, we consider the case of a TE_z wave $\boldsymbol{H}^{\mathrm{inc}} = H_0 e^{-j(\beta_{1x}x+\beta_{1y}y)}\hat{z}$ incident on a half-space interface with an absorber at $x = 0$. Importantly, the (fictitious) absorber has *both* electrical (σ) and magnetic (σ^*) loss. The fields on the incident ($x < 0$) side, region 1, are the usual free-space fields:

$$\boldsymbol{H}_1 = H_0(1 + \Gamma\, e^{j2\beta_{1x}x})\, e^{-j(\beta_{1x}x+\beta_{1y}y)}\hat{z}$$

$$\boldsymbol{E}_1 = \left[-\frac{\beta_{1y}}{\omega\epsilon_1}\left(1+\Gamma\, e^{j2\beta_{1x}x}\right)\hat{x} + \frac{\beta_{1x}}{\omega\epsilon_1}\left(1-\Gamma\, e^{j2\beta_{1x}x}\right)\hat{y}\right] H_0\, e^{-j(\beta_{1x}x+\beta_{1y}y)}$$

The fields on the transmitted ($x > 0$) side are:

$$\boldsymbol{H}_2 = H_0 \tau\, e^{-j(\beta_{2x}x+\beta_{2y}y)}$$

$$\boldsymbol{E}_2 = \left[-\frac{\beta_{2y}}{\omega\epsilon_2\left(1+\frac{\sigma}{j\omega\epsilon_2}\right)}\hat{x} + \frac{\beta_{2x}}{\omega\epsilon_2\left(1+\frac{\sigma}{j\omega\epsilon_2}\right)}\hat{y}\right] H_0\tau\, e^{-j(\beta_{2x}x+\beta_{2y}y)}$$

Here, Γ and τ are the usual plane-wave reflection and transmission coefficients at the interface. These equations follow simply from the Maxwell equations, if a (fictitious) magnetic current and hence loss term are included in Faraday's law; they are generalizations of the case discussed by Balanis in [4, Section 5.4.2].

Similarly, the dispersion relationships are:

$$\beta_{1x} = k_1 \cos\theta_i, \qquad \beta_{1y} = k_1 \sin\theta_i, \qquad \forall x < 0$$
$$\beta_{2x} = \sqrt{k_2^2 \left(1 + \frac{\sigma}{j\omega\epsilon_2}\right)\left(1 + \frac{\sigma^*}{j\omega\mu_2}\right) - (\beta_{2y})^2}, \qquad \forall x > 0 \tag{3.44}$$

with $k_i = \omega\sqrt{\epsilon_i \mu_i}$, $i = (1, 2)$.

Enforcing continuity of the tangential fields at the interface, $x = 0$, one obtains:

$$\Gamma = \frac{\frac{\beta_{1x}}{\omega\epsilon_1} - \frac{\beta_{2x}}{\omega\epsilon_2(1+\sigma/j\omega\epsilon_2)}}{\frac{\beta_{1x}}{\omega\epsilon_1} + \frac{\beta_{2x}}{\omega\epsilon_2(1+\sigma/j\omega\epsilon_2)}}$$
$$\tau = 1 + \Gamma$$
$$\beta_{2y} = \beta_{1y} = k_1 \sin\theta_i$$

For normal incidence ($\theta_i = 0$), this simplifies to:

$$\Gamma = \frac{\eta_1 - \eta_2}{\eta_1 + \eta_2}$$

with

$$\eta_1 = \sqrt{\frac{\mu_1}{\epsilon_1}}, \qquad \eta_2 = \sqrt{\frac{\mu_2(1 + \sigma^*/j\omega\mu_2)}{\epsilon_2(1 + \sigma/j\omega\epsilon_2)}}$$

Now, the core idea: *set* $\mu_2 = \mu_1$, $\epsilon_2 = \epsilon_1$ *and, further, enforce*

$$\frac{\sigma^*}{\mu_1} = \frac{\sigma}{\epsilon_1} \Rightarrow \sigma^* = \sigma\mu_1/\epsilon_1 = \sigma(\eta_1)^2 \tag{3.45}$$

Then, $k_1 = k_2$, $\eta_1 = \eta_2$, and thus we obtain perfect absorption: $\Gamma = 0$. Also, very importantly,

$$\beta_{2x} = \left(1 + \frac{\sigma}{j\omega\epsilon_1}\right) k_1 = k_1 - j\sigma\eta_1 \tag{3.46}$$

and the transmitted fields in region 2 are

$$\boldsymbol{E}_2 = \eta_1 H_0\, e^{-jk_1 x} e^{-\sigma\eta_1 x}\, \hat{y}$$
$$\boldsymbol{H}_2 = H_0\, e^{-jk_1 x} e^{-\sigma\eta_1 x}\, \hat{z}$$

In summary, note the following important features of this solution:

- At normal incidence, there is *no* reflection at the interface: hence (at this angle at least) we have a *perfectly matched layer (PML)*.
- The transmitted wave in the absorber has the same velocity as in region 1, but *attenuates* in the normal direction.
- Although lossy, the absorbing material is *dispersionless* (that is, the wave speed is independent of frequency).

3.3.3 Berenger's split field PML formulation

The previous fictitious absorber exhibits PML behavior *only at normal incidence*; its properties degrade rapidly off-normal. As discussed in the introductory comments, Berenger recognized that an additional degree of freedom would permit a match off-normal *as well*. He did this by "splitting" the transverse fields into two orthogonal components, for example $H_z = H_{zx} + H_{zy}$ in his notation. Associated with these were *two* components[11] of σ^* (σ_x^* and σ_y^*) and similarly, *two* components of σ (σ_x and σ_y).

Applying this to our previous two-dimensional TE problem, instead of the usual three equations in E_x, E_y and H_z – for example, as in Eqs. (3.5)–(3.7) – we now have four:

$$j\omega\epsilon_2\left(1+\frac{\sigma_y}{j\omega\epsilon_2}\right)E_x = \frac{\partial(H_{zx}+H_{zy})}{\partial y} \tag{3.47}$$

$$j\omega\epsilon_2\left(1+\frac{\sigma_x}{j\omega\epsilon_2}\right)E_y = -\frac{\partial(H_{zx}+H_{zy})}{\partial x} \tag{3.48}$$

$$j\omega\mu_2\left(1+\frac{\sigma_x^*}{j\omega\mu_2}\right)H_{zx} = -\frac{\partial E_y}{\partial x} \tag{3.49}$$

$$j\omega\mu_2\left(1+\frac{\sigma_y^*}{j\omega\mu_2}\right)H_{zy} = \frac{\partial E_x}{\partial y} \tag{3.50}$$

Introducing the variables

$$s_k = (1+\sigma_k/j\omega\epsilon_2), \qquad s_k^* = \left(1+\sigma_k^*/j\omega\mu_2\right), \qquad k = x, y \tag{3.51}$$

it may be shown that:

$$H_z = H_0\tau\, e^{-j\sqrt{s_x s_x^*}\beta_{2x}x - j\sqrt{s_y s_y^*}\beta_{2y}y} \tag{3.52}$$

$$E_x = -H_0\tau\frac{\beta_{2y}}{\omega\epsilon_2}\sqrt{\frac{s_y^*}{s_y}}\, e^{-j\sqrt{s_x s_x^*}\beta_{2x}x - j\sqrt{s_y s_y^*}\beta_{2y}y} \tag{3.53}$$

$$E_y = H_0\tau\frac{\beta_{2x}}{\omega\epsilon_2}\sqrt{\frac{s_x^*}{s_x}}\, e^{-j\sqrt{s_x s_x^*}\beta_{2x}x - j\sqrt{s_y s_y^*}\beta_{2y}y} \tag{3.54}$$

with

$$(\beta_{2x})^2 + (\beta_{2y})^2 = (k_2)^2 \tag{3.55}$$

Clearly, these can be discretized using the central-differenced leapfrog Yee approach.

The phase-matching condition at the interface requires that the propagation constants in the y-direction are identical; this can be achieved if $s_y s_y^* = 1$, or equivalently $\sigma_y = \sigma_y^* = 0$. Thus, $\beta_{2y} = \beta_{1y} = k_1 \sin\theta_i$. Further, the H-field reflection coefficient may be shown to be:

$$\Gamma = \frac{\dfrac{\beta_{1x}}{\omega\epsilon_1} - \dfrac{\beta_{2x}}{\omega\epsilon_2}\sqrt{\dfrac{s_x^*}{s_x}}}{\dfrac{\beta_{1x}}{\omega\epsilon_1} + \dfrac{\beta_{2x}}{\omega\epsilon_2}\sqrt{\dfrac{s_x^*}{s_x}}}, \qquad \tau = 1+\Gamma \tag{3.56}$$

[11] In retrospect, *this* was the crucial idea, and the split field simply a mathematical device to accomplish this: clearly this defines an anisotropic medium of some type.

Now, let $\epsilon_1 = \epsilon_2$, $\mu_1 = \mu_2$, and $s_x = s_x^*$. This is equivalent to $k_1 = k_2$, $\eta_1 = \sqrt{\mu_1/\epsilon_1} = \sqrt{\mu_2/\epsilon_2} = \eta_2$ and $\sigma_x/\epsilon_1 = \sigma_x^*/\mu_1$. Thus, from Eq. (3.55), $\beta_{1x} = \beta_{2x}$, and from Eq. (3.56), $\Gamma = 0$. The resultant TE$_z$ field transmitted into the PML is then:

$$H_z = H_0\, e^{-j\beta^{1x}x - j\beta_{1y}}\, e^{-\sigma_x \eta_1 \cos\theta_i x} \tag{3.57}$$

and similar expressions for E_y and E_z.

These have the same behavior as the previous normal-only PML, but attenuate *without dispersion* for *all* incident angles.

These results are so important that we will highlight them again in summary form:

- Theoretically, the PML absorbs incident waves of *all* polarizations, at *all* frequencies, and at *all* angles of incidence.
- Further, the wave transmitted into the PML has the *same* wave speed as the incident wave, the *same* characteristic impedance, but *attenuates* (potentially rapidly) in the normal distance.
- All that is needed to implement the absorber is to modify the FDTD update equations in the PML region. (Again, in retrospect what is required is the ability to handle a certain type of lossy anistropic material; this at heart is why the update equations need to be modified.)
- Although perhaps not immediately clear from the above, a "corner region" of a mesh can be treated by simply overlapping an x-attenuating and a y-attenuating PML. This had long been a very troublesome problem with analytical ABCs.

We have already discussed the alacrity with which Berenger's idea was adopted in the FDTD community; within a few months, the PML had been extended to three dimensions by Katz *et al.* [10]; Berenger himself also extended his formulation to three dimensions [11].

3.3.4 The FDTD update equations for a PML

With the theoretical background in place, we turn our attention to implementing and then testing a split field PML. The time domain equivalents of Eqs. (3.47)–(3.50) are

$$\left(\epsilon_2 \frac{\partial}{\partial t} + \sigma_y\right) E_x = \frac{\partial (H_{zx} + H_{zy})}{\partial y} \tag{3.58}$$

$$\left(\epsilon_2 \frac{\partial}{\partial t} + \sigma_x\right) E_y = -\frac{\partial (H_{zx} + H_{zy})}{\partial x} \tag{3.59}$$

$$\left(\mu_2 \frac{\partial}{\partial t} + \sigma_x^*\right) H_{zx} = -\frac{\partial E_y}{\partial x} \tag{3.60}$$

$$\left(\mu_2 \frac{\partial}{\partial t} + \sigma_y^*\right) H_{zy} = \frac{\partial E_x}{\partial y} \tag{3.61}$$

Compared to Eqs. (3.5)–(3.7), the loss terms bring a slight complication: we require the value of the electric field, for instance, at a half timestep, e.g. $E_x\left(i + \frac{1}{2}, j, n + \frac{1}{2}\right)$, a point at which it is not available. (Note that this problem is due to the presence of loss, and not specifically because of the PML – even a normal material with finite electrical conductivity presents this problem.) A method widely used with success is the "semi-implicit"[12] approximation: the required value is computed as the arithmetic average of the previous (known) value and the as-yet-to-be-computed value, i.e.

$$E_x\left(i+\tfrac{1}{2}, j, n+\tfrac{1}{2}\right) = \frac{E_x\left(i+\tfrac{1}{2}, j, n+1\right) + E_x\left(i+\tfrac{1}{2}, j, n\right)}{2} \tag{3.62}$$

Using this approximation, and otherwise proceeding as before, the result is the following set of update equations:

$$H_{zx}(i,j,n) = D_{a_{Hzx}}(i,j) \cdot H_{zx}(i,j,n-1) - D_{b_{Hzx}}(i,j) \cdot [E_y(i+1,j,n) - E_y(i,j,n)] \tag{3.63}$$

$$H_{zy}(i,j,n) = D_{a_{Hzy}}(i,j) \cdot H_{zy}(i,j,n-1) + D_{b_{Hzy}}(i,j) \cdot [E_x(i,j+1,n) - E_x(i,j,n)] \tag{3.64}$$

$$E_x(i,j,n+1) = C_{a_{Ex}}(i,j) \cdot E_x(i,j,n) + C_{b_{Ex}}(i,j) \cdot [H_z(i,j,n) - H_z(i,j-1,n)] \tag{3.65}$$

$$E_y(i,j,n+1) = C_{a_{Ey}}(i,j) \cdot E_y(i,j,n) - C_{b_{Ey}}(i,j) \cdot [H_z(i,j,n) - H_z(i-1,j,n)] \tag{3.66}$$

where we have combined[13] the H field in Eqs. (3.65) and (3.66):

$$H_z(i,j,n) = H_{zx}(i,j,n) + H_{zy}(i,j,n) \tag{3.67}$$

and the material constants are defined as

$$C_{a_{Ex}}(i,j) = \frac{1 - \frac{\sigma_y(i,j)\Delta t}{2\epsilon_2(i,j)}}{1 + \frac{\sigma_y(i,j)\Delta t}{2\epsilon_2(i,j)}} \tag{3.68}$$

$$C_{b_{Ex}}(i,j) = \frac{\frac{\Delta t}{\epsilon_2(i,j)\Delta y}}{1 + \frac{\sigma_y(i,j)\Delta t}{2\epsilon_2(i,j)}} \tag{3.69}$$

$$C_{a_{Ey}}(i,j) = \frac{1 - \frac{\sigma_x(i,j)\Delta t}{2\epsilon_2(i,j)}}{1 + \frac{\sigma_x(i,j)\Delta t}{2\epsilon_2(i,j)}} \tag{3.70}$$

$$C_{b_{Ey}}(i,j) = \frac{\frac{\Delta t}{\epsilon_2(i,j)\Delta x}}{1 + \frac{\sigma_x(i,j)\Delta t}{2\epsilon_2(i,j)}} \tag{3.71}$$

[12] The FDTD method is an explicit method; "future" values are computed entirely from "present" and "past" ones. The approach discussed here uses a "future" value as unknown in the update equation, albeit itself, and hence the name "semi-implicit."

[13] This is slightly more convenient to code. However, note that the split fields must be retained, and updated as usual before the next iteration.

$$D_{a_{Hzx}}(i,j) = \frac{1 - \frac{\sigma_x^*(i,j)\Delta t}{2\mu_2(i,j)}}{1 + \frac{\sigma_x^*(i,j)\Delta t}{2\mu_2(i,j)}} \tag{3.72}$$

$$D_{b_{Hzx}}(i,j) = \frac{\frac{\Delta t}{\mu_2(i,j)\Delta x}}{1 + \frac{\sigma_x^*(i,j)\Delta t}{2\mu_2(i,j)}} \tag{3.73}$$

$$D_{a_{Hzy}}(i,j) = \frac{1 - \frac{\sigma_y^*(i,j)\Delta t}{2\mu_2(i,j)}}{1 + \frac{\sigma_x^*(i,j)\Delta t}{2\mu_2(i,j)}} \tag{3.74}$$

$$D_{b_{Hzy}}(i,j) = \frac{\frac{\Delta t}{\mu_2(i,j)\Delta y}}{1 + \frac{\sigma_y^*(i,j)\Delta t}{2\mu_2(i,j)}} \tag{3.75}$$

As usual with an FDTD equation set, there are subtle differences between the otherwise very repetitive equations which one must be careful to code correctly. In particular, note that σ_x is associated with the E_y update (and vice versa), whereas σ_x^* and σ_y^* are associated with the H_{zx} and H_{zy} updates respectively.

3.3.5 PML implementation issues

One issue which one needs to decide upon when implementing an FDTD PML code is whether the PML update equations are going to be used throughout the entire computational domain, or whether different code will be written for each section. (By simply setting the conductivities to zero, the PML reduces to the usual update equations; alternatively, the electrical conductivity may be retained if required, etc.) The former has the advantage of being far simpler – and corner regions are very simply catered for automatically – but it does increase the memory requirement. The latter is far more tedious to code and the potential for coding error is much higher, but it is more memory efficient. In 2D, the overhead is only 33% in the non-PML regions, and since 2D FDTD codes are in any case not especially memory intensive, it is almost certainly better to use the PML update equations throughout. In 3D, however, the overhead is 100% in the non-PML regions, and the decision is not quite so straightforward. Bear in mind though that the PML works so well that the absorbing boundary can be brought quite close to the scatterer, reducing the memory required in any case.

Remember also that the exact positions of the material parameters are implied but not explicitly stated in the Eqs. (3.68)–(3.75); for example, in Eqs. (3.68) and (3.69), $\sigma_y(i,j)$ must be evaluated at $([i-\frac{1}{2}]\Delta_x, [j-1]\Delta_y)$, the position of the relevant E_x field component; similarly, in Eqs. (3.70) and (3.71), $\sigma_x(i,j)$ must be evaluated at $([i-1]\Delta_x, [j-\frac{1}{2}]\Delta_y)$; and in Eqs. (3.72)–(3.75), $\sigma_x^*(i,j)$ and $\sigma_y^*(i,j)$ must be evaluated at $([i-\frac{1}{2}]\Delta_x, [j-\frac{1}{2}]\Delta_y)$. (Note that H_{zx} and H_{zy} are located at the *same* grid point, the usual H_z location.) This implies of course that σ_x, σ_y, and the pair $\sigma_x^*; \sigma_y^*$ are always evaluated a half-grid point apart. Since the usual polynomial scaling results in quite rapidly changing conductivities, this is an important point to bear in mind for a high-performance PML.

Theoretically, the PML can be made as thin as desired by simply making the material extremely lossy. In practice, the FDTD discretization, with the accompanying half-cell offset, produces some "numerical" reflection. To ameliorate this, practical PML schemes use a number of FDTD cells to implement the absorber, with a "graded" loss profile, increasing from zero loss at the PML/free space interface to some maximum value at the boundary of the grid. A widely used profile is polynomial grading; for a PML of thickness d, the value of σ_x at depth x is

$$\sigma_x = (x/d)^m \sigma_{x,\max} \tag{3.76}$$

where $\sigma_{x,\max}$ is the maximum value attained at $x = d$. Typical practical PMLs are five to ten FDTD cells thick, with a polynomial-order loss profile from two to four.

When discretized in an FDTD mesh, the discretization error produces a filtering effect, which produces some frequency dependence – typically low frequencies are not absorbed as well as higher frequencies.

Thus far, nothing has been said about suitable values for σ. An extensive series of numerical experiments has demonstrated that an optimal choice of this parameter for polynomial grading is

$$\sigma_{x,\max} = \frac{0.8(m+1)}{\eta \Delta s} \tag{3.77}$$

Usually, the external walls are treated as PECs for simplicity, i.e. the relevant tangential field is set to zero.

When implementing a PML, one needs to think carefully about the slight lack of symmetry in FDTD grids. As an example, consider σ_y in the layer of the cells with, on the one hand, $j = 1$ and on the other, $j = N_y$. Setting the tangential fields (E_x) to zero, the result is that in the layer of cells with $j = N_y$, there is no field, since the relevant E_x field component is "below" the last cell (in the geometry of Fig. 3.1). Thus, the value of σ_y in cell layer $j = 1$ actually corresponds to that in cell layer $j = N_y - 1$, rather than $j = N_y$. Also, once σ_y has been computed, it is tempting to find σ_y^* using $\sigma_y^* = \eta^2 \sigma_y$, but, as we have already commented above, this is subtly incorrect due to the $\Delta s/2$ offset between electric and magnetic field points.

Coding hints – testing a PML

The first test to run with a PML is a free-space test: set all the conductivities to zero, which effectively reduces the PML to free space. Errors in the update equations will often quickly make themselves apparent without having to worry about whether conductivity profiles have been set correctly, for instance.

3.3.6 Results for a split field PML

The PML performs so well as an absorber that trying to identify the reflection visually, as we did with the first-order ABC earlier, is impossible. The correct approach to testing

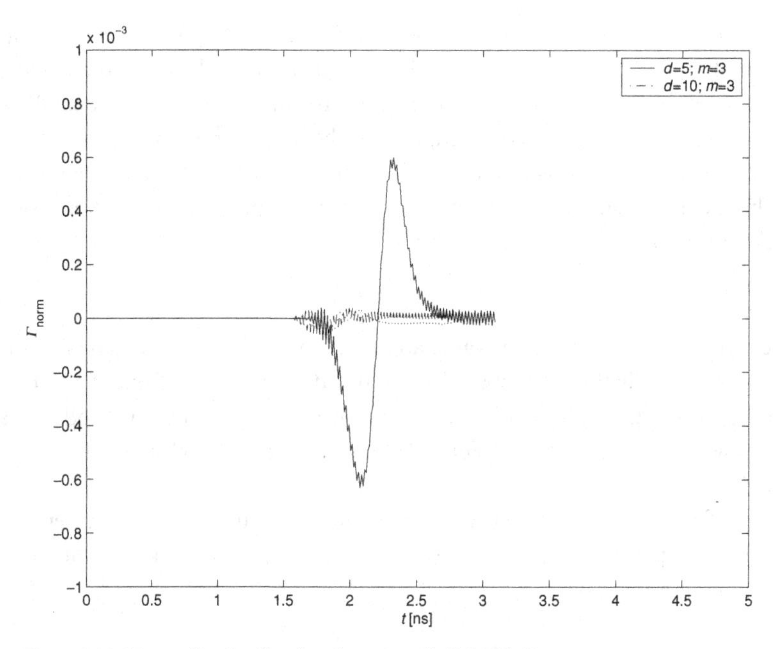

Figure 3.16 Normalized reflection from a split field PML.

a PML (or indeed any ABC) is to run two simulations, with identical discretization and source: one with the ABC under test, and another with a rather larger computational space. The signal is then compared at a point near the ABC. In this case, a 200×200 simulation was compared with a 400×400 simulation. The two signals cannot be distinguished on a graph, so on Fig. 3.16, the difference between the signals is shown – this is the reflection. Note the vertical scale. This has also been normalized by the signal peak, and further time-gated to remove double reflections, etc. When expressed in dBs in Fig. 3.17, the results are *deeply* impressive: the five cell thick, third-order polynomial grading PML has a maximum reflection of around −65 dB; the ten cell thick PML improves this to −85 dB.

Prior to the Berenger PML, the best ABCs were challenged to produce reflection coefficients significantly less than around −50 dB. As we have seen, the Berenger PML offers astounding performance – broadband reflection coeffecients *far* less than this are easily achieved, and with care (for example, optimized conductivity profile, double precision), absorptions of the order of −100 dB and significantly less have been obtained. The FDTD is in a position to out-perform very careful measurements; as mentioned earlier, sophisticated anechoic chambers have dynamic ranges of around 70 dB but there is little prospect of dramatic improvements there.

3.3.7 Drawbacks of the Berenger PML

It may seem curmudgeonly to offer any criticism at all of such an innovation, but despite its superb performance, the PML has some drawbacks, especially in three dimensions.

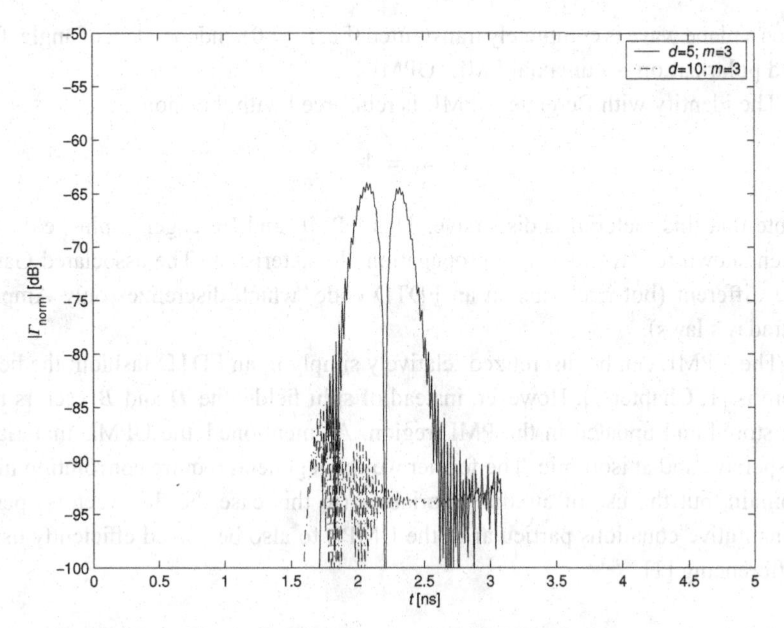

Figure 3.17 Normalized reflection from a split field PML, in dB.

For 3D formulations, the PML requires that *all* field components be split, doubling the memory requirements in the absorbing region; with a 5–10 cell thick layer in 3D this can become a significant overhead. (Other formulations, such as the UPML, do not split the field but instead require the $\boldsymbol{D}$ and $\boldsymbol{B}$ fields to be stored as well in each cell, so the overhead is the same.) The Berenger PML is non-Maxwellian; the field splitting is a mathematical artifact which works very well but leaves niggling questions about physical reality. These drawbacks led to the investigation of other equivalent formulations, aiming to reproduce the superb performance of the PML with a (potentially) physical realizable material. This is also important for applications in FEM codes, where the split-field formalism has no counterpart. Two approaches have emerged: the uniaxial anisotropic absorber and the stretched coordinate formulations. Although our implementation is the original split field one, we will briefly outline these other approaches.

3.3.8 Uniaxial absorber theory

A uniaxial material has the following tensor characterization:

$$\overline{\overline{\epsilon}} = \epsilon_1 \begin{bmatrix} a & 0 & 0 \\ 0 & b & 0 \\ 0 & 0 & b \end{bmatrix}, \qquad \overline{\overline{\mu}} = \mu_1 \begin{bmatrix} c & 0 & 0 \\ 0 & d & 0 \\ 0 & 0 & d \end{bmatrix} \tag{3.78}$$

with $\boldsymbol{D} = \overline{\overline{\epsilon}}\boldsymbol{E}$ and $\boldsymbol{B} = \overline{\overline{\mu}}\boldsymbol{H}$. It has been shown that if the tensors are chosen as follows:

$$\overline{\overline{\epsilon}} = \epsilon_1 \overline{\overline{s}}, \quad \overline{\overline{\mu}} = \mu_1 \overline{\overline{s}}, \quad \overline{\overline{s}} = \begin{bmatrix} s_x^{-1} & 0 & 0 \\ 0 & s_x & 0 \\ 0 & 0 & s_x \end{bmatrix}$$

then a plane wave is completely transmitted (i.e. $\Gamma = 0$), independent of angle, frequency and polarization – a uniaxial PML (UPML).

The identity with Berenger's PML is reinforced with the choice:

$$s_x = 1 + \frac{\sigma_x}{j\omega\epsilon} \tag{3.79}$$

Note that this material is dispersive. This UPML and Berenger's split field PML have been shown to have the same propagation characteristics. The associated Gauss' laws are different (but irrelevant in an FDTD code, which discretizes only Ampère's and Faraday's laws).

The UPML can be discretized relatively simply in an FDTD fashion; the best source here is [1, Chapter 7]. However, instead of split fields, the $\boldsymbol{D}$ and $\boldsymbol{B}$ vectors must also be stored and updated in the PML region. As mentioned, the UPML material is both dispersive and anisotropic. The former would in general require convolution in the time domain, but the use of auxiliary variables (in this case the flux vectors) permits the constitutive equations particular to the UMPL to also be solved efficiently using finite differencing [1].

3.3.9 Stretched coordinate theory

Another formulation shown to be equivalent is the "stretched coordinate" theory. The Cartesian coordinates (x, y, z) are mapped into complex space using

$$\tilde{x} \to \int_0^x s_x(x')\,dx' \tag{3.80}$$

and similarly y and z. Partial derivatives then become:

$$\frac{\partial}{\partial \tilde{x}} = \frac{1}{s_x}\frac{\partial}{\partial x} \tag{3.81}$$

and these are carried into the Maxwell equations. Stretched coordinates have been useful in extending the PML to cylindrical and spherical coordinate systems. The theory also admits a particularly efficient implementation in the FDTD; see again [1, Chapter 7]. Auxiliary computational variables are required; however, unlike the split-field and UPML formulations, where an efficient and simple implementation requires storage of auxiliary variables throughout the computational volume, these can stored in only the PML region. The dispersive nature of the materials requires convolution, which in this case is implemented efficiently using a recursive formulation, and this approach is hence known as CPML.

3.3.10 Further reading on PMLs

An excellent description by Gedney may be found in [1, Chapter 7]. The treatment presented here is based on this approach. Gedney well summarizes the fervor with which the FDTD community adopted, expanded and generalized Berenger's work, and provide an extremely useful unified view of the original split field formulation, the

UMPL and the stretched coordinate viewpoints, with a consistent notation. The original paper by Berenger remains interesting reading [5]. There are a very large number of papers on the subject of PML and the FDTD; the interested reader is referred to the extensive list of references in [1, Chapter 7]. One paper which is worth highlighting is Wittwer and Ziolkowski's contribution [12], since this discusses a number of practical issues in PML implementation.

3.3.11 Conclusions on the PML

Berenger's PML (and the related UPML) came close to putting the ABC "industry" out of business, at least in the FDTD community. Using the Berenger PML, a numerical absorber for the FDTD with essentially arbitrarily good performance can be produced. This has been extended to terminating conductive and/or dispersive regions, as well as half-spaces [1, Chapter 7]. There are still some detail issues to consider – although the basic formulation has been done, *details* for the PML in other coordinate systems are not always readily available.

The PML has some computational overhead and does complicate a code to some extent, whether one uses the split field, UPML or stretched coordinate (CPML) formulations.

It should be commented that such superb absorption is not always required, and a simple ABC is sometimes sufficient, especially if combined with time-gating.

A final comment: the issue of high-performance numerical absorbers in FEM codes is *not* such a closed topic; however, a number of recent papers have addressed this successfully; see for instance [13, 14].

3.4 The 3D FDTD algorithm

Extending the two-dimensional algorithm to three dimensions is straightforward from the viewpoint of the update equations. However, there is no analogy to the TM and TE modes, and all six field components must be updated. The field components are located on the full Yee cell. Again, the field components are offset in both space and time. Details are available in a number of texts. A good introduction is available in [2, Chapter 11]. For a very comprehensive study of the FDTD method, including state-of-the art material, refer to [1]. We will discuss the 3D FDTD algorithm for the free-space case, and show a simple application, which will later be revisited by the finite element method in Chapter 11.

Before doing this, the greatly increased computational cost associated with adding another dimension should be noted. The algorithm is now $\mathcal{O}(N)^4$, or $\mathcal{O}(k_{\max}d)^5$–$\mathcal{O}(k_{\max}d)^{5.5}$. Halving the mesh size increases the runtime by a factor of 16, doubling the frequency, by between 32 and 45 or so (when numerical dispersion is correctly controlled as discussed previously).

In 3D, memory also starts becoming a serious issue; the storage requirements for the six field components (times two, for past and present) and the material arrays (in double precision) become $144 N_x \times N_y \times N_z$ bytes. A computational volume with 100 cells on a side will require 144 MB. This will run on most contemporary personal computers (depending obviously on the amount of memory installed), but just doubling this to 200 cells in each direction increases the memory requirement to well over 1 GB. This is within the scope of most PCs at the time of writing, but doubling this again will most likely exceed available capacity, and will certainly exceed the address space of 32 bit operating systems.[14] Double precision is unnecessary for many applications, and one can save storage by storing an integer index rather than the material arrays as done in the 2D example; similarly, in many applications, such as that to be developed shortly, the fields can be overwritten immediately, approximately halving storage requirements; but even so, the storage requirement grows very rapidly.

It is for these reasons that the development of efficient ABCs was so crucial as the enabling technology which permitted widespread adoption of the FDTD. Highly efficient ABCs permit one to place the scatterer very close to the boundary, and one can also obtain scattered fields very close to the boundary without unphysical reflections corrupting the fields.

3.4.1 The Yee cell in 3D

The Yee cell and the FDTD algorithm as proposed by Yee [15] has proven extraordinarily robust. In Chapter 1, a number of the attractive features of the algorithm were discussed, primarily from a code-development viewpoint. From a mathematical viewpoint, there are some additional features which deserve mention, not least that the algorithm implicitly enforces the divergence condition in Gauss's laws, Eq. (1.3) and (1.4) (assuming that the initial conditions are correctly specified). It has also recently been shown that the algorithm satisfies some deep-lying theoretical requirements for discretization of Maxwell's equations on dual grids [16]. From the perspective of high-performance computing, the matrix-free nature of the algorithm and structured mesh has made the FDTD algorithm the easiest of all the major CEM algorithms to parallelize efficiently.

As with our 2D cell in Fig. 3.1, the original Yee cell positions magnetic field vectors in the center of the faces of the cube, with electric field vectors on the edges, as shown in Fig. 3.18. (Taflove chooses a different convention, interchanging the role of the field vectors [1, Fig. 3.1]); either convention is valid, but as boundary conditions are most often specified on $\boldsymbol{E}$ fields, it would seem convenient to have the electric fields located on the edges, as in Yee's original scheme which we use here.) Although the Yee cell is a very convenient way to visualize the interleaved position of the field vectors, it is important to appreciate that Fig. 3.18 shows fields in some cases associated with a number of indices – for instance, there are three E_x fields shown – and that what is shown on the sketch is actually cell $i-1, j, k$.

For the FDTD algorithm, we associate fields with spatial indices and time steps following Yee's choice, as shown in Table 3.1.

[14] At the time of writing, 64 bit operating systems for PCs were available, but some way from being standard.

Table 3.1 Spatial and temporal location of fields in 3D FDTD algorithm

Field component	x	y	z	t
$E_x(i,j,k,n)$	$x_{i+1/2}$	y_j	z_k	$t_{n+1/2}$
$E_x(i,j,k,n)$	x_i	$y_{j+1/2}$	z_k	$t_{n+1/2}$
$E_x(i,j,k,n)$	x_i	y_j	$z_{k+1/2}$	$t_{n+1/2}$
$H_x(i,j,k,n)$	x_i	$y_{j+1/2}$	$z_{k+1/2}$	$t_{n+1/2}$
$H_y(i,j,k,n)$	$x_{i+1/2}$	y_j	$z_{k+1/2}$	$t_{n+1/2}$
$H_z(i,j,k,n)$	$x_{i+1/2}$	$y_{j+1/2}$	z_k	$t_{n+1/2}$

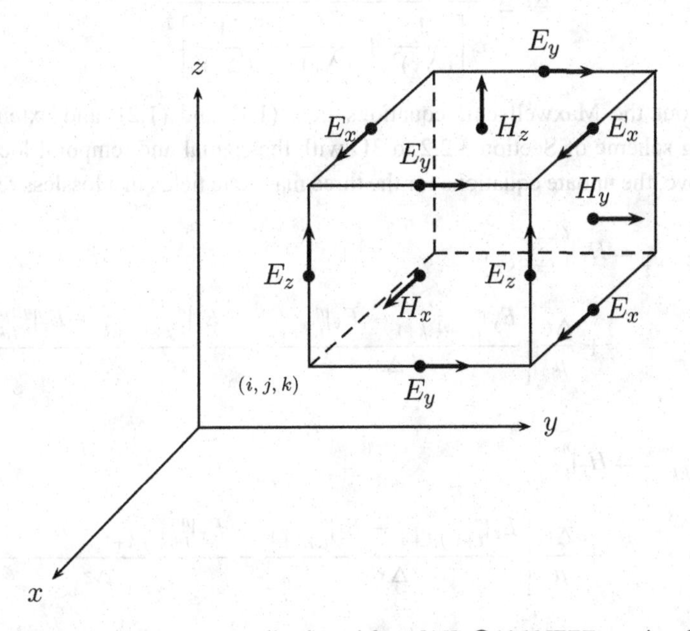

Figure 3.18 The 3D Yee cell, adapted from [15], ©1966 IEEE, reprinted with permission.

The associated grid points are defined by

$$x_i = (i-1)\Delta x, \qquad i = 1, 2, \ldots, N_x \tag{3.82}$$

$$y_j = (j-1)\Delta y, \qquad j = 1, 2, \ldots, N_y \tag{3.83}$$

$$z_k = (k-1)\Delta z, \qquad k = 1, 2, \ldots, N_z \tag{3.84}$$

$$t_n = (n-1)\Delta t, \qquad n = 1, 2, 3, \ldots \tag{3.85}$$

and half-grid points by

$$x_{i+1/2} = (i-1/2)\Delta x, \qquad i = 1, 2, \ldots, N_x \tag{3.86}$$

$$y_{j+1/2} = (j-1/2)\Delta y, \qquad j = 1, 2, \ldots, N_y \tag{3.87}$$

$$z_{k+1/2} = (k - 1/2)\Delta y, \qquad k = 1, 2, \ldots, N_y \tag{3.88}$$

$$t_{n+1/2} = (n - 1/2)\Delta t, \qquad n = 1, 2, 3, \ldots \tag{3.89}$$

These will immediately be recognized as the 3D extension of the relevant equations in Section 3.2.2. Once again, this provides a spatial grid with the appropriate electric fields circulating around each magnetic field and offset in time by $\Delta t/2$. The spatial mesh sizes are given by $\Delta x = X/\mathcal{N}_x$; $\Delta x = Y/\mathcal{N}_y$ and $\Delta x = Z/\mathcal{N}_z$, where $\mathcal{N}_x = N_x - 1$ is the number of Yee cells in the x-direction, one less than the number of x nodes (and similarly for y and z). The time step is of course restricted by the Courant limit:

$$\Delta t \leq \frac{1}{c\left[\dfrac{1}{(\Delta x)^2} + \dfrac{1}{(\Delta y)^2} + \dfrac{1}{(\Delta z)^2}\right]^{\frac{1}{2}}} \tag{3.90}$$

Writing out the Maxwell curl equations Eqs. (1.1) and (1.2), and extending the differencing scheme of Section 3.2.2 to 3D with the spatial and temporal locations as defined above, the update equations for the three magnetic fields in a lossless region are:

$$H_x\Big|_{i,j+\frac{1}{2},k+\frac{1}{2}}^{n+\frac{1}{2}} = H_x\Big|_{i,j+\frac{1}{2},k+\frac{1}{2}}^{n-\frac{1}{2}} + \frac{\Delta t}{\mu}\left[\frac{E_y\big|_{i,j+\frac{1}{2},k+1}^{n} - E_y\big|_{i,j+\frac{1}{2},k}^{n}}{\Delta z} - \frac{E_z\big|_{i,j+1,k+\frac{1}{2}}^{n} - E_z\big|_{i,j,k+\frac{1}{2}}^{n}}{\Delta y}\right] \tag{3.91}$$

$$H_y\Big|_{i+\frac{1}{2},j,k+\frac{1}{2}}^{n+\frac{1}{2}} = H_y\Big|_{i+\frac{1}{2},j,k+\frac{1}{2}}^{n-\frac{1}{2}} + \frac{\Delta t}{\mu}\left[\frac{E_z\big|_{i+1,j,k+\frac{1}{2}}^{n} - E_z\big|_{i,j,k+\frac{1}{2}}^{n}}{\Delta x} - \frac{E_x\big|_{i+\frac{1}{2},j,k+1}^{n} - E_x\big|_{i+\frac{1}{2},j,k}^{n}}{\Delta z}\right] \tag{3.92}$$

$$H_z\Big|_{i+\frac{1}{2},j+\frac{1}{2},k}^{n+\frac{1}{2}} = H_z\Big|_{i+\frac{1}{2},j+\frac{1}{2},k}^{n-\frac{1}{2}} + \frac{\Delta t}{\mu}\left[\frac{E_x\big|_{i+\frac{1}{2},j+1,k}^{n} - E_x\big|_{i+\frac{1}{2},j,k}^{n}}{\Delta y} - \frac{E_y\big|_{i+1,j+\frac{1}{2},k}^{n} - E_y\big|_{i,j+\frac{1}{2},k}^{n}}{\Delta x}\right] \tag{3.93}$$

and for the electric fields,

$$E_x\Big|_{i+\frac{1}{2},j,k}^{n+1} = E_x\Big|_{i+\frac{1}{2},j,k}^{n} + \frac{\Delta t}{\varepsilon}\left[\frac{H_z\big|_{i+\frac{1}{2},j+\frac{1}{2},k}^{n+\frac{1}{2}} - H_z\big|_{i+\frac{1}{2},j-\frac{1}{2},k}^{n+\frac{1}{2}}}{\Delta y} - \frac{H_y\big|_{i+\frac{1}{2},j,k+\frac{1}{2}}^{n+\frac{1}{2}} - H_y\big|_{i+\frac{1}{2},j,k-\frac{1}{2}}^{n+\frac{1}{2}}}{\Delta z}\right] \tag{3.94}$$

```
H_x = H_x + Delta_t/mu * (diff(E_y,1,3)/Delta_z - diff(E_z,1,2)/Delta_y);
H_y = H_y + Delta_t/mu * (diff(E_z,1,1)/Delta_x - diff(E_x,1,3)/Delta_z);
H_z = H_z + Delta_t/mu * (diff(E_x,1,2)/Delta_y - diff(E_y,1,1)/Delta_x);
```

Figure 3.19 MATLAB code stub for updating H in 3D.

$$E_y|^{n+1}_{i,j+\frac{1}{2},k} = E_y|^{n}_{i,j+\frac{1}{2},k} + \frac{\Delta t}{\varepsilon}\left[\frac{H_x|^{n+\frac{1}{2}}_{i,j+\frac{1}{2},k+\frac{1}{2}} - H_x|^{n+\frac{1}{2}}_{i,j+\frac{1}{2},k-\frac{1}{2}}}{\Delta z} - \frac{H_z|^{n+\frac{1}{2}}_{i+\frac{1}{2},j+\frac{1}{2},k} - H_z|^{n+\frac{1}{2}}_{i-\frac{1}{2},j+\frac{1}{2},k}}{\Delta x}\right] \tag{3.95}$$

$$E_z|^{n+1}_{i,j,k+\frac{1}{2}} = E_z|^{n}_{i,j,k+\frac{1}{2}} + \frac{\Delta t}{\varepsilon}\left[\frac{H_y|^{n+\frac{1}{2}}_{i+\frac{1}{2},j,k+\frac{1}{2}} - H_y|^{n+\frac{1}{2}}_{i-\frac{1}{2},j,k+\frac{1}{2}}}{\Delta x} - \frac{H_x|^{n+\frac{1}{2}}_{i,j+\frac{1}{2},k+\frac{1}{2}} - H_x|^{n+\frac{1}{2}}_{i,j-\frac{1}{2},k+\frac{1}{2}}}{\Delta y}\right]. \tag{3.96}$$

Here, we have adopted the more compact sub- and superscript notation of [1, 17], rather than that of Eqs. (3.15)–(3.17), as otherwise the equations become very lengthy, but have still explicitly indicated the half-steps.

Regarding coding, the necessity of using vectorized operations for execution speed has already been highlighted in Section 3.2.2. MATLAB has an even more efficient method of forming differences than was used there; this is the `diff` command [17], and an application to coding Eqs. (3.91)–(3.93) efficiently is shown in Fig. 3.19. (This code was developed independently, but such is the nature of FDTD code, in particular written in an environment such as MATLAB, that it looks extremely similar to code in [17, Chapter 5].) The first argument of the function is the matrix to be differenced, the second is the order of differentation, and the third is the dimension along which the differencing is performed. Note that the appropriate spatial step ($1/\Delta_z$ and $1/\Delta_y$ respectively, in this case) must still be explicitly included to form the approximation of the relevant partial differential. This code also demonstrates the use of in-place operations, with H_x being overwritten immediately the new value is computed. (In MATLAB, one should be aware that temporary copies of the field arrays are created during such operations, which can result in more memory usage than intended.)

When coding the FDTD, the dimensions of the matrices representing the field components will usually be slightly different. The reason is simply that the $\boldsymbol{H}$ fields are positioned in the center of cell faces (in both our and the original Yee cells), whereas the $\boldsymbol{E}$ fields are positioned on edges. As mentioned, it is usually convenient to have tangential $\boldsymbol{E}$ fields postitioned on the outer boundaries of the region, as very often, a PEC boundary is applied there (in which the relevant field component is simply zeroed). For a cubical region with $\mathcal{N}_x \times \mathcal{N}_y \times \mathcal{N}_z$ Yee cells the dimensions are shown in Fig. 3.20. Looking at one in detail, we see that E_x, which is positioned in the middle of each edge in the x-direction, will have only $\mathcal{N}_x$ samples in that direction, but clearly $\mathcal{N}_{y+1}$ and

```
E_x = zeros(N_x,   N_y+1,N_z+1);
E_y = zeros(N_x+1,N_y,   N_z+1);
E_z = zeros(N_x+1,N_y+1,N_z);
H_x = zeros(N_x+1,N_y,   N_z);
H_y = zeros(N_x,   N_y+1,N_z);
H_z = zeros(N_x,   N_y,   N_z+1);
```

Figure 3.20 MATLAB code stub dimensioning field arrays. Note that N_x in the code corresponds to the number of Yee cells in the x-direction, $\mathcal{N}_x$ in the text, and similarly for N_y and N_z).

```
E_x(:,2:N_y,2:N_z) = E_x(:,2:N_y,2:N_z) + Delta_t/epsilon * ...
  (diff(H_z(:,:,2:N_z),1,2)/Delta_y - diff(H_y(:,2:N_y,:),1,3)/Delta_z);
E_y(2:N_x,:,2:N_z) = E_y(2:N_x,:,2:N_z) + Delta_t/epsilon * ...
  (diff(H_x(2:N_x,:,:),1,3)/Delta_z - diff(H_z(:,:,2:N_z),1,1)/Delta_x);
E_z(2:N_x,2:N_y,:) = E_z(2:N_x,2:N_y,:) + Delta_t/epsilon * ...
  (diff(H_y(:,2:N_y,:),1,1)/Delta_x - diff(H_x(2:N_x,:,:),1,2)/Delta_y);
```

Figure 3.21 MATLAB code stub for updating E in 3D.

$\mathcal{N}_{z+1}$ in the y- and z-directions respectively. One has to be careful to ensure that the dimensions in code such as Fig. 3.19 are compatible (here, they are); for the E_x update, suitable code is given in Fig. 3.21. All these points are obvious once coded, but can cause some headaches when starting out.

3.4.2 An application: determining the resonant frequencies of a PEC cavity

Developing an application similar to the 2D TE scattering problem discussed earlier in this chapter is quite a lengthy procedure, but there is a very simple application which requires only elementary boundary conditions and a very simple source. This is the approximate determination of the eigenvalues of a cavity with PEC walls. This problem will be worked in detail in Chapter 11, using finite elements, but the FDTD algorithm can also be applied to this problem. A random source is injected as an initial value, the timestepping proceeds for a sufficiently long time, and a Fourier transform is then taken of the field(s). Peaks in the frequency response indicate the presence of resonant modes in the cavity. As such, this is not truly an eigenanalysis procedure, but it can provide a good approximation of the resonant frequencies.

As usual, there are a few details that need to be taken into account when doing this. Firstly, both TM and TE modes are potentially present (a more detailed discussion of these will be found in Chapter 11). It is possible to excite only one of the two mode sets by suitable choice of initial condition; for instance, an initial H_z field excites only the TE modes. (This is the computational analogy of the methods used in practice for feeding cavities.) For a general analysis, one needs to sample the fields at a number of points, as it is possible to inadvertently sample the field at a node (zero, or null) of certain modes.

An example of such an analysis with a randomly excited H_z field (accomplished in this case using the MATLAB command $\texttt{H\_z}(:,:,:) = \texttt{rand}(\texttt{N\_x}, \texttt{N\_y}, \texttt{N\_z}+1) - 0.5$; the last term attempts to produce zero DC component) is given in Fig. 3.22 for a cavity $1 \times 0.5 \times 0.75\ \text{m}^3$. The dotted lines on the figure indicate the first four eigenfrequencies, which in this case are the TE_{101}, $\text{TE}_{011}/\text{TE}_{201}$, TE_{111} and TE_{102} modes. (Note that the second and third modes are degenerate; see Table 11.2 for more detail.) It is notable that

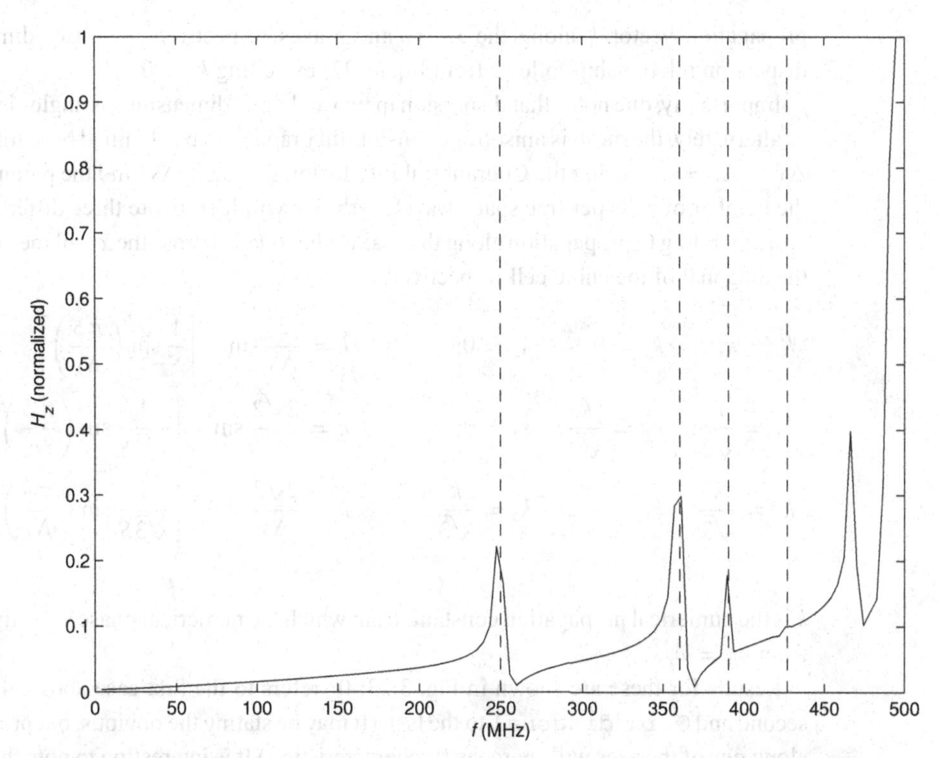

Figure 3.22 An example of the FDTD eigenanalysis of a cubical PEC cavity. The dotted lines are the analytically determined TE eigenvalues.

the fourth discrete eigenfrequency (corresponding to the TE_{102} mode) is hardly detected; this is because the sampling location was in the middle of the cavity, corresponding to a null in the modal pattern of this mode. It should also be commented that some numerical experimentation, requiring a number of runs, was needed to obtain the result shown in this figure; many of the initial random distributions do not provide as clear a result. So, in its basic, very simple, form the method is not especially reliable. The post-processing can be greatly improved using methods from signal processing such as the application of Padé approximations; the interested reader is referred to [17, Chapter 4] for further discussion.

3.4.3 Dispersion in two and three dimensions

The dispersion relation in three dimensions is given by

$$\frac{\sin^2(\omega\Delta t/2)}{(c\Delta t)^2} = \frac{\sin^2(k_x\Delta x/2)}{(\Delta x)^2} + \frac{\sin^2(k_y\Delta y/2)}{(\Delta y)^2} + \frac{\sin^2(k_z\Delta z/2)}{(\Delta z)^2} \tag{3.97}$$

An succinct account of the derivation of this may be found in [17, Chapter 5]; there is a similar treatment in [1, Chapter 4]. k_x, k_y and k_z are the components of the propagation vector $\boldsymbol{k}$ along the x-, y- and z-axes respectively. The two-dimensional dispersion relationship follows from Eq. (3.97) by setting $k_z = 0$.

Importantly, one notes that dispersion in two and three dimensions is angle-dependent, or alternately, the mesh is anisotropic. Insightful graphs can be obtained by setting $\Delta x = \Delta y = \Delta z = \Delta s$, using the Courant stability factor $S = c\Delta t/\Delta s$, and the parameter N_λ, the number of cells per free space wavelength. We will investigate three different cases, corresponding to propagation along the x-axis, diagonally across the xy-plane, and along the diagonal of the cubic cell respectively:

$$k_x = k; \quad k_y = 0; \quad k_z = 0; \qquad \tilde{k} = \frac{2}{\Delta s}\sin^{-1}\left[\frac{1}{S}\sin\left(\frac{\pi S}{N_\lambda}\right)\right] \tag{3.98}$$

$$k_x = \frac{k}{\sqrt{2}}; \quad k_y = \frac{k}{\sqrt{2}}; \quad k_z = 0; \qquad \tilde{k} = \frac{2\sqrt{2}}{\Delta s}\sin^{-1}\left[\frac{1}{\sqrt{2}S}\sin\left(\frac{\pi S}{N_\lambda}\right)\right] \tag{3.99}$$

$$k_x = \frac{k}{\sqrt{3}}; \quad k_y = \frac{k}{\sqrt{3}}; \quad k_z = \frac{k}{\sqrt{3}}; \qquad \tilde{k} = \frac{2\sqrt{3}}{\Delta s}\sin^{-1}\left[\frac{1}{\sqrt{3}S}\sin\left(\frac{\pi S}{N_\lambda}\right)\right] \tag{3.100}$$

$\tilde{k}$ is the numerical propagation constant, from which the numerical phase velocity follows from $\tilde{v}_p = \omega/\tilde{k}$.

Results for these are shown in Fig. 3.23; 0° refers to the first case above, 45° to the second and `Grid diagonal` to the last. (It may be stating the obvious, but propagation along *any* of the axes will share the 0° characteristics.) It is interesting to note that for the case where $S = 1/\sqrt{3}$, there is no dispersion when operating at the Courant limit; this is again the "magic time step" encountered in 1D analysis, and Fig. 2.27 explains this. The curve for the 45° case stops at $N_\lambda \approx 0.33$; actually, the argument of the $\sin^{-1}$ function exceeds one, and the resulting propagation constant becomes complex for larger values of N_λ. There is an extensive theory now around this; see [1, Chapter 4]. Also notable on the curves is how rapidly phase velocity error (and hence phase error) increases as the mesh density decreases beyond a certain point.

It is interesting to compute the discretization level required for a specified error. Assuming one is operating at the Courant limit, by performing a Taylor series expansion on Eq. (3.98) the following expression is obtained for the numerical phase velocity (see Problem P3.2 at the end of the chapter):

$$\tilde{v}_p = \frac{\omega}{k} = c\left(1 - \frac{k^2h^2}{36} + \mathcal{O}\left(k^4h^4\right)\right) \tag{3.101}$$

Note that h is used here in place of Δs, conforming to more general usage in numerical analysis. Note also that this confirms the second-order accuracy of the method. For a 1% error in phase, one can easily show (Problem P3.3) that this requires around 10.5 cells per wavelength. The requirement on group velocity is tighter:

$$\tilde{v}_p = \frac{\partial\omega}{\partial k} = c\left(1 - \frac{k^2h^2}{12} + \mathcal{O}\left(k^4h^4\right)\right) \tag{3.102}$$

and the same level of error requires around 18 cells per wavelength.

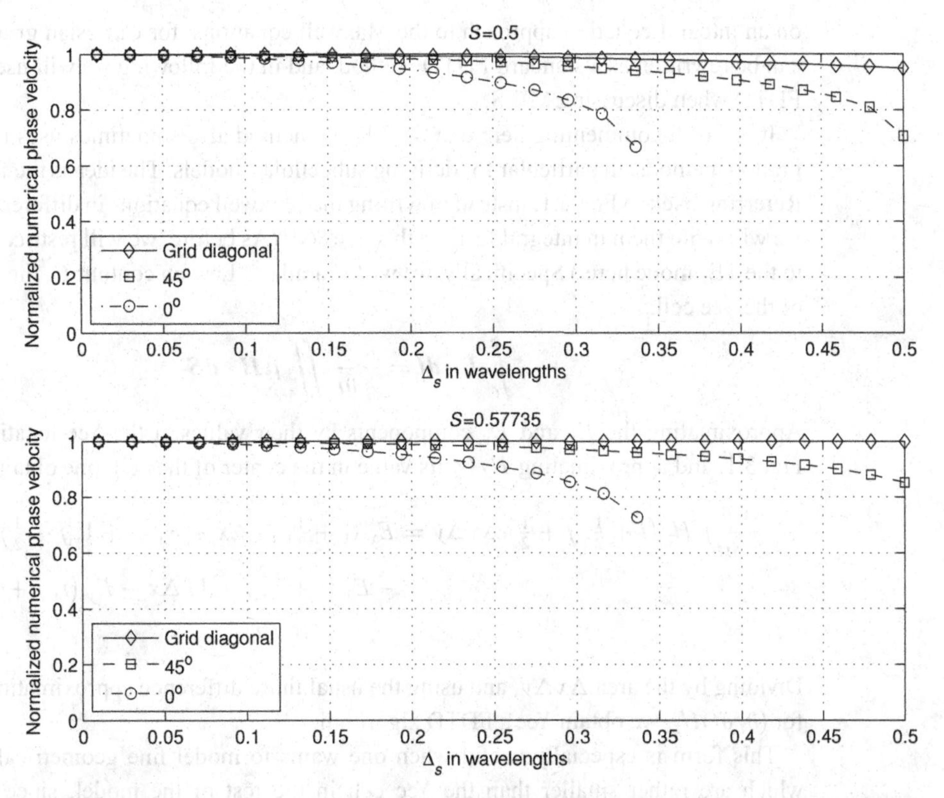

Figure 3.23 Dispersion curves for the 3D FDTD, comparing three different propagation angles. The top graph is for $S = 0.5$, i.e. 86.6% of the 3D Courant limit, the lower for $S = 1/\sqrt{3}$, i.e. the Courant limit.

It is very important to appreciate that phase error accumulates across a domain. The absolute phase error over a fixed length L is

$$\mathrm{err}_{\mathrm{phase}} = (\tilde{k} - k)L = \left(\frac{\omega}{c(1 - (kh)^2/36 + \cdots)} - \frac{\omega}{c} \right) L = \approx \frac{k^3 h^2 L}{36} \tag{3.103}$$

which implies that cell size must scale with frequency as $\omega^{-1.5}$ to keep the error constant, and hence the number of cells in each dimension scales with $\omega^{1.5}$. This explains the scalings for the FDTD given in Chapter 1.

3.5 Commercial implementations

Perhaps the most well-known commercial implementations of the FDTD are XFDTD and CST MICROWAVE STUDIO^{TM} (MWS). The former is an implementation of the standard FDTD. The latter is actually a suite of codes, including a transient solver which uses the finite integration technique (FIT) [18, 19]; its predecessor was known as MAFIA and one may still encounter reference to this in the literature. Although apparently based on an integral equation approach to the Maxwell equations, for Cartesian grids the FIT can be rewritten as a standard FDTD method, and in the following we will use the term FDTD when discussing MWS.

It is worth commenting here that the FDTD method also sometimes uses finite integration methods, in particular for deriving subcellular models. The idea is the following. Referring back to Fig. 3.1, instead of writing the Maxwell equations in differential form, we will write them in integral form in this Yee cell. (As before, we will restrict ourselves to the TE_z mode here.) Specifically, we write Faraday's Law on contour C, the boundary of the Yee cell:

$$\oint_C \boldsymbol{E} \cdot \boldsymbol{dl} = -\frac{\partial}{\partial t} \iint_A \mu \boldsymbol{H} \cdot \boldsymbol{dS} \tag{3.104}$$

Approximating the E_x and E_y components by their values at the Yee locations as in Fig. 3.1, and approximating H_z by its value in the center of the cell, one obtains:

$$\begin{aligned} -\frac{\partial}{\partial t}\mu H_z\left(i+\tfrac{1}{2}, j+\tfrac{1}{2}\right)\Delta x\,\Delta y = {} & E_x\left(i+\tfrac{1}{2}, j\right)\Delta x + E_y\left(i+1, j+\tfrac{1}{2}\right)\Delta y \\ & - E_x\left(i+\tfrac{1}{2}, j+1\right)\Delta x - E_y\left(i, j+\tfrac{1}{2}\right)\Delta y \end{aligned} \tag{3.105}$$

Dividing by the area $\Delta x \Delta y$, and using the usual finite difference approximation in time for $(\partial/\partial t)H_z$, we obtain Yee's FDTD algorithm.

This form is especially useful when one wants to model fine geometrical features which are rather smaller than the Yee cell in the rest of the model, since the field behavior can be taken into account when performing the integral. (As a simple example, the quasi-static $1/r$ nature of the magnetic field near a thin wire is used to incorporate thin wires.) These are generally known as local subcell models.[15] Typical examples include thin sheets, better approximations of curved boundaries, thin wires and thin cracks.

3.5.1 An introductory example – a waveguide "through"

The following is the first use of a commercial code in this book – in this case, MWS – and we will use this to highlight some important points about using an unfamiliar simulation tool.

Firstly, most packages nowadays ship with good documentation, usually with some form of "Getting Started" manual, or some variant on this theme, and time spent working through this type of manual is time very well spent indeed. Most simulators have some features and functions which are not immediately obvious, even if one is familiar with the method implemented, and the introductory manuals will often highlight these and save much time and subsequent frustration.

Secondly, even with the very best user interfaces – and MWS has a very impressive one – modelling complex three-dimensional geometries is not straightforward. One

[15] Another term often used is partially filled cells. Subcell is also sometimes used to describe submeshing, a method whereby a cell is divided into a number of smaller cells to improve accuracy.

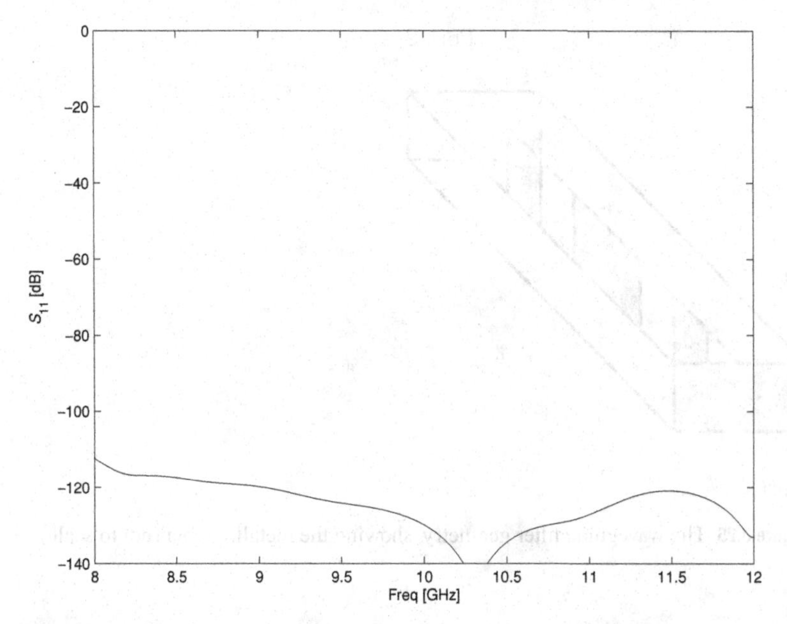

Figure 3.24 An MWS simulation of an empty piece of waveguide, showing an extremely low reflection coefficient as expected.

needs to try out simpler structures first, before attempting to model some complex device, quite possibly of unknown performance. Although MWS is at heart an FDTD code, the mesh is very largely invisible to the user. Model creation proceeds by defining geometrical primitives, which are then combined into more complex structures, before finally adding electrical parameters such as ports, field monitors, etc. (One exception in MWS is the electrical and/or magnetic properties of materials, which are defined as needed during model building; in some other packages, this is only done once the geometrical model is finished.)

So, with the notes of caution in mind, before analyzing a real device, the first structure which we will simulate is an empty piece of waveguide. We will do this at X band (8.2–12.4 GHz), using a piece 40 mm long. (This is long enough to test the model without requiring a significant runtime.) In MWS, we create the waveguide using either of the pre-defined waveguide "templates." (Templates simplify generating particular types of frequently used models; in the case of a waveguide, for instance, the exterior region is set to PEC.) Then, the "brick" primitive is used to generate the length of waveguide (the standard cross-section inside dimensions are 22.86 mm × 10.16 mm). Finally, the "pick face" function is used twice, to assign waveguide ports to each end of the length of waveguide. Since the waveguide is empty, the magnitude of the transmission coefficient should be unity, and the reflection coefficient zero. A result is shown in Fig. 3.24.[16] The reflection coefficient is less than −100 dB across the band, showing excellent performance, and giving confidence in basic modelling and simulation setup.

[16] Throughout this book, results have generally been plotted from MATLAB, using data computed by the relevant program, so as to provide some visual unity. Most programs provide a command to export data to some type of neutral file format.

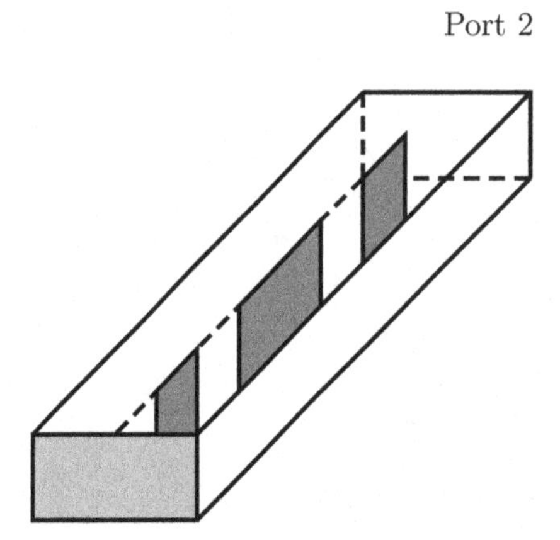

Figure 3.25 The waveguide filter geometry, showing the metallic septa (not to scale).

3.5.2 A waveguide filter

With some confidence that one has basic modelling skills with a particular package, one can turn to more interesting and challenging problems. Again, we will use a waveguide example, but now a more complex double-pole filter. The following example was originally designed by Meyer and van der Walt [20]. This X-band waveguide filter consists of three metal septa along its center, normal to the broad walls of the waveguide. The smaller septa are each 6.556 mm in length, and the longer is 16.788 mm. The inter-septa spacing is 12.148 mm. The septa are 0.2 mm thick. See Fig. 3.25 for a sketch of the filter.

When dealing with waveguide discontinuities, one of the first things one must note is that only the dominant waveguide mode should be present at the ports. In this case, an extra section of empty guide, 23.9 mm, was added, but any similar value would be acceptable. (The evanescent modes dampen exponentially, and at 10 GHz, the guide wavelength is around 40 mm, so the above length is around one-half a guide wavelength, more than sufficient.)

The modelling process in MWS is very similar to that already discussed in the previous introductory example, although here the "waveguide filter" template is chosen. (This sets some internal analysis parameters which are optimized for highly resonant structures.) The septa inside the waveguide are added quite easily using the "brick" primitive. (There are various ways of doing this; using the "working coordinate system" – a local coordinate system which can be easily repositioned – which the package supports can simplify this.)

In this case, the initial results were not especially accurate. The reason is that a filter relies on resonances and anti-resonances for its operation, and these must be computed extremely accurately for good overall accuracy. MWS offers an adaptive mesh facility,

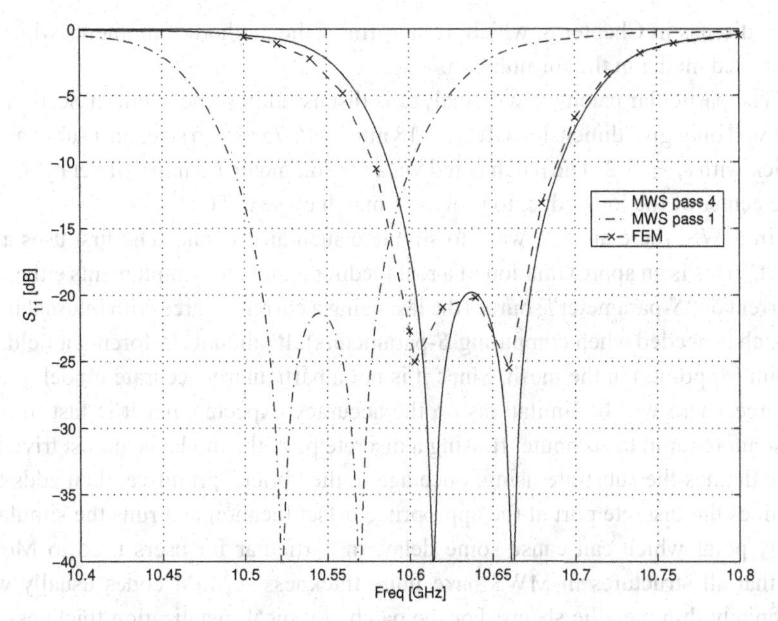

Figure 3.26 An MWS simulation of the waveguide filter in the text.

which automatically refines the mesh in regions it determines. Using this option provides a much more accurate result in this case. In Fig. 3.26, three results are shown: MWS pass 1 is the result after one solution; MWS pass 4 is the result after four adaptive passes have been undertaken; and the FEM results were computed using FEMFEKO, an experimental FEM program that will be described in Chapter 12, using complete second-order vector elements.[17] Clearly, the FEM results and pass 4 are in excellent agreement. For this filter, measured data are also available; the measured center frequency was 10.47 GHz. This is an example of the difference we have already discussed between the approximate field problem (which these two different techniques have solved with great accuracy, the difference in center frequency being less than 0.1%) and the actual problem (both analyses differ from the measured result by about 2%); the difference is very likely due to manufacturing tolerances.

3.5.3 A microstrip patch antenna

FDTD codes can also be applied to antennas, provided a suitable ABC is available. MWS offers a PML-based ABC; as we have seen, this is a very accurate mesh truncation technique. An "antenna on planar substrate" template is available, although for an accurate model we will have to work a little harder. One important point which one must bear in mind is that with the FDTD, the substrate will not be of infinite extent, unless we use a suitable boundary condition to simulate this. This is different to the simulations we

[17] More details on the FEM simulation may be found in Section 12.5; this solution had an average edge length of 3.0 mm, with 4968 tetrahedral elements and 41 526 degrees of freedom.

will discuss in Chapter 8, which use a form of the method of moments which includes stratified media in the formulation.

The particular patch we will analyze is discussed in some detail in Section 8.2; here, we will only give dimensions. It is 31.18 mm $\times$ 46.75 mm in size, on a substrate 2.87 mm thick with $\epsilon_r = 2.2$. The patch is fed via a pin (diameter 1.3 mm), offset by 8.9 mm from the center of the long edge, to provide a match close to 50 Ω.

In MWS, there are two ways to simulate such an antenna. The first uses a "discrete port." This is an approximation of a real feeding region, and implements either a voltage, current or "S-parameter" source (the last being a current source with internal impedance, which is needed when computing S-parameters). It amounts to forcing a field value at a point (or points) in the mesh. Since it is not a particularly accurate model of a physical source, there will be limitations on the accuracy expected, but it is fast to model and also more rapid to compute. If using a discrete port, the model is almost trivial to build: one defines the substrate using, once again, the "brick" primitive, then adds the patch, defines the discrete port at the appropriate offset location and runs the simulation. The only point which can cause some delay, in particular for users used to MoM codes, is that all structures in MWS have finite thickness – MoM codes usually work with infinitely thin metallic sheets. For the patch, a typical metalization thickness would be 25 μm, although the value is really not critical. MWS uses an elegant subcell model, known as the perfect boundary approximation [21] so that thin metal sheets do *not* have to comprise a full FDTD cell.

A more accurate model of the patch uses a coaxial feed and waveguide port. One way to do this is to add explicitly a ground plane of PEC of finite thickness, in which the coaxial feed will be embedded. (For reasons of internal code operation, MWS recommends that the length of the coaxial feed should be several times the thickness of the substrate; in this case, a length of 10 mm was chosen.) When adding the coaxial feed pin, one needs to be careful, since one is adding structures in regions where material already exists. In this case, it is easiest first to add the outer dielectric coaxial region cutting through the ground plane (and to use the same dielectric filler as the substrate material) and then to add the PEC inner conductor, which extends to become the feed pin. Although in general different materials cannot be defined in the same geometrical region,[18] MWS permits PECs and dielectrics to coexist, but the region is effectively treated as perfectly conducting.

An alternative approach is to use a thin ground plane, and construct a coaxial cable on the reverse side.

Results for two such models are compared to a FEKO computation in Fig. 3.27. There are actually three MWS results in the plot. Model one used the discrete port approach, and a 100 mm $\times$ 100 mm substrate, using open boundaries on the substrate sides, an open boundary with additional space above the patch, and an electric boundary on the ground plane. Model two used the same substrate and boundary treatment, but a full

[18] One makes use of various Boolean operations to combine, intersect, etc. such overlapping regions to resolve this.

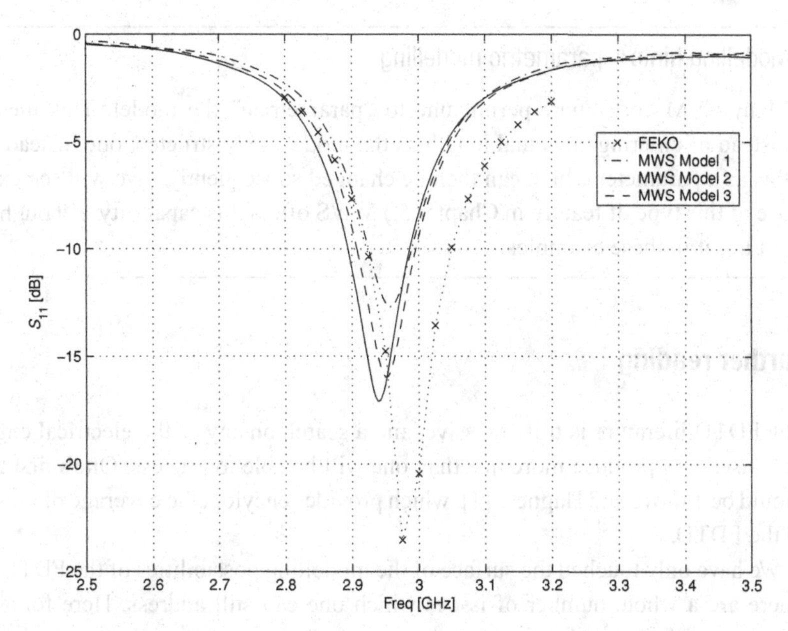

Figure 3.27 An MWS simulation of a microstrip patch antenna. FEKO results are also shown for comparison.

coaxial feed model.[19] Model three used the same coaxial feed model and boundary treatment, but with a smaller substrate, 50 mm × 50 mm in size; the results are very similar to those of model two, indicating that the open boundary is simulating an infinite substrate quite well. All models were also run through the adaptive meshing process. The agreement between all three models and the FEKO computation is good; the discrete port model indicates the least good match, but the results are still quite usable. The difference in center frequencies between all four analyses is less than 1%. As we will see frequently in this book, this is a commonly encountered limit in CEM for resonant antenna models, unless tremendous care is taken with the model. (It is worth commenting that manufacturing and material tolerances will often render this moot in any case.)

Modelling hints – open boundaries and MWS

MWS has two types of open boundaries, both simulated using the PML, and the difference between them is subtle. Although we did not discuss this, PMLs can also terminate a region with two different materials; see, for example, [1, Section 7.11]. An *open boundary* in MWS places a PML at the plane indicated, permitting the code effectively to continue the substrate indefinitely. An *open boundary* (*add space*) does much the same, but adds some additional (free) space first; hence, this will *not* produce an infinite substrate.

[19] The outer diameter of the coaxial feed, i.e. the region penetrating the ground plane, was chosen to give $Z_0 = 50\,\Omega$. For a coaxial cable of course, $Z_0 = \frac{60}{\sqrt{\epsilon_r}}\ln(b/a)$, with b and a the outer radius and inner radius respectively. In this case, with the same dielectric constant as the substrate, the outer radius was 2.24 mm.

> **Modelling hints – parametric modelling**
>
> Many CEM codes now permit one to "parametrize" the model. This means that instead of entering an actual length as the model is constructed, one instead defines this as a parameter which can then be changed subsequently. (We will see extensive use of this type of feature in Chapter 5.) MWS offers this capability, although we did not use it in these examples.

3.6 Further reading

The FDTD literature is truly massive, and a search on any of the electrical engineering databases will produce more hits than one will be able to process. One's first reference should be Taflove and Hagness [1], which provides encylopedic coverage of most aspects of the FDTD.

We have only touched the surface of the modelling possibilities of the FDTD method. There are a whole number of issues which one can still address. Here follows just a selection of these:

- Our 2D example already indicated that the rectangular cells of the standard FDTD method may not approximate curved geometries very well. Methods of improving fine geometrical detail are generally known as "subcell" models, and usually rely on an equivalent formulation of the FDTD in terms of Faraday's and Ampère's laws, as briefly introduced in Section 3.5. See [1, Chapter 10] for more on this topic. Thin wires are another type of structure which do not fit into the Yee grid very well. Bingle, the present author and Cloete describe a formulation incorporating finitely conducting wires in [22].
- When dealing with wideband pulses, one should appreciate that many materials cannot be represented accurately by a fixed value of ε_R. Again, elegant methods have been developed for dealing with materials with frequency-dependent material parameters; this is discussed in detail in [1, Chapter 9].
- For larger scatterers, it is extremely inefficient to try to position a field point in the far field. Formulations are available to compute the far field from a near field time domain computation, which permits one to use a much smaller mesh. See [1, Chapter 14] for details.
- Non-linear problems can *only* be addressed using time domain methods. A considerable amount of work has been done using the FDTD for such materials, including work at optical frequencies. FDTD codes have also been hybridized with circuit simulators to include non-linear devices (e.g. diodes). [1, Chapters 9 and 15] addresses these issues.
- We have discussed one-, two- and three-dimensional formulations of the FDTD. There is another interesting formulation, suitable for rotationally symmetric structures: the body of revolution FDTD. (This has been described as a two-and-a-half dimensional formulation; the full three-dimensional fields are computed, but using a

two-dimensional grid for each Fourier mode present – for some problems, only one such mode is needed.) A discussion of this may be found in [1, Chapter 12]. The present author and Ziolkowski also used this formulation for modelling optical wave phenomena; in [23], we presented the formulation. Rather importantly, the correct numerical stability criterion (the Courant limit) for this case is also given in this paper.

- The FDTD can also be used for handling periodic structures. The present author, Smith and van Tonder used this for modelling frequency selective surfaces [24]. The treatment by Maloney and Kesler [1, Chapter 13] provides an up-to-date account of the formulations available in this context.
- Another type of boundary condition of interest is the complementary operator. Ramahi has worked extensively on this, and a summary may be found in [1, Chapter 6]. Work also continues on other types of ABCs for the FDTD; see, for instance, [25].
- A recently (re-)discovered algorithm, the alternating direction implicit (ADI) formulation of the FDTD method, permits one to exceed the Courant limit, but retain stability. The ADI-FDTD method does pose some other challenges [26].

3.7 Conclusions

Our treatment of the FDTD method, which started out in the previous chapter with a very simple 1D transmission line problem, solved essentially in the frequency domain, continued in this chapter with a quite sophisticated 2D simulation, incorporating wide-band pulses, absorbing boundary conditions, and a physical analysis of scattering in the resonance regime[20] in both the time and frequency domains, and finished with some examples computed using the commercial package MWS.

We have also looked at computational issues, both runtime and memory, which impact on our ability to perform useful FDTD simulations. Berenger's PML has been introduced, and its extraordinary performance demonstrated. The 3D FDTD was briefly outlined; theoretically, there are no new issues to understand, but in practice writing a 3D code is challenging, since it needs to be very efficient in order to handle realistic problems (in 2D, far less optimal code can still be useful). Furthermore, for good results one should ideally use some of the more advanced FDTD approaches, in particular subcellular models and better modelling of curved boundaries. Unless one is fortunate enough to have access to an existing 3D FDTD code, such codes are generally best left to experts unless one has a very specific application in mind. The commercial code we discussed, MWS, provides a powerful implementation of the FDTD, offering (amongst other advanced modelling features) thin sheets, and a method called "perfect boundary approximation" which is essentially a type of subcell formulation improving geometrical modelling. It also features a user interface which at the time of writing was state-of-the-art. Other commercial FDTD codes are also available.

[20] The region in which the dimension(s) of the scatterer are on the order of several wavelengths at most.

The FDTD has truly become the workhorse of CEM computation over the last decade – even when it is not necessarily the best technique to use! In the next chapter, we introduce the method of moments, which is a very powerful method for dealing with highly conducting structures, and often more efficient for these applications than the FDTD method.

References

[1] A. Taflove and S. Hagness, *Computational Electrodynamics: the Finite Difference Time Domain Method*. Boston, MA: Artech House, 3rd edn., 2005.

[2] W. L. Stutzman and G. A. Thiele, *Antenna Theory and Design*. New York: Wiley, 2 edn., 1998.

[3] A. Ishimaru, *Electromagnetic Wave Propagation, Radiation and Scattering*. Engelwood Cliffs, NJ: Prentice-Hall, 1991.

[4] C. A. Balanis, *Advanced Engineering Electromagnetics*. New York: Wiley, 1989.

[5] J.-P. Berenger, "A perfectly matched layer for the absorption of electromagnetic waves," *J. Comput. Phys.*, **114**, 185–200, October 1994.

[6] W. C. Chew and W. H. Weedon, "A 3D perfectly matched medium from modified Maxwell's equations with stretched coordinates," *Microwave Opt. Technol. Guided Wave Lett.*, **7**, 599–604, September 1994.

[7] C. M. Rappaport, "Perfectly matched absorbing boundary conditions based on anisotropic lossy mapping of space," *IEEE Microwave Guided Wave Lett.*, **5**, 90–92, 1995.

[8] Z. S. Sacks, D. M. Kingsland, R. Lee and J. F. Lee, "A perfectly matched anisotropic absorber for use as an absorbing boundary condition," *IEEE Trans. Antennas Propagat.*, **43**, 1460–1463, December 1995.

[9] R. W. Ziolkowski, "Time-derivative Lorentz materials and their utilization as electromagnetic absorbers," *Phys. Rev. E, pt. B*, **55**, 7696–7703, 1997.

[10] D. S. Katz, E. T. Thiele and A. Taflove, "Validation and extension to three dimensions of the Berenger (PML) absorbing boundary condition for FD-TD meshes," *IEEE Microwave Guided Wave Lett.*, **4**, 268–270, August 1994.

[11] J.-P. Berenger, "Three-dimensional perfectly matched layer for the absorption of electromagnetic waves," *J. Comput. Phys.*, **127**, 363–379, 1996.

[12] D. C. Wittwer and R. W. Ziolkowski, "How to design the imperfect Berenger PML," *Electromagnetics*, **16**, 465–468, 1996.

[13] T. Rylander and J. Jin, "Perfectly matched layer in three dimensions for the time-domain finite-element method applied to radiation problems," *IEEE Trans. Antennas Propagat.*, **53**, 1489–1499, April 2005.

[14] D. B. Davidson and M. M. Botha, "Evaluation of a spherical PML for vector FEM applications," *IEEE Trans. Antennas Propagat.*, **55**, 494–498, February 2007.

[15] K. Yee, "Numerical solution of initial boundary value problems involving Maxwell's equation in isotropic media," *IEEE Trans. Antennas Propagat.*, **AP-14**, 302–307, May 1966.

[16] B. He and F. Teixeira, "Geometric finite element discretization of Maxwell equations in primal and dual spaces," *Phys. Lett. A*, **349**, pp. 1–14, 2006.

[17] A. Bondeson, T. Rylander and P. Ingelström, *Computational Electromagnetics*. New York, NY: Springer Science, 2005.

[18] T. Weiland, "A discretization method for the solution of Maxwell's equations for six-component fields," *Electronics Comm. (AEÜ)*, **31**, 116–120, 1977.

[19] T. Weiland, "Time domain electromagnetic field computation with finite difference methods," *Int. J. Numerical Modelling*, **9**, 295–319, 1996.

[20] P. Meyer and P. W. van der Walt, "Design of narrowband E-plane waveguide filters," in *Proceedings of the 1988 SAIEE 2nd. Joint AP-MTT Symposium*, Pretoria, South Africa, pp. 26.1–26.11, August 1988.

[21] B. Krietenstein, R. Schuhmann, P. Thoma and T. Weiland, "The perfect boundary approximation technique facing the big challenge of high precision field computation," in *Proceedings of the XIX International Linear Accelerator Conference (LINAC 98)*, Chicago, IL, pp. 860–862, 1998.

[22] M. Bingle, D. B. Davidson and J. H. Cloete, "Scattering and absorption by thin metal wires in rectangular waveguides: FDTD simulation and physical experimants," *IEEE Trans. Microwave Theory Tech.*, **50**, 1621–1627, June 2002.

[23] D. Davidson and R. W. Ziolkowski, "Body-of-revolution finite-difference time-domain modelling of space-time focusing by a three-dimensional lens," Special Issue on 3D Scattering, *J. Opt. Soc. Am. A*, **11**, pp. 1471–1490, April 1994.

[24] D. B. Davidson, A. G. Smith, and J. J. van Tonder, "The analysis, measurement and design of frequency selective surfaces," in *Proceedings of the 10th International Conference on Antennas and Propagation*, **1**, pp. 1.156–1.160. Edinburgh: IEE, April 1997.

[25] R. E. Díaz and I. Scherbatko, "A simple stackable re-radiating boundary condition (rRBC) for FDTD," *IEEE Antennas Propagat. Mag.*, **46**, 124–130, February 2004.

[26] S. W. Staker, C. L. Holloway, A. U. Bhobe and M. Piket-May, "Alternating-direction implicit (ADI) formulation of the finite-difference time-domain (FDTD) method: algorithm and material dispersion implementation," *IEEE Trans. Electromagn. Compat.*, **45**, 156–166, May 2003.

[27] D. M. Pozar, *Microwave Engineering*. New York: Wiley, 2nd edn., 1998.

Problems and assignments

Problems

P3.1 Derive Eqs. (3.98)–(3.100).

P3.2 Derive Eq. (3.101). (*Hint*: start with Eq. (3.97) for the case $k_x = k, k_y = 0, k_z = 0$, at the Courant limit, $c\Delta t/\Delta s = 1/\sqrt{3}$. Perform a Taylor series expansion on both sides. Collect terms and simplify, using $c\Delta t/\Delta s = 1/\sqrt{3}$ and $\omega/k = c$ as needed, to obtain the result.)

P3.3 Using Eq. (3.101), verify that a 1% error in phase velocity requires around 10.5 cells per wavelength. Similarly, using using Eq. (3.102), verify the 18 cell requirement for group velocity.

Assignments

A3.1 Repeat the TE_z scattering analysis discussed in this chapter using longer (in time) pulses and shorter pulses. Explain the time domain results obtained with each of these.

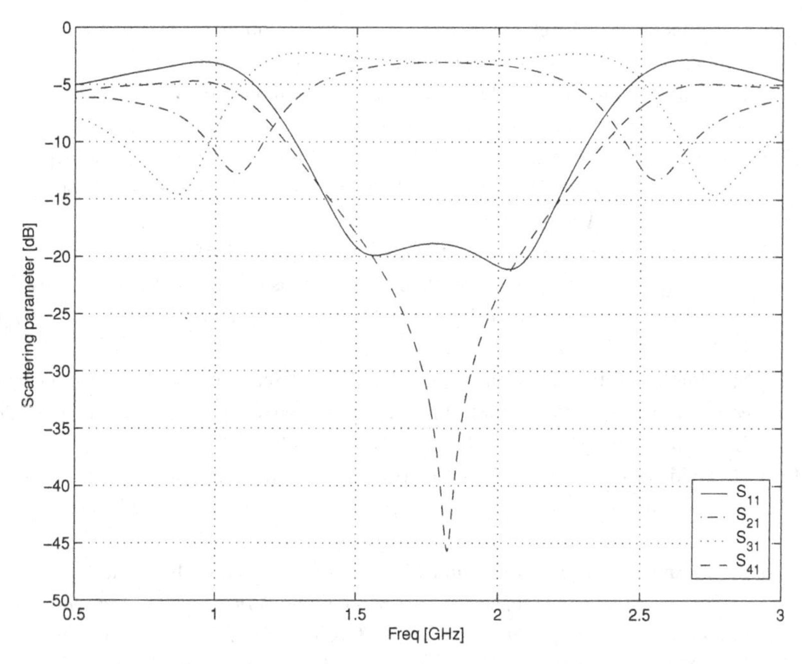

Figure 3.28 A MWS simulation of a rat-race hybrid coupler in microstrip.

Keep the grid size at 800×400 and $M = 1024$ so that runtimes remain minutes rather than hours.

A3.2 Modify the code to compute TM_z scattering from a cylinder. Does the TM_z polarization also show creeping waves?

A3.3 Finally, extend the code (either TM or TE) to use the PML ABC. Since one needs to verify the PML, this is quite time consuming.

A3.4 A ring hybrid, or *rat-race*, is a four-port device which functions as a 180° hybrid. (These are discussed in some detail in Section 11.5.) Descriptions may be found in many books on microwave circuits, such as [27, Section 7.8]. Using a commercial package, predict the behavior of such a device fabricated in microstrip.

Partial solution

The device was designed to operate at 1.8 GHz. One must first obtain the dimensions for the microstrip; this was based on the example in [27, p. 163]. The substrate was chosen as $d = 1.27$ mm thick, $\epsilon_r = 2.2$. For the 50 Ω feedlines, the strip width to substrate thickness ratio W/d is 3.0981, hence $W = 3.91$ mm. For the 70.7 Ω components in the rat-race, W/d works out as 1.768, hence $W = 2.24$ mm. The effective dielectric constant in the 70.7 Ω section is 1.82 (it is slightly dependent on W/d). Hence, at 1.8 GHz, a quarter-wavelength in the dielectric is 30.8 mm. The average radius of the ring is thus 29.4 mm.

The four ports were modelled as discrete ports on the ends of sections of 50 Ω feedline approximately 15 mm long. In this case, the standard planar coupler template was used, and the space on top is five times the substrate thickness, as recommended by a MWS tutorial. In this case, however, open boundaries should be used (apart obviously from the ground plane).

At the design frequency, the results show the expected good match at port 1, 3 dB coupling to ports 2 and 3, and some 45 dB of isolation with respect to port 4 (Fig. 3.28).

4 A one-dimensional introduction to the method of moments: modelling thin wires and infinite cylinders

4.1 Introduction

The method of moments – MoM – was one of the first numerical methods to achieve widespread acceptance in electronic engineering for the analysis of antennas and scatterers. It is generally defined as a method for reducing an integro-differential equation to a set of linear equations. The origins of the method are old: as was already indicated in Chapter 1, some of the early work was done over a century ago. One of the widely used integral equation formulations still used for the analysis of thin wires (that due to Pocklington) was first presented in 1897 (although he used a series expansion method, rather than the modern segmentation approach). The first publications in the antenna and propagation professional literature were in the early 1960s, and some of the canonical papers (those of Harrington, Richmond, Mei and Andreasen) appeared at much the same time as Yee's paper. The specific name "method of moments" was introduced by Harrington in his early work, and the name caught on quickly; this was perhaps unfortunate, since the name has a slightly different meaning in contemporary applied mathematics. In that field, and also fields such as computational mechanics, the term "method of weighted residuals" is generally used for what has become known as the MoM in radio-frequency engineering. Another term widely used in other fields of engineering is "boundary element method"; for highly conducting structures, this term and the MoM as used in electromagnetics are synonymous.[1]

Primarily for two reasons, the MoM rapidly achieved widespread acceptance. Firstly, to a generation of engineers and scientists trained on analytical methods, along the lines of Harrington's classic text on time-harmonic fields [1][2] – which in turn was based on methods of mathematical physics, as expounded by Stratton [2] and Morse and Feshbach [3] – the method was clearly based on sound electromagnetic theory and more generally, methods of mathematical physics, in particular variational calculus (which was then in widespread use). Secondly, because the method discretized *only* the metallic wires or surfaces of the antennas, it was far more efficient than methods such as the FDTD for analyzing the relatively small – typically resonance regime – antenna structures which were then the main topic of research. (As we have seen, the FDTD requires the discretization of all space surrounding the antenna or scatterer.) Furthermore, many

[1] We return to the topic of nomenclature in the penultimate section of this chapter.

[2] Originally published in 1961, but reprinted since.

problems then of current research interest could be solved used the MoM in a reasonable time – this was far less true of the FDTD, whose requirements for memory and computer time could generally not be accommodated on 1960s era computers.

In this chapter, we will present an introduction to the MoM, starting with an extremely simple electrostatic example. Again, as with the FDTD, the simple physics and geometry permit us to illustrate a number of core ideas without becoming overwhelmed by implementation details. Following this, we will extend the discussion to electrodynamics. Thin-wire modelling uses locally one-dimensional basis functions, but for general wire geometries, one must of course take the full three-dimensional geometry into account, and hence writing one's own MoM program for any reasonably interesting engineering problem is well beyond the scope of an introductory book of this nature. Fortunately, there are some excellent commercial implementations of the MoM, as well as one very useful public domain code; these are the topics of Chapter 5. Also in this chapter, we will consider scattering from highly conducting cylinders of infinite length. Although a problem in two-dimensional space, the MoM permits one to reduce this to a one-dimensional problem along the boundary of the cylindrical cross-section, illustrating the power of the MoM in reducing problem dimensionality.

4.2 An electrostatic example

The problem we will address as an illustration of the MoM is the charge distribution $\rho(z)$ on a perfectly conducting straight thin wire, of radius $\rho = a$, charged to a potential V volts relative to ground. It is based on an example presented in [4, Chapter 12]. The wire could, for instance, be charged by induction. It is important to note that this is the *opposite* of typical work in introductory courses in electromagnetics, where $\rho(z)$ is given and one must then establish the potential (and hence field). Given $\rho(z)$, $V(\boldsymbol{r})$ (and hence $\boldsymbol{E} = -\nabla V$) is easily found:

$$V(\boldsymbol{r}) = \frac{1}{4\pi\epsilon_0}\int_V \frac{\rho(\boldsymbol{r}')}{R(\boldsymbol{r},\boldsymbol{r}')} dV' \tag{4.1}$$

with

$$R(\boldsymbol{r},\boldsymbol{r}') = |\boldsymbol{r} - \boldsymbol{r}'| = \sqrt{(x-x')^2 + (y-y')^2 + (z-z')^2}$$

The primed coordinates ($\boldsymbol{r}'(x', y', z')$) are those of the source point. The field point coordinates are $\boldsymbol{r}(x, y, z)$.

However, our problem now is, given the voltage on the wire, to establish how the charge distributes itself. (In passing, we note that it cannot be a uniform distribution; the charges near the ends would clearly experience an unbalanced electrostatic force which would push them towards the ends of the wire.) This falls into the general class of *inversion* problems, and cannot generally be solved in *closed form*, i.e. analytically. A numerical approach is the only general solution method for such problems.

Before we proceed further, some terminology: Eq. (4.1) is known as an *integral equation*; the part inside the integral operator is frequently called the *kernel*. The function $V(\boldsymbol{r})$ is the *forcing function*. Two other concepts that are central to this theory is that the physical environment surrounding the radiator/scatterer (in this case, free space, i.e. an infinitely large and empty vacuum) *and* the boundary conditions are all included in the formulation. This is what permits the MoM to solve typical antenna problems (at least those involving perfect or highly conducting conductors) very efficiently. We will later encounter *Green functions*; it is these that effectively take the environment surrounding the structure into account, but they are only available for a very limited number of environments.

The critical idea is that Eq. (4.1) is valid *everywhere* – including on the wire itself, *where* $V(x, y, z)$ *is known*. This is the *boundary condition* (BC) for the problem. The idea that we will pursue to solve this problem is to approximate the charge by a number of simple functions, of unknown amplitude, which we will then find by assembling a matrix equation representing the geometry of the model and the BCs in discrete form.

4.2.1 Some simplifying approximations

Before we proceed further with the MoM solution of this problem, we will make a number of assumptions, which will considerably simplify the solution process:

- Equation (4.1) contains a volumetric integral. If we assume that the wire is a perfect electrical conductor (PEC), the charge is restricted to the surface and becomes a *surface charge* $\rho_s(z, \rho = a, \phi)$. (Note that we use cylindrical coordinates here, and that ρ refers both to the radius in this coordinate system and to the charge. The meaning will be clear from the context.)
- Secondly, we will simplify the geometry, by assuming a $\hat{z}$-directed wire.
- Thirdly, we will assume that the charge distribution is *uniform* in the circumferential direction, i.e. we can simply write $\rho_s(z, \rho = a, \phi) = \rho_s(z, \rho = a)$. This permits us to approximate further the *surface* charge $\rho_s(z, \rho = a)$ by an equivalent *line* charge, $\rho_l(z) = 2\pi a \rho_s(z, \rho = a)$, placed on the $\hat{z}$-axis.

Using these approximations, the integral equation Eq. (4.1) becomes

$$V(z, \rho = a) = \frac{1}{4\pi\epsilon_0} \int_0^{\ell} \frac{\rho_\ell(z')}{R(z, z')} dz' \tag{4.2}$$

with (a is the wire radius)

$$\begin{aligned} R(z, z') &= \sqrt{[(x - x')^2 + (y - y')^2] + (z - z')^2} \\ &= \sqrt{a^2 + (z - z')^2} \end{aligned}$$

Note that we now write $V(z, \rho = a)$, rather than $V(\boldsymbol{r})$, since V is restricted to be on the wire surface (where the boundary is) and is rotationally invariant by assumption.

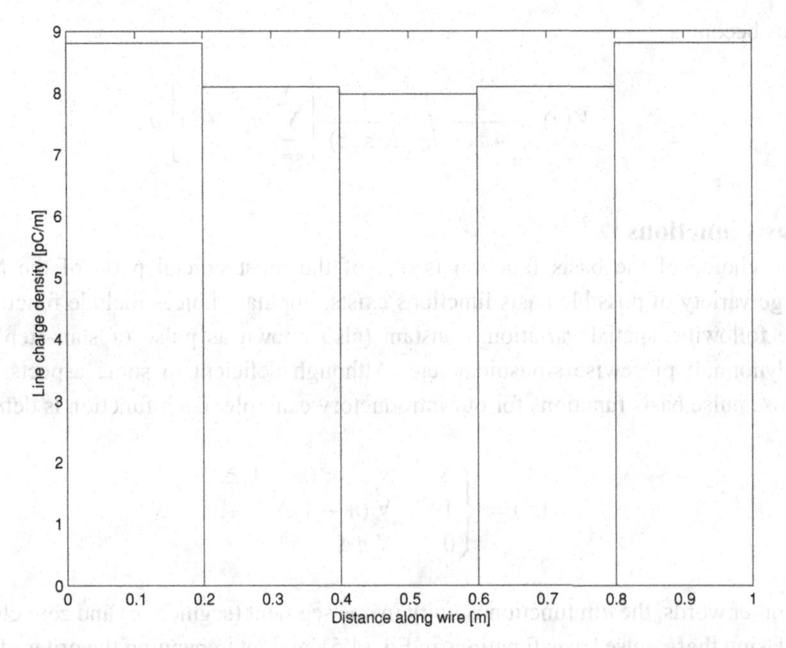

Figure 4.1 Equivalent line charge density for an $N = 5$ segment MoM solution using piecewise constant basis functions. $L = 1$ m, $V = 1$ V, $a = 0.001$ m.

4.2.2 Approximating the charge

Up to this point, the approximations have been in the mathematical formulation (the integral equation). Now, we introduce the MoM as a method of approximately solving this equation. The wire, of length L, is broken up into N segments, using $N + 1$ nodes, defined as follows:

$$z_n = (n-1)\Delta, \qquad n = 1, 2, \ldots, N+1 \tag{4.3}$$

$$\Delta = \frac{L}{N} \tag{4.4}$$

In the following, "segment n" will mean the segment located between z_n and z_{n+1}.

The charge is approximated as

$$\rho(z') \approx \sum_{n=1}^{N} a_n h_n(z') \tag{4.5}$$

Here, a_n are unknown (but constant) coefficients, and $h_n(z')$ are *basis functions* – also often known as *expansion functions*. (Many texts use $f_n(z')$, but we want to reserve f and g for a specific purpose, discussed later in this chapter.) An example, with $N = 5$, is shown in Fig. 4.1. (Note that this is the solution obtained *after* the procedure to be discussed has been performed and the unknown coefficients obtained.) Equation (4.2) thus becomes

$$V(z) = \frac{1}{4\pi\epsilon_0}\int_0^{\ell}\frac{1}{R(z,z)}\left[\sum_{n=1}^{N} a_n h_n(z')\right] dz' \tag{4.6}$$

Basis functions

The choice of the basis function is one of the most crucial parts of the MoM. A large variety of possible basis functions exists. Popular choices include functions with the following spatial variation: constant (also known as pulse or stair-step); linear; polynomial; piecewise sinusoidal; etc. Although deficient in some aspects, we will chose pulse basis functions for our introductory example. Each function is defined as

$$h_n(z') = \begin{cases} 0 & \forall\, z' < (n-1)\Delta \\ 1 & \forall\, (n-1)\Delta \leq z' \leq n\Delta \\ 0 & \forall\, n\Delta < z' \end{cases} \tag{4.7}$$

In other words, the nth function is unity in one segment (segment n) and zero elsewhere.

Using these pulse basis functions in Eq. (4.5), and interchanging the order of integration and summation, one obtains:

$$4\pi\epsilon_0 V(z) = a_1\int_0^{\Delta}\frac{h_1(z')}{R(z,z')}dz' + a_2\int_{\Delta}^{2\Delta}\frac{h_2(z')}{R(z,z')}dz' + \cdots + a_N\int_{(N-1)\Delta}^{N\Delta}\frac{h_N(z')}{R(z,z')}dz' \tag{4.8}$$

This is one equation in N unknowns, viz. $\{a_1, a_2, \ldots, a_N\}$. To obtain a unique solution, one requires N equations, or *constraints*.[3]

4.2.3 Collocation

To provide these N constraints, we enforce (match) the boundary condition at N points along the wire, z_m; this is also described as *testing* (sampling) $V(z, \rho = a)$. This method is called *collocation* or *point-matching*. It is convenient to locate these points in the middle of each segment, in between the nodes:

$$z_m = (m - 1/2)\Delta, \qquad m = 1, 2, \ldots, N \tag{4.9}$$

Note that *unlike* the FDTD method, this "half-point" offset has no adverse effect on the accuracy of the method, and is not essential to its implementation; sampling points at other locations within the segment would also work, this is merely convenient.

[3] Strictly speaking, these must be *linearly independent* equations.

Sampling Eq. (4.6) at each of these N points, the following set of N equations is obtained:

$$4\pi\epsilon_0 V(z_1) = a_1 \int_0^{\Delta} \frac{h_1(z')}{R(z_1, z')} dz' + \cdots + a_N \int_{(N-1)\Delta}^{N\Delta} \frac{h_N(z')}{R(z_1, z')} dz'$$

$$\vdots$$

$$\vdots$$

$$4\pi\epsilon_0 V(z_N) = a_1 \int_0^{\Delta} \frac{h_1(z')}{R(z_N, z')} dz' + \cdots + a_N \int_{(N-1)\Delta}^{N\Delta} \frac{h_N(z')}{R(z_N, z')} dz' \tag{4.10}$$

4.2.4 Solving the system of linear equations

The above set of equations is a *system* of linear equations. At this point, it is important to appreciate that the original *integral equation* inversion problem has now been reduced to a *matrix equation* inversion problem. It can be written as

$$\{V\} = [Z]\{I\} \tag{4.11}$$

sometimes known as generalized network parameters. Square braces indicate a matrix, curled braces a vector. The relevant entries are:

$$V_m = 4\pi\epsilon_0 V(z_m)$$
$$I_n = a_n$$
$$Z_{mn} = \int_{(n-1)\Delta}^{n\Delta} \frac{1}{[(z_m - z')^2 + a^2]^{1/2}} dz' \tag{4.12}$$

The n subscript refers to source points; m refers to testing (sampling) points. Symbolically, the solution is

$$\{I\} = [Z]^{-1}\{V\} \tag{4.13}$$

However, a linear system, usually written in the form $[A]\{x\} = \{b\}$, is *almost never* solved by inverting the matrix explicitly. Instead, the matrix $[A]$ is factored into the product of lower and upper triangular matrices:

$$[A] = [L][U] \tag{4.14}$$

Hence $[L][U]\{x\} = \{b\}$. An auxiliary vector $\{z\} = [U]\{x\}$ is introduced, and then $[L]\{z\} = \{b\}$ is solved by *forward substitution* to yield $\{z\}$; finally, $\{x\}$ is solved from $\{z\} = [U]\{x\}$ using *backward substitution*. (This process, an extension of Gaussian elimination, is generally covered in introductory undergraduate courses in numerical analysis.)

There are a number of reasons for pursuing this rather than direct inversion of the matrix; the most important is that solving a linear system using LU-factorization has a cost $\sim\mathcal{O}(N^3)$, whereas inverting a matrix costs at least twice this, since following the factorization N forward and backward substitutions are required, each of cost $\sim\mathcal{O}(N^2)$.

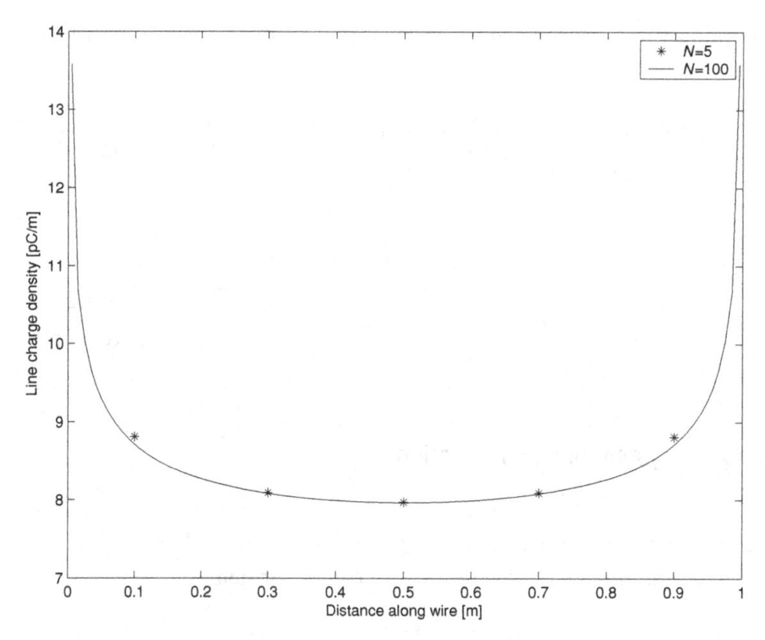

Figure 4.2 Comparison of 5-segment and 100-segment solutions.

Before the matrix equation can be solved, however, there is still one issue to attend to. In Eq. (4.12), the term Z_{mn} is given as an integral over the nth segment. This usually has to be done numerically using *quadrature* (numerical integration). In this *specific* case, analytical results are available [4, p. 674]:

$$Z_{mn} = \begin{cases} 2\ln\left(\frac{\Delta/2+\sqrt{a^2+(\Delta/2)^2}}{a}\right) & \forall m = n \\ \ln\left(\frac{d^+_{mn}+\sqrt{(d^+_{mn})^2+a^2}}{d^-_{mn}+\sqrt{(d^-_{mn})^2+a^2}}\right) & \forall m \neq n \text{ but } |m-n| \leq 2 \\ \ln\frac{d^+_{mn}}{d^-_{mn}} & \forall |m-n| > 2 \end{cases} \tag{4.15}$$

with

$$\begin{aligned} d^+_{mn} &= l_m + \Delta/2 \\ d^-_{mn} &= l_m - \Delta/2 \\ l_m &= \sqrt{[(m-n)\Delta]^2 + a^2} \end{aligned} \tag{4.16}$$

The last parameter is the distance between the mth matching point and the center of the nth source segment.

4.2.5 Results and discussion

Results are shown in Figs. 4.1 and 4.2. In Fig. 4.1, the piecewise constant nature of the basis function has been explicitly shown (the `bar` command in MATLAB provides a simple way of doing this). In Fig. 4.2, one observes that the $N = 5$ solution is surprisingly accurate, although of course it does not correctly predict the behavior of the charge at the ends of the wire.

A number of approximations have been made in this development. These include the following, with the implications indicated:

- An equivalent line charge was assumed. This relied on a *rotationally symmetric* charge distribution. For a thin wire, this is generally a very good approximation.
- The ends of the wire were ignored; for instance, was the wire a hollow or solid tube? Again, for thin wires, this is a reasonable approximation.
- In the collocation process, the integrals (which represent the boundary conditions) were only exactly enforced at N discrete points. In between these points, the potential will depart from the specified value. Fortunately, using more (i.e. smaller) segments will reduce the impact of this.
- The specific basis function that was chosen – constant – is *discontinuous* at segment ends. Since we were approximating charge, which is continuous, this is non-physical. This is clearly evident in Fig. 4.1. (Again, the impact of this can be mitigated using smaller segments.)
- We assumed that the surface of the wire was perfectly conducting, so that the wire was an equipotential surface. For most good conductors, this is a very good approximation.

The reason that we are discussing these in detail is that all these comments also apply to electrodynamics.

4.3 Thin-wire electrodynamics and the MoM

With these basics behind us, electrodynamics (or *full-wave* behavior) can now be investigated. The ideas of incident and scattered field decomposition are important here. Other than this, and the more complex equations, we will find the overall process very similar indeed.

4.3.1 The electrically thin dipole

The problem that we now want to solve is the current distribution $I(z)$ on a straight thin wire. It is assumed here that the basics of the dipole radiator have already been studied. In such introductory courses on electrodynamics, some *assumption* is generally made regarding the distribution. For very short dipoles, a linear or even constant approximation of current can yield quite good results, and for the typical resonant dipole, the widely assumed sinusoidal distribution also produces useful results. However, the most obvious information which cannot be thus obtained is the *reactance* of the dipole.

Although the overall process is very similar to the electrostatic charge distribution problem just worked out, there are two important differences. Firstly, the boundary condition: for a perfect electric conductor, the boundary condition is

$$\boldsymbol{E}_{\tan} = 0 \tag{4.17}$$

We will use the incident/scattered field decomposition method. This was already introduced with the FDTD. To revise this briefly: since the Maxwell equations are linear, the fields may be decomposed into an *incident* field $\boldsymbol{E}^{\text{inc}}$ and a *scattered* field $\boldsymbol{E}^{\text{scat}}$. The overall field, called the *total* field $\boldsymbol{E}^{\text{tot}}$, is then

$$\boldsymbol{E}^{\text{tot}} = \boldsymbol{E}^{\text{inc}} + \boldsymbol{E}^{\text{scat}} \tag{4.18}$$

By definition, the incident field is the field which would exist if the scatterer were absent. As an example, if the incident field is a plane wave, propagating in the x-direction, in free space, with a z-polarized electric field, the expressions for the incident fields are:

$$\boldsymbol{E}^{\text{inc}} = e^{-jkx}\hat{z} \tag{4.19}$$

$$\boldsymbol{H}^{\text{inc}} = -\frac{1}{\eta_0}e^{-jkx}\hat{y} \tag{4.20}$$

As usual, $\eta_0 = \sqrt{\mu_0/\varepsilon_0}$ is the wave impedance of free space, and $k = 2\pi/\lambda_0$ is the wavenumber. It is of interest to compare these expressions to those used in Section 3.2.3. The main difference is of course that these expressions are *frequency domain* ones. (A rather more minor difference is that the electric field is polarized in the $\hat{z}$-direction rather than $\hat{y}$.) On the surface of a PEC wire, $\boldsymbol{E}_{\text{tot}} = 0$. The boundary condition on the surface of the wire thus becomes

$$\boldsymbol{E}^{\text{inc}} = -\boldsymbol{E}^{\text{scat}} \tag{4.21}$$

As indicated above, $\boldsymbol{E}^{\text{inc}}$ typically has a simple form. The scattered fields, $\boldsymbol{E}^{\text{scat}}$, must be computed from the surface current.

In general, the electric field can be computed from the magnetic vector potential $\boldsymbol{A}$ and electric scalar potential Φ as

$$\boldsymbol{E} = -j\omega\boldsymbol{A} - \nabla\Phi \tag{4.22}$$

It will be recalled that various *gauges* can be applied to these potentials.[4] The Lorenz[5] gauge is widely used in this context:

$$\nabla \cdot \boldsymbol{A} = -j\omega\mu_0\epsilon_0\Phi \tag{4.23}$$

Applied now to the $\hat{z}$-directed surface current source, and assuming that the wire is in free space, so that ϵ, μ and the wavenumber k have the usual values in vacuum[6] this becomes

$$\frac{\partial A_z}{\partial z} = -j\omega\mu_0\epsilon_0\Phi \tag{4.24}$$

Hence,

$$E_z^{\text{scat}}(r) = -j\frac{1}{\omega\mu_0\epsilon}\left(k^2 A_z + \frac{\partial^2 A_z}{\partial z^2}\right) \tag{4.25}$$

[4] The potentials are not unique, and contain elements of arbitrariness, which the gauging resolves.

[5] More properly attributed to L. Lorenz than H. Lorentz.

[6] This formulation is actually valid in any linear, isotropic and uniform medium, with μ and ϵ taking the appropriate values. For simplicity, we show only the free-space case.

with

$$A_z = \frac{\mu_0}{4\pi} \int_{-l/2}^{l/2} \int_0^{2\pi} J_z(\phi', z') \frac{e^{-jkR}}{R} a \, d\phi' \, dz' \tag{4.26}$$

We have used the "free-space Green function", here ($\psi(z, z') = e^{-jkR}/4\pi R$), which gives the resulting magnetic vector potential for a current element.[7] R is the distance from source to field point coordinates. Substituting Eq. (4.26) in Eq. (4.25), and integrating over the source region, one obtains

$$E_z^{\text{scat}}(r) = \frac{1}{j\omega\epsilon_0} \int_0^{2\pi} \int_{-l/2}^{l/2} \left[\frac{\partial^2 \psi(z, z')}{\partial z^2} + k^2 \psi(z, z') \right] J_z(\phi', z') a \, d\phi' \, dz' \tag{4.27}$$

Note that the differentiation in Eq. (4.25) has been taken inside the integral operator. This is valid since the differentiation is with respect to the field point coordinates, and the integration is over the source points.

At this stage, the unknown is still the $\hat{z}$-directed (by assumption) surface current $J_z(\phi', z')$. For sufficiently thin wires, this can be reduced to the *Pocklington* equation, first introduced in 1897:

$$\begin{aligned} E_z^{\text{scat}}(r) &= \frac{1}{j\omega\epsilon_0} \int_{-l/2}^{l/2} \left[\frac{\partial^2 \psi(z, z')}{\partial z^2} + k^2 \psi(z, z') \right] I_z(z') \, dz' \\ &= -E_z^i(r) \end{aligned} \tag{4.28}$$

This equation is obtained by assuming that (as for the electrostatic case), we locate the filament on the axis and enforce the boundary condition on the surface (the reciprocal case is sometimes more convenient in deriving this). Although it looks fairly straightforward, the presence of the second derivative of z inside the integral kernel, acting on the Green function, makes this non-trivial to implement. A useful further simplification can be made if the wire is assumed to be *very* thin ($a \ll \lambda$):

$$\int_{-l/2}^{l/2} I_z(z') \frac{e^{-jkR}}{4\pi R^5} \left[(1 + jkR)(2R^2 - 3a^2) + (kaR)^2 \right] dz' = -j\omega\epsilon_0 E_z^i(\rho = a) \tag{4.29}$$

with a the wire radius and $R = \sqrt{a^2 + (z - z')^2}$. This is now a convenient form to program. It appears in numerous texts (for example, [4, p. 720]) and appears to have been first introduced by Richmond [5] (reprinted in [6]).

Further discussion on these and other integral equations (such as Hallén's) may be found in [4, 7].

[7] A more detailed discussion of Green functions is deferred to Chapter 7.

Before solving this numerically, recall that we are assuming the following:

- Circumferential currents are negligible.
- The axial current $I(z')$ does not vary circumferentially. (This is *not* the same as the first assumption!)
- As for the electrostatic case, we locate the filament on the axis and enforce the boundary condition on the surface, or the reciprocal case.

The reason that we offset the source filament and testing surface (or vice versa) is, as in the electrostatic case, to avoid the singularity present at $z = z'$. Although approximate, this method works well for thin wires. As for the static case, the kernel is not singular, but for small a can become more nearly so than in the electrostatic case – the R^5 term in the denominator of Eq. (4.29) is largely responsible – and more sophisticated treatments are frequently used. The problem usually occurs with the "self" term (the element of $[Z]$ with $m = n$). The usual remedy is to subtract a term with the same order of singularity but which can be integrated analytically, and then to integrate numerically the difference between the singular term and the remainder, since this is usually quite well behaved. Examples of this type of treatment of singular integrals will be discussed in Chapter 7 (although in a slightly different context).

Approximating the current

The same idea is used for approximation of the current as we used for charge, namely some sort of discrete approximation using a set of functions of known shape but unknown amplitude. The most widely used basis functions are pulse (piecewise constant, as used for the electrostatic problem), triangular (piecewise linear) and piecewise sinusoidal. An especially convenient form arises when piecewise sinusoidal basis functions are chosen. In this case, for a wire ℓ in length, lying on the z-axis from $-\ell/2$ to $\ell/2$, the nodes are defined as

$$z_n = -\ell/2 + n\,\Delta, \qquad n = 0, 1, \ldots, N \tag{4.30}$$

$$\Delta = \frac{\ell}{N} \tag{4.31}$$

assuming that all N segments are the same length Δ.

For a $\hat{z}$-directed interior segment,[8] the piecewise sinusoidal basis function is expressed as

$$h_n(z) = \begin{cases} \dfrac{\sin k(z - z_{n-1})}{\sin k(z_n - z_{n-1})}, & \forall z_{n-1} \le z \le z_n \\ \dfrac{\sin k(z_{n+1} - z)}{\sin k(z_{n+1} - z_n)}, & \forall z_n \le z \le z_{n+1} \end{cases} \tag{4.32}$$

Note that here we have permitted segments to be potentially different in length, in which case Eq. (4.31) must be modified accordingly. With the above numbering scheme for an

[8] At the ends of the wire, special half-basis functions are required, to be discussed shortly.

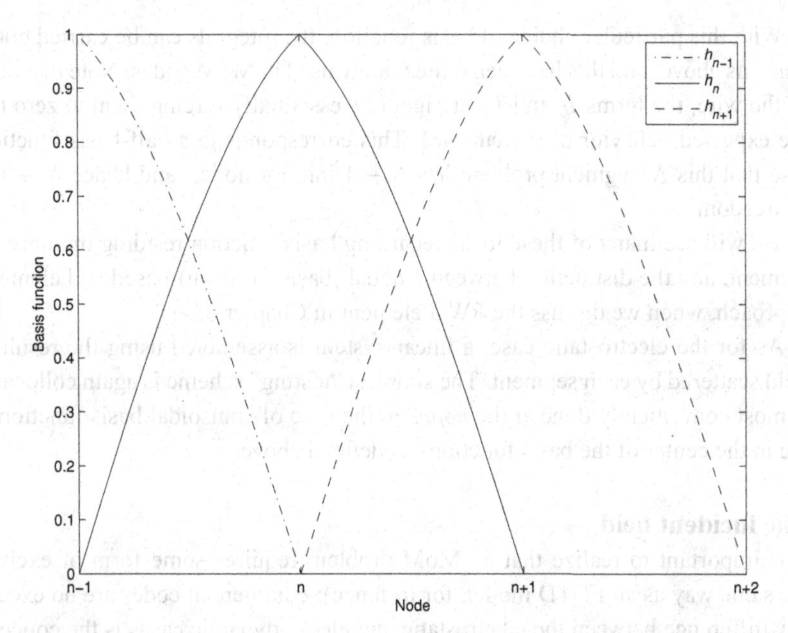

Figure 4.3 Basis functions h_{n-1}, h_n and h_{n+1}. Note that only the second half of h_{n-1} is shown.

N-segment problem (numbering the first node as 0, and the last as N), basis function n is associated with nodes $n-1$, n and $n+1$. The coefficient (also known as the degree of freedom) I_n is associated with the current at node n (as the basis function is zero at nodes $n-1$ and $n+1$). Three such basis functions are plotted in Fig. 4.3. Note the overlapping nature of the basis functions. The functions plotted in this figure are for a segmentation $\Delta = \lambda/5$; usually, finer segments are required. As Δ becomes smaller, the piecewise sinusoid increasingly closely resembles a piecewise linear function.

However, *segment* (or *element*) n is associated with $z_{n-1} \le z \le z_n$. In terms of the approximation of current on the nth segment, the current is approximated as a linear superposition of the two basis functions defined on segment n:

$$I_z(z) \approx I_n \frac{\sin k(z - z_{n-1})}{\sin k\Delta_n} + I_{n-1} \frac{\sin k(z_n - z)}{\sin k\Delta_n} \qquad \forall z_{n-1} \le z \le z_n \tag{4.33}$$

where Δ_n is the length of segment n.

It may be shown (see end-of-chapter problem P4.1) that the $\hat{z}$-directed scattered field from the nth basis function of Eq. (4.32) is given by

$$\boldsymbol{E}_z^{\text{scat}} = -j30 \left[\frac{e^{-jkR_{n-1}}}{R_{n-1} \sin k(z_n - z_{n-1})} - \frac{e^{-jkR_n} \sin k(z_{n+1} - z_{n-1})}{R_n \sin k(z_n - z_{n-1}) \sin k(z_{n+1} - z_n)} + \frac{e^{-jkR_{n+1}}}{R_{n+1} \sin k(z_{n+1} - z_n)} \right] \tag{4.34}$$

The lengths R_{n-1}, R_n, and R_{n+1} are respectively the distances from nodes $n-1$, n and $n+1$ to the field point.

With this particular choice of basis function, the integrals can be carried out *analytically*, as above, and this has been quite widely used in MoM codes. Note that at the ends of the wire, the terms I_0 and I_N are ignored, essentially forcing them to zero (which is the expected behavior of the current). This corresponds to a half-basis function. Note also that this N-segment problem has $N-1$ interior nodes, and hence $N-1$ degrees of freedom.

We will see many of these ideas regarding basis function residing on more than one element, and the distinction between a nodal (basis function)-based and element-based approach, when we discuss the RWG element in Chapter 6.

As for the electrostatic case, a linear system is assembled using the results for the field scattered by each segment. The simplest "testing" scheme is again collocation: this is most conveniently done at the *nodes* in the case of sinusoidal basis functions, which are in the center of the basis functions as defined above.

The incident field

It is important to realize that an MoM problem requires some form of excitation (in the same way as an FDTD model, for instance); commercial codes are no exception. A key difference between the electrostatic and electrodynamic cases is the concept of the incident field, as already outlined, which provides this excitation. For an incident plane wave, peak value E_0 V/m normally incident on the z-directed dipole (along the x-axis in this case), the expression is

$$E_z^{\text{inc}} = E_0 \, e^{-jkx} \tag{4.35}$$

as already discussed.

For an antenna problem, a very simple form of feed is the "delta-gap"; in this case

$$E_z^{\text{inc}} = \pm V/\delta \tag{4.36}$$

for an impressed voltage of V at the terminals of the antenna and gap length δ (quite often, the length of the segment). This source is also sometimes placed at the node between segments. The sign depends on the convention adopted regarding voltage. For the basis functions discussed, the direction of positive current flow is from node n to $n+1$.

More realistic models are available, such as the "frill" source. This models a coaxial line, whose center conductor becomes a monopole and whose outer conductor opens into an infinite ground plane. In this case, the electric field on the axis of the $\hat{z}$-directed monopole is (again, similar comments pertain regarding the sign):

$$E_z^{\text{inc}} = \pm \frac{V}{2\ln(b/a)} \left(\frac{e^{-jkR_1}}{R_1} - \frac{e^{-jkR_2}}{R_2} \right) \tag{4.37}$$

with

$$R_1 = \sqrt{z^2 + a^2}$$
$$R_2 = \sqrt{z^2 + b^2} \tag{4.38}$$

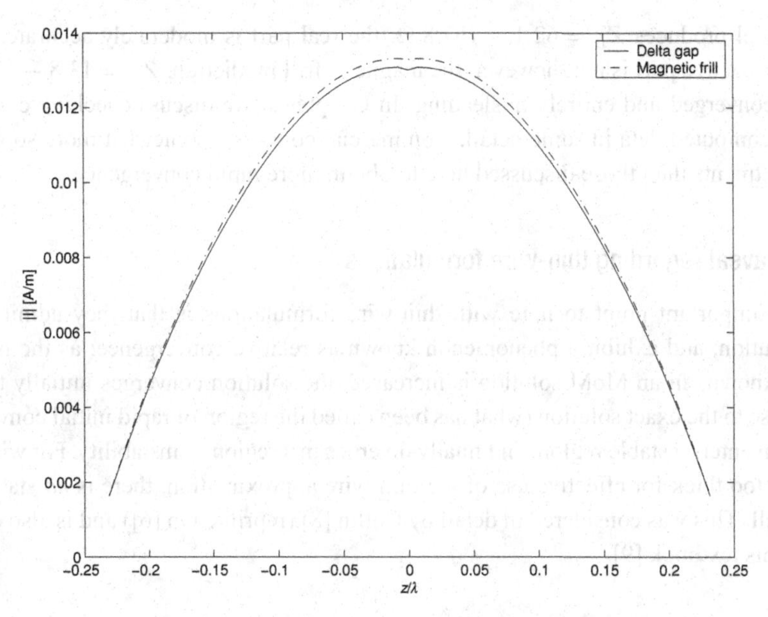

Figure 4.4 Current on a resonant dipole computed with the MoM using piecewise sinsuoidal basis functions and collocation. $L = 0.47\lambda$, $a = 0.005\lambda$, with $N = 60$ segments.

where a and b are the inner and outer radii of the coaxial feedline. V is the terminal voltage. Usually, this is used an an equivalent model, in which case a is the radius of the wire and b is then chosen as some reasonable value – often the equivalent characteristic feedline impedance $Z_0 = 60 \ln(b/a)$ is chosen as $50\,\Omega$, i.e. $b \approx 2.3a$. It is worth commenting that the current (and hence antenna terminal impedance) is very little affected by this value.

Some computed results

An example for the current distribution on a thin resonant dipole ($L = 0.47\lambda$, $a = 0.005\lambda$) computed using the MoM is shown in Fig. 4.4. This MoM code, implementing the theory in this chapter in MATLAB, uses piecewise sinsuoidal basis functions and collocation. Results are shown for both the delta-gap and magnetic frill sources, using $N = 60$ segments. The impedance computed with the former was $Z_L = 76.7 + j4.7\,\Omega$, and for the latter, $Z_L = 74.8 + j8.2\,\Omega$. Considering the relative simplicity of the approximation, this agreement is excellent. An even better comparison is to look at the magnitude of the reflection coefficient Γ; a $75\,\Omega$ system is appropriate here (and was also used for the equivalent coaxial radius in the frill model); the results are -29.7 dB and -25.3 dB respectively. Anyone who has ever tried to measure the reflection coefficients of antennas will be aware that such agreement is more than satisfactory.

However, this computed result appears better than it actually is! What is *not* shown on Fig. 4.4 is that the magnetic frill source converges very slowly; 60 segments corresponds to a sampling density of around 120 per wavelength, approximately an order of magnitude times the usual rule-of-thumb for full-wave MoM codes. Using $N = 6$, the delta-gap model produces $Z_L = 62.1 - j67.8\ \Omega$; the real part is moderately accurate, although the reactive part is not; however, the magnetic frill prediction, $Z_L = 13.8 - j15.1\ \Omega$, is unconverged and entirely misleading. In Chapter 5, we discuss checking convergence of computed data in some detail. Commercial codes use somewhat more sophisticated treatments than those discussed here to obtain more rapid convergence.

4.3.2 A caveat regarding thin-wire formulations

An important point to note with thin-wire formulations is that they admit no exact solution, and exhibit a phenomenon known as relative convergence: as the number of unknowns in an MoM solution is increased, the solution converges initially to a value close to the exact solution (what has been called the region of rapid initial convergence), then enters a stable region, and finally diverges in a region of instability. For wires which are too thick for effective use of the thin-wire approximation, there is *no* stable region at all. This was considered in detail by Collin [8] (reprinted in [6]) and is also discussed in his textbook [9].

4.4 More on basis functions

Suitable basis functions were the topic of research for many years, and in this section, some details are provided of two other solutions which have been widely adopted. Firstly, it is appropriate to provide some background on a public domain code, NEC-2, which for many years was the workhorse of MoM computation. The basis function used by NEC-2 had some particularly elegant features. Following this, some more details are provided on piecewise linear basis functions, which are also very popular.

4.4.1 The numerical electromagnetic code (NEC) – method of moments

It would be inappropriate in a book of this nature not to include some discussion of NEC, or NEC-2 in the case of the public domain version. This code has a long lineage, with its genesis in a code called BRACT (released in 1970), which was developed by contractors MBA Associates, primarily for US Air Force applications. The code that eventually became NEC started as the AMP (Antenna Modelling Program), first released in 1974, again with US military funding. A discussion of the theoretical background and a number of applications for what is clearly this code (although unnamed in the article) may be found in [10], available in the collection [6].

NEC-1 was released in 1977, and NEC-2 in 1981. NEC-2 became, and still is, the most widely used public domain MoM code.[9] NEC-3 was an intermediate version, and saw only limited distribution; NEC-4 was released in 1992 and was the last major release of the code, which is no longer being actively developed further. Until recently,

[9] Whether it was indeed the intention of the US government to make the code public domain is still not entirely clear, but this became the de facto situation by the 1990s.

NEC-4 was still US Military Restricted technology,[10] although all NEC-4 functionality is now available in commercial codes (most prominently FEKO). The various NEC codes were developed at the Lawrence Livermore National Laboratory, one of the major US government research laboratories. Here, we will focus on NEC-2, owing to its ready availability; despite its even more venerable age, it is still a useful tool and quite widely used as a benchmark.

NEC-2 incorporates the Pocklington integral equation formulation for thin wires, as well as a treatment for closed conducting surfaces (the magnetic field integral equation, which will be discussed in Chapter 6). It includes support for a number of features very useful in modelling wire antennas, including: non-radiating networks (e.g. transmission lines); lumped element loading; perfectly or highly conducting wires; incident plane-wave or voltage sources; and treatments of perfect or imperfect grounds. The last included the Sommerfeld formulation for half-spaces; this will be discussed in Chapter 7. It can compute induced currents and charges; near- and far-fields (electric or magnetic); radar cross-section; antenna impedance (and admittance); gain and directivity; and antenna to antenna coupling. It can exploit symmetry of rotation or reflection.

NEC-2 was primarily developed for wire antenna modelling, and many of the problems which have been reported with NEC-2 arose because users tried to use it for modelling surfaces via meshes of wires. Although one can obtain useful answers with careful work with this approach, it is not the purpose for which the code was primarily designed. Provided that NEC-2 is used within its limits, it is still a very useful code.

4.4.2 NEC basis functions

Much of the success of NEC was due to the basis function used. (In the following discussion, NEC and NEC-2 will be used interchangeably; this theory is also applicable to NEC-4.) A highly desirable requirement of a good basis function is that it should satisfy physical requirements of current and charge continuity. This implies that both the current and its first derivative should be continuous. NEC makes the usual thin-wire approximations, viz. transverse currents are negligible; the circumferential variation of current is negligible; current can be represented by a filament on the wire axis; and the boundary conditions on the tangential electric field are only enforced axially, so the basis function is one-dimensional, as in our preceding discussion in this chapter. In developing the basis function, the following interpolation function is first introduced for segment j:

$$I_j = A_j + B_j \sin k(s - s_j) + C_j \cos k(s - s_j), \qquad \forall\ |s - s_j| < \Delta_j/2 \tag{4.39}$$

The parameter s is a local coordinate along the length of the wire, with s_j the value of s at the center of segment j. Δ_j is the length of segment j. This is based on a function originally proposed by Yeh and Mei [11] (reprinted in [6]). Although this is quite often described loosely as the basis function, this is *not* entirely correct. The full basis function is rather more complex. Each NEC basis function spans at least three segments: central, left (minus) and right (plus), supporting interpolation functions of the form of Eq. (4.39)

[10] Since 2003, NEC-4 has been available for a very modest license fee for users in most countries.

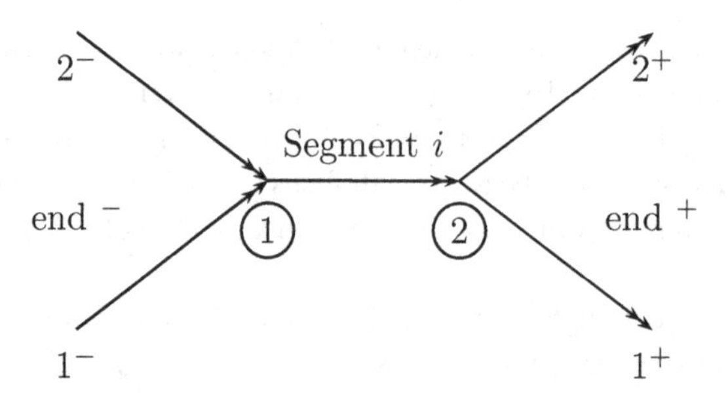

Figure 4.5 Segments covered by the ith basis function.

on *each* segment, f_i^0, $f_{i_j}^-$ and $f_{i_j}^+$ respectively. The double subscript is used to identify the jth segment connected to segment i. Figure 4.5 shows the situation for a wire segment with two wire segments connecting to the left and two to the right of the central segment. (In this case, the basis function "centered" on segment i spans all five segments.) For a straight wire, with only one segment on the left and one on the right, one can drop the double subscript and the basis function comprises interpolation functions f_i^0, f_i^- and f_i^+ associated with it (nine unknowns in total) – *each* interpolating as Eq. (4.39). For a wire junction as in Fig. 4.5, there are contributions from five segments (and hence 15 unknowns).

The unknowns are now reduced to one per segment by the following constraints:

(1) The current must go to zero at outer edges of connected segments.
(2) The derivative of the current must go to zero at outer edges of connected segments.
(3) The current must be continuous at a segment junction.
(4) At a segment junction, the charge must satisfy a condition known as the *Wu–King* condition; it is continuous for a straight, uniform wire.

These conditions are then enforced on *each individual* basis function – these are sufficient (but not necessary) conditions to ensure current and charge continuity, since the final approximation of current is a linear sum of these basis functions. This was a crucial insight.

For example, these constraints for a segment in a straight wire are as follows:

(1) One from end 1^- and one from end 1^+.
(2) Again, one from end 1^- and one from end 1^+.
(3) Two (one at each end of the central segment).
(4) Four (one each from the segments connected to the $-$ and $+$ ends, two from the central segment itself).

This amounts to ten constraints. From Eq. (4.39), there are three unknowns per interpolation function, and three such functions, making nine unknowns. A charge-related parameter at the segment junctions provides two additional ("invisible") unknowns, producing eleven unknowns per wire segment (more details on this are given below). The ten constraints are then applied to yield one unknown per segment, which is arbitrarily chosen as $-A_i^0$, i.e. the coefficient associated with the constant part of the interpolation function centered on segment i. The details of this process are quite lengthy, and are available in [12].

The advantage of this formulation is that it can be generalized to handle multi-wire connections. Although it appears complex (and indeed the implementation is non-trivial), it is handled entirely within the code and the user is unconcerned with the details.

NEC-2 can also handle junctions involving wires of different radii. The so-called Wu–King condition (an attempt to enforce the continuity of scalar potential, which is the correct quasi-static continuity condition) is applied at each junction:

$$\left.\frac{\partial I(s)}{\partial s}\right|_{\text{at junction}} = \frac{Q}{\ln(2/ka) - \gamma} \tag{4.40}$$

In this expression, $\gamma = 0.5772$, Euler's constant. Q is an unknown related to charge: it is constant for all wires at a junction and is the "invisible" unknown in the previous discussion.

4.4.3 Piecewise linear basis functions

The NEC-2 basis function is very useful for modelling wire antennas, but is difficult to apply when the structure to be modelled comprises large amounts of conducting surfaces. We will discuss effective methods for modelling surfaces in Chapter 6; at present, all we need to know is that the usual basis function for this is piecewise linear. Hence, such basis functions are very convenient for models including both wires and surfaces. The formulation is very similar to that of Eq. (4.33):

$$h_n(z) = \begin{cases} \dfrac{I_{n-1}(z_n - z) + I_n(z - z_{n-1})}{\Delta} & \forall z_{n-1} \leq z \leq z_n \leq \Delta \\ 0 & \text{otherwise} \end{cases} \tag{4.41}$$

As with the piecewise sinusoid, the basis function consists of two parts, with two associated (and unknown) coefficients I_{n-1} and I_n; again, it is often convenient to reinterpret the function as spanning two segments, with one associated coefficient I_n. This idea is very useful at wire junctions.

4.4.4 Junction treatments with piecewise linear basis functions

The NEC junction treatment is sophisticated, but a simpler approach first introduced by Chao and Strait in 1970 is worth mentioning, since it is still quite widely used. The only place the proof appears to have been published is a report for a government research laboratory [13, pp. 22–25] and given that these are frequently rather difficult to obtain, even when unlimited distribution was approved as was the case here, it is worth briefly deriving their approach. A description of the method (without proof) appears in [14, Chapter 4]. Chao and Strait used a slightly more complex variant of the piecewise linear function, with an interior node in each segment to permit better approximation of curved wires; here, we use straight segments.

Consider a three-wire junction at node n, as shown in Fig. 4.6. (The method works for any number of wires, but this keeps things simple. The general case is outlined at

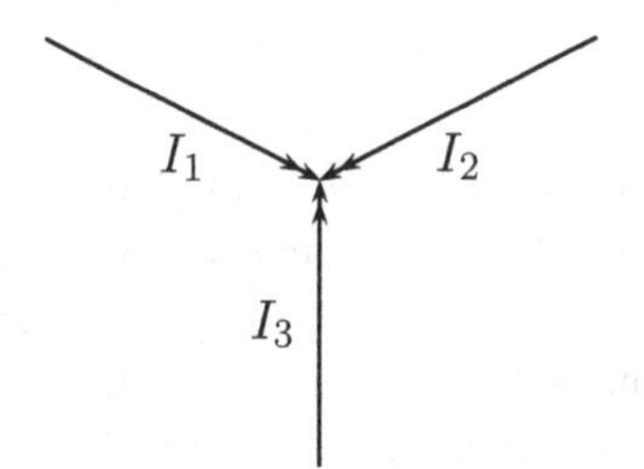

Figure 4.6 A three-wire junction.

the end of the discussion.) Firstly, we introduce a "half-triangle" function of the form

$$h_n(s) = \begin{cases} \dfrac{I_n(s - s_{n-1})}{\Delta} & \forall\, s_{n-1} \leq s \leq s_n \\ 0 & \text{otherwise} \end{cases} \tag{4.42}$$

This is simply half the basis function defined in Eq. (4.41), but with z replaced by s, a local distance parameter along each wire. In this discussion, it is convenient if $s = 0$ corresponds to the end of each wire away from the junction, with s increasing as one approaches the junction. There are three currents to consider: the current on wire 1, just before the junction; and the same for wires 2 and 3. Note that, at these points, the only basis functions contributing to the current are these half-triangle functions. We will call the corresponding coefficients I_1, I_2 and I_3. (We will not include the node n in the notation since it is unnecessary here.) With only two wires, it is sufficient to set $I_1 = -I_2$ and Kirchoff's current law is automatically satisfied. (The negative sign is due to the convention on s adopted above.) With three wires, one possibility is to allow the MoM procedure to include I_1, I_2 and I_3, and then impose the additional constraint

$$I_1 + I_2 + I_3 = 0 \tag{4.43}$$

However, this often results in a constraint equation with very different magnitudes to the usual impedance matrix elements.

The approach suggested by Chao and Strait is to consider each half-triangle coefficient as the sum of two components, hence:

$$\begin{aligned} I_1 &= I_1' + I_1'' \\ I_2 &= I_2' + I_2'' \\ I_3 &= I_3' + I_3'' \end{aligned} \tag{4.44}$$

Further, they propose that:

$$\begin{aligned} I_1'' &= -I_3' \\ I_2'' &= -I_1' \\ I_3'' &= -I_2' \end{aligned} \tag{4.45}$$

What this implies is that these basis functions are simply the usual piecewise linear basis functions, spanning both the last segment of the relevant wire and *overlapping* by one segment onto the next wire: that is wire 1 overlaps onto wire 2, wire 2 onto wire 3, and

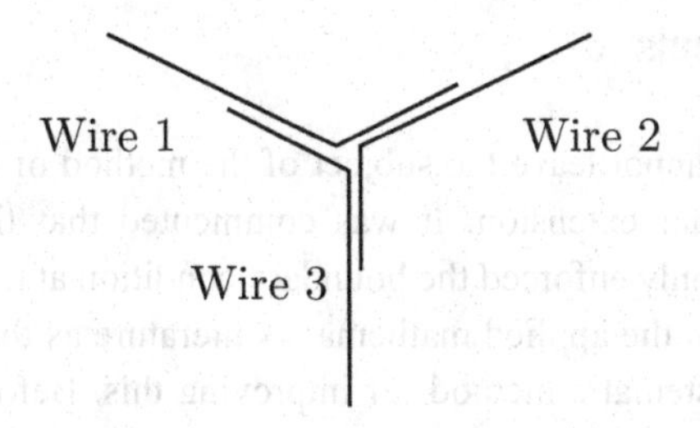

Figure 4.7 The three-wire junction, with the wires overlapped.

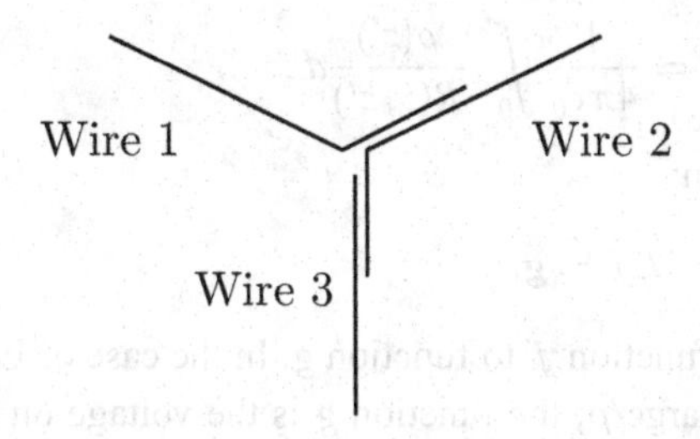

Figure 4.8 The final junction treatment, with two overlapped wires, and one not.

wire 3 onto wire 1. This is shown in Fig. 4.7. (Our previous comment regarding sign convention applies here too.) Substituting Eq. (4.45) into Eq. (4.44),

$$\begin{aligned} I_1 &= I_1' - I_3' \\ I_2 &= I_2' - I_1' \\ I_3 &= I_3' - I_2' \end{aligned} \tag{4.46}$$

one notes that this choice identically satisfies Eq. (4.43) for any values of I_1', I_2' and I_3'. A unique solution is obtained by arbitrarily choosing one of the degrees of freedom; it is convenient to set $I_3' = 0$. This yields

$$\begin{aligned} I_1 &= I_1' \\ I_2 &= I_2' - I_1' \\ I_3 &= -I_2' \end{aligned} \tag{4.47}$$

A little thought shows that this implies that we overlap wire 1 onto wire 2, wire 2 onto wire 3, but do *not* overlap wire 3 onto wire 1, as in Fig. 4.8. For a general N wire junction, the procedure is to overlap wire n onto wire $n + 1$, but not wire N onto wire 1. Each of these overlapped wires is then treated with the usual MoM procedure as an open wire, with zero current at the end, as is the one non-overlapped wire.

This is a somewhat cruder approximation than in NEC-2, since it satisfies *only* Kirchoff's current law, and not the continuous scalar potential. However, for junctions involving wires of the same or similar radius it works satisfactorily. It also incorporates an element of arbitrariness, since which wire is not to be overlapped can be chosen at will. Finally, note that this procedure also works with piecewise sinusoidal basis functions.

4.5 The method of weighted residuals

Even at an introductory level, one cannot leave the subject of the method of moments without introducing a very important extension. It was commented that the point-matched procedure which was used only enforced the boundary condition at the sample points. A method generally known in the applied mathematics literature as the *method of weighted residuals* provides a systematic method for improving this. Before we do this, some notation needs to be introduced first. We return to Eq. (4.2), repeated here for convenience:

$$V(z, \rho = a) = \frac{1}{4\pi\epsilon_0}\int_0^{\ell} \frac{\rho(z')}{R(z, z')} dz' \tag{4.48}$$

and introduce *linear operator* notation:

$$\mathcal{L}f = g \tag{4.49}$$

where $\mathcal{L}$ is the operator which maps function f to function g. In the case of Eq. (4.48), for instance, the function f is the charge ρ; the function g is the voltage on the wire; and the linear operator $\mathcal{L}$ is

$$\mathcal{L} = \frac{1}{4\pi\epsilon_0}\int_0^{\ell} \frac{1}{R(z, z')} (\cdot)\, dz' \tag{4.50}$$

The bracketed dot is used as a place-holder for the function on which this operator acts. Using this notation, the previous development then produces

$$\mathcal{L}\sum_{n=1}^{N} a_n h_n = g \tag{4.51}$$

where, as before, f has been approximated using the basis functions, viz.

$$f \approx \sum_{n=1}^{N} a_n h_n$$

Using point-matching, the $N \times N$ linear system can be obtained by testing the above at N test points. But now, instead of doing this, we form the *residual* as:

$$\mathcal{R} = \mathcal{L}\sum_{n=1}^{N} a_n h_n - g \tag{4.52}$$

This residual is the difference between the approximate solution and the actual solution. (At the risk of belaboring the obvious, if this was one of the very rare problems which can be solved exactly using the MoM procedure, then the residual would be zero.) The point-matching procedure forces this residual to zero at N discrete points. A better approach would be to try to obtain some type of average value of the residual over the domain of the problem (the length of the wire in this case), and set this to zero. One can do this in a quite general fashion by introducing the idea of a *weighting function*, which is multiplied by the residual (and hence the name, method of weighted residuals) and integrated over the domain. The weighting function (also often known as a testing function) is also usually expressed as some type of finite series:

$$w = \sum_{m=1}^{M} w_m \tag{4.53}$$

In this case, the equality is appropriate, since we are not approximating this function. Note also that there are no unknown coefficients. Symbolically, the weighted residual method becomes

$$\int_L \mathcal{R} \sum_{m=1}^{M} w_m dz = \int_L \sum_{m=1}^{M} w_m \mathcal{L} \sum_{m=1}^{N} a_n h_n - \int_L \sum_{n=1}^{M} w_n g = 0 \tag{4.54}$$

Usually, the number of basis functions (N) and the number of weighting functions (M) are equal. Because this integration process frequently defines an *inner product*, an equivalent notation frequently encountered is

$$\langle w_m, \mathcal{L} a_n h_n \rangle = \langle w_m, g \rangle \tag{4.55}$$

This is of course the bracket notation widely used in quantum mechanics, for the matrix algebra formulation of Heisenberg. We will not pursue this further, other than to note that the reason for this analogy is that both classical electromagnetics and quantum mechanics are at heart field theories.

It is easy to show that the method of weighted residuals produces a matrix equation, of the same form as Eq. (4.11), repeated here:

$$\{V\} = [Z]\{I\} \tag{4.56}$$

except that the matrix entries are now

$$\begin{aligned} Z_{mn} &= \langle w_m, \mathcal{L} h_n \rangle \\ V_m &= \langle w_m, g \rangle \\ I_n &= a_n \end{aligned} \tag{4.57}$$

In addition to the question of which type of basis functions to adopt, one now can also choose a variety of weighting functions. This matter has been quite extensively researched. In practice, however, there are two very popular choices. The *Galerkin* procedure uses the same basis and weighting functions. The collocation method, which we have already studied, uses Dirac delta functions, which of course reduce to just testing the operator at the sample points.

Before concluding this section, one or two points which can (and have) caused confusion in the past should be highlighted. Firstly, the inner product implied above for two functions f and g defined on domain $\mathcal{D}$ is

$$\langle f, g \rangle = \int_{\mathcal{D}} f g \, dV \tag{4.58}$$

For real valued functions, the operation thus defined satisfies the mathematical requirements of an inner product. However, for complex valued functions, it defines a *symmetric* rather than *inner* product, and in this case, the Galerkin procedure requires weighting functions which are the complex conjugate of the basis functions. (The symmetric product defines a quantity known as *reaction* in electromagnetic theory [7, Section 10.7; 4, Section 7.6].) A valid inner product for complex-valued functions is

$$\langle f, g \rangle = \int_{\mathcal{D}} f g^* \, dV \tag{4.59}$$

where g^* is the complex conjugate of g. In this case, the basis and weighting functions are identical in the Galerkin procedure. Heated debates have arisen over this in the literature; mathematically, it is important, because functions and operators defined within the framework of a proper inner product (and also with some additional properties) are known to be elements of Hilbert and/or Sobolev *spaces*, which confer various properties, important with regard to error analysis and convergence studies, on the problem. In practical engineering applications, the difference is usually unimportant, as the basis and weighting functions are usually real-valued, but one place where it *is* important is when complex-valued functions are used, as is the case with body of revolution formulations.

On a different topic, the use of Dirac delta functions to derive the collocation approach from the method of weighted residuals has been criticized by some writers [15]. The core of this criticism is the observation that functions such as these are only properly defined in a distributional sense (i.e. under an integral sign). Again, whilst valid from a theoretical viewpoint, in practice the collocation method stands on its own merits, does not need to be derived thus, and is often a very effective formulation.

One final point we can now explain – the origin of the name "method of moments." Again, consider a one-dimensional problem, such as the electrostatic one we started the chapter with. If we use a method of weighted residuals approach, but select as weighting functions the set $\{z, z^2, z^3, \ldots\}$ we form the *moments* of the residual. In applied mathematics, the method of moments is this specific form of the method of weighted residuals. Harrington chose it as the generic name for method of weighted residuals approaches in electromagnetics, and the name stuck. (In [16], he explained that when first working with the method, he tried to avoid introducing new jargon, and that the name method of moments had previously been used by the Russian mathematicians Kantorovich and Akilov.) Arguably, it may not have been the best choice of name, but four decades of usage in computational electromagnetics have established it so firmly as to be beyond debate. One will also sometimes find the term *boundary element method* used instead of MoM; usually, these terms are identical, although we caution that volumetric MoM formulations are available which are *not* boundary, but rather volume, element methods. (We briefly discuss volume elements in Chapter 6.)

4.6 Scattering from infinite cylinders

Scattering from infinite cylinders is another venerable application of the method of moments which very well illustrates some of its core features. A formulation will be presented here which draws heavily on [17, Chapter 2]. It will provide an ideal opportunity to compare the MoM with the FDTD treatment of this problem in the previous chapter, in Section 3.2. These include the development of the relevant integral equation on the boundary of the scatterer – the electric field integral equation (EFIE) in this case; the solution of the resulting integral equation using pulse basis functions and collocation; and special treatment for singular terms. For cylinders of infinite length, the problem can of course be decomposed into transverse magnetic and transverse electric modes, resulting in a scalar problem. Initially, we will study the TM_z case, as it is the easier of the two, using E_z as the working variable. Also, we will again use the decomposition of the total field into an incident and scattered field, $\boldsymbol{E}^{\text{tot}} = \boldsymbol{E}^{\text{inc}} + \boldsymbol{E}^{\text{scat}}$, as done in both the previous chapter and this one. In particular, we will study a plane wave, originating from ϕ^{inc}:

$$E_z^{\text{inc}} = E_0 e^{jk(x\cos\phi^{\text{inc}} + y\sin\phi^{\text{inc}})} \tag{4.60}$$

In passing here, note that the usage of ϕ^{inc} is not unique. Some sources define this angle as the angle $\boldsymbol{k}$ (the wavenumber vector) makes with the reference coordinate system (the $\hat{x}$-axis, in this case). The usage here is consistent with [4, p.579], for instance, and is also described as the angle of arrival. This definition has the advantage that for *monostatic* scattering (source and receiver co-located, as is the case with most radar systems), the relevant angle for the scattered field $\phi^{\text{scat}} = \phi^{\text{inc}}$.

4.6.1 General derivation of surface integral equation operators

With the field decomposed into incident and scattered fields, the latter may be found from Maxwell's equations as solutions to

$$\nabla^2 \boldsymbol{E}^{\text{scat}} + k^2 \boldsymbol{E}^{\text{scat}} = j\omega\mu_0 \boldsymbol{J} - \frac{\nabla\nabla\cdot\boldsymbol{J}}{j\omega\epsilon_0} + \nabla\times\boldsymbol{K} \tag{4.61}$$

with $\boldsymbol{J}$ and $\boldsymbol{K}$ equivalent sources. For problems with only equivalent electric sources (more soon), $\boldsymbol{K}$ is zero, and vice-versa.

A solution of this (for $\boldsymbol{K} = 0$) is found in terms of the magnetic vector potential $\boldsymbol{A}'$ as

$$\boldsymbol{E}^{\text{scat}} = \frac{\nabla\nabla\cdot\boldsymbol{A}' + k^2\boldsymbol{A}'}{j\omega\epsilon_0} \tag{4.62}$$

Note that the magnetic vector potential $\boldsymbol{A}'$ is defined in terms of the magnetic field $\boldsymbol{H} = \nabla\times\boldsymbol{A}'$ here; in many texts (including later chapters of this one), it is more conventionally defined in terms of the flux $\boldsymbol{B} = \mu\boldsymbol{H} = \nabla\times\boldsymbol{A}$, and we have used the prime to indicate this difference. (In a homogeneous medium, μ is constant and cancels when fields are computed, so it is of no matter.)

The vector potential $\boldsymbol{A}'$ is the solution of

$$\nabla^2\boldsymbol{A}' + k^2\boldsymbol{A}' = -\boldsymbol{J} \tag{4.63}$$

which is found in terms of the relevant *Green* function G:

$$\boldsymbol{A}' = \boldsymbol{J} * G \tag{4.64}$$

where $*$ denotes two-dimensional convolution and the two-dimensional Green function is

$$G = \frac{1}{4j} H_0^2(k|\rho|) \tag{4.65}$$

with H_0^2 the zero-order Hankel function of the second kind. Green functions are discussed in more detail in Chapter 7. Note that when $\boldsymbol{J}$ is z-directed, $\boldsymbol{A}'$ is too.

When a PEC scatterer is present, the following integro-differential operator follows for the scattered (and hence incident) field in terms of the magnetic vector potential:

$$\hat{n} \times \boldsymbol{E}^{\text{inc}} = -\hat{n} \times \boldsymbol{E}^{\text{scat}} = -\hat{n} \times \left. \frac{\nabla\nabla \cdot \boldsymbol{A}' + k^2 \boldsymbol{A}'}{j\omega\epsilon_0} \right|_S \tag{4.66}$$

This equation is applied *on* the PEC surface S and explicitly enforces the boundary condition $\hat{n} \times \boldsymbol{E}^{\text{tot}} = \hat{n} \times (\boldsymbol{E}^{\text{inc}} + \boldsymbol{E}^{\text{scat}}) = 0$ on surface S and is known as the *electric field integral equation*. Importantly, such *surface integral equation operators* all reduce the dimensionality of the problem by one. In this case, a two-dimensional problem is reduced to a one-dimensional problem, along the contour of the PEC scatterer. It is worth reiterating that this reduction in dimensionality, and the automatic inclusion of the far-field radiation condition via the Green function, are major strengths of the surface integral formulations, and MoM solutions thereof. The major drawback is that this comes at the cost of computing global interactions (required by the Green function), and hence the resulting MoM impedance matrix is full.

4.6.2 The EFIE for TM scattering

For this case, illustrated in Fig. 4.9, only the $E_z(x, y)$, $H_x(x, y)$ and $H_y(x, y)$ components are present, and hence only J_z, permitting the problem to be worked as a one-dimensional scalar problem in terms of $E_z(x, y)$ and J_z. (Note that the figure shows a right circular cylinder, positioned about the origin, but the formulation can handle cylinders of arbitrary cross-section and location – see also the footnote in Section 3.2.4 regarding the mathematical sense of "cylinder" here.) For the TM polarization, $\nabla \cdot \boldsymbol{J} = 0$ and also $\nabla \cdot \boldsymbol{A}' = 0$. (This may not be obvious immediately. To show $\nabla \cdot \boldsymbol{J} = 0$, use the current continuity equation, viz. $\nabla \cdot \boldsymbol{J} = -j\omega\nabla \cdot \epsilon\boldsymbol{E}$; since the electric field is only z-directed and does not vary in z, its divergence is zero.) Hence Eq. (4.66) simplifies to

$$E_z^{\text{inc}}(t) = jk\eta A_z(t) \tag{4.67}$$

with

$$A_z(t) = \int J_z(t) \frac{1}{4j} H_0^{(2)}(kR) dt' \tag{4.68}$$

$$R = \sqrt{[x(t) - x(t')]^2 + [y(t) - y(t')]^2}$$

where t is a parametric variable denoting the position around the contour of the cylindrical surface. (If the cylinder is a right circular cylinder of radius a, centered on the origin, as in Fig. 4.9, than the obvious parametrization is $x = a\cos\phi;\ y = a\sin\phi$.) This is now a

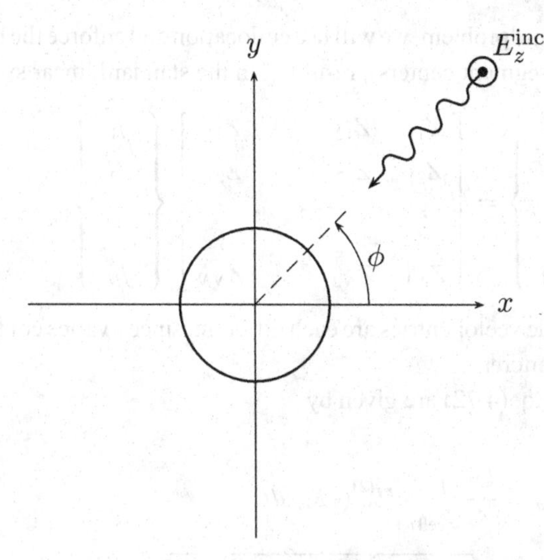

Figure 4.9 A highly conducting cylinder illuminated by an incident TM_z wave.

one-dimensional integral equation, with $J_z(t)$ the unknown to be found – and the MoM can be applied.

One might ask why, given the presence of a highly conducting cylinder, the free space Green function may be used in the MoM? The answer to this is the application of the equivalence principle, which permits the interior problem (the cylinder in this case) to be replaced by a more convenient material (free space, in this case) to give an equivalent exterior problem. This is discussed in Section 6.5 in more detail. For now, it is sufficient to note that, in this formulation, the currents computed are the real currents that could be measured with a current probe. (This is *not* always the case.)

4.6.3 MoM solution of EFIE for TM scattering

As with the electrostatic problem addressed earlier in this chapter, we will use pulse basis functions:

$$p_n(t) = \begin{cases} 1 & \text{if } t \in \text{cell } n \\ 0 & \text{otherwise} \end{cases} \tag{4.69}$$

The axially directed current J_z will be approximated with N such functions as

$$J_z(t) \approx \sum_{n=1}^{N} j_n p_n(t) \tag{4.70}$$

Substitute Eq. (4.69) into Eq. (4.67) to find

$$E_z^{\mathrm{inc}}(t) \approx jk\eta \sum_{n=1}^{N} j_n \int_{\mathrm{cell}\ n} \frac{1}{4j} H_0^{(2)}(kR)dt' \tag{4.71}$$

Again, as with the electrostatic problem, we will use collocation and enforce the boundary condition at N points (the segment centers), resulting in the standard linear system:

$$\begin{Bmatrix} E_z^{\text{inc}}(t_1) \\ E_z^{\text{inc}}(t_2) \\ \vdots \\ E_z^{\text{inc}}(t_N) \end{Bmatrix} = \begin{bmatrix} Z_{11} & Z_{12} & \cdots & Z_{1N} \\ Z_{21} & Z_{22} & \cdots & Z_{2N} \\ \vdots & & & \\ Z_{N1} & Z_{N2} & \cdots & Z_{NN} \end{bmatrix} \begin{Bmatrix} j_1 \\ j_2 \\ \vdots \\ j_N \end{Bmatrix} \tag{4.72}$$

In this case, the left-hand side vector entries are each different, since t varies continuously along the contour of the scatterer.

The Z matrix entries in Eq. (4.72) are given by

$$Z_{mn} = \frac{k\eta}{4} \int_{\text{cell } n} H_0^{(2)}(kR_m)dt' \tag{4.73}$$

$$R_m = \sqrt{[x_m - x(t')]^2 + [y_m - y(t')]^2}$$

where (x_m, y_m) is the center of the mth segment. When the segment (cell) size $\ll \lambda$, this can be approximated using single-point evaluation:

$$Z_{mn} \approx \frac{k\eta}{4} w_n H_0^{(2)}(kR_{mn})dt' \quad \forall\ m \neq n \tag{4.74}$$

$$R_{mn} = \sqrt{[x_m - x(t')]^2 + [y_m - y(t')]^2}$$

where w_n is the width of segment n.

For $m = n$, the *self-impedance* term, the approximate formula cannot be used since the Hankel function is singular (infinite) for $R_{mn} = 0$. A small argument power series expansion of the Hankel function can be used to give an approximate result:

$$Z_{mm} \approx \frac{k\eta w_m}{4} \left\{1 - j\frac{2}{\pi}[ln\left(\frac{\gamma k w_m}{4}\right) - 1\right\} \tag{4.75}$$

$$\gamma = 1.781\,072\,418\ldots$$

This is derived as follows. The small argument expansion is:

$$H_0^2(x) \approx \left(1 - \frac{x^2}{4}\right) - j\left\{\frac{2}{\pi}\ln\left(\frac{\gamma x}{2}\right) + \left[\frac{1}{2\pi} - \frac{1}{2\pi}\ln\left(\frac{\gamma x}{2}\right)\right]x^2\right\} + \mathcal{O}(x^4) \tag{4.76}$$

Assuming that the total curvature of each cell is small enough that the cell may be considered flat, the dominant terms in the expansion are retained to produce

$$\int_{\text{cell } m} H_0^{(2)}(kR_m)dt' \approx 2\int_0^{w_m} \left[1 - j\frac{2}{\pi}\ln\left(\frac{\gamma k u}{2}\right)\right] du$$

$$= w_m - j\frac{2}{\pi} w_m \left[\ln\left(\frac{\gamma k w_m}{4}\right) - 1\right] \tag{4.77}$$

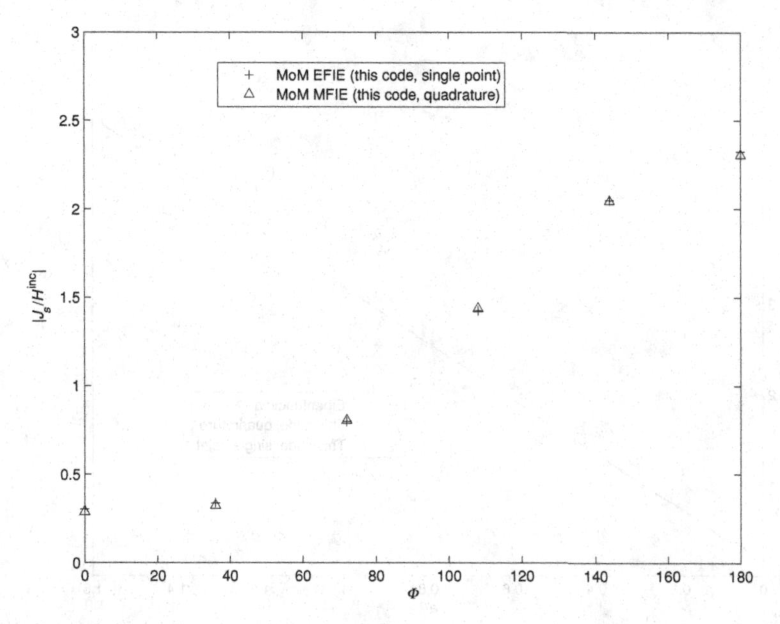

Figure 4.10 Surface current induced by a plane wave incident on a highly conducting infinite cylinder. $ka = 1$, $N = 10$, $\phi^{\text{inc}} = 180°$.

using

$$\int \ln x = x \ln x - x \tag{4.78}$$

4.6.4 Coding in MATLAB for right circular PEC cylinder

Coding is very similar to the static example. MATLAB directly supports evaluation of the Hankel function – it is called here as `besselh(0,2,k*R_mn)`. Remember that the "voltage" matrix V has different entries; the mth entry of $[V]$ is found using Eq. (4.60) evaluated at the center of segment m.

Figure 4.10 shows an MoM solution. Note that $\phi = 180°$ corresponds to the back-scatter direction. In the high-frequency (optical) limit, $J_s \approx 2H^{\text{inc}}$ at that angle. Here, $J_s \approx 2.3H^{\text{inc}}$ due to the finite size of the cylinder. The quadrature result in the figure was computed using a very simple ten-point subdivision of each segment.

4.6.5 Post-processing: echo width and radar cross-section

Radar cross-section (RCS) is very important parameter for radar systems analysis, and a full definition and discussion may be found in Section 6.4.1. In two dimensions, the analogous concept is echo width, or scattering width, σ_w, defined as

$$\sigma_w = \lim_{\rho \to \infty} 2\pi\rho \frac{|E^{\text{scat}}|^2}{|E^{\text{inc}}|^2} \tag{4.79}$$

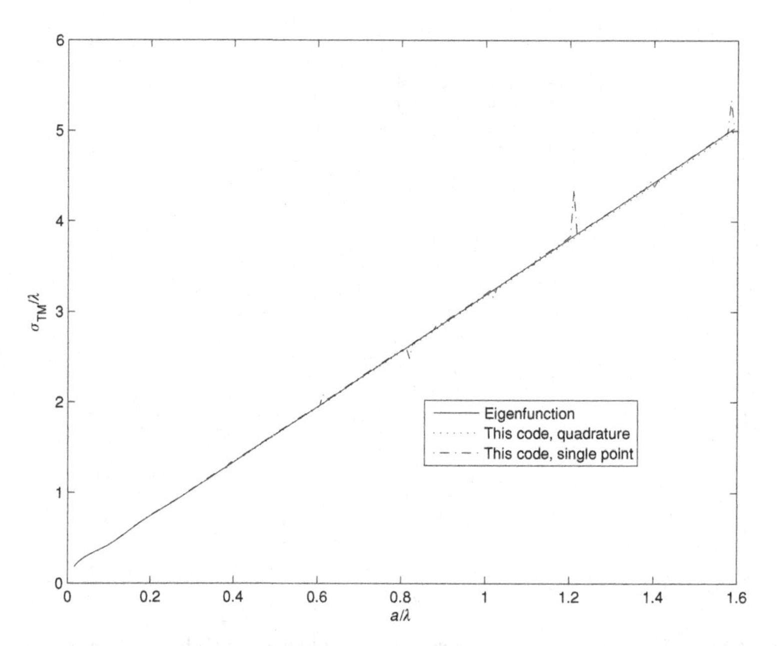

Figure 4.11 The monostatic echo width of PEC cylinder, TM incidence, as a function of frequency.

Importantly, for a cylinder of long but finite length ℓ, RCS and echo width are related as

$$\sigma \approx \sigma_w \frac{2\ell^2}{\lambda} \tag{4.80}$$

For a PEC cylinder, the scattering width is

$$\sigma_{\text{TM}}(\phi, \phi^{\text{inc}}) = \frac{k\eta^2}{4} \left| \int J_z(t')e^{jk[x(t')\cos\phi + y(t')\sin\phi]}\, dt' \right|^2 \tag{4.81}$$

assuming an incident E_z field of unit amplitude (i.e. $E_0 = 1$ in Eq. (4.60)). From the MoM solution, using a simple single-point quadrature rule, one finds that

$$\sigma_{\text{TM}}(\phi, \phi^{\text{inc}}) = \frac{k\eta^2}{4} \left| \sum_{n=1}^{N} j_n w_n e^{jk[x_n \cos\phi + y_n \sin\phi]} \right|^2 \tag{4.82}$$

Using the convention of Eq. (4.60), for monostatic scattering $\phi = \phi^{\text{inc}}$. Results for a typical calculation are given in Fig. 4.11. Whilst the results exhibit excellent agreement almost everywhere, careful study shows a number of "spikes," especially for the single-point evaluation scheme. This is a manifestation of a problem known as interior resonance, which will be addressed briefly in Section 6.10.

The eigenfunction solution to which results are compared is given by [4, Eq. (11-102)]:

$$\sigma_{\mathrm{TM}}(\phi, \phi^{\mathrm{inc}} = \pi) = \frac{2\lambda}{\pi}\left|\sum_{n=0}^{\infty} \epsilon_n \frac{J_n(ka)}{H_n^{(2)}(ka)} \cos(n\phi)\right|^2 \tag{4.83}$$

with

$$\epsilon_n = \begin{cases} 1 & \forall\, n = 0 \\ 2 & \text{otherwise} \end{cases}$$

A general discussion of such solutions may be found in Section 6.4.1.

4.6.6 Discussion, and the Fredholm alternative

Some notes are in order at this stage. Firstly, the off-diagonal entries can be evaluated using *quadrature*; more accurate results are usually obtained. Secondly, when approximating a circular cylinder with flat segments, one can choose to have the center of the segments lie on the circle, or the ends of the segments, or somewhere in between. The results shown here located the phase center on the circumference, but better results may be obtained by setting the perimeter of the flat segment approximation to equal the actual circumference of the cylinder. This can be done by "tweaking" the radius a of the cylinder in the code.

In Chapter 6, we return to the subject of integral equations. As will be mentioned there, they are can be classified as *Fredholm* integral equations of the first or second type. The EFIE is a type one equation, and tends to be ill-conditioned. We mention the existence of another possibility – the magnetic field integral equation – MFIE (Fredholm type two, also known as the *Fredholm alternative*). An MFIE equation may be derived for the problem of TE_z scattering. It is slightly more complex than the TM_z EFIE case, and we will not pursue it in detail here. An excellent treatment of this is available in [17, Chapter 2], on which the TM_z EFIE presented here is also based. Briefly, it may be shown that when a PEC scatterer is present, the following integro-differential operator can be derived for the scattered (and hence incident) field in terms of the magnetic vector potential:

$$\hat{n} \times \boldsymbol{H}^{\mathrm{inc}} = -\hat{n} \times \boldsymbol{H}^{\mathrm{scat}} = \boldsymbol{J}_s - \hat{n} \times \nabla \times \boldsymbol{A}'\big|_{S+} \tag{4.84}$$

When specialized to the two-dimensional TE case, the MFIE becomes

$$\begin{aligned} H_z^{\mathrm{inc}}(t) &= -J_t(t) - \hat{z} \cdot \nabla \times \boldsymbol{A}'\big|_{S+} \\ &= -J_t(t) - \frac{\partial A'_y}{\partial x} - \frac{\partial A'_x}{\partial y}\bigg|_{S+} \end{aligned} \tag{4.85}$$

where

$$\begin{aligned} A'(t) &= \int \hat{t}(t')J_t(t)\frac{1}{4j}H_0^{(2)}(kR)dt' \\ R &= \sqrt{[x(t) - x(t')]^2 + [y(t) - y(t')]^2} \end{aligned} \tag{4.86}$$

where t is a parametric variable denoting the position around the contour of the cylindrical surface, and the unit tangent vector is defined as

$$\hat{t}(t) = \hat{x}\cos\Omega(t) + \hat{y}\sin\Omega(t) \tag{4.87}$$

$\Omega(t)$ is the orientation parameter of each segment, i.e. the angle with respect to the x-axis.

Comparing this development with the EFIE, Eq. (4.66) and the TM case, Eq. (4.67), is instructive. Obviously, the MFIE equation only contains one spatial derivative, as opposed to two in the EFIE, so the degree of singularity in general is lower;[11] however, as we will see in Section 6.3, there are ways to work around this. More subtly, the unknown, $\boldsymbol{J}_s$, appears both in the kernel of the integral, and on its own. For these reasons, the MFIE is much more *well posed*, and would appear to be the formulation of choice. Unfortunately, the MFIE only works for closed scatterers. In practice, this is unfortunately often a very serious limitation, and despite its attractive features, the MFIE has not been the formulation of choice for general-purpose codes. (A rectangular patch MFIE formulation is available in the public domain code NEC, described in Chapter 5, but, as noted, the theoretical limitations of the MFIE, combined with the restricted geometrical modelling possible with rectangular patches, severely limit its application.)

The reason that the EFIE, but not the MFIE, is valid for open structures is as follows. The boundary condition ($\boldsymbol{n} \times \boldsymbol{E} = 0$) from which the EFIE is derived remains valid when the surface becomes infinitesimally thin. The current $\boldsymbol{J}_s$ in that case becomes the superposition of the electric currents on both sides of the shell. However, the boundary condition from which the MFIE is derived is not valid in that limit. The MFIE is derived from the BC $\boldsymbol{n} \times \boldsymbol{H} = J_s$, and this is a special case of the general BC $\boldsymbol{n} \times (\boldsymbol{H}_1 - \boldsymbol{H}_2) = \boldsymbol{J}_s$. For a closed body, one of these fields vanishes, but for an open body, they are both non-zero; hence the MFIE is invalid in this case.

4.7 Further reading

Although elegant theoretically, the MoM is probably the most difficult formulation of those presented in this book to implement accurately and efficiently. In the next chapter, we will turn our attention to the use of commercial codes, and not attempt to develop the simple codes presented in this chapter further. For those intending to develop codes themselves, the MoM is surprisingly badly served by textbooks for applications significantly more advanced than the introductory level treatment presented here, and the following notes may be of use.

Firstly, one still needs to refer to some of the original papers on the topic – there is no MoM equivalent of the books by Silvester and Ferrari or Jin on the FEM [15, 18] or

[11] In this particular case, as $\nabla \cdot \boldsymbol{A}' = 0$, the second-order derivative in the TM_z EFIE formulation is absent.

Taflove on the FDTD [19]. In this context, the original paper by Pocklington [20] is both still available in specialized libraries, and still interesting reading, although it will be of little help in developing an MoM code.

An historical aside – H. C. Pocklington

Reading scientific papers from this age can be a little humbling for modern researchers. At the same meeting of the Cambridge Philosophical Society where Pocklington presented his work (25 October 1897), a paper by C. T. R. Wilson on his cloud chamber was presented. At other meetings of that year, numerous papers appear by J. J. Thomson. 1897 was of course the year that Thomson announced the discovery of the electron at the 30 April, 1897 meeting of the Royal Institution – although he called it a corpuscle at that time. Pocklington was a fellow of St. John's College, Cambridge, and during a sabbatical visit to Cambridge the present author tried to obtain more details about his life. Sadly, no photograph or any other information about him was available, unlike Thomson, who went on to become Master of Trinity College, Cambridge, one of the most prestigious positions at that University, as well as of course winning the Nobel prize. Both Trinity and St. John's have a proud tradition of scientific accomplishment, Trinity numbering Newton and Maxwell amongst its fellows in addition to Thomson, and St. John's Dirac.

The collection of reprints edited by Miller *et al.* is very useful in this context, almost two decades after publication [6]. It contains a number of seminal papers, many of which have been referenced in this chapter, as well as a translation from the original German of an important basic theoretical paper by Maue [21], dating back to 1949, which derived what have become known as the electric/magnetic field integral equations, discussed in Chapter 6. The original text by Harrington [22], although reprinted on several occasions and still very widely referenced, is not particularly useful when implementing complex RF simulation codes since its focus is more on basic concepts. However, several important chapters in the now hard-to-find [23], such as [24], are of considerable interest when implementing complex wire codes, and this still appears to be the only comprehensive derivation available of the magnetic field integral equation as generally used; this work generalized some aspects of Maue's original derivation. Another hard-to-find reference with useful information on MoM procedures for arbitrarily oriented wire antennas is [25]. In this context, Moore and Pizer's monograph [14] was useful in its time, but unfortunately has never been revised and may be difficult to locate. Finally, another useful source on this topic, which should be far easier to obtain, is the theory manual for NEC-2 [12]. Good introductory treatments of the MoM for antenna applications are available in [4, 7, 26], which provide a somewhat more extended coverage of the subject than in this chapter; however, these are by no means fully comprehensive treatments. The only extended text on the MoM is Wang's [27], and the book has some material which has dated quickly, specifically in the context of a controversy then raging in the literature about iterative methods. Peterson *et al.*'s book [17] has a good theoretical treatment of canonical problems, but as with the introductory MoM treatment in the antenna textbooks mentioned above (and also Wang's volume), it does not deal with the complexities of arbitrarily oriented wire antennas, providing only a brief overview of the topic.

Finally, the question of the convergence of the MoM has proven far from trivial; a brief discussion may be found in Appendix C.

4.8 Conclusions

Although highly simplified, the theory discussed in this first chapter on the MoM is at the core of very complex and powerful MoM programs such as NEC-2 and FEKO. The former uses collocation, with a variant of the sinusoidal basis function as discussed; the latter uses a Galerkin formulation with piecewise linear basis functions, also as discussed. Extensions to arbitrarily oriented wire antennas rapidly become complex, due to the presence of different components of the electric field (set up by the arbitrarily oriented currents) which need to be taken into account. Highly (as opposed to perfectly) conducting metallic structures can also be addressed with very similar theory. NEC-2 was one of the first codes to incorporate a large number of such facilities; modern commercial codes such as FEKO incorporate all these, as well as many other powerful analysis capabilities.

In the next chapter, we will look specifically at the use of FEKO and NEC-2 for wire antenna modelling. Following this, we return to more theoretical topics, considering modelling highly conducting surfaces in Chapter 6, as well as hybrid formulations to reduce the computational cost of this, and we conclude our study of the MoM in Chapters 7 and 8 with a discussion of Green functions, stratified media formulations and the Sommerfeld potentials. In Chapter 12, we will introduce a very powerful hybrid of the MoM with the finite element method, which permits a very efficient solution of certain classes of problems.

References

[1] R. F. Harrington, *Time-Harmonic Electromagnetic Fields*. New York: McGraw-Hill, 1961.

[2] J. A. Stratton, *Electromagnetic Theory*. New York: Mc-Graw Hill, 1941. Reprinted by IEEE, 2007.

[3] P. M. Morse and H. Feshbach, *Methods of Theoretical Physics*. New York: McGraw-Hill, 1953.

[4] C. A. Balanis, *Advanced Engineering Electromagnetics*. New York: Wiley, 1989.

[5] J. H. Richmond, "A wire-grid model for scattering by conducting bodies," *IEEE Trans. Antennas Propagat.*, **14**, 782–786, November 1966.

[6] E. K. Miller, L. Medgyesi-Mitschang and E. H. Newman, eds., *Computational Electromagnetics: Frequency Domain Method of Moments*. New York: IEEE Press, 1992.

[7] W. L. Stutzman and G. A. Thiele, *Antenna Theory and Design*. New York: Wiley, 2nd edn., 1998.

[8] R. E. Collin, "Equivalent line current for cylindrical dipole antennas and its asymptotic behavior," *IEEE Trans. Antennas Propagat.*, **32**, 200–204, February 1984.

[9] R. E. Collin, *Antennas and Radiowave Propagation*. New York: McGraw-Hill, 1985.

[10] S. Gee, E. K. Miller, A. J. Poggio, E. S. Selden and G. J. Burke, "Computer techniques for electromagnetic scattering and radiation analyses," in *IEEE. Internat. Electromgn. Compat. Symp. Rec.*, pp. 122–131, 1971.

[11] Y. S. Yeh and K. K. Mei, "Theory of conical equiangular-spiral antennas Part 1 – numerical techniques," *IEEE Trans. Antennas Propagat.*, **15**, pp. 634–639, September 1967.

[12] G. J. Burke and A. J. Poggio, "Numerical electromagnetics code (NEC) – method of moments; Part I: Program description – theory." Lawrence Livermore National Laboratory, CA, UCID 18834, January 1981.

[13] H. H. Chao and B. J. Strait, "Computer programs for radiation and scattering by arbitrary configurations of bent wires." Syracuse University, Report number AFCRL-70-034, September 1970.

[14] J. Moore and R. Pizer, eds., *Moment Methods in Electromagnetics Techniques and Applications*. Letchworth, Hertfordshire: Research Studies Press, 1986.

[15] P. P. Silvester and R. L. Ferrari, *Finite Elements for Electrical Engineers*. Cambridge: Cambridge University Press, 3rd edn., 1996.

[16] R. F. Harrington, "Origin and development of the method of moments for field computation," in *Computational Electromagnetics: Frequency-Domain Method of Moments* (E. K. Miller, L. Medgyesi-Mitschang and E. H. Newman, eds.), pp. 43–47. New York: IEEE Press, 1992.

[17] A. F. Peterson, S. L. Ray and R. Mittra, *Computational Methods for Electromagnetics*. Oxford & New York: Oxford University Press and IEEE Press, 1998.

[18] J.-M. Jin, *The Finite Element Method in Electromagnetics*. New York: Wiley, 2nd edn., 2002.

[19] A. Taflove and S. Hagness, *Computational Electrodynamics: the Finite Difference Time Domain Method*. Boston, MA: Artech House, 3rd edn., 2005.

[20] H. C. Pocklington, "Electrical oscillations in wires," *Cambridge Philos. Soc. Proc.*, **9**, 324–332, 1897.

[21] A. E. Maue, "Toward formulation of a general diffraction problem via an integral equation," in *Computational Electromagnetics: Frequency-Domain Method of Moments* (E. K. Miller, L. Medgyesi-Mitschang and E. H. Newman, eds.), pp. 7–14. New York: IEEE Press, 1992.

[22] R. F. Harrington, *Field Computation by Moment Methods*. Malabar, FL: Krieger, 1982. (Reprint of 1968 edition.)

[23] R. Mittra, edn., *Computer Techniques for Electromagnetics*. Oxford: Pergamon, 1973.

[24] A. J. Poggio and E. K. Miller, "Integral equation solutions of three dimensional scattering problems," in *Computer Techniques for Electromagnetics* (R. Mittra, ed.). Oxford: Pergamon, 1973.

[25] W. A. Imbriale, "Applications of the Method of Moments to thin-wire elements and arrays," in *Numerical and Asymptotic Techniques in Electromagnetics* (R. Mittra, ed.). Berlin: Springer-Verlag, 1975.

[26] C. A. Balanis, *Antenna Theory: Analysis and Design*. New York: Wiley, 2nd edn., 1997.

[27] J. J. H. Wang, *Generalized Moment Methods in Electromagnetics*. New York: Wiley, 1991.

[28] W. L. Stutzman and G. A. Thiele, *Antenna Theory and Design*. New York: Wiley, 1981.

Problems and assignments

Problems

P4.1 This problem outlines the derivation of Eq. (4.34).[12] Firstly, note the following identities, which are easily verified by performing the indicated differentations:

$$\frac{\partial \psi(z,z')}{\partial z} = -\frac{\partial \psi(z,z')}{\partial z'} \tag{4.88}$$

$$\frac{\partial^2 \psi(z,z')}{\partial z} = -\frac{\partial^2 \psi(z,z')}{\partial z'^2} \tag{4.89}$$

Now, integrate the first term in Eq. (4.28) by parts twice using Eq. (4.89), substitute the result into Eq. (4.28), and use Eq. (4.88) to obtain

$$\begin{aligned} E_z^{\text{scat},1}(r) = {} & \frac{j}{\omega\epsilon_0}\left[\frac{dI(z')}{dz'}\psi(z,z') + I(z')\frac{\partial \psi(z,z')}{\partial z}\right]_{z=z_{n-1}}^{z=z_n} \\ & + \frac{1}{j\omega\epsilon_0}\int_{z_{n-1}}^{z_n}\left[\frac{d^2 I(z')}{dz'^2} + k^2 I(z')\right]\psi(z,z')dz' \end{aligned} \tag{4.90}$$

for the first half of the basis function (indicated by the superscript 1). With the piecewise sinusoidal function of Eq. (4.32), show that the term in brackets in the integrand is zero, obtaining thus the contribution from the first half of the basis function after subsitution of Eq. (4.32) into the above as

$$E_z^{\text{scat},1}(r) = \frac{j}{4\omega\epsilon_0}\left[\frac{k\cos k(z'-z_{n-1})}{\sin k(z_n - z_{n-1})} + \frac{\sin k(z'-z_{n-1})}{\sin k(z_n - z_{n-1})}\frac{\partial}{\partial z}\right]\frac{e^{-jkr}}{r}\Bigg|_{z'=z_{n-1}}^{z'=z_n} \tag{4.91}$$

Combined now with the contribution from the second half, show finally that Eq. (4.34) follows. (*Hints:* Be very careful to distinguish between z and z'. Note also that is not necessary to find $\frac{\partial}{\partial z}\left(\frac{e^{-jkr}}{r}\right)$ explicitly, as the relevant terms from the first and second parts of the basis function cancel.)

Assignments

A4.1 Using the theory presented in Section 4.2, develop an MoM code for a charged wire. Use pulse (piecewise constant) basis functions and collocation, so that Eqs. (4.15) and (4.16) are applicable. Replicate Fig. 4.2.

A4.2 Using the theory presented in Section 4.3, develop a thin-wire MoM code for a $\hat{z}$-directed dipole. Use sinusoidal weighting functions and collocation, so that Eq. (4.34)

[12] A detailed derivation of this was given in the first edition of Stutzman and Thiele's antenna text [28, p. 330], but was removed from the second edition [7].

is applicable. Use both the delta-gap and magnetic frill source models, and replicate Fig. 4.4.

A4.3 Using the theory presented in Section 4.6, develop an MoM EFIE code for TM scattering from an infinitely long right circular cylinder. Use pulse (piecewise constant) basis functions and collocation, so that Eqs. (4.74) and (4.75) are applicable. Replicate Fig. 4.10.

5 The application of the FEKO and NEC-2 codes to thin-wire antenna modelling

5.1 Introduction

With the theoretical background now established, one is in a position to start using commercial and public domain MoM programs intelligently. In this chapter, we will discuss primarily the application of the commercial code FEKO for antenna modelling, but will also discuss the use of the public domain code NEC-2[1] in this regard. Other than FEKO, few commercial programs (other than some proprietary NEC-2 extensions) provide good support for modelling thin-wire antennas, the topic of this chapter; such antennas are still very widely used indeed. For commercial programs, material is usually available to assist novice users to get started with the codes.[2] Hence we will not describe the basic concepts of entering the geometry of the problem, including the source, and specifying parameters such as operating frequency and radiation patterns, since these vary from program to program, indeed quite often from release to release, and are usually quite well documented by the suppliers. However, in the case of NEC-2, some comments are in order.

NEC-2 is a "card driven" program, dating back to the days of "decks" of punched cards. A NEC model is described by a geometry file, usually with a `.nec` extension. An example is given in Fig. 5.1. If using NEC in this form, one *must* obtain a copy[3] of the user manual [1]. Each line in this file describes either a geometrical element or an analysis operation; the first two lines are simply comments; the third line `GW` is a straight wire, with a *tag* of 1 in this case (a tag is a number referring to the particular wire, and is used to simplify later references), divided into 41 segments, with (x, y, z) coordinates of the first end $(0, 0, -0.25)$, of the second end $(0, 0, 0.25)$ and radius 0.005. All dimensions are in meters by default. The fourth line `GE` indicates that the geometry section has ended. The fifth line `FR` specifies the frequency; the sixth, `EX`, specifies a voltage-source excitation on the 21st segment of the wire with tag 1; and the penultimate line, `XQ`, executes the program, computing input impedances and (possibly) radiation patterns. The final line `EN` ends the "deck."

[1] Again, as in the previous chapter, we will use NEC-2 and NEC interchangeably in this chapter. All the comments made are equally applicable to NEC-4.

[2] In the case of FEKO, a "Getting Started" manual is provided.

[3] This has been made available on the internet. See Appendix E for a list of websites which can assist in this regard.

```
CM Dipole Example
CE Start of geometry
GW  1  41  0.000000  0.000000 -0.250000  0.000000  0.000000  0.250000   0.00500
GE  0    0
FR  0   51    0     0 250.00000 2.0000000
EX  0    1   21    00    1.00000     0.00000
XQ  0
EN
```

Figure 5.1 A sample NEC input file.

> **Code tip – using NEC-2**
>
> NEC-2 is *only* the computational engine, originally written in one of the earlier versions of FORTRAN, which performs the MoM computations as specified in the input file, and writes data to an output file. *No* graphical support is provided at all. An entire industry grew up providing such support; some packages are fully featured commercial products with major additional computational features, such as SuperNEC. The freeware package, Wiregrid for Windows, is unfortunately no longer supported nor available.
>
> Although not clear from Fig. 5.1, the column spacing can be *crucial* – i.e. the x coordinate of end 1 *must* be entered between columns 11 and 20 for some versions of NEC-2. There are many slightly different versions of the code, compiled by different authors, and the earlier versions had limited parsing ability on data files. Later versions relaxed this, and also permitted the use of commas to demarcate data fields. One is well advised to get one of the many GUI interfaces mentioned above, since otherwise preparing a NEC-2 data file can be very frustrating indeed.
>
> An advantage of the NEC-2 open-source mode of operation is that it lends itself to use in a variety of applications – optimization, for instance – since it is relatively easy to generate NEC-2 input files automatically, and using tools such as `grep`, the output file can be parsed for the required output parameters. However, this is *not* an operation recommended for beginners. In some cases the code has even been partially or entirely rewritten in other languages – part of the present author's doctoral dissertation was an implementation in a language called Occam, to permit efficient parallelization of the code [2].

FEKO was also influenced by NEC; at the time of writing FEKO still referred to "cards" in the input file. The actual input file used by FEKO has a `.fek` extension, and consists of lines of data, usually preceded by a two-letter label. (It is either in ASCII or binary format; the former is advantageous when generating geometry files on a PC for running on a more powerful computer such as a workstation or even supercomputer.) However, this is a very difficult format for users to comprehend. Earlier versions of FEKO were usually run from a PREFEKO file (with extension `.pre`). PREFEKO is a type of *scripting* language which generates the `.fek` file from elementary geometrical and other primitives. The current version of FEKO, in common with codes such as MWS, discussed in Chapter 3 and HFSS, discussed in Chapter 11, also offers an entirely graphical geometry-enter process, CADFEKO. It is very attractive for users, but does not always offer the same fine control as the scripting approach (and the two can be combined when necessary). The `.pre` file geometrical description has been retained in the present text, and is still supported in FEKO, although users might well elect to rather use CADFEKO when running these examples.

The code FEKO

This code had its genesis in the doctoral work of Jakobus at the University of Stuttgart in Germany during the early 1990s. It is an acronym of the German name: "**FE**ldberechnung bei **K**örpern beliebiger **O**berfläche," which translates as *field computations involving bodies of arbitrary shape.* It incorporates a powerful MoM treatment using piecewise linear triangular functions for metallic structures – both wires and surfaces, the latter using the RWG basis functions which will be discussed in detail in the next chapter. It also supports the MoM treatment of dielectric structures, using either *surface* or *volumetric* treatments. A unique feature of FEKO is the approximate hybrid treatment available using *physical optics*. We will discuss many of these topics in Chapter 6. FEKO also offers the Sommerfeld treatment for stratified media (the topic of Chapters 7 and 8) and the fast multipole method (see Chapter 6). Recent additions have included support for the FEM/MoM hybrid formulation (see Chapter 12). FEKO is available across a wide range of platforms, including supercomputers; some benchmarking results will be presented in Chapter 6. The code ships with a powerful CAD GUI, as well as support for a scripting language. The code is widely used worldwide. A version with restricted capabilities (sometimes called FEKO Lite) is available at no cost. See Appendix E for contact details.

Historical note – other thin-wire codes

MININEC is another program which one quite frequently sees mentioned. The name is slightly misleading, since it implies that it is a stripped-down version of NEC-2 – this was indeed the original intent of authors Rockway and Logan when the project was first mooted in 1980. However, it evolved into an entirely separate implementation, using a different formulation, and different basis functions (in the current version, triangular ones). See Appendix E for contact details.

Wire (also known as Thin Wire) was a program originally developed by Richmond at Ohio State University; it still has a loyal following there and versions have been made publicly available. See [3, Appendix F] for more details of the code in its 1989 incarnation, WIRE89, with a FORTRAN listing.

5.2 An introductory example: the dipole

No matter what numerical technique has been used – MoM, FDTD, FEM – one of the first things to check is that the solution is indeed *converged*. What we mean by this is that, after a certain point, refining the mesh (making segment size smaller, for a simple MoM problem) does not change the solution. (In Chapter 3, Section 3.2.7, we saw how making Δ smaller improved the quality of the solution by comparison to the analytical result.) To investigate this we will study the half-wavelength dipole. A note is in order here: this term can cause confusion for newcomers in antenna engineering, since what is usually meant is the wavelength at which the dipole exhibits its first resonance – i.e. has no reactive part of the impedance. This is usually equivalent to the wavelength at which the reflection coefficient is minimized in a typical 50 Ω or 75 Ω system, since the real part of the input impedance is generally on the order of 50–70 Ω and changes far less rapidly than the reactance at resonance. It generally occurs at somewhere between $0.46\lambda \sim 0.49\lambda$, depending on the dipole thickness.

Modelling hints – convergence studies using FEKO

In FEKO, there is unfortunately no simple way to undertake a convergence study by creating multiple structures in one file, and one needs to change the discretization manually in the PREFEKO file, run PREFEKO again, and also of course re-run FEKO.[a] There are various ways of proceeding from here, but probably the easiest is to save the output file (`out`) after each run with a distinctive name, and then use the `Import` option in the FEKO graphical postprocessor to read the data in from each file.

[a] It is possible to do this with OptFEKO, the FEKO optimizer, but this is beyond the scope of the present discussion.

The result of such a convergence study is shown in Fig. 5.2. The default reference impedance of 50 Ω was used to create these plots.[4] All produce a minimum reflection coefficient of around −15 dB except for the coarsest mesh (−14 dB); interpolating a little, the frequency of this varies from 292 MHz (5 segments) through 281 MHz (11 segments) and 278 MHz (21 segments) to 276 MHz (41 segments). The five-segment model has a segment size of just under $\lambda/10$, which is about the largest segment length that should be used in thin-wire modelling, certainly near a source. FEKO will issue a warning or an error if the segmentation is grossly inadequate. NEC, however, does *not* – many of the preprocessors now available provide this functionality, another reason that it is strongly recommended to use one!

The obvious course is now to proceed with further refining of the mesh (81 segments, etc.), but for subtle theoretical reasons, this is not wise. The problem is that the FEKO solution is based on the *thin-wire* approximation, discussed in Chapter 4. With a large number of segments, each segment becomes very short, and although the wire overall may indeed be thin, this is no longer true for a particular segment. FEKO issues a warning if the ratio of segment length to radius is less than around 3.3, and an error if

[4] FEKO offers the ability to *load* sources – this is not the same as setting the reference impedance Z_0.

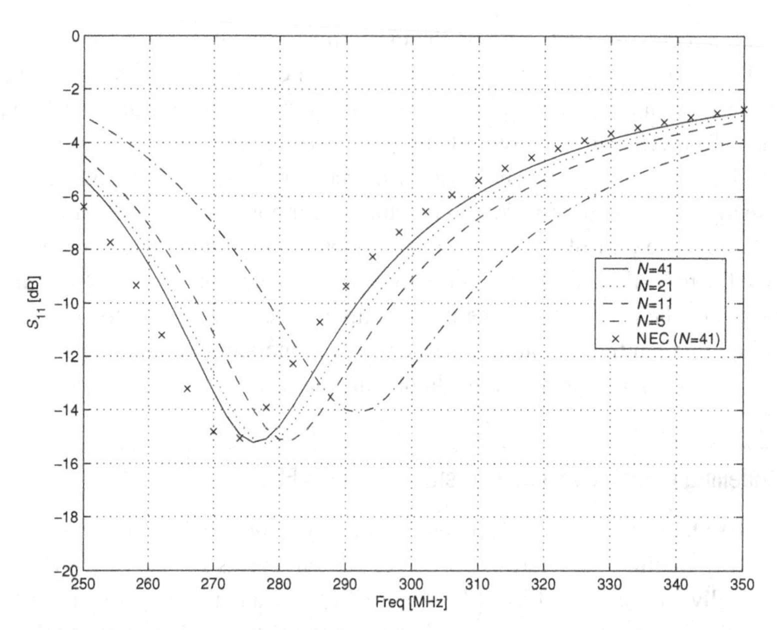

Figure 5.2 Results of convergence study for a dipole of length 0.5 m, radius 0.005 m. `A1` feed model.

this is less than 1. (The developers of NEC suggest an even more conservative ratio of around 8 as a preferred lower bound [1, p. 4].) Indeed, our 41-segment model actually violated this, with a ratio of 2.5. If one opens the output file and views the warnings, one will observe that a warning was indeed issued with the 41-segment model. (FEKO computes the ratio as radius to segment length, so the values reported in the file are the inverse of those in this discussion.)

The difference in resonant frequencies between the 21 and 41 segmentation runs is under 1%. It is important to note that resonant frequencies predicted numerically are often in error, typically by some few percent; indeed, this is perhaps the *least* accurate physical parameter computed by the MoM (and other numerical methods). This is especially true of thin-wire structures, but is generally true of resonant devices. To illustrate this further, we also show a result computed using NEC-2 in Fig. 5.2. NEC-2 predicts a center frequency of around 273 MHz using 41 segments, as opposed to the 276 MHz of the corresponding FEKO computation, an error also of the order of around 1%. NEC-2 uses different basis functions and a collocation approach, whereas FEKO uses piecewise linear basis functions and the Galerkin formulation, so one cannot expect the NEC-2 and FEKO results to be identical. To improve this further, one will need a more sophisticated source model for both codes [5] and one should be aware that this is about the level of accuracy for this parameter which can be expected from standard thin-wire codes.

[5] One such approach uses a quasi-static MoM model first to establish the incident field, which is usually *assumed* in such MoM models, and then uses this in the full-wave solution. One also needs to treat end-caps carefully. The best source on this is [4], whose results were also supported by careful measurements.

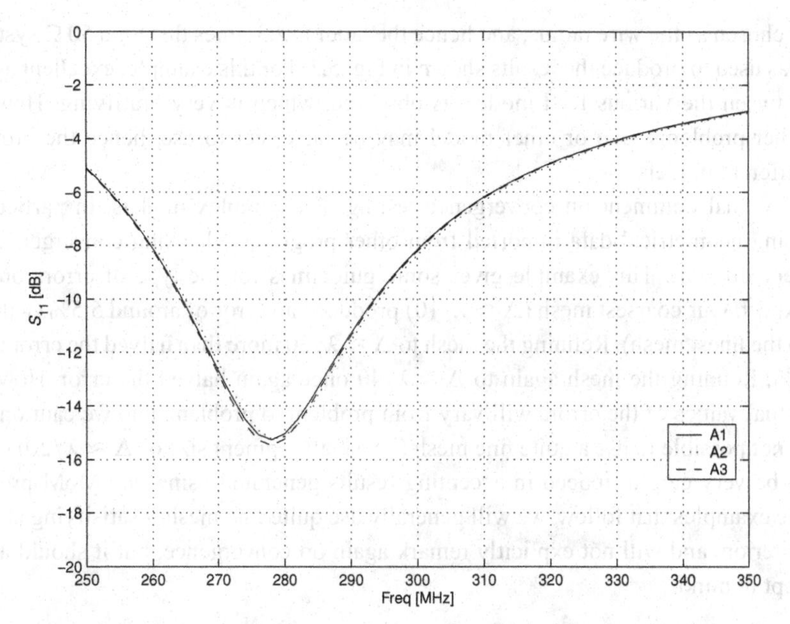

Figure 5.3 Comparison of different sources using 20 or 21 segments: voltage gap on segment (A1); voltage gap at node between segments (A2); magnetic frill feed (A3).

FEKO offers other methods for driving dipoles, and it is worth looking at them briefly. The A1 model essentially replaces a segment with a region of impressed electric field. It is important to note that this is done *within the code*!

Modelling hints – feed points for wire antennas

Many new users of MoM codes – FEKO, NEC-2, etc. – try to create a dipole from two wires, with a gap in the middle for the feed. *This is incorrect!* The correct approach is to specify a feed on an existing segment. In the region of the feed, the current is of course *displacement* current, rather than *conduction* current; it is effectively the former which the MoM is approximating in the feed region, but it still needs a segment (even though it is fictitious) and its associated expansion function in order to do this.

The other feed models for thin-wire structures offered by FEKO are the A2 and A3 models. The former uses a very thin gap between two nodes. The latter models a coaxial feed; it is derived by considering the TEM fields in a coaxial cable feeding a monopole against a very large ground plane, as discussed in Section 4.3. In Fig. 5.3, the results obtained by applying these three different feed models to this dipole are shown. Twenty-one segments were used for the A1 and A3 sources, and 20 for the A2 source. (Because the A2 source models a feed *at a node* rather than on a segment, the model requires an even number of segments for this case in order to place the feed at the dipole center.) For the A3 source, an equivalent inner and outer radius must be specified; usually, the former is chosen as the wire radius, and hence the latter is 2.3 times this for a 50 Ω system. This was used to produce the results shown in Fig. 5.3. For this example, excellent agreement between the various feed models is observed, which is very gratifying. However, for other problems, one or other model may be far easier to use, hence the provision of different models.

A final comment on convergence testing. For complex models, in particular ones using geometrical data imported from other programs, checking convergence may be very difficult. This example gives some guidelines for the type of errors one should expect. Our coarsest mesh ($\Delta \approx \lambda/10$) produced an error of around 5.5% (with respect to the finest mesh). Refining the mesh to $\Delta \approx \lambda/20$ more than halved the error to around 2%. Refining the mesh again to $\Delta \approx \lambda/40$ once again halved the error. However, the actual values of the errors will vary from problem to problem, and we caution that if it is not possible to use a quite fine mesh (i.e. small segment size of $\Delta \approx \lambda/20$) one needs to be very careful indeed in accepting results generated using *any* MoM program. In the examples that follow, we will generally use quite fine meshes satisifying at least this criterion, and will not explicitly remark again on convergence, but it should always be kept in mind.

Code tips – structural versus control cards in NEC

NEC differentiates between two different types of cards, namely structural and control cards. The former define actual metallic segments and patches, either via the direct creation of a wire or surface, or via operations on structural elements such as copying or reflection. The latter control parameters such as the location of the excitation, operating frequencies, grounds, near- and far-fields requested, etc.

Note that a NEC file requires at least one card which triggers execution, such as a field computation. The `XQ` card is a convenient way of forcing execution otherwise.

FEKO also distinguishes cards in a similar fashion, using the terms geometry and control cards respectively.

5.3 A wire antenna array: the Yagi–Uda antenna

In the preceding section, we discussed how to specify feed models, as well as the importance of checking that the analysis has converged. However, the thin-wire half-wavelength dipole is not a very stimulating engineering design on its own. A much more interesting example is an *array* of dipoles. Two well-known examples here are the Yagi–Uda antenna[6] and the log-periodic antenna, invented at the University of Illinois Urbana-Champaign during the 1950s. Design tables are available for both antennas, and some are reproduced in [5, 6]. The main difference is that the former is a narrowband, moderately high-gain structure, but with only one element (the driven element) fed;

[6] S. Uda is credited with the original design in 1926; the first English language publication was by his professor, H. Yagi, in 1927 [5, p. 188].

Table 5.1 Design data for a six-element Yagi array, wire radius $a = 0.00425\lambda$, using Viezbicke's results

Element	Length (in wavelengths)	Spacing (in wavelengths)
Reflector	0.482	−0.2
Driven	0.475	N/A
D_1	0.428	0.25
D_2	0.420	0.25
D_3	0.420	0.25
D_4	0.428	0.25

Spacing is relative to the previous element.

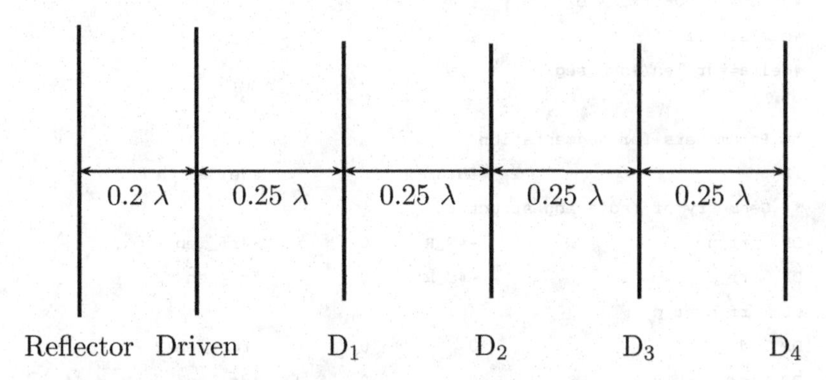

Figure 5.4 The six-element Yagi array described in the text.

the latter is a wideband structure, somewhat lower in gain, with all the elements fed in parallel via a transmission line network. Both are very widely used for VHF and UHF communication, as well as TV reception from terrestrial broadcasts. (Satellite transmissions are in the microwave band and a high-gain dish is generally used.) As an example, we will analyze a simple Yagi–Uda array, with one *reflector*, one *driven element* and four *director* elements. This is illustrated in Fig. 5.4.

We use the design data of Viezbicke, available in [5, Section 5.4] or in [6, Section 10.3.3]. Viezbicke's design process usually consists of two stages: firstly, establish the director and reflector lengths for the prototype Yagi [5, Table 5.4]; secondly, compensate for the actual wire radius using [5, Fig. 5–37]. By using the wire diameter $d = 2a = 0.0085\lambda$ of the prototype given in [5, Table 5.4], no compensation is required. These tables do not give the length of the driven element; this is usually the resonant dipole length in free space [5, p. 190]. (This can be established from standard results, for instance [5, Table 5.2]: for $L \approx 0.5\lambda$, $L/2a \approx 59$, the required shortening is about 5%, i.e. 0.475λ.) Hence our design is as summarized in Table 5.1. Director 1 is closest to the driven element. Extracts from the FEKO `.pre` file are given in Fig. 5.5; a NEC-2 data file is shown in Fig. 5.6.

Results for the reflection coefficient and the H-plane pattern at 291 MHz (the actual resonant frequency) are given in Figs. 5.7 and 5.8 respectively. The simulation indicates

```
.
.
#freq_o = 300.0e6          ** center frequency in Hertz
#lam_o = #c0/#freq_o   ** wavelength in metre,  #c0 = speed of light in vacuum

#rf_len = 0.482*#lam_o   ** Reflector
#dr_len = 0.475*#lam_o  ** driven element
#d1_len = 0.428*#lam_o
#d2_len = 0.420*#lam_o
#d3_len = 0.420*#lam_o
#d4_len = 0.428*#lam_o
#S_R = 0.2*#lam_o
#S_D = 0.25*#lam_o
#diam = 0.0085*#lam_o
#num_seg=21
#delta=#dr_len/#num_seg

** Parameters for segmentation
IP                          #diam/2                #delta
** Geometry of radiating structure
DP    rf_n                  -#S_R      0           -#rf_len/2
DP    rf_p                  -#S_R      0           #rf_len/ 2
BL    rf_n rf_p
DP    dr_n                  0          0           -#dr_len/2
DP    dr_p                  0          0           #dr_len/ 2
BL    dr_n dr_p
DP    d1_n                  1*#S_D     0           -#d1_len/2
DP    d1_p                  1*#S_D     0           #d1_len/2
BL    d1_n d1_p
DP    d2_n                  2*#S_D     0           -#d2_len/2
DP    d2_p                  2*#S_D     0           #d2_len/2
BL    d2_n d2_p
DP    d3_n                  3*#S_D     0           -#d3_len/2
DP    d3_p                  3*#S_D     0           #d3_len/2
BL    d3_n d3_p
DP    d4_n                  4*#S_D     0           -#d4_len/2
DP    d4_p                  4*#S_D     0           #d4_len/2
BL    d4_n d4_p
** End of geometric input
EG    1    0    0    0    0
.
.
.
```

Figure 5.5 Part of a PREFEKO file for the six-element Yagi array illustrating the use of user-defined variables and scaling.

```
CM 6 element Yagi
CE Start of geometry
GW1,21,-0.200000,0.000000,-0.241000,-0.200000,0.000000,0.241000,0.00425
GW2,21,0.000000,0.000000,-0.237500,0.000000,0.000000,0.237500,0.00425
GW3,19,0.250000,0.000000,-0.214000,0.250000,0.000000,0.214000,0.00425
GW4,19,0.500000,0.000000,-0.210000,0.500000,0.000000,0.210000,0.00425
GW5,19,0.750000,0.000000,-0.210000,0.750000,0.000000,0.210000,0.00425
GW6,19,1.000000,0.000000,-0.214000,1.000000,0.000000,0.214000,0.00425
GE  0    0
FR  0   51    0    0 275.00000 1.0000000
EX  0    2   11   00    1.00000    0.00000
XQ  0
EN
```

Figure 5.6 A NEC-2 file for the six-element Yagi array. This file uses the comma-delimited format.

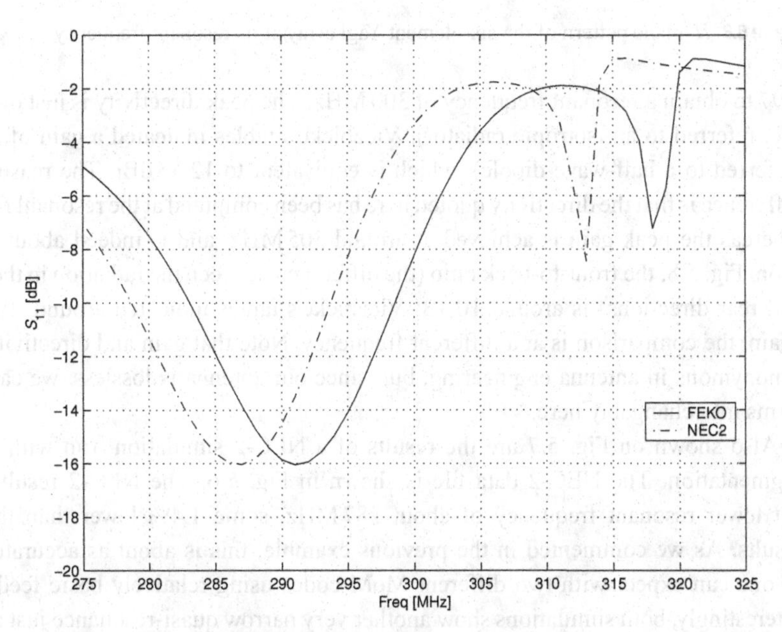

Figure 5.7 Reflection coefficient of the six-element Yagi array.

around a 5% −10 dB impedance bandwidth (the range of frequencies for which $|S_{11}|$ is less than −10 dB, corresponding to VSWR ≤ 2), which is as expected for a thin-wire structure. (These results were obtained for a segment length of around $\lambda_0/40$ at the center frequency.) The resonant frequency is 291 MHz, some 3% lower than the design frequency. Since quite fine segmentation has been used, this is probably a real effect, and were one to build this antenna, all the dimensions should be scaled by a factor of

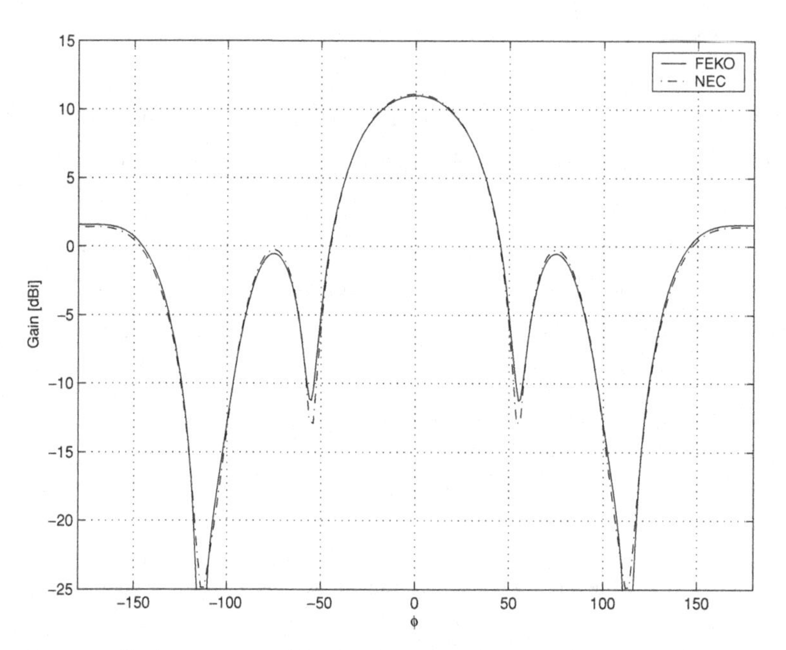

Figure 5.8 H-plane pattern of the six-element Yagi array at its resonant frequency.

0.97 to obtain a resonant frequency of 300 MHz. The peak directivity is just over 11 dBi (i.e. referred to an isotropic radiator). Viezbicke's tables indicated a gain of 10.2 dBd (referred to a half-wave dipole), which is equivalent to 12.35 dBi. The reason for the difference is that the directivity quoted here has been computed at the resonant frequency, whereas the peak gain is achieved at around 305 MHz, and is indeed about 12.3 dBi. From Fig. 5.8, the front-to-back ratio (the difference between the radiation in the forward and rear directions) is around 10 dB; Viezbicke's tables indicated around, 19 dB, but, again, the comparison is at a different frequency. Note that gain and directivity are not synonymous in antenna engineering, but, since our antenna is lossless, we can use the terms interchangably here.

Also shown on Fig. 5.7 are the results of a NEC-2 simulation, run with a similar segmentation. The NEC-2 data file is shown in Fig. 5.6. The NEC-2 results show a yet lower resonant frequency of about 287 MHz, some 1.4% lower than the FEKO results. As we commented in the previous example, this is about as accurate a result as one can expect with two different MoM codes using relatively basic feed models. Interestingly, both simulations show another very narrow quasi-resonance just above the design frequency.

Figure 5.8 also shows the NEC radiation pattern predictions (the NEC results are computed at 287 MHz, the resonant frequency computed by NEC); we use these to illustrate an important point, namely the far-field radiation patterns are not as sensitive a parameter as the input impedance, and hence excellent agreement with other codes can usually be expected. (Agreement with measurements tends to be less satisfactory; frequently, the problem lies with the experimental setup, for instance problems with the feed cables interfering with the patterns.)

We did not explicitly perform a convergence check, since we are using a fine discretization with around 40 segments per wavelength, but of course the comments in our introductory dipole section apply. Due to the relatively thick dipoles in use, one cannot refine the mesh further without starting to violate the thin-wire assumptions.

Aside from the lower center frequency – which, as we commented above, is easily fixed in practice (or indeed in simulation) by scaling – our six-element Yagi array works moderately satisfactorily. Now, we are in a position to evaluate quickly the effect of having to use a different wire radius, etc., as is quite probable in an actual design. This however might degrade the performance of the antenna. We might also not be satisfied with the front-to-back ratio, for instance, and wish to improve this. This leads into the field of optimization, which FEKO supports, although we will not pursue this further here.

Modelling hints – using user-defined variables and scaling

When developing a general-purpose model, it is often useful to specify dimensions in terms of λ_0, which makes it very easy to change the operating frequency. Also, all the dimensions are given in terms of user-defined variables, so that if we want to change the design of the antenna (perhaps by optimization), we have already done a lot of the work. An example of this is shown in Fig. 5.5, which shows part of the PREFEKO file exploiting user-defined variables. Some other commercial codes, such as MWS, have similar abilities. Connected to this is scaling: a popular use of this is to permit microwave structures to be entered in millimeters. Whilst NEC-2 does support scaling, it does not support user-defined variables.

Modelling hints – wire radius versus diameter

Here is an important point to note, which even experienced users forget from time to time: wire thicknesses in FEKO and NEC-2 are specified in terms of *radius*, whereas especially older texts in antenna design often use *diameter*. Accidentally confusing these is a common source of error; to make things worse, the simulation will often still appear to work, but the results produced are usually subtly incorrect.

5.4 A log-periodic antenna

The Yagi–Uda example highlighted a number of points, but in a sense was simply an extension of the dipole problem, since the additional wires – the reflector and the directors – were passive, and it was just a case of adding these into the `.pre` file. The problem we will now investigate, however, brings some new points, with regard to both FEKO modelling and antenna engineering. It also serves as an introduction to some ideas in wideband antennas.

The log-periodic (log-p) antenna consists of a number of wire dipoles, but unlike the Yagi–Uda antenna, they are *all* fed (by means of a transmission line, which provides a parallel feed). Also, each element is smaller than and more closely spaced to its predecessor; the *ratio* is constant, and τ is the design parameter which specifies this. With dipole lengths L_n and spacing d_n, this is defined as

$$\tau = \frac{L_{n+1}}{L_n} = \frac{d_{n+1}}{d_n} \tag{5.1}$$

The other parameter which defines a log-periodic array is the *spacing factor* σ, defined as

$$\sigma = \frac{d_n}{2L_n} \tag{5.2}$$

One can also compute α, the angle of the wedge bounding the dipole arms of the log-p, from these parameters:

$$\alpha = 2 \arctan \left(\frac{1 - \tau}{4\sigma} \right) \tag{5.3}$$

A value of τ close to 1 indicates a log-p with a very slow expansion, i.e. long overall length, but also higher gain. The design of a log-p is typically a trade-off between length, gain and impedance match. Most design data are based on tables originally published by Carrel in 1961; subsequent research has improved these tables and a typical set are presented in [5, Section 6.7]. We will base our FEKO simulation on [5, Example 6.2].

To summarize this briefly for readers without ready access to this reference, the design specification is for a 6.5 dB gain antenna over the VHF-TV and FM broadcast bands, which span the frequency range 54–216 MHz (a 4:1 bandwidth). From the design tables, $\tau = 0.822$ and $\sigma = 0.149$ are selected to satisfy the gain requirement. The lowest frequency determines the length of the longest element, usually chosen as $\lambda_{\text{max}}/2$, or 2.78 m in this case. Elements are then placed until an element shorter than $\lambda_{\text{min}}/2$ is produced. In this case, nine elements are required. The tabulated data are for a dipole radius 1/250 of the dipole length, clearly varying from element to element. The characteristic impedance of the transmission line is 100 Ω. The design is summarized in Table 5.2 and illustrated in Fig. 5.9.

To implement this in FEKO, there are several approaches that can be taken. The first is simply to create nine wires. A better approach is to use the `!!FOR ... !!NEXT` loop structure, as illustrated in Fig. 5.10. It will also be noted that we construct the elements from four points: two at each end, but also two very close to the center. We do the latter for two reasons. Firstly, there is then always a segment at the center of the element to feed, no matter what the segment length. Secondly, we use the label (`LA`) card (the equivalent of a tag in NEC) to attach a unique label to these central segments; this makes connecting these fed segments (which represent the terminals of the elements) via a transmission line much easier. This is the next step to consider.

Table 5.2 Design data for a nine-element log-periodic array

Element	Length (in meters)	Spacing to next element (in meters)
1	2.78	0.828
2	2.29	0.682
3	1.88	0.560
4	1.54	0.459
5	1.27	0.378
6	1.04	0.310
7	0.858	0.256
8	0.705	0.210
9	0.579	–

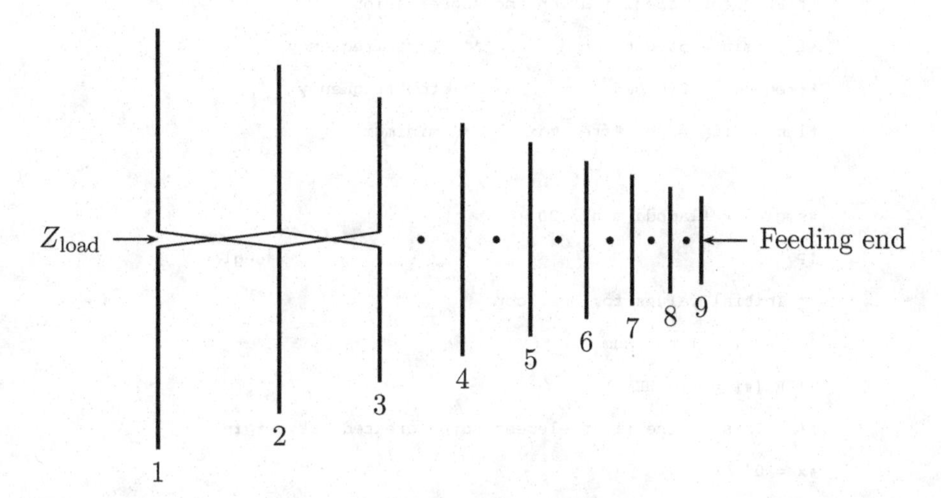

Figure 5.9 The nine-element log-periodic array described in the text. The details of the crossed feed are only shown for the largest three elements, but repeat to the end of the array. Also shown is the feeding end, as well as the position for a possible terminating load, as discussed in the text.

Modelling hints – using iteration loops and conditional execution structures in PREFEKO

Many antennas consist of repeated components, and PREFEKO has a very useful feature to implement this, namely the `!!FOR ... !!NEXT` loop (iteration) structure. This is illustrated in Fig. 5.10. We have used another useful feature as well, namely the `!!IF ... !!THEN ... !!ELSE` conditional. Note that d_n is computed from the *current* length, and is computed before we update (reduce) the length for the next execution of the loop.

NEC-2 has no such functionality – the closest NEC-2 gets is the coordinate transformation GM card, which allows one to copy, translate or rotate parts of the geometry.

```
** Analysis of a 9-element logarithmic periodic antenna.

** Some definitions for the geometry

#sigma = 0.149        ** scaling factor for spacing [eqn.6.83,S&T]

#tau = 0.822          ** scaling factor for elements [eqn.6.85,S&T]

#len = 2.78           ** length of element (initially L_1)

#rad = #len/250.0     ** radius of first element: L/2a = 125

#Zline = 100          ** transmission line impedance

#Zload = 100          ** load impedance at the last element (set to very large value
                         if not present)

#num =9               ** number of elements


** Frequency specification and segmentation

#freq_min = 50.0e6           ** start frequency

#freq_max = 250.0e6          ** stop frequency

#lambda_min = #c0/#freq_max   ** minimum


#seglen = #lambda_min / 20

IP                                                 #seglen

** Initial values for the loop

!!FOR #i = 1 to #num

!!IF (#i = 1) THEN

**   This is the first element to be created, at origin

#x = 0

!!ELSE

**   Other elements spaced logarithmically

#x = #x+#d

!!ENDIF

**   Create the wire with the correct radius, use a unique

**   label #i for the center segment

#z = #seglen   ** ensure that just one segment at the center

DP   P1                    #x        0         -#len/2.0

DP   P2                    #x        0         -#z/2.0

DP   P3                    #x        0         #z/2.0

DP   P4                    #x        0         #len/2.0

LA   0
```

Figure 5.10 PREFEKO file for the nine-element log-periodic array.

```
BL    P1    P2                       #rad
LA    #i
BL    P2    P3                       #rad
LA    0
BL    P3    P4                       #rad
** Compute inter-element spacing to next element. Note that d_n is the spacing
   between elements
** L_n and L_n+1 and must be computed using current length.
#d = 2.0*#sigma*#len
**   Now apply scaling for next element (shorter)
#len = #len*#tau
#rad = #rad*#tau
!!NEXT
** End of the geometry
EG   1    0    0    0    0


** Create all the transmission lines (again a loop is very useful)
!!FOR #i = 1 to #num-1
**   Extra shunt admittance at the first element
!!IF #i=1 THEN
#YS = 1 / #Zload
!!ELSE
#YS = 0
!!ENDIF
**   Define the transmission line from label #i to label #i+1 (crossed)
TL    1    #i    #i+1 1        -1        #Zline              #YS
!!NEXT
** Excitation by a voltage source at the last (shortest) element
FR        2                  #freq_min                #freq_max
A1   0    #num                1         0
** Vertical radiation pattern - gain
FF   1    1    1    1        90         0
** Vertical radiation pattern - directivity
FF   1    1    1    0        90         0
EN
```

Figure 5.10 (*cont.*)

We also have to consider how to interconnect the radiating elements. The obvious way is to connect wires to the elements to form a transmission line explicitly. However, this is not a very efficient way of handling the problem. Transmission lines are non-radiating structures, and can be succinctly described using two-port circuit theory. FEKO incorporates this feature, implemented using the `TL` card. (This functionality is also available within NEC-2, with the same name.) We need eight of these transmission lines; a subtle design point is that the transmission lines are *crossed*, i.e. reverse phase, from element to element; this is done to compress the overall length of the antenna. (In NEC, such crossed lines are specified by using a negative characteristic impedance.) These are also implemented using a loop. Finally, the transmission lines of log-periodic antennas are often terminated with a resistive load (usually equal to the transmission line characteristic impedance, 100 Ω in this case) to improve the impedance match. This is done here via the special handling of the last transmission line, which adds a shunt (parallel) admittance of $1/100$ S to the feed segment of the last antenna.

In NEC, the absence of user-defined variables, loops, etc., means that we have no option other than to compute the values explicitly and enter them by hand, either into a NEC file directly, or using a preprocessor. An example of a NEC file for this log-periodic array is given in Fig. 5.11.

Code tip – some useful NEC functions

In Fig. 5.11, two cards `PT` and `PL` are used which offer useful functionality. The former is used for selectively or entirely suppressing outputting of the currents, which is, perhaps unfortunately, the NEC default, since this otherwise inflates output files with data which are rarely used. The latter produces an extra data file (the specific name varies from implementation to implementation) with radiation patterns or currents suitable for later plotting.

Results for the reflection coefficient and the gain of the log-periodic array are given in Figs. 5.12 and 5.13 respectively. Results computed with NEC-2 using a similar segment length are given for some of the parameters, and excellent agreement is noted. Also indicated on Fig. 5.12 is the reflection coefficient level corresponding to a VSWR of 2, widely used as a specification for antenna impedance. (A VSWR ≤ 2 actually corresponds to $|S_{11} \leq -9.54|$ dB, as indicated, but $|S_{11} \leq -10|$ dB is often used instead for convenience.) It will be noted how the use of the terminating resistance improves the impedance match; the antenna has $|S_{11}| < -10$ dB over almost the entire band in this case. Without the terminating resistance, the reflection coefficient varies far more over the frequency band, sometimes lower, but also sometimes unacceptably high. Another point to note is that the log-p array must be fed from the *shorter* end; if fed from the longer end, the long dipoles are excited (but not very effectively) so that there is too little power at the higher frequencies to radiate properly from the shorter dipoles.

On Fig. 5.13, both *gain* and *directivity* (also sometimes known as directive gain) are given. To revise these terms briefly, the former indicates how well the antenna focusses

```
CM 9 element log-p
CE Start of geometry
GW1,47,0.000000,0.000000,-1.390000,0.000000,0.000000,1.390000,0.01110
GW2,39,0.828400,0.000000,-1.142600,0.828400,0.000000,1.142600,0.00910
GW3,33,1.509400,0.000000,-0.939200,1.509400,0.000000,0.939200,0.00750
GW4,27,2.069200,0.000000,-0.772000,2.069200,0.000000,0.772000,0.00620
GW5,23,2.529300,0.000000,-0.634600,2.529300,0.000000,0.634600,0.00510
GW6,19,2.907500,0.000000,-0.521600,2.907500,0.000000,0.521600,0.00420
GW7,15,3.218400,0.000000,-0.428800,3.218400,0.000000,0.428800,0.00340
GW8,13,3.474000,0.000000,-0.352500,3.474000,0.000000,0.352500,0.00280
GW9,11,3.684100,0.000000,-0.289700,3.684100,0.000000,0.289700,0.00230
GE  0    0
PT -1
PL  3    1    0    1
TL  1   24    2   20 -100.0000   0.00000   0.01000   0.00000   0.00000   0.00000
TL  2   20    3   17 -100.0000   0.00000   0.00000   0.00000   0.00000   0.00000
TL  3   17    4   14 -100.0000   0.00000   0.00000   0.00000   0.00000   0.00000
TL  4   14    5   12 -100.0000   0.00000   0.00000   0.00000   0.00000   0.00000
TL  5   12    6   10 -100.0000   0.00000   0.00000   0.00000   0.00000   0.00000
TL  6   10    7    8 -100.0000   0.00000   0.00000   0.00000   0.00000   0.00000
TL  7    8    8    7 -100.0000   0.00000   0.00000   0.00000   0.00000   0.00000
TL  8    7    9    6 -100.0000   0.00000   0.00000   0.00000   0.00000   0.00000
EX  0    9    6   00    1.00000   0.00000
FR  0  101    0    0  50.00000 2.0000000
RP  0    1    1 1000  90.00000   0.00000   0.00000   0.00000   0.00000         0
EN
```

Figure 5.11 NEC file for the nine-element log-periodic array.

power spatially, relative to the power *delivered to* it; the latter indicates how well the antenna focusses power spatially, relative to the power *radiated by* it. Clearly, if the antenna has any loss, the two will not be identical, and the difference on Fig. 5.13 is due to the losses in the termination. We have traded off a better impedance match for a slightly poorer gain. (At the very top of the band, we are slightly under the 6.5 dB gain design specification. To improve this, we would have to repeat the design using a longer array, i.e. with more elements, but we will leave this as an exercise.) A final point: because the transmission line has a characteristic impedance of 100 Ω, it is tempting to use this as the impedance level when computing S_{11}, etc. However, one should recall

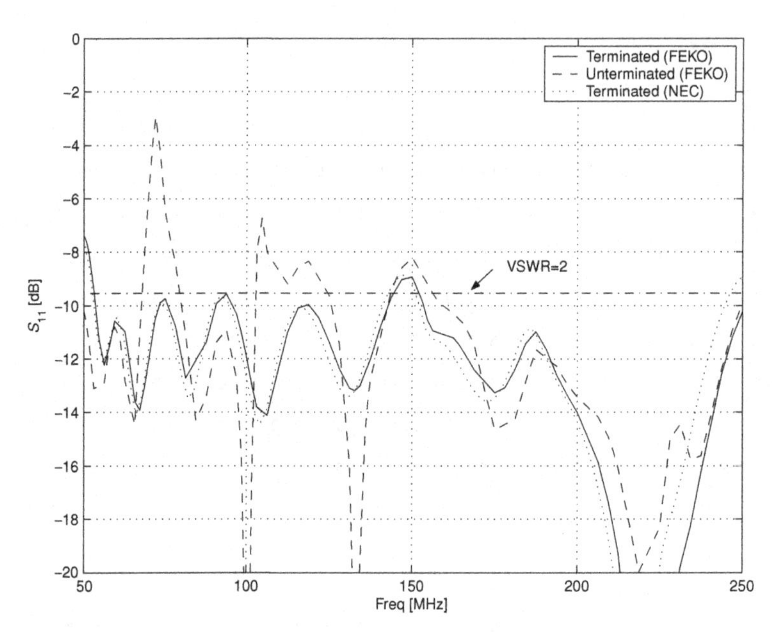

Figure 5.12 Reflection coefficient of the nine-element log-periodic antenna in the text, for both resistively terminated and unterminated cases.

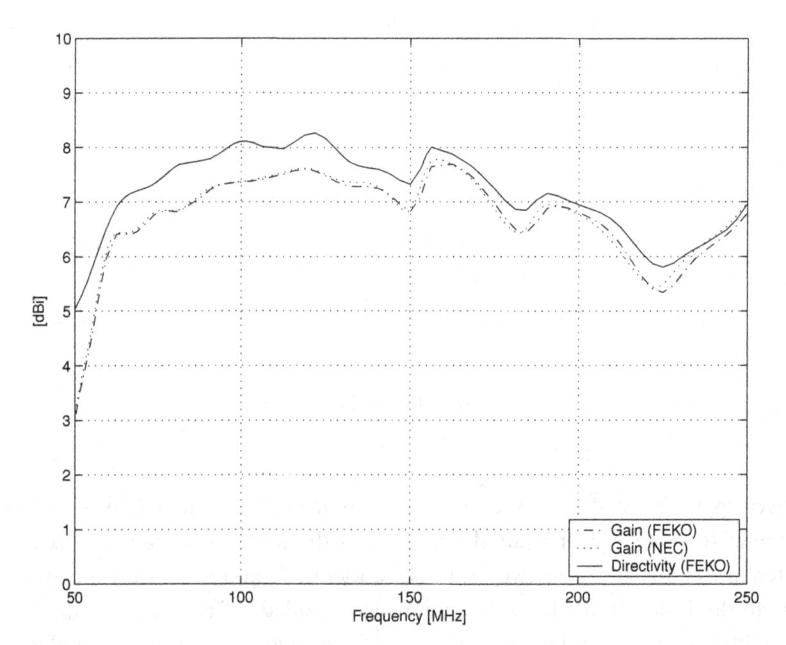

Figure 5.13 A comparison of gain and directivity for the nine-element resistively terminated log-periodic antenna. The gains computed by FEKO and NEC lie essentially on top of one another.

that this line is in parallel with the radiating dipole(s), with an impedance of typically $50 \sim 70\,\Omega$. The net result is that this antenna is quite well matched to a $50\,\Omega$ system, which is the FEKO default. Note also that we only compute the gain at one angle, in the direction along the axis of the antenna. A log-p is an end-fire antenna, and radiates in the direction from longest to shortest element.

This example also introduces another feature which FEKO supports, namely the use of *adaptive frequency sampling* (this is not supported by NEC). This example is sufficiently complex (227 wire segments as discretized) that FEKO takes a noticeable amount of time, typically a second or two, to compute each frequency point. However, the data change rapidly over frequency, requiring a *lot* of points; to obtain good results with uniformly spaced frequency points over the frequency band of interest, one would need at least 100 points, preferably more. FEKO has the ability to determine where to place frequency points in a *non-uniform* fashion, as well as then intelligently interpolating the data by using what is termed a *model-based parameter* representation. We use the defaults for this option in the frequency card in the PREFEKO file in Fig. 5.10.

Modelling hints – gains in dB or actual values

Be very careful when plotting gains for these relatively low-gain structures; the gains in dB or in actual value are quite similar numerically, and it is easy to plot the wrong dataset, especially when exporting data!

5.5 An axial mode helix antenna

Helical antennas are another interesting type of antenna. The axial mode helix was invented by Kraus at Ohio State University in 1946, and his textbook on antennas is a mine of information on the subject [7]. Their bandwidth ratio is given theoretically as approximately 1.78 [5, Section 6.2.2]. (A wideband antenna is conventionally defined as one where this ratio exceeds 2, so the helix is close to being "wideband.") Details are also available in [6, Section 10.3.1]. It is also a wire structure, but *unlike* all the previous antennas we have analyzed, which all relied on a *standing wave* on some part of the structure, this is a *travelling wave* antenna, at least in its axial mode of operation, which is the most common mode of employment. The circumference of the antenna is chosen such that currents on opposite sides of the antenna (which would radiate fields out of phase due to the winding of the helix, if the currents were in phase) are delayed by a half-wavelength, so that the resulting radiation is now in phase again along the axis of the helix (and hence the name, axial mode). The radiation is circularly polarized, with the sense of the winding, i.e. a right(left)-hand wound helix generates right(left)-hand polarization. Compared to other candidates, the axial mode helix is quite compact – the helical structure permits a lot of wire to be contained in a moderately small volume – and the design is very popular in the UHF band, especially for satellite communication.

(A closely related structure, namely the normal mode helix, is very popular for mobile telephones. It radiates almost isotropically.)

FEKO provides the HE card, which greatly simplifies creation of a helix. Indeed, all that is required other than this card is to add a short segment below the helix to feed it with, and to add a ground plane of some type underneath it. A ground plane of around 0.75λ on each side is usually adequate [5, Section 6.2.2].

To create a ground plane, one can use a mesh of wires – and indeed this was a very widely used method with NEC-2. However, FEKO supports the creation and meshing of surfaces. A simple method of defining a surface is using the parallelogram card (BP). This surface is then meshed using triangles.

Modelling hints – connecting wires and plates

Here is something to be careful of. The obvious approach when grounding the helical wire is to generate one surface in the plane $z = 0$, where the feed segment terminates. However, this usually will not work properly! The reason is that FEKO, and indeed any MoM code, needs the *nodes* defining the segments on the wire and the triangular segments on the surface to coincide. Many new users overlook this and it is a frequently encountered fault. In the PREFEKO file, we have generated only a quarter of the ground; this of course includes a point at the origin, where the feed segment connects. We then use geometrical symmetry in two planes ($x = 0$ and $y = 0$) first to create half the ground plane, and then to create the entire ground plane. (The PREFEKO file supplied does this in first the $x = 0$ plane, then the $y = 0$ plane, but the order is actually irrelevant in this example.) FEKO also permits users to specify *internal nodes* in polygonal plates, using the PM card, which makes it easier to make sure that wires correctly connect to nodes on surfaces.

One final point regarding creating the geometry. FEKO also offers a ground plane (BO) card, and this would appear to be very useful. However, one needs to read the "fine print" in this case. This card uses a reflection coefficient *approximation*; i.e. the *fields* radiated by the structure are imaged in the ground plane, but the ground plane is *not* taken into account when the currents are computed by the MoM. As such, it is very useful for antennas some distance above a ground plane, where the currents are indeed hardly changed by the presence of the ground, but entirely *inappropriate* for an antenna fed right against a ground, as the helix is. A careful reading of the user manual cautions that segments should not connect to the ground, but does not describe in detail why this ground plane would be incorrect in this application.

A detailed design example is given in [5, Example 6.2]. The antenna is to operate in the microwave band, with center frequency 8 GHz. The circumference of the antenna, C, is specified as $0.92\lambda = 34.5$ mm. (It will be noted that the scaling card is also used in the PREFEKO file to permit all dimensions to be entered in millimeters, which is far more convenient than meters in this frequency range.) The pitch angle α is chosen as $13°$ (a value based on prior design experience). The spacing between turns, S, works out at 7.96 mm, and the antenna has $N = 10$ turns. With the 1.78 bandwidth ratio and center frequency of 8 GHz, the lower and upper frequencies are 5.75 GHz and 10.25 GHz respectively. The PREFEKO file is given in Fig. 5.14, and Fig. 5.15 shows a FEKO model of the antenna.

Radiation patterns at the lower, center and upper frequencies are shown in Fig 5.16 with a ground plane 1.5λ on a side, somewhat larger than the minimum recommended. The gain at 8 GHz is exactly 13 dBi, somewhat higher than the 10.5 dBi gain predicted by the approximate formula [5, Eq. (6-34)]

$$G \approx 6.2 \left(\frac{C}{\lambda}\right)^2 N \frac{S}{\lambda} \tag{5.4}$$

Commensurate with this increased gain, the half-power (HP) beamwidth of 40° is somewhat smaller than that predicted by the approximate formula [5, Eq. (6-33)]

$$HP \approx \frac{65^\circ}{\frac{C}{\lambda}\sqrt{N \frac{S}{\lambda}}} \tag{5.5}$$

of 48°. It must be emphasized that these are approximate empirical formulas, so some differences are to be expected. Kraus provides another formula [7, Eq. (7), p. 235] for directivity, which he describes as more realistic:

$$D \approx 12 \left(\frac{C}{\lambda}\right)^2 N \frac{S}{\lambda} \tag{5.6}$$

Using this formula yields a gain of around 13.3 dBi, almost exactly as simulated. (Since the antenna is essentially lossless, we are again using gain and directivity interchangeably.)

From Fig. 5.16, the gain at the lower frequency is almost 3 dB less than at the center frequency, and the pattern is starting to show some "squint"; the main beam has moved slightly to the left. At the upper frequency band, the gain has increased and the main beam has narrowed (which may or may not be acceptable, depending on the design requirements).

Impedance results are shown in Fig. 5.17. (These data were generated using adaptive frequency sampling.) It will be noted that the antenna is largely resistive across most of the frequency band. However, towards the lower end of the band, the otherwise smooth impedance curves break down. This type of behavior is *not* predicted by the simple description of operation as a travelling wave antenna [5]. Measured data by Baker [7, Fig. 8–73], who worked on helix arrays with Kraus, indicate almost exactly the same impedance behavior at around 0.7 of the center frequency, with the reflection coefficient suddenly increasing dramatically from less than −20 dB to −2 dB or worse over a very small frequency change. (Baker's helix was not precisely the same as the one simulated here, hence the frequency at which this effect occurs is slightly different.) The reason is that the axial mode ceases effective operation quite abruptly; [7, Fig. 8–34] provides more information on this, in particular via the phase velocity.

In the region near the design frequency, the resistance and reactance values are well behaved, as shown in Fig. 5.18. An approximate formula for the input resistance of the

```
** A 10-turn helical antenna
**
** Variables
**   Optional scaling factor (set to 0.001 for geometrical data
**   coordinates etc. defined in cm instead of metres)
#scaling = 0.001
**   Frequency and wavelength
#freq = 8e9           ** frequency in Hertz
#freq_min=5.75e9
#freq_max=10.25e9
#lam = #c0/#freq      ** wavelength in metre,  #c0 = speed of light in vacuum
#lam_mm = #lam/#scaling
#circum = 34.5        ** helix cicumference
#h_rad = #circum/(2*#pi)
#h_len = 79.6         ** helix length
#gnd = 1.5*#lam_mm
** Parameters for segmentation
#seg_rad = #lam_mm/100 ** radius of the wire segments
#seg_len = #lam_mm/20  ** maximum length of wire segments
#tri_len = #lam_mm/10  ** maximum size of triangles
IP                           #seg_rad  #tri_len  #seg_len

** Quarter of ground plane
DP   G1                      0.0       0.0       0.0
DP   G2                      #gnd/2    0.0       0.0
DP   G3                      #gnd/2    #gnd/2    0.0
DP   G4                      0.0       #gnd/2    0.0
BP   G1   G2   G3   G4
** Generate rest of ground - imaged first in x=0, then y=0 planes.
SY   1    1    0    0
SY   1    0    1    0
** Helix
DP   ZERO                    0.0       0.0       0.0
DP   A1                      0.0       0.0       2*#seg_len
DP   B1                      0         0.0       #h_len
DP   C1                      #h_rad    0.0       2*#seg_len
HE   A1   B1   C1   0        10
** Wire
LA   1
BL   A1   ZERO
```

Figure 5.14 PREFEKO file for the 10-turn helix.

```
** Apply the scaling factor
SF   1                        #scaling
** End of geometric input
EG   1    0    0    0    0
** Voltage gap excitation at segment just above ground
A1   0    1                   1.0

** Note: using adaptive frequency sampling permits only
** ONE of the following analysis options:
** ** Set the frequency card for adaptive frequency sampling.
** FR        2                    #freq_min           #freq_max
**
** ** Trigger execution, no patterns.
** FF    0
**
** Set discrete frequency for radiation patterns.
FR   3    0                   #freq_min           #freq_max
**   Radiation pattern
FF   1    181   1    0        -90       0         1.0
FF   1    181   1    0        -90       90        1.0

** End
EN
```

Figure 5.14 *(cont.)*

axial mode helix is

$$R \approx 140C/\lambda\,\Omega \tag{5.7}$$

At 8 GHz, this gives a value of $\approx 129\,\Omega$, whereas FEKO indicates a value closer to $170\,\Omega$. It should be noted that the above formula is to be regarded only as an approximation, so the FEKO result is very credible. Also giving confidence in the FEKO results is the approximately linear increase in resistance, at least in the central part of the frequency band. In practice, such an antenna would probably be fed via an impedance matching transformer, probably with a 3:1 ratio. As such, a reference impedance of $Z_0 = 150\,\Omega$ is appropriate when plotting the reflection coefficient, which is shown in Fig. 5.19.

To evaluate this antenna fully for an actual design exercise, one should also check the axial ratio of the polarization, since this is an important parameter when designing circularly polarized antennas. This information is also available in the `.out` file, but may require some manipulation to present graphically. More details are available in [7].

In summary, the helix performs well from around 6.2 GHz to at least 10.75 GHz, in terms of impedance match (S_{11} less than -10 dB, assuming a 3:1 impedance transformer for a $50\,\Omega$ system, as above) and offering reasonable pattern behavior. This is a bandwidth ratio of 1.73. The gain at the center frequency agrees very well with Kraus's

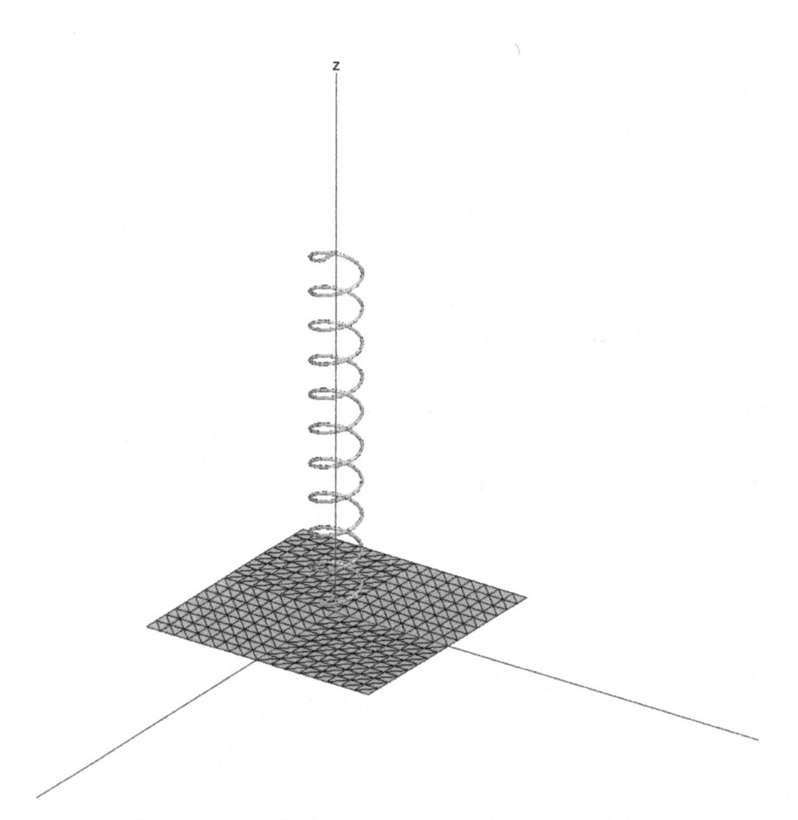

Figure 5.15 The FEKO model of the axial mode helix antenna discussed in the text.

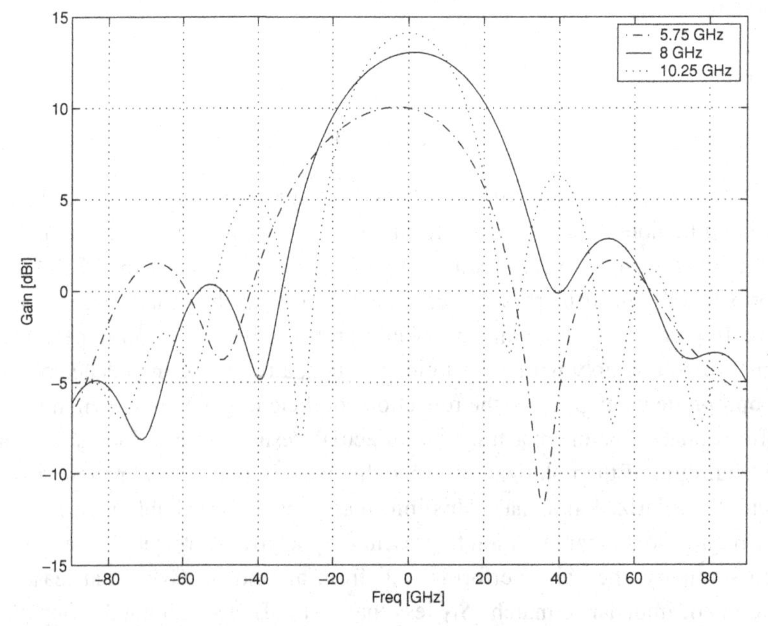

Figure 5.16 The gain of the axial mode helix antenna at the lower, center and upper ends of its operating band.

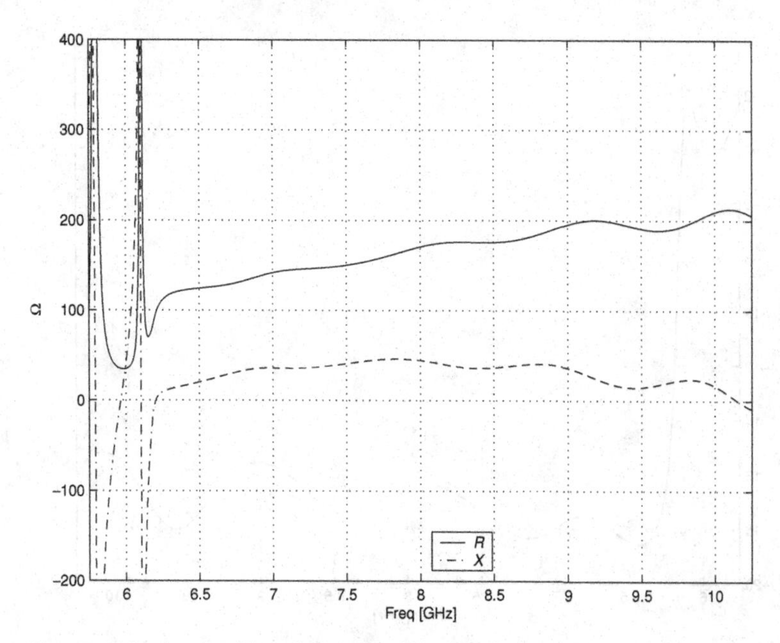

Figure 5.17 Resistance (R) and reactance (X) of the helix antenna across the entire operating band.

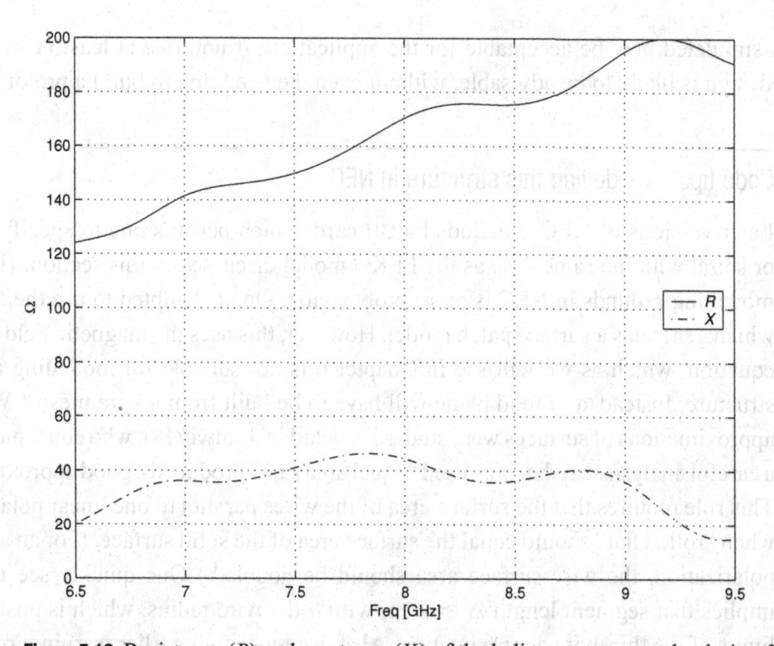

Figure 5.18 Resistance (R) and reactance (X) of the helix antenna near the design frequency.

improved formula, and at the lower end of operation, the reflection coefficient shows the same behavior as measured data for a similar (but not identical) helix. The empirical design formulas give reasonable guidelines for gain and half-power beamwidth, but the numerical simulation provides much more accurate data. In an actual design, the helix

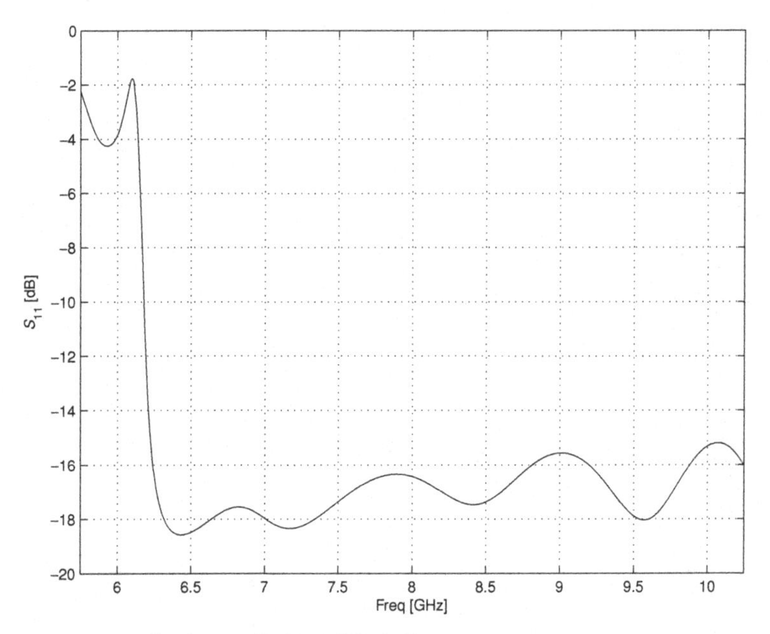

Figure 5.19 Reflection coefficient of the helix.

as simulated may be acceptable for the application; if not, one at least is aware that a redesign is likely to be advisable, without even the need first to build a prototype.

Code tips – modelling this structure in NEC

Later versions of NEC-2 included a `GH` card, which permits one to specify a helix or spiral with the same ease as the FEKO model discussed in this section. However, modelling grounds in NEC is more problematic. One is tempted to use the `SP` card, which generates a surface patch model. However, this uses the magnetic field integral equation, which as we will see in Chapter 6 is not suitable for modelling an open structure. Instead, a ground plane will have to be built from a wire mesh.[a] Wiregrid approximations of surfaces were studied in detail by Ludwig [8], who confirmed using a careful analysis that the long-used "equal area rule" produced a good approximation. This rule requires that the surface area of the wires parallel to one linear polarization when "rolled flat" should equal the surface area of the solid surface. (For an arbitrary polarization, the wire surface area should be doubled.) One quickly see that this implies that segment length $\Delta \approx 2\pi a$, with a the wire radius, which is pushing the limits of the thin-wire approximation. Also, we repeat our earlier warning: one must be very careful to ensure that the helix wire and wires representing the ground plane actually connect.

[a] FEKO includes a `WG` card to do this, although due to its surface meshing capabilities, one will probably not use this too often.

5.6 A Wu–King loaded dipole

Thus far, all the antennas discussed in this chapter were assumed to consist of perfectly conducting wires. (The log-periodic antenna included a terminating resistance, which was introduced to improve the impedance match, although the elements were still assumed to be perfectly conducting.) In practice, the conductance of the metals traditionally used for constructing wire antennas (aluminum, steel, etc.) is sufficiently high that this is an excellent assumption. In this example, however, we are going to study an antenna deliberately loaded with resistance – the *Wu–King* resistively loaded dipole. This antenna, first described in [9, 10], has a continuous resistive loading. In practice, this can be made either using thin tubular sections of varying radius and material [9], or by approximating the continuous loading by discrete resistors [11].

Wu and King showed that if the loading on a dipole, half-length h, had the following form:[7]

$$Z(z) = \frac{\eta_0 \Psi}{2\pi h(1 - |z|/h)} \; [\Omega/\text{m}] \tag{5.8}$$

then the current had the following approximate form:

$$I(z) \sim h(1 - |z|/h)\, e^{-jk_0|z|} \tag{5.9}$$

This is clearly a *travelling wave*. By comparison, on the usual half-wave resonant PEC dipole, the current has the *standing wave* form $\sin[k(h - |z|)]$.

The dimensionless parameter Ψ is complex valued, and a function of the electrical dimensions of the antenna. It is usually approximated by its DC value, Ψ_0. It must be computed numerically; typical values are from just under 10 for moderately thick dipoles to around 20 for very thin ones.

We will study the loaded dipole described by Maloney and Smith [11]; for their antenna, the ratio of half-height to radius h/a was 65.8, and $\Psi_0 = 7.79$. For convenience, we will work with $h = 0.25$ m, so that the unloaded PEC dipole resonates close to 300 MHz.

In FEKO, loading can be accomplished using several different cards: `LD`, `LS` and `LP`. The first implements *distributed* loading, in Ω/m, which is what we need here. (The other two cards implement lumped loads in series and parallel respectively.) FEKO loads segments via their label number, and hence one needs to label each segment on the dipole separately. (A FEKO label is the equivalent of a NEC tag.) One way of doing this is shown in Figs. 5.20 and 5.21, where the dipole is first built from individual segments, and then loading is applied to each of these.

The reflection coefficient of the Wu–King dipole is compared to a PEC (unloaded) one of the same dimensions in Fig. 5.22. In these results, two values of loading are shown: the "high" value is as in Eq. (5.8), the "low" value is as given in their original paper, with an 8 instead of 2 in the denominator. The Wu–King dipole has a rather high input impedance (given approximately by $60\Psi_0$), so $Z_0 = 300\ \Omega$ was used when computing

[7] Note the major corrections in [10]; the corresponding expression [11, Eq. (1)] is correct.

```
** A resistively loaded (Wu-King profile) dipole.
** As in Maloney and Smith, IEEE T-AP, May 1993 p.668-676.

** Variables
**   Frequency and wavelength
#lam = 1.00          ** wavelength in metre
#freq = #c0/#lam     ** frequency in Hertz
#h = #lam/4          ** half-height of antenna [m]
#seg_rad = #h/65.8 ** radius of wire
#f_l = 200e6
#f_u = 600e6
#psi_0 = 7.79        ** Wu-King parameter
#eta_0 = sqrt(#mu0/#eps0)

** Parameters for segmentation
#seg_ln = #lam/40     ** nominal length of wire segments
IP                             #seg_rad            #seg_ln
#num_sg2 = ceil(#h/#seg_ln)      ** segments on each dipole half (excl. source)
#num_sg = 2*#num_sg2+1           ** to ensure odd number of overall segments
#delta = 2*#h/#num_sg            ** actual length of wire segments
** Geometry of radiating structure
** Has to be constructed with two loops and a special source segment, since a
   separate label
** is required for each segment

** Construct center (source) segment
#lab = #num_sg
LA   #lab
DP   A                         0.0        0.0        -#delta/2
DP   B                         0.0        0.0        #delta/2
BL   A    B
** Construct upper half
#ell1 = #delta/2
#ell2 = #delta/2+#delta
!!for #ii = 1 to #num_sg2
#lab = #ii
LA   #lab
DP   A                         0.0        0.0        #ell1
DP   B                         0.0        0.0        #ell2
BL   A    B
#ell1 = #ell1+#delta
#ell2 = #ell2+#delta
!!next
.
.
.
```

Figure 5.20 PREFEKO file for the Wu–King loaded dipole, geometry.

```
** Construct lower half
#ell1 = -#delta/2
#ell2 = -#delta/2-#delta
!!for #ii = 1 to #num_sg2
#lab = #ii+#num_sg2
LA   #lab
DP   A                            0.0       0.0       #ell1
DP   B                            0.0       0.0       #ell2
BL   A    B
#ell1 = #ell1-#delta
#ell2 = #ell2-#delta
!!next
** End of geometric input
EG   1    0    0    0    0
```

Figure 5.20 (*cont.*)

```
.
.
.
** Load the structure - again, a loop structure is used.
** Load source segment
#load = #psi_0*#eta_0/(8*#pi*#h)
#lab = #num_sg
LD   #lab                         #load  ** Loss
** Upper half and lower half at same time now:
!!for #ii = 1 to #num_sg2
#z = #ii*#delta
#load = #psi_0*#eta_0/(8*#pi*#h*(1-#z/#h))
#lab = #ii
LD   #lab                         #load  ** Loss
#lab = #ii+#num_sg2
LD   #lab                         #load  ** Loss
!!next

** Set the frequency
FR   41   0                       #f_l                #f_u
** Voltage gap excitation at a segment
#lab = #num_sg
A1   0    #lab                    1.0
** Calculate surface currents for current display
OS   1         1
EN
```

Figure 5.21 PREFEKO file for the Wu–King loaded dipole, loading.

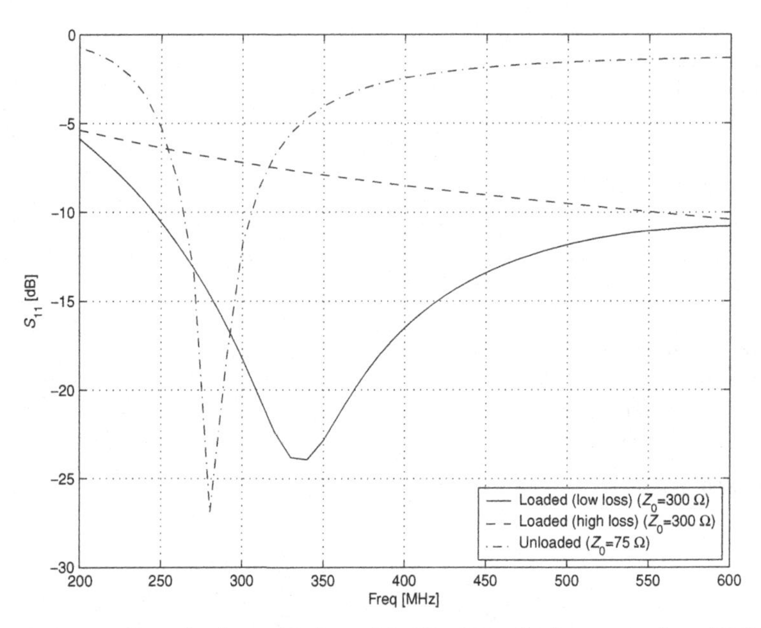

Figure 5.22 The reflection coefficient of the Wu–King dipole compared to a PEC dipole.

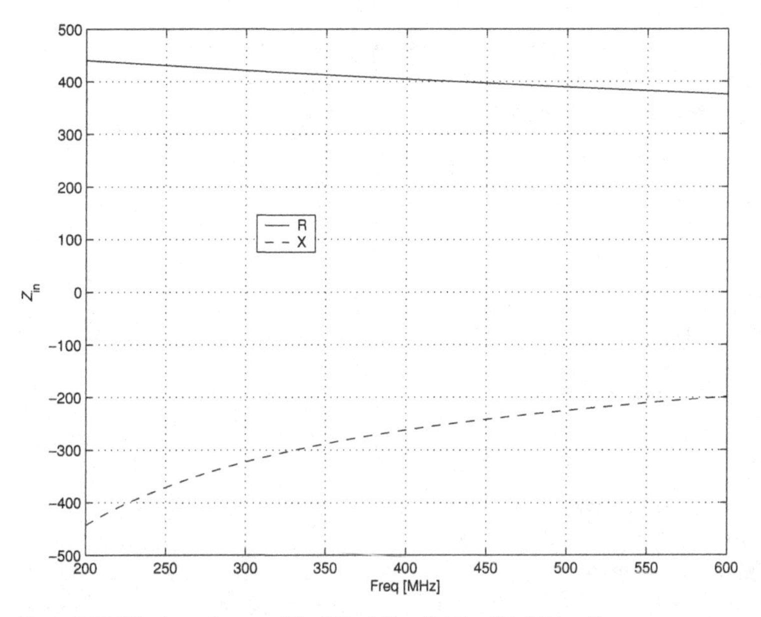

Figure 5.23 The impedance of the Wu–King dipole, "high" loading.

the reflection coefficient (for the PEC dipole, $Z_0 = 75\ \Omega$ was used). The loaded dipole clearly has a *much* larger impedance bandwidth, and is indeed a wideband antenna. The rather poor result for the higher loading is due to a large, but slowly varying, reactive component, as shown in Fig. 5.23; this could be removed by adding a tuning component in an actual application, but this has not been done here. Figure 5.24 shows the

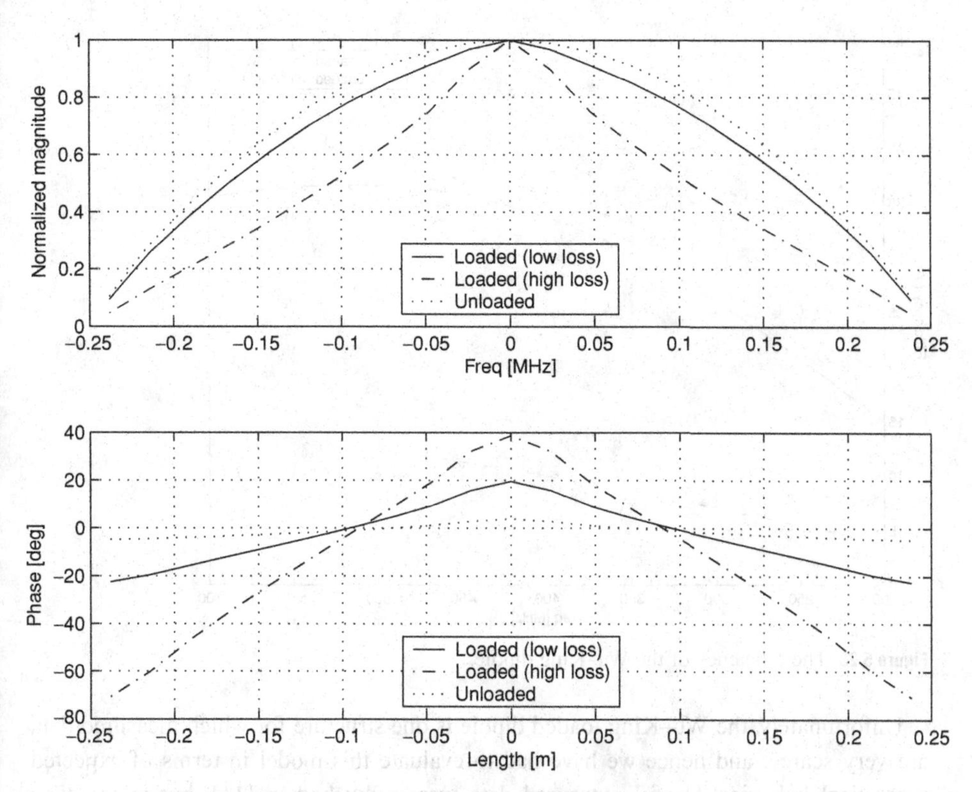

Figure 5.24 The current (normalized magnitude and phase) on the Wu–King and PEC dipoles.

current distributions along both loaded and unloaded dipoles at 280 MHz, the resonant frequency of the unloaded dipole. (The magnitudes have been normalized; the higher impedance of the loaded dipole results of course in smaller values of current.) The loaded dipole with the higher loading clearly supports a travelling wave, with a phase difference along the dipole arm of a little more than the 90° predicted by Eq. (5.9) for $h \approx \lambda_0/4$, and with an almost linear current distribution, also as predicted. The phase for the unloaded dipole is almost constant, as one would expect from a standing wave distribution. The results for the lower loading are somewhere in between the pure standing wave of the unloaded dipole and the pure travelling wave of the dipole with higher loading.

The wide bandwidth is, however, bought at a price: efficiency. Wu and King originally predicted a theoretical efficiency of 50% for $h = \lambda_0/4$, but FEKO shows a *much* lower efficiency of around 7% at 300 MHz (Fig. 5.25). In a subsequent correction [10], Wu and King drastically revised their calculation, predicting a very similar value to the FEKO computation. The result for the lower loading case is around 23%, rather better. Interestingly and serendipitously, the original (incorrect) result by Wu and King provides generally better antenna performance, certainly in terms of reflection coefficient and efficiency, even though the current is not a pure travelling wave.

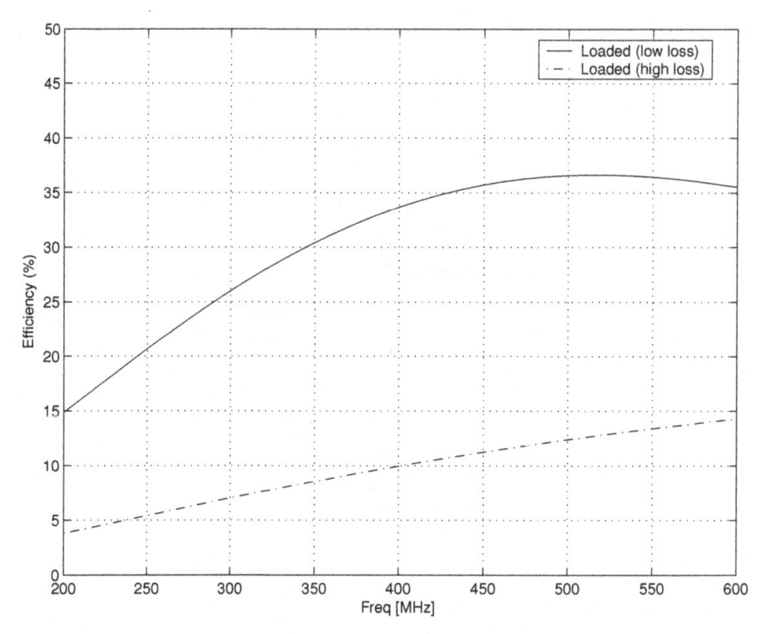

Figure 5.25 The efficiency of the Wu–King dipole.

Unfortunately, the Wu–King loaded dipole is one structure for which measured data are very scarce, and hence we have had to evaluate this model in terms of expected theoretical behavior. Useful measured data were published in [11], but in the *time domain*. (Although FEKO has a time domain option, it is only available for scattering problems.)

Before leaving this structure, a fundamental point should be noted about wide-band antennas. The definition of this is inherently a frequency domain concept, and one should be careful to differentiate between a wideband antenna on the one hand, and a *non-dispersive* antenna on the other. The former type of antenna works well over a wide range of operating frequencies; the latter can radiate actual time domain pulses without distortion (obviously, it will also be wideband). A little thought about this from the viewpoint of the Fourier transform shows that this translates to requirements on not just constant magnitude response, but also phase linearity. Many wideband antennas (such as spirals and log-periodics) are dispersive because different frequencies radiate from different parts of the structure. We will not pursue this further, but will mention in closing that the loaded dipole exhibits limited dispersion, and because of this is widely used in time domain antenna systems despite its low efficiency.

Code tips – modelling this structure in NEC

The `LD` card provides the same functionality as in FEKO, but the absence of user-defined variables in NEC means that one will have to compute the loading manually at each segment, so this will be a tedious structure to model in NEC.

5.7 Conclusions

In this chapter, we have discussed modelling thin-wire antennas using FEKO and NEC-2. Starting with a very simple dipole example, we progressed to more complex antennas, including Yagi–Uda and log-periodic dipole antennas, an axial mode helix and a loaded dipole. The helix example also introduced the use of *surface* modelling. We highlighted a number of points which one must be careful with; perhaps the most crucial is to check that the solution is converged (but also not over-converged, due to the limits of the thin-wire approximation). We also emphasized the importance of validation, that is, checking computed results in some way. Historically, comparison to measured data or an analytical solution has been the most convincing method of validation. Nowadays, comparisons with data computed using other codes and/or formulations are increasingly widely used and accepted, and we have directly compared FEKO and NEC-2 results on several occasions, noting that one cannot expect exact agreement. (It has also been commented in this context that measured data must also be used with discretion.) A number of features supported by FEKO (but not NEC-2) which simplify antenna modelling were introduced, including iteration and conditional execution. Several other FEKO and NEC-2 features were also discussed, including the use of labels/tags, transmission lines, and various types of grounds. We also took the opportunity afforded by numerical simulation to improve an antenna design, by adding a terminating resistance to a log-periodic antenna and evaluating the change in antenna performance.

Properly used, within its region of validity, we have seen that the thin-wire formulation is both accurate and very efficient computationally. Having completed this chapter, the reader should feel far more confident in modelling a wide range of wire antennas using tools such as FEKO and NEC-2.

During this chapter, we very briefly touched on the modelling of surfaces in Section 5.5. This is an important part of many antenna designs – and also for scattering problems – and in the next chapter we will comprehensively discuss the modelling of surfaces and volumes using the MoM, as well as the attendant problems of high computational cost.

References

[1] G. J. Burke and A. J. Poggio, "Numerical electromagnetics code (NEC) – method of moments; Part III: User's guide." Lawrence Livermore National Laboratory, CA, UCID 18834, January 1981.

[2] D. B. Davidson, "Parallel processing revisited: a second tutorial," *IEEE Antennas Propagat. Mag.*, **34**, 9–21, October 1992.

[3] J. J. H. Wang, *Generalized Moment Methods in Electromagnetics*. New York: Wiley, 1991.

[4] D. J. Janse van Rensburg and D. A. McNamara, "On quasi-static source models for wire dipole antennas," *Microwave Optical Technol. Lett.*, **3**, 396–398, November 1990.

[5] W. L. Stutzman and G. A. Thiele, *Antenna Theory and Design*. New York: Wiley, 2nd edn., 1998.

[6] C. A. Balanis, *Antenna Theory: Analysis and Design*. New York: Wiley, 2nd edn., 1997.
[7] J. D. Kraus and R. J. Marhefka, *Antennas for All Applications*. Boston, MA: McGraw-Hill, 3rd edn., 2002.
[8] A. C. Ludwig, "Wire grid modelling of surfaces," *IEEE Trans. Antennas Propagat.*, **35**, 1045–1048, September 1987.
[9] T. T. Wu and R. W. P. King, "The cylindrical antenna with nonreflecting resistive loading," *IEEE Trans. Antennas Propagat.*, **13**, 369–373, May 1965.
[10] T. T. Wu and R. W. P. King, "Correction: the cylindrical antenna with nonreflecting resistive loading," *IEEE Trans. Antennas Propagat.*, **13**, 998, November 1965.
[11] J. G. Maloney and G. S. Smith, "A study of transient radiation from the Wu-King resistive monopole – FDTD analysis and experimental measurements," *IEEE Trans. Antennas Propagat.*, **41**, 668–676, May 1993.

6 The method of moments for surface modelling

The helix antenna discussed in the previous chapter used a new type of element to model *surfaces*. The theory underlying this is described in this chapter. The basic theory is quite complex, and general implementations are especially challenging. However, by choosing a suitable problem, it proves possible to undertake a limited implementation of a three-dimensional scattering problem, using a basis function defined on a triangular patch known as the RWG element. This is named after Rao, Wilton and Glisson, who introduced the element in their classic 1982 paper [1]. It represented a new type of element, the *vector* or *edge-based* element, and a closely related class of element was also under development for finite element applications at that time, although it would be some years before the connection was fully appreciated. (This will be pursued in more detail in the later coverage of the FEM.) The RWG element underlies the surface treatment of modern codes such FEKO (although not NEC), and some examples of using existing codes (in particular FEKO) to compute scattering from more general surfaces will further illustrate this.

We will also see that not only can perfectly (or highly) conducting structures be efficiently modelled using surface currents, but also homogeneous dielectric and/or magnetic regions, using fictitious equivalent currents. (We will even briefly describe how *in*homogeneous bodies can be modelled using volumetric currents, but note at the outset that this is not one of the strong points of the MoM.) Modelling surfaces is far more computationally expensive than modelling wires, and some methods for reducing the computational cost will also be discussed. These include a hybrid of the MoM and physical optics, and the general class of fast methods, including both those based on the FFT and the fast multipole method. We will also discuss the use of parallel processing.

6.1 Electric and magnetic field integral equations

Following the same lines as the Pocklington equation (Chapter 4), integral equations in either the magnetic or electric fields can be derived for problems with currents flowing on surfaces. In this overview section, only the results will be presented; subsequently, an electric field integral equation will be derived in detail. One integral equation couples the incident electric field to the induced surface current, and is known as the electric field integral equation (EFIE):

$$\hat{n} \times \boldsymbol{E}^{\mathrm{inc}}(\boldsymbol{r}) = \hat{n} \times \int_S \left[jk\eta \boldsymbol{J}_S(\boldsymbol{r}')G(\boldsymbol{r}, \boldsymbol{r}') + \frac{\eta}{jk}\{\nabla'_s \cdot \boldsymbol{J}_S(\boldsymbol{r}')\}\nabla' G(\boldsymbol{r}, \boldsymbol{r}') \right] dS', \qquad \forall\, \boldsymbol{r}, \boldsymbol{r}' \in S \tag{6.1}$$

The ∇' operator implies differentiation in the *source* coordinates. $\hat{n}$ is the unit vector on the surface S. $G(\boldsymbol{r}, \boldsymbol{r}')$ is the scalar free-space Green function given by

$$G(\boldsymbol{r}, \boldsymbol{r}') = \frac{e^{-jkR}}{4\pi R} \tag{6.2}$$

$$R = |\boldsymbol{r} - \boldsymbol{r}'| \tag{6.3}$$

Equation (6.1) is valid for both closed and open surfaces. In the latter case, $\boldsymbol{J}_S$ is the sum of surface currents on both sides of the sheet.

There are other, related forms of the EFIE. We will shortly study the EFIE mixed potential integral equation – MPIE, which explicitly retains charge as an unknown. From the continuity equation,[1] charge is of course connected to current, and this is exploited in the MPIE formulation. We will also see an example of the MPIE in Chapter 7.

The other integral equation couples the incident magnetic field to the induced surface current, and is known as the magnetic field integral equation (MFIE):

$$\frac{1}{2}\boldsymbol{J}_S(\boldsymbol{r}) = \hat{n} \times \boldsymbol{H}^{\mathrm{inc}}(\boldsymbol{r}) + \hat{n} \times \oint_S \boldsymbol{J}_S(\boldsymbol{r}') \times \nabla' G(\boldsymbol{r}, \boldsymbol{r}')dS', \qquad \forall\, \boldsymbol{r}, \boldsymbol{r}' \in S \tag{6.4}$$

This is valid *only* for closed surfaces. (The reason was outlined at the end of Section 4.6.) It is interesting to note that if we neglect the surface integral, what remains is the physical optics approximation, $\boldsymbol{J}_S(\boldsymbol{r}) = 2\hat{n} \times H^{\mathrm{inc}}(\boldsymbol{r})$, of which more later.

The integrals in the above should be interpreted as the principal value of the integral. (The principal value of an integral with a singularity at $\boldsymbol{r}_0$ is essentially the value of the integral with a δ neighborhood around $\boldsymbol{r}_0$ removed; then the limit as $\delta \to 0$ is found.) In both these equations, the presence of singularities raises delicate issues and requires careful treatment. The simple expedient of slightly offsetting field and source points as was done with the one-dimensional wire problem (in that case, by treating the source as a filament on the wire axis, but still imposing the boundary condition on the surface of the wire) can still be done, although in this case one offsets the quadrature points corresponding to source and field points rather than concentrating the source elsewhere. As will be seen, surprisingly good results can be obtained with such an off-set point treatment.

Mathematically, the EFIE is a *Fredholm integral equation of the first kind* – the unknown is present only in the kernel. The MFIE is a Fredholm integral equation of the *second kind* – the unknown is present both inside *and* outside the kernel. The reason for the difference is due to the boundary condition. The EFIE and MFIE are both derived

[1] The time rate of change of charge is the negative of the divergence of the current.

from the Statton–Chu formula, which states that for points *on* the surface, the following relations hold [2, p. 172]:

$$E^{\text{inc}}(r) + \text{PV}\int_S E_s\, dS' = \frac{1}{2}E(r)$$
$$H^{\text{inc}}(r) + \text{PV}\int_S H_s\, dS' = \frac{1}{2}H(r) \qquad (6.5)$$

The PV here reminds us that these are the principal values of the relevant integrals. In the equations above, E_s and H_s are not directly the scattered fields, but rather the kernels which are integrated to obtain these (the full expressions may be found in [2, p. 173]). For a PEC, the boundary condition on $\hat{n} \times E$, i.e. tangential E, is of course zero, whereas $\hat{n} \times H$ is the surface current J_s; hence the different nature of the two integral equations. For more details, see [2, Section 12.3].

Mathematically, it is well known that Fredholm type two integral equations are generally more *well posed* – this motivated much work using the MFIE. (Put simply, a well-posed problem is one whose solution is not strongly dependent on the physics and geometry of the problem.) However, the requirement for a *closed* surface S is frequently a problem in applied CEM work, with the result that the EFIE is usually preferred in practical codes. Finally, linear combinations of the EFIE and MFIE have also been used; not surprisingly, this method is known as the combined field integral equation (CFIE). The CFIE will not be discussed here.

Because the EFIE and MFIE are both quite complex, it is convenient to introduce a simplifying notation. As an example, for the EFIE, the right-hand side of Eq. (6.1), which represents the scattered field, is often written in the following shorthand:

$$\mathbf{E}_s = \mathcal{L}_J^E \mathbf{J}^{MM}$$

$\mathcal{L}$, which represents all the mathematical operations to be performed on the current J, is known as an *operator* – it is an extension of the concept of a function.

A mathematical aside – functions, functionals and operators

A function, of course, maps a number (integer, real or complex) to another number; a functional maps a function to a number; and an operator maps a function to another function. (We will encounter functionals in Chapter 9.) The Fourier transform is a commonly encountered example of an operator: it maps a function of time to a function of frequency (or more generally, from one domain to the corresponding spectral domain). Operator notation will be used subsequently in this chapter.

6.2 The Rao–Wilton–Glisson (RWG) element

When dealing with surfaces using the MoM, two matters need attention. The first is that we need to split the geometry up into small elements. The simplest approach, and the first one explored historically in codes such as NEC-2, was to use square

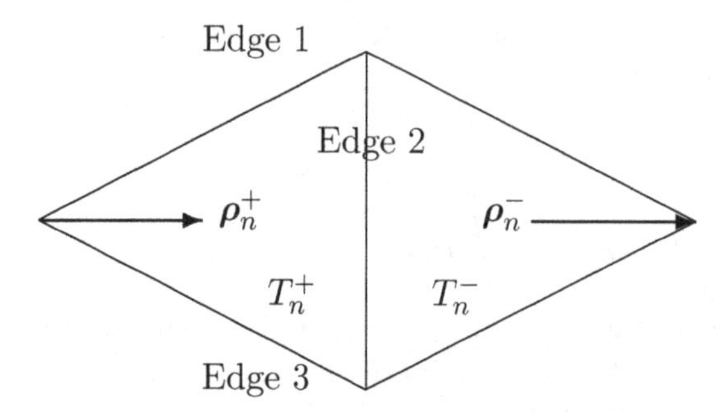

Figure 6.1 The two connected triangles T_n^+ and T_n^-, sharing a common edge, which support a RWG basis function.

(or rectangular) patches. However, for general two-dimensional geometries, triangular elements are better for approximating the geometry, and this is the approach which most modern codes (including FEKO) use. The second matter is that the physical parameter being approximated, $\boldsymbol{J}_S$, is now two dimensional. The basis function must also incorporate this.

In this context, a very widely used basis function for the triangular patch was introduced by Rao, Wilton and Glisson in 1982 [1]. The basis function is often known simply as the RWG element. Subsequent work led to the realization that this basis function is very closely related to the edge-based elements widely used in contemporary finite element analysis. We will return to this later in this book when we address finite elements.

The basis function includes some new features which have not yet been encountered in this book. Most importantly, the basis function is *vector* in nature, which means that the individual scalar components (e.g. J_x, J_y and J_z), can only be recovered with some manipulation. The essential idea is to enforce current continuity over an *edge* of a patch. The interpolation function used to achieve this is the following:

$$\boldsymbol{f}_n(\boldsymbol{r}) = \begin{cases} \dfrac{l_n}{2A_n^+}\,\boldsymbol{\rho}_n^+ & \forall \boldsymbol{r} \text{ in } T_n^+ \\ \dfrac{l_n}{2A_n^-}\,\boldsymbol{\rho}_n^- & \forall \boldsymbol{r} \text{ in } T_n^- \\ 0 & \text{otherwise} \end{cases} \tag{6.6}$$

Figure 6.1 defines the vectors $\boldsymbol{\rho}_n^+$ and $\boldsymbol{\rho}_n^-$. It is important to note these position vectors are defined *with respect to the relevant free vertex*, rather than a global origin. Note that the basis function is defined over two adjoining triangles T_n^+ and T_n^-, which share a common edge. A_n^+ is the area of triangle T_n^+ (and similarly A_n^-). l_n is the length of the shared edge. The vector $\boldsymbol{\rho}_n^+$ is the vector position within triangle T_n^+, with the left-hand node of T_n^+ as origin; similarly, $\boldsymbol{\rho}_n^-$ is the (negative of the) vector position, with the right-hand node of T_n^- as origin. There exists a coordinate system known as *simplex coordinates* which makes the study of interpolation functions on triangles much simpler, and is widely used in finite element analysis; these vectors can be written rather simply in that coordinate system, a topic studied in some detail in Chapter 10. The terms $l_n/2A_n^-$

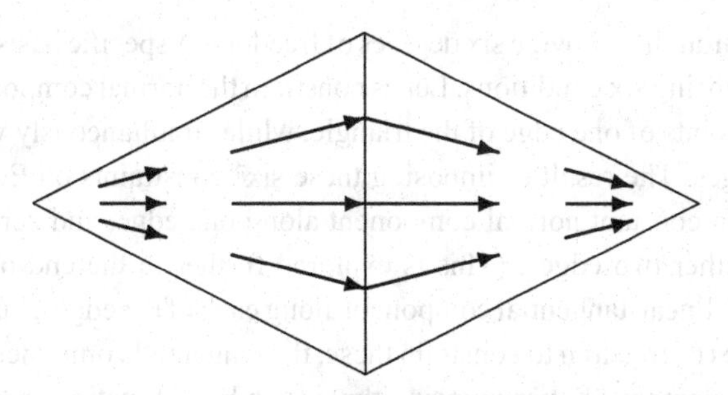

Figure 6.2 A vector plot of the RWG basis functions.

and $l_n/2A_n^+$ are normalizing constants. The convention of [1] regarding subscripts and superscripts is used here; the former refer to edges, and the latter to faces.

The resulting current interpolation is shown in Fig. 6.2. The following points may not be immediately apparent. Firstly, it should be noted that this basis function has *no* component normal to the upper or lower sides of either of the triangles, but only to the central (shared) edge. Without more detailed theoretical analysis, the following is not obvious, and is stated without proof here:[2] the current crossing this shared edge is linearly interpolated in the tangential direction (i.e. along the edge) and interpolated as a constant normal to (i.e. across) the edge. This latter value is usually the "degree of freedom" (the unknown value of current) which is associated with this basis function; the current associated with this edge is thus approximated as $\boldsymbol{J}_n(\boldsymbol{r}) \approx I_n \boldsymbol{f}_n(\boldsymbol{r})$. Note that all these terms are expressed in terms of the local coordinates on the triangle; again, the conversion to Cartesian coordinates is most readily performed using simplex coordinates.

What of the current flowing across the two *other* edges? To approximate these, one defines *additional* basis functions on each of the other two connected triangles; thus on any one triangle, there are *three* such basis functions, with three associated unknowns, which are the normal components of current on each edge. Within the element, it should be appreciated that the *total* current is thus approximated by the sum of these three basis functions. With the edges numbered as on Fig. 6.1, the total current on triangle T_n^+ is given by

$$\boldsymbol{J}(\boldsymbol{r}) \approx I_1 \boldsymbol{f}_1(\boldsymbol{r}) + I_2 \boldsymbol{f}_2(\boldsymbol{r}) + I_3 \boldsymbol{f}_3(\boldsymbol{r}), \qquad \forall \boldsymbol{r} \text{ in } T_n^+ \tag{6.7}$$

At the risk of repetition, note that the basis functions carry the vector information; the unknowns for which the code solves (I_1, I_2, etc.) are just complex-valued scalars.

There are other methods of deriving the RWG element which provide additional insight. One is the following [3, Section 9.13]. Consider a vector basis function for a triangular element in the xy-plane, expressed as a linear polynomial:

$$\boldsymbol{B}(x, y) = (A + Bx + Cy)\hat{x} + (D + Ex + Fy)\hat{y} \tag{6.8}$$

[2] Again, because this RWG basis function is so intimately related to the edge-based Whitney function of finite element analysis, we postpone detailed mathematical analysis of this class of element until Chapter 10.

The six coefficients A through F provide six degrees of freedom. A specific basis function can be generated by imposing six conditions. Let us constrain the normal component of $\boldsymbol{B}$ to be unity at both endpoints of one edge of the triangle, while simultaneously vanishing along the other two edges. The result of imposing these six[3] constraints on Eq. (6.8) is a vector function having constant normal component along one edge, and zero normal component along the other two edges. (This is explored further in the end-of-chapter problem.) There is also a linear-tangential component along each of the edges, but as there are no additional degrees of freedom to constrain these, the tangential component may be discontinuous between elements. Superimposing three such basis functions provides an approximating function with continuous normal components across element boundaries, but potentially discontinuous tangential components. The degrees of freedom are then associated with the normal component at the appropriate edge.

In passing, it worth noting that basis functions of this type, which have finite divergence, continuous normal and possibly discontinuous tangential components are known as *divergence-conforming* elements. In later chapters on the finite element method, *curl-conforming* elements will be studied, which are closely related. The RWG elements are also known as constant normal/linear tangential (CN/LT) elements; again, in the later context of finite elements, constant tangential/linear normal (CT/LN) elements will be encountered.

6.3 A mixed potential electric field integral equation for electromagnetic scattering by surfaces of arbitrary shape

In this section, we will formulate a "mixed potential" electric field integral equation for electromagnetic scattering by surfaces of arbitrary shape, which we will then proceed to solve with a Galerkin formulation, using RWG elements as introduced in the preceding section. Our formulation and implementation directly parallels the original Rao *et al.* reference [1]. A Galerkin formulation can be computationally costly, due to the integration required over elements containing both source and field points. As proposed in [1], we will use approximations to the integration to reduce computation time.

6.3.1 The electric field integral equation (EFIE)

The EFIE was given without derivation earlier in this chapter, and it is useful to develop it from first principles in this case. We will follow the development, and largely the notation, of [1] very closely in the following. Here, S will be the surface of an open or closed perfectly conducting scatterer, with unit normal $\hat{n}$. $\boldsymbol{E}^{\text{inc}}$ is the incident field, which induces currents on surface S. As already noted, if a thin, open surface is being modelled, $\boldsymbol{J}$ is the vector sum of the surface currents on opposite sides of the scattering surface S. Hence, the normal component of $\boldsymbol{J}$ must vanish on boundaries of S, a property we will

[3] Constraining the basis function to having only a normal component at each node, which is additinally constrained to unity, constitutes two conditions.

implicitly exploit in the MoM solution. As before, the scattered electric field $\boldsymbol{E}^{\text{scat}}$ can be computed from the surface current $\boldsymbol{J}$ and surface charge σ from

$$\boldsymbol{E}^{\text{scat}} = -j\omega \boldsymbol{A} - \nabla\Phi \tag{6.9}$$

The magnetic vector potential is defined by

$$\boldsymbol{A}(\boldsymbol{r}) = \frac{\mu}{4\pi}\int_S \boldsymbol{J}\frac{e^{-jkR}}{R}dS' \tag{6.10}$$

and the scalar potential as

$$\Phi(\boldsymbol{r}) = \frac{1}{4\pi\epsilon}\int_S \sigma\frac{e^{-jkR}}{R}dS' \tag{6.11}$$

As usual, $k = \omega\sqrt{\mu\epsilon} = 2\pi/\lambda$, with λ the wavelength in the homogeneous region exterior to S with permeability μ and permittivity ϵ. (Usually, the formulation is applied in free space, $\mu = \mu_0$ and $\epsilon = \epsilon_0$, but this is not required.) $R = |\boldsymbol{r} - \boldsymbol{r}'|$ is the distance between an arbitrarily located observation point $\boldsymbol{r}$ and a source point $\boldsymbol{r}'$ on S. Both $\boldsymbol{r}$ and $\boldsymbol{r}'$ are defined with respect to a global coordinate origin $\mathcal{O}$.

The surface charge density σ is related to the surface divergence of $\boldsymbol{J}$ (the divergence in the plane of the surface) through current continuity:

$$\nabla_s \cdot \boldsymbol{J} = -j\omega\sigma \tag{6.12}$$

An integro-differential equation for $\boldsymbol{J}$ (and σ, related as above) is obtained by enforcing the boundary condition on the total field, $\hat{n} \times (\boldsymbol{E}^{\text{inc}} + \boldsymbol{E}^{\text{scat}}) = 0$ on S, resulting in

$$-\boldsymbol{E}^{\text{inc}}_{\text{tan}} = (-j\omega\boldsymbol{A} - \nabla\Phi)_{\text{tan}}, \quad \boldsymbol{r} \text{ on } S \tag{6.13}$$

Equation (6.13), together with Eqs. (6.10)–(6.12), constitute the mixed potential electric field integral equation. The presence of derivatives on the current in Eq. (6.12), and on the scalar potential in Eq. (6.13), will require care with the selection of basis and testing functions in the MoM development.

6.3.2 The RWG basis function revisited

The mixed-order basis function defined in Eq. (6.6) is the basis of the MoM formulation; as a Galerkin formulation will be used, the RWG basis function will serve as both basis and testing functions. In Eq. (6.12), the (surface) divergence of the current, and hence basis function, must be computed. The surface charge density associated with the RWG basis function is

$$\nabla_s \cdot \boldsymbol{f}_n = \begin{cases} \dfrac{l_n}{A_n^+} & \boldsymbol{r} \text{ in } T_n^+ \\ \dfrac{l_n}{A_n^-} & \boldsymbol{r} \text{ in } T_n^- \\ 0 & \text{otherwise} \end{cases} \tag{6.14}$$

This is not an obvious result. It may be derived using simplex coordinates (to be discussed in Chapter 10); see [3, Section 9.13] for the derivation.

Equation (6.14) shows that the charge density is constant within each triangle (which is to be expected, given that the current is approximated to mixed first order, and that the linear terms absent from the RWG element are precisely those which make no contribution to the divergence thereof). The charge is thus approximated by a *pulse doublet*. The charge will be discontinuous from element to element, but this will not pose problems in the MoM formulation.

6.3.3 The MoM formulation

As already discussed, a Galerkin formulation is adopted. The surface current is approximated as

$$\boldsymbol{J} \approx \sum_{n=1}^{N} I_n \boldsymbol{f}_n(\boldsymbol{r}) \tag{6.15}$$

with N the number of interior edges. The basis functions $\boldsymbol{f}_n$, $n = 1, 2, \ldots, N$ are as defined in Eq. (6.6). As noted, the normal component of $\boldsymbol{J}$ is zero, so on open edges no basis functions need be – or indeed are – defined. Importantly, with the basis function as chosen, *the coefficients I_n in Eq.* (6.15) *give directly the normal component of current flowing over the nth edge*. The testing functions are the same RWG functions of Eq. (6.6), $\boldsymbol{f}_m$. Note that we use n and m to distinguish between basis functions (source points) and testing functions (field points) respectively, as is usual MoM practice.

The symmetric product, already discussed in Chapter 4, is defined as usual as

$$< \boldsymbol{f}, \boldsymbol{g} >= \int_S \boldsymbol{f} \cdot \boldsymbol{g} dS \tag{6.16}$$

Equation (6.13) is thus tested with $\boldsymbol{f}_m$, $m = 1, 2, \ldots, N$, yielding

$$< \boldsymbol{E}^{\text{inc}}, \boldsymbol{f}_m >= j\omega < \boldsymbol{A}, \boldsymbol{f}_m > + < \nabla\Phi, \boldsymbol{f}_m > \tag{6.17}$$

Because the expansion for the surface charge σ in Φ is the pulse doublet of Eq. (6.14), discussed in the preceding section, applying the gradient operator directly to σ is not advisable (it results in Dirac delta functions). Instead, an approach often used in both MoM and FEM formulations will be adopted, which is to move a differential operator from the source term to the testing function. Using a surface vector calculus identity, the last term in Eq. (6.17) can be rewritten as

$$< \nabla\Phi, \boldsymbol{f}_m >= -\int_S \Phi \nabla_S \cdot \boldsymbol{f}_m dS \tag{6.18}$$

Thus the differential is moved to the testing function, which has appropriate first-order terms to yield a finite result. (The manipulation of the surface divergence in Eq. (6.18) involves some points not immediately obvious; details of this type of operation may be found in Chapter 7.)

With Eq. (6.14), the integral in Eq. (6.18) may be written and approximated as follows:

$$\int_S \Phi \nabla_S \cdot \boldsymbol{f}_m dS = \ell_m \left(\frac{1}{A_m^+} \int_{T_m^+} \Phi dS - \frac{1}{A_m^-} \int_{T_m^-} \Phi dS \right) \approx \ell_m \left[\Phi(\boldsymbol{r}_m^{c+}) - \Phi(\boldsymbol{r}_m^{c-}) \right] \tag{6.19}$$

In the above, the average of Φ over each triangle is approximated by the value of Φ at the triangle centroid; $c+$ and $c-$ refer to the centroids of the $+$ and $-$ triangles respectively. Similar approximations are applied to the vector potential and incident field terms in Eq. (6.17), yielding

$$\left\langle \left\{ \begin{matrix} \boldsymbol{E}^{\text{inc}} \\ \boldsymbol{A} \end{matrix} \right\}, \boldsymbol{f}_m \right\rangle = \ell_m \left[\frac{1}{2A_m^+} \int_{T_m^+} \left\{ \begin{matrix} \boldsymbol{E}^{\text{inc}} \\ \boldsymbol{A} \end{matrix} \right\} \cdot \boldsymbol{\rho}_m^+ dS + \frac{1}{2A_m^-} \int_{T_m^-} \left\{ \begin{matrix} \boldsymbol{E}^{\text{inc}} \\ \boldsymbol{A} \end{matrix} \right\} \cdot \boldsymbol{\rho}_m^- dS \right] \approx \frac{\ell_m}{2} \left[\left\{ \begin{matrix} \boldsymbol{E}^{\text{inc}}(\boldsymbol{r}_m^{c^+}) \\ \boldsymbol{A}(\boldsymbol{r}_m^{c^+}) \end{matrix} \right\} \cdot \boldsymbol{\rho}_m^{c+} + \left\{ \begin{matrix} \boldsymbol{E}^{\text{inc}}(\boldsymbol{r}_m^{c^-}) \\ \boldsymbol{A}(\boldsymbol{r}_m^{c^-}) \end{matrix} \right\} \cdot \boldsymbol{\rho}_m^{c-} \right] \tag{6.20}$$

where the integral over each triangle has been eliminated by approximating $\boldsymbol{E}^{\text{inc}}$ (or $\boldsymbol{A}$) by its value at the triangle centroid. As in Eq. (6.6), $\rho_m^{\pm}$ refer to the position vector with reference to the free vertex in triangles $+$ or $-$. With Eqs. (6.18)–(6.20), Eq. (6.17) now becomes

$$j\omega\ell_m \left[\boldsymbol{A}(r_m^{c^+}) \cdot \frac{\boldsymbol{\rho}_m^{c+}}{2} + \boldsymbol{A}(r_m^{c^-}) \cdot \frac{\boldsymbol{\rho}_m^{c-}}{2} \right] + \ell_m \left[\Phi(r_m^{c^-}) - \Phi(r_m^{c^+}) \right] = \ell_m \left[\boldsymbol{E}^{\text{inc}}(r_m^{c^+}) \cdot \frac{\boldsymbol{\rho}_m^{c+}}{2} + \boldsymbol{E}^{\text{inc}}(r_m^{c^-}) \cdot \frac{\boldsymbol{\rho}_m^{c-}}{2} \right] \tag{6.21}$$

Approximating the integration over the testing element (triangles $m+$ and $m-$) as above does of course result in an approximate implementation of the Galerkin formulation. The movitation is as follows. The surface integrals must be computed numerically, using quadrature. Integrating carefully only over the source triangles (to be discussed) as in this approximate implementation already results in significant computational cost; were the integrations to be performed over both source and testing elements, the computational cost would be far higher. For many applications, this tradeoff of accuracy versus computational cost (and hence runtime) is acceptable, but we will return to this issue later. One major deficiency of the approximation that should be noted is that, in general, this approach does not yield a symmetrical impedance matrix, which a Galerkin formulation should.[4]

This approximation also admits another interpretation, also mentioned in the original paper. One can define a "razor-blade" testing function, along the line connecting triangle centroids; approximating that integral with the integrands evaluated at the centroids,

[4] If the mesh has certain symmetry properties, the impedance matrix produced by this appoximation may happen to be symmetric, but this is not true in general.

the result is identical (within a constant). This interpretation has been explored in other work on the topic.

6.3.4 Derivation of the matrix entries

Substituting the current expansion of Eq. (6.15) into Eq. (6.21) yields the usual system of $N \times N$ linear equations of an MoM formulation. (Note that in Section 4.5, and indeed more commonly in MoM formulations, the current was expanded, substituted into the linear operator, and only then was the weighting function defined, but since the operator is linear, the order of discretization is actually immaterial.) As usual – see Eqs. (4.11) and (4.56) – this is written as

$$\{V\} = [Z]\{I\} \tag{6.22}$$

where $[Z]$ is an $N \times N$ matrix, and $\{V\}$ and $\{I\}$ are column vectors of length N. Elements of $[Z]$ and $\{V\}$ are given by

$$Z_{mn} = \ell_m \left[j\omega \left(\boldsymbol{A}^+_{mn} \cdot \frac{\boldsymbol{\rho}^{c+}_m}{2} + \boldsymbol{A}^-_{mn} \cdot \frac{\boldsymbol{\rho}^{c-}_m}{2} \right) + \Phi^-_{mn} - \Phi^+_{mn} \right] \tag{6.23}$$

$$V_m = \ell_m \left(\boldsymbol{E}^+_m \cdot \frac{\boldsymbol{\rho}^{c+}_m}{2} + \boldsymbol{E}^-_m \cdot \frac{\boldsymbol{\rho}^{c-}_m}{2} \right) \tag{6.24}$$

where

$$\boldsymbol{A}^\pm_{mn} = \frac{\mu}{4\pi} \int_S \boldsymbol{f}_n(\boldsymbol{r}') \frac{e^{-jkR^\pm_m}}{R^\pm_m} dS' \tag{6.25}$$

$$\Phi^\pm_{mn} = -\frac{1}{4\pi j\omega\epsilon} \int_S \nabla'_s \cdot \boldsymbol{f}_n(\boldsymbol{r}') \frac{e^{-jkR^\pm_m}}{R^\pm_m} dS' \tag{6.26}$$

$$R^\pm_m = |\boldsymbol{r}^{c\pm}_m - \boldsymbol{r}'| \tag{6.27}$$

and

$$\boldsymbol{E}^\pm_m = \boldsymbol{E}^{\text{inc}}(\boldsymbol{r}^{c\pm}_m) \tag{6.28}$$

Note that the signs on the scalar potential in Eqs. (6.23) and (6.26) derive from Eq. (6.18).

For plane wave incidence, one sets

$$\boldsymbol{E}^{\text{inc}}(\boldsymbol{r}) = (E_\theta \hat{\theta}_0 + E_\phi \hat{\phi}_0) e^{j\boldsymbol{k}\cdot\boldsymbol{r}} \tag{6.29}$$

with the propagation vector $\boldsymbol{k}$ given by

$$\boldsymbol{k} = k(\sin\theta_0 \cos\phi_0 \hat{x} + \sin\theta_0 \sin\phi_0 \hat{y} + \cos\theta_0 \hat{z}) \tag{6.30}$$

Here, $\hat{\theta}_0$ and $\hat{\phi}_0$ define the angle of arrival of a plane wave in spherical coordinates. (For a uniform plane wave of unit magnitude normally incident on the xy-plane, as in the examples to be considered in this chapter, $\boldsymbol{E}^{\text{inc}}(z = 0)$ may be set to unity.)

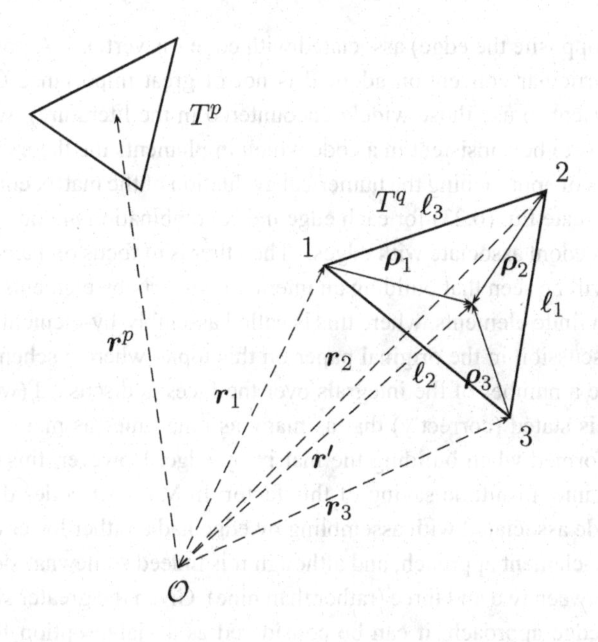

Figure 6.3 Local coordinates, nodes and edges for source triangle T^q and observation point in testing triangle T^p. Note that all vectors $\boldsymbol{r}$ are defined w.r.t. global origin $\mathcal{O}$, but all vectors $\boldsymbol{\rho}$ are defined w.r.t. the relevant local free vertex. After [1, Fig. 4], ©1982 IEEE, reprinted with permission.

This completes the derivation of matrix elements, which must be computed by numerical integration (quadrature). Doing this efficiently is the subject of the next section.

6.3.5 Numerical approximation of the matrix entries

Before proceeding further, one needs to establish both local and global numbering schemes for edges, nodes and triangles/elements – we use the last two interchangeably here. Such a numbering scheme is a theme which will recur in the context of finite elements. Associated with each triangle are, obviously, three nodes and three edges. These are illustrated in Fig. 6.3. The numbering scheme used in [1] is given here:[5]

Edge	Edge nodes	Associated free vertex
edge 1	2;3	1
edge 2	1;3	2
edge 3	1;2	3

Note that this is *not* compatible with the local edge numbering scheme used in finite elements (see Section 10.8.3); however, it is advantageous in the present context; the

[5] This is not explicitly stated there, but can be inferred from the figures and discussion.

free vertex (i.e. that opposite the edge) associated with edge i is vertex i. A point worth noting is that the particular convention adopted is not of great importance (although it is certainly convenient to use those widely encountered in the literature); what *is* of cardinal importance is to be consistent in a code which implements the theory!

There are two ways of approaching the numerical evaluation of the matrix entries. The first is to directly evaluate Eq. (6.23) for each edge index combination m and n. (Recall that the degrees of freedom associate with edges.) The other is to focus on face-pairs, or elements. (Later, it will be seen that building an interaction matrix by elements is a very common approach in finite elements, where this is called assembly-by-elements.) There is a quite lengthy discussion in the original paper on this topic, where a scheme which permits one to re-use a number of the integrals over the faces is discussed (we outline this later). There, it is stated (correctly) that as many as nine times as many integrals may need to be performed when building the matrix by edge. However, this does not necessarily translate into a runtime saving of this factor. In MATLAB codes developed by the author, the code associated with assembling by edge had a rather lower overhead than the assembly-by-element approach, and although it is indeed somewhat slower, the factor was usually between two and three (rather than nine). Given the greater simplicity of the assembly-by-edge approach, it can be considered as a viable option for a first prototype.

To evaluate Eq. (6.23), we need to evaluate the magnetic vector and scalar potentials due to each of the three basis functions residing on an element (more accurately, one of the two halves of the basis function, which spans the $+$ and $-$ elements as in Fig. 6.1). Each of these functions is proportional to the position vector $\boldsymbol{\rho}_i$,

$$\boldsymbol{\rho}_i = \pm(\boldsymbol{r}' - \boldsymbol{r}_i), \qquad i = 1, 2, 3 \tag{6.31}$$

the relevant position vector defined with respect to free vertex i, and is associated with edge i in the above edge and vertex numbering scheme. (Subsequently, data-structures needed for an actual code will be discussed; clearly, some general global numbering scheme for edges and elements will be needed for arbitrary scatterers.) Substituting Eqs. (6.6) into Eq. (6.23), we need to evaluate the magnetic vector potential,

$$\boldsymbol{A}_i^{pq} = \frac{\mu}{4\pi} \int_{T^q} \left(\frac{\ell_i}{2A^q}\right) \boldsymbol{\rho}_i \frac{e^{-jkR^p}}{R^p} dS' \tag{6.32}$$

and the electric scalar potential

$$\Phi_i^{pq} = -\frac{1}{4\pi j\omega\epsilon} \int_{T^q} \left(\frac{\ell_i}{A^q}\right) \frac{e^{-jkR^p}}{R^p} dS' \tag{6.33}$$

associated with the ith basis function on face q observed at the centroid of face p. In Eqs. (6.32) and (6.33),

$$R^p = |\boldsymbol{r}^{cp} - \boldsymbol{r}'| \tag{6.34}$$

where $\boldsymbol{r}^{cp}$ is the position vector of the centroid of face p. (We recall that $\boldsymbol{r}^{cp}$ is defined with respect to the global origin $\mathcal{O}$.)

When implementing the formulation, it is important to bear in mind the convention that subscripts associate with edges, superscripts with faces.

The integrals above must be evaluated using quadrature. To replicate the results which will be presented shortly, the six-point rule given in Table 10.4 is adequate; results were very similar to those computed using a 12-point rule (not shown). When $p = q$, the source point and field point coincide. In general, a rigorous singularity extraction scheme is required. Perhaps surprisingly, quite good results can be obtained by the simple expedient of not evaluating the function at the singular point. This is accomplished naturally using this six-point rule, as it does not have a node point at the centroid, where the field point is located. This point has been made by other workers who have implemented this formulation [4, 5]. Rigorous handling of the singularity will be discussed shortly.

The coordinates given in Table 10.4 are simplex, or normalized area coordinates. These will be discussed in Section 10.4 in detail. For now, it is sufficient to note that these are an alternate way of locating a point P located at $\boldsymbol{r}'$ inside a triangle. Such a point divides a triangle into three areas, A_1, A_2 and A_3 (A_i is the area of the triangle formed by P and the nodes connecting edge i; see Fig. 6.3). The ratios of these areas to the total area of the triangle provide an alternate, *linearly dependent*, description of point P, as

$$\xi = \frac{A_1}{A^q}, \qquad \eta = \frac{A_2}{A^q}, \quad \zeta = \frac{A_3}{A^q} \tag{6.35}$$

Clearly, since the three sub-triangles sum to the full triangle, $A_1 + A_2 + A_3 = A$, which implies that

$$\xi + \eta + \zeta = 1 \tag{6.36}$$

hence the linear dependence of this description. It is also important to note that these coordinates provide both a local description $(\xi; \eta; \zeta)$, which is independent of rotation and/or translation of the triangle, and also an alternate description in terms of Cartesian coordinates:

$$\boldsymbol{r}' = \xi \boldsymbol{r}_1 + \eta \boldsymbol{r}_2 + \zeta \boldsymbol{r}_3 \tag{6.37}$$

where $\boldsymbol{r}_1$, $\boldsymbol{r}_2$ and $\boldsymbol{r}_3$ are the Cartesian coordinates of the triangle vertices, as in Fig. 6.3. Equation (6.37) provides the required formula to convert the nodal points given in the quadratures tables in Table 10.4 to the Cartesian coordinates required to find $\boldsymbol{r}_i$ and hence R^p in Eq. (6.34) for each quadrature point in T^q (and observation point in T^p).

For the evaluation of the surface integrals over T^q in Eqs. (6.32) and (6.33), this can be formulated elegantly in simplex coordinates. It can be shown (see, for instance, [6, Appendix 1]) that the surface integrals transform as follows:

$$\int_{T^q} g(\boldsymbol{r}) dS = 2A^q \int_0^1 \int_0^{1-\eta} g\left[\xi \boldsymbol{r}_1 + \eta \boldsymbol{r}_2 + (1 - \xi - \eta)\boldsymbol{r}_3\right] d\xi\, d\eta \tag{6.38}$$

With Eqs. (6.31), (6.34), (6.37) and (6.38), Eqs. (6.32) and (6.33) may now be written as

$$A_i^{pq} = \pm \frac{\mu \ell_i}{4\pi} \left(\boldsymbol{r}_1 I_\xi^{pq} + \boldsymbol{r}_2 I_\eta^{pq} + \boldsymbol{r}_3 I_\zeta^{pq} - \boldsymbol{r}_i I^{pq} \right) \tag{6.39}$$

and

$$\Phi_i^{pq} = \mp \frac{\ell_i}{j2\pi\omega\epsilon} I^{pq} \tag{6.40}$$

where

$$I^{pq} = \int_0^1 \int_0^{1-\eta} \frac{e^{-jkR^p}}{R^p} d\xi d\eta \tag{6.41}$$

$$I_\xi^{pq} = \int_0^1 \int_0^{1-\eta} \xi \frac{e^{-jkR^p}}{R^p} d\xi d\eta \tag{6.42}$$

$$I_\eta^{pq} = \int_0^1 \int_0^{1-\eta} \eta \frac{e^{-jkR^p}}{R^p} d\xi d\eta \tag{6.43}$$

$$I_\zeta^{pq} = I^{pq} - I_\xi^{pq} - I_\eta^{pq} \tag{6.44}$$

Each of these can be evaluated using quadrature, as already discussed.

It is these integrals that can be re-used when computing integrals over a face. For interior triangles, there will be three basis functions (with local number i) on each face. The only i dependence in the scalar potential, Eq. (6.33), is ℓ_i, hence I^{pq} only needs be computed once. More subtly, the i dependence in the vector potential $\boldsymbol{A}_i^{pq}$, Eq. (6.32), is in terms of $\boldsymbol{r}_i$ (vertex i) in Eq. (6.39), hence Eqs. (6.41)–(6.44) may be (re-)used to compute $\boldsymbol{A}_1^{pq}$, $\boldsymbol{A}_2^{pq}$ and $\boldsymbol{A}_3^{pq}$ without recomputing the integrals.

6.3.6 Coding issues

Coding the formulation developed in this section is a non-trivial undertaking, and it ranks as one of the three most complex codes undertaken in this book, along with the Sommerfeld MoM code to be discussed in Section 7.6 and the mixed second-order tetrahedral element three-dimensional FEM code discussed in Chapter 11. Although the formulation is somewhat simpler mathematically than the Sommerfeld formulation to be addressed in the next chapter, the triangular elements add geometrical complexity. In this context, the suggestions made at the end of Chapter 7 regarding coping with coding complexity apply equally to this project.

The triangular discretization of the scatterer is precisely the same meshing problem which will be discussed in Section 10.8.3 in the context of the two-dimensional FEM formulation and code, and the scheme developed there is equally applicable to this MoM code. Readers are referred to that section for more detail. Similarly, the datastructures which are required are in general very similar to those required by the FEM in the same section. Both the present MoM code and the FEM code have edge-based degrees of freedom, although in the present (MoM) case, these are the normal components of surface currents across the edges, and in the FEM case they are tangential electric fields along the edges. Furthermore, both formulations have only non-zero degrees of freedom on triangle edges interior to the mesh.

The MoM does need some geometrical data which the FEM code does not. One of these is the set of $\boldsymbol{\rho}_m^{c\pm}$ vectors, giving the centroids in *local* Cartesian coordinates (i.e. with reference to the appropriate free vertex as local origin). Further, one needs a datastructure indicating, for a given edge, which element is associated with the "plus" triangle T_n^+, and which with the "minus" T_n^-, with reference to Fig. 6.1. It may be stating the obvious, but note that a particular element may represent T_n^+ for one edge, but T_n^- for another.

When coding, it is important to distinguish carefully between vectors defined using Cartesian coordinates with respect to the global origin $\mathcal{O}$, such as $\boldsymbol{r}$; vectors defined using "local" Cartesian coordinates, i.e. defined with respect to the relevant free vertex, such as $\boldsymbol{\rho}^{\pm}$; and quantities defined in simplex (area) coordinates, such as the quadrature points on a triangle.

When using quadrature tables, note the coding hints discussion regarding such tables in Section 11.2.4; the weights given in Table 10.4 must be normalized by $1/2$ when applied to Eqs. (6.41)–(6.44), as this is the area of the "unitary" triangle associated with the simplex coordinates.

It is also useful to appreciate that the Galerkin process is in essence computing the mutual coupling between two edge-based basis functions, each defined on two triangles; hence four triangles are involved in the computation of any one matrix entry. For instance, $\boldsymbol{A}_{mn}^{\pm}$, as in Eq. (6.25), comprises two terms, each computed using Eqs. (6.39) and (6.41)–(6.44), and is the result of integrating over the potentials, weighted by the appropriate basis function, over "source" triangles T_q^+ and T_q^- associated with edge n, for an observation point located at the centroid of either "testing" triangles T_p^+ or T_p^- associated with edge m. These terms are then summed in Eq. (6.23), representing the two-point (centroid) approximation of the integration over the testing triangles T_p^+ and T_p^-. It is useful to keep this in mind when coding.

When post-processing, it is very useful indeed to recall that each degree of freedom represents the normal component of current on that edge. Vertical and horizontal cuts of a "regular" mesh such as that shown in Fig. 10.10 yield the distribution of the normal component of current along the cut-line directly; this is exploited in Figs. 6.4 and 6.5, where the meshes were set up with this in mind. For more general visualization applications, it is necessary to compute the current at a point as the sum of the three RWG vector basis functions multiplied by the relevant coefficient, as in Eq. (6.7). Very similar post-processing operations are discussed later in the context of vector basis functions for the FEM, see Section 10.8.

6.3.7 Verification

To verify the formulation, two results given in [1] will be replicated. The first shows the dominant component of current on a 0.15λ square flat PEC plate, with a uniform plane wave, comprising an x-directed electric field normally incident on the plate [1, Fig. 5]; the second is the same, but for a 1.0λ square plate [1, Fig. 6]. Results computed with a code implementing this theory are compared to the original results in [1] in Figs. 6.4 and 6.5; [Lit] in the figure legend refers to [1]. In both, cut AA is a centered vertical cut (constant x), and cut BB is a centered horizontal cut (constant y), as in the original. (For a discussion of the physics illustrated by the different problems, see [1]; the first plate is electrically small, the second is in the resonant regime.) The mesh sizes for the problems

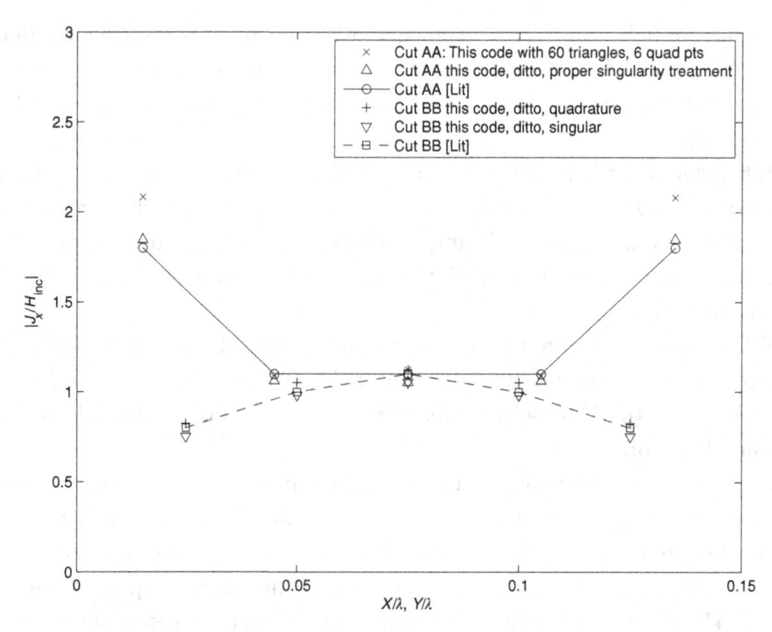

Figure 6.4 Dominant component of current on a 0.15λ square flat plate. ([Lit] refers to [1].)

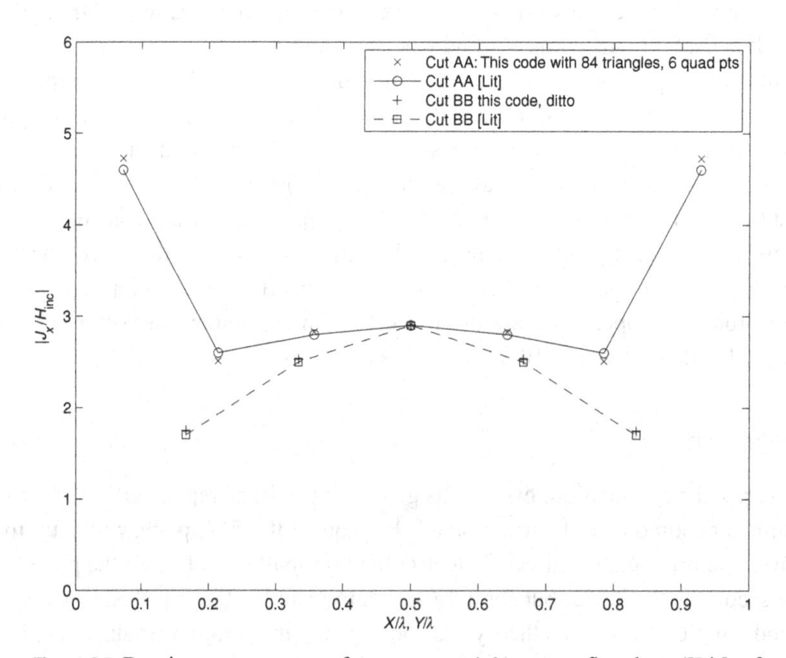

Figure 6.5 Dominant component of current on a 1.0λ square flat plate. ([Lit] refers to [1].)

are as in the original reference, viz. 6×5 rectangular cells in x and y respectively, each split into two triangles, for the former, and $6 \times 7 \times 2$ for the latter. This permits straightforward extraction of the normal current components on vertical and horizontal cuts, as discussed at the end of the preceding section. In Fig. 6.5, the agreement is excellent; in Fig. 6.4, some minor discrepancies are noted. This is an electrically smaller problem; it is due to the approximate handling of the singular self-term. This is shown in Fig. 6.4, where the singularity has also been handled rigorously using the method of [7].

This affords a clear illustration of the difference between verification and validation. These figures confirm that the formulation has been correctly implemented – i.e. the code implementing it has been verified. However, it says nothing about how well the formulation actually models the real physical problem. Fortunately, validation was addressed to a considerable extent in the [1], hence verification suffices here. (At the risk of repetition, validation of a general formulation is in essence an open-ended issue, as new applications will sometimes throw up a discrepancy requiring attention.)

6.3.8 Discussion

The above provides a simple working code based on the RWG formulation for scattering problems. Radiation problems can also be addresssed; see, for instance, [4] for a discussion of including antenna sources. For a more rigorous implementation, a modern singularity handling scheme should be used. Traditionally, singularity subtraction schemes (as will be used in Chapter 7) have been used in MoM codes of this nature; the idea is to split the singular kernel into a singular component which can be integrated analytically, and a slowly varying remainder which can be integrated numerically. More modern approaches use singularity cancellation, using a change of variables chosen such that the Jacobian of the transformation cancels the singularity; the integral is then performed purely numerically on the transformed domain. When applied to a triangular domain, the singularity is placed at a vertex, and the transformation stretches the vertex into a line, so that the transformed domain is rectangular, and softens the singularity in the process. This permits quadrature using a Cartesian product of Gauss–Legendre rules to be applied. The singularity cancellation scheme due to Khayat and Wilton [7] using an arcsinh transformation is exact, relatively straightforward to implement, and can be recommended. (There is a widely used earlier scheme due to Duffy which only approximately cancels the singularity.) Ironically, near-singular functions are actually more difficult to integrate; this has been addressed further in subsequent work by those authors [8]. Graglia and Lombardi have also made recent contributions in this regard [9].

In codes such as FEKO, more sophisticated treatments are used.[6] A problem with the RWG two-point approximation of the testing process is that it becomes unstable for low frequencies, an issue explained in the next paragraph. Strict "razor-blade" testing had been proposed using closed paths on the surfaces connecting the vertices of adjacent triangular patches (i.e. using line integrals for the weighting, as discussed

[6] The author gratefully acknowledges information from Dr. Jakobus, FEKO product manager, in this regard.

earlier). This was also originally implemented in FEKO, but was changed to an adaptive weighting formulation involving surface integrals to further improve accuracy for low-frequency problems (such as a car body at 10 MHz) and to reduce runtime for high-frequency problems. Depending on the frequency and the distance of source and observer triangles, and also on the source/observer type, e.g. metallic/dielectric or triangle/wire interactions, etc., different weighting algorithms are used. Furthermore, parameters in these algorithms are changed adaptively, such as the number of points for Gaussian quadratures, etc.

In the previous paragraph, issues with the low-frequency stability of the MoM scheme were mentioned. The problem is more fundamental than just the particular approach taken to discretize it (the RWG basis functions in this case), although more accurate evaluation of the matrix entries apparently assists; it is inherent in the EFIE formulation, unless special measures are taken. Consider the matrix entries, Eq. (6.23). The vector potential contribution scales as ω; the scalar potential, Eq. (6.26), as $1/\omega$. Clearly, as $\omega \to 0$, there is a problem: the condition number of the matrix will tend to infinity. Two approaches have been pursued recently. The first is the use of what are widely called "loop-star" basis functions; the aim is to model the solenoidal (zero divergence) component of the current correctly. Since solenoidal currents form closed loops, the loop basis functions are precisely that, and are associated with vertices (nodes) in the mesh. The theory underlying the approach is discussed in some detail in [10], which also discusses earlier work on the topic; a somewhat more succinct description was given in [11], which also describes how the loop-star functions may be obtained from the RWG ones. The method has also been described in the context of a Helmholtz decomposition or Helmholtz splitting [11] – this splits a field into its solenoidal and irrotational (zero curl) components. However, this can only be performed approximately in a numerical scheme based on div-conforming elements such as the RWG ones. (In [10], it is questioned whether the term should be used at all in this context; nonetheless, it appears to provide a useful perspective.) The loop-star approach is closely related to tree-cotree decomposition techniques in the FEM, which also suffers from low-frequency instability. The other approach pursued more recently is the use of Calderon preconditioners; a recent paper reviewing the contributions to date and proposing a multiplicative preconditioner is [12].

6.4 Some examples of surface modelling

6.4.1 Scattering from a sphere

One of the classical problems of analytical electromagnetics was that of scattering from a sphere. Early work on this was done in the nineteenth century by Lord Rayleigh (John William Strutt, 1842–1919), who has lent his name to the general field of scattering from electrically small objects. For electrically small spheres, Lord Rayleigh showed that scattering was proportional to the fourth power of frequency; this permitted him to explain the color of the sky. For electrically large spheres, the scattering cross-section is simply the cross-sectional area of the sphere. In between these extremes, the *resonant regime* is encountered, where energy creeping around the surface of the sphere results in constructive and destructive interference. The process of electromagnetic scattering will be recalled from Chapter 3.

A brief historical aside – why *is* the sky (usually) blue?

The color of the sky is due to the presence of the earth's atmosphere. On the moon or in space, the sky appears black. For our present purposes, we can view the atmosphere as consisting of a large number of small particles and molecules in suspension. These are considerably smaller than the wavelength of visible light (approximately 400 to 700 nm), so that the scattering from each particle is proportional to $1/\lambda^4$, as in the text. Hence, the scattering from the violet (short-wavelength) end of the spectrum is almost an order of magnitude larger than that from the red (long-wavelength) end. The spectral irradiance of sunlight – see for example [13, Fig. 7.49] – which peaks near the wavelength of blue light, 470 nm, and varies by about 30% over the visible spectrum, makes the overall calculation slightly more complex. It is this scattered radiation which colors the sky blue. (It is worth noting in passing that the scattered light is also polarized, although we will not pursue this here.) At sunset, however, the radiation has to pass through much more of the atmosphere, and the blue scatters out completely, leaving the red sunset. When there is dust in the air, this exacerbates the effect, leading to spectacular sunsets. More details of this may be found in many texts, such as [14, Chapter 12] and [13, Chapter 7]. The latter has a particularly insightful discussion, and also provides extensive historical background on this topic.

The echo width of a three-dimensional target is also known as its radar cross-section (RCS). It is usually abbreviated σ. The RCS is defined as follows:

$$\sigma(\theta, \phi, f) = \lim_{R\to\infty} 4\pi R^2 \frac{|E^{\text{scat}}|^2}{|E^{\text{inc}}|^2} \tag{6.45}$$

where R is the distance to the target. The dimensions of the RCS are square meters, since it is in essence an equivalent area. Frequently, results are given in dB form, and quite often normalized to 1 m^2, in which case the symbol dBsm is often used. The RCS of a target is in general a function of orientation and frequency, and this has been explicitly indicated above. Note that this definition is entirely equivalent to

$$\sigma(\theta, \phi, f) = \lim_{R\to\infty} 4\pi R^2 \frac{P^{\text{scat}}}{P^{\text{inc}}} \tag{6.46}$$

The RCS is a far-field parameter; once the surface currents are have been found using the MoM, the radiated fields may be computed in a straightforward fashion using standard antenna theory.

As a simple example of a scattering problem, we will now study the RCS of a sphere. A highly conducting sphere with a radius of 5 cm will be chosen; this is the typical size of anti-personnel landmines (although of course these are generally buried, and

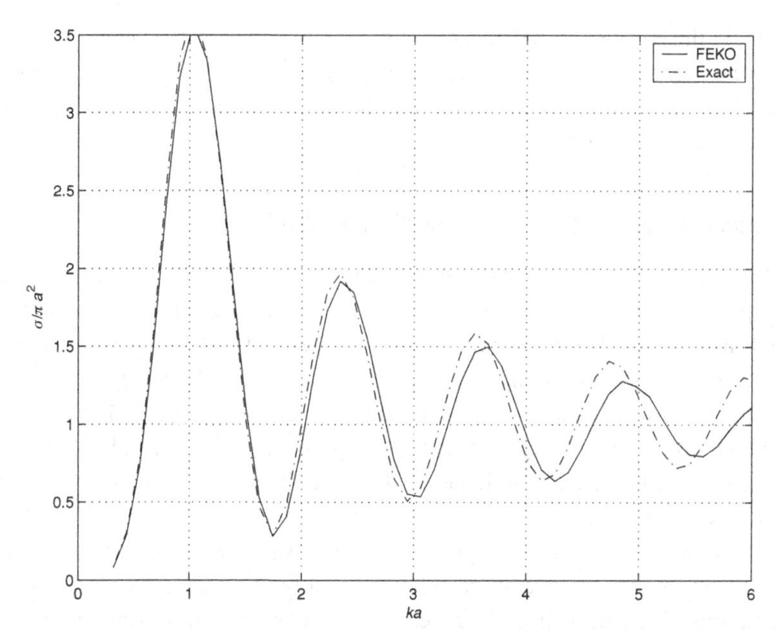

Figure 6.6 Normalized RCS of a PEC sphere plotted against circumference in wavelengths ka.

also unfortunately usually made largely of non-metallic materials to make detection even more difficult). We expect the interesting resonance interactions to occur when the circumference of the sphere is of the order of a wavelength, hence $\lambda \approx 2\pi a$. This corresponds to a frequency of around 1 GHz. Running the simulation from 300 MHz to 6 GHz should produce some interesting results.

Note that we are only going to investigate the *back-scatter* from the sphere; hence, only one RCS angle is required (the same one the field is incident from). The results of the analysis are shown in Fig. 6.6. The RCS has been normalized by the high-frequency limit πa^2 to illustrate more clearly the different scattering regimes; note how the RCS initially climbs steeply (this is the Rayleigh scattering regime), then oscillates sharply through the resonance regime, before finally converging to the high-frequency limit. The horizontal axis has also been normalized, by plotting $ka = (2\pi/\lambda)a$ (the sphere circumference in wavelengths). Note the peak as expected at $ka = 1$. Also shown on this plot is the exact analytical solution, computed as a sum of spherical Hankel functions [15, Eq. (11–247), p. 657] – more on this shortly. When compared with the exact solution, we note that the accuracy with which the resonances are computed decreases as the frequency increases.

If we were to analyze this problem over a rather larger frequency band, we would find that eventually, the result should converge to the high-frequency limit. We cannot do this with the present file, because our discretization will not be sufficiently fine for frequencies much beyond 6 GHz. Refining the discretization will result in far longer execution times. However, some thought about the problem shows that we can use symmetry to generate a more efficient solution. The incident electric field is $\hat{x}$ polarized,

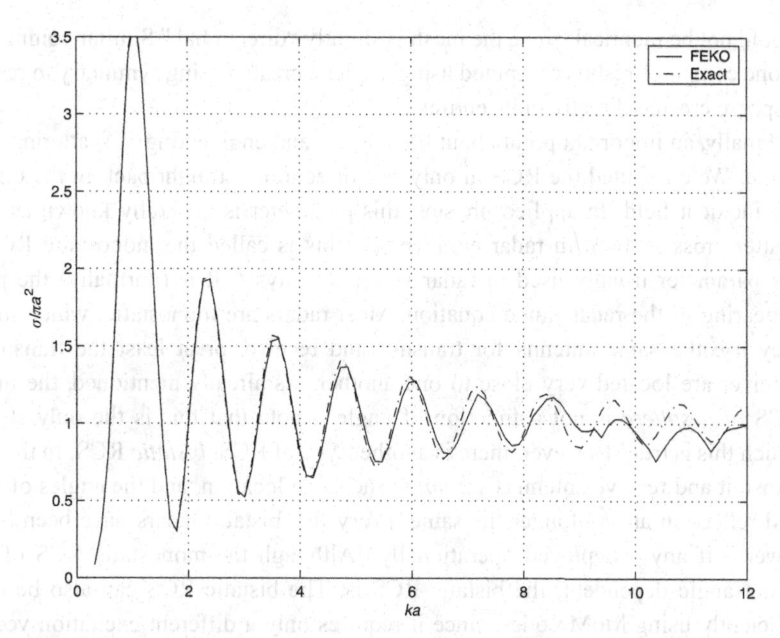

Figure 6.7 Normalized RCS of a PEC sphere plotted against circumference in wavelengths ka. Results were computed exploiting symmetry.

travelling in the $-\hat{z}$-direction. As such, there is a plane of electric symmetry in the plane $x = 0$. Similarly, there is a plane of magnetic symmetry in the plane $y = 0$. Finally, there is a plane of geometrical symmetry in the plane $z = 0$. (In this last plane, the geometry is symmetrical, but not the excitation.) Results for a wider frequency range computed using symmetry are shown in Fig. 6.7. Note the improvement in the resonances when compared to Fig. 6.6. However, at the high-frequency end, the mesh is too coarse even with this model, as is clear by comparison with the analytical solution.

Modelling hints – modelling spheres

All meshers generate some approximation of the actual spherical surface; in the case of FEKO, the triangular mesh is inscribed within the sphere. (FEKO provides the KU card to generate a spherical section or a sphere, which makes modelling the sphere very straightforward.) The model can be improved by using a slightly larger radius, chosen to provide the same surface area as the sphere. Conveniently, FEKO computes the surface area of the triangles; for the first model, the area was 0.030 96 m^2, whereas the surface of a 0.05 m radius sphere should be 0.03142 m^2. Increasing the radius by 1.007, the square root of this ratio, should provide a slightly better model.

A couple of closing comments on this study would be in order. Firstly, because a sphere is rotationally symmetric, we could have used a field incident from *any* angle. The choice of the $\hat{x}$-polarized field, travelling in the $-\hat{z}$-direction, was however convenient. Note that if a different incident field were used, results would (or should) be very similar, but would not be identical, since the mesh is slightly "directional." Similar comments apply if one compares results computed using a sphere created using symmetry to results from a sphere created directly in its entirety.

Finally, an important point about the physics and engineering of scattering should be made. We computed the RCS in only one direction – straight back in the direction of the incident field. In applied physics, this parameter is generally known as the back-scatter cross-section; in radar engineering, this is called the monostatic RCS, and is the parameter usually used in radar systems analysis. It is (normally) the parameter appearing in the radar range equation. Most radars are monostatic, which means that they use the same antenna for transmit and receive, or at least the transmitter and receiver are located very close to one another. As already mentioned, the monostatic RCS *of a sphere* is not a function of angle – note that this is the only structure of which this is true! However, there is another type of RCS, *bistatic* RCS. In this case, the transmit and receive antennas are *not* in the same location, and the angles of incidence and reflection are no longer the same. (Very few bistatic radars have been built, even fewer – if any – deployed operationally.) Although the monostatic RCS of a sphere is not angle dependent, the bistatic RCS is. The bistatic RCS can also be computed efficiently using MoM codes, since it requires only a different excitation vector to be computed.

6.4.2 The analytical solution

The exact solution of scattering from a PEC sphere, plotted in Figs. 6.6 and 6.7, is one of the classic analytical solutions in electromagnetics, dating back to the turn of the previous century. Nonetheless, despite the venerable status of the solution, there are some points which are worth making about it, and indeed about analytical solutions in general.

A brief historical aside – Mie scattering

The analytical solution for scattering from a PEC sphere was originally derived by Mie and published in 1908, and the solution bears his name to this day. Debye undertook a very similar study, published in 1909. For details, see [16, p. 415]; for elegant sketches of the fields for the first four modes, reproduced from Mie's paper, see [16, p. 567]. Stratton's book has fortunately been reprinted by the IEEE. The derivation may also be found in somewhat more recent texts, such as [17, Chapter 6], and a particularly detailed derivation is given in [15, Section 11.8].

The monostatic RCS is given by the following expression:

$$\sigma = \frac{\lambda^2}{4\pi} \left| \sum_{n=1}^{\infty} \frac{(-1)^n (2n+1)}{\hat{H}_n^{(2)\prime}(ka) \hat{H}_n^{(2)}(ka)} \right|^2 \tag{6.47}$$

with a the radius of the sphere and k the free-space wavenumber. The function $\hat{H}_n^{(2)}(ka)$ is the alternative spherical Hankel function. It is related to the regular cylindrical Hankel function by [15, p. 938]

$$\hat{H}_n^{(2)}(x) = \sqrt{\frac{\pi x}{2}} H_{n+1/2}^{(2)}(x) \tag{6.48}$$

The prime in $\hat{H}_n^{(2)'}(ka)$ indicates differentiation with respect to the argument.

This would really appear to be a relatively straightforward formula to implement. MATLAB provides only the regular cylindrical Hankel function, but the scaling required by Eq. (6.48) is very easy to implement. For FORTRAN implementations, routines are available in [18, Chapter 6], although one will need to build the Hankel function from its constitutive Bessel functions of the first and second kinds, viz. $H_p^{(2)}(x) = J_p(x) - jY_p(x)$. The derivative requires some simple manipulation to evaluate, using the rule for the differentiation of products applied to Eq. (6.48), and the standard identity [15, p. 936]

$$\frac{d}{dx}[H_p^{(2)}(\alpha x)] = -\alpha H_{p+1}^{(2)}(\alpha x) + \frac{p}{x} H_p^{(2)}(\alpha x) \tag{6.49}$$

to obtain:

$$\hat{H}_n^{(2)'}(x) = \frac{1}{2}\sqrt{\frac{\pi}{2x}} H_{n+1/2}^{(2)}(x) + \sqrt{\frac{\pi x}{2}} \left[-H_{n+3/2}^{(2)}(x) + \frac{n+\frac{1}{2}}{x} H_{n+1/2}^{(2)}(x) \right] \tag{6.50}$$

Hence, Eq. (6.47) can be implemented within a few lines of code. However, one needs to be cautious! Routines to compute Bessel functions (by which we include Hankel functions) are not bulletproof. In particular, when the argument (ka in this case) or the order (n) becomes very large, the results lose accuracy. Good implementations should warn of such problems: MATLAB, for instance, provides five different error flags, ranging from warnings of possible loss of precision to outright error messages and not returning a numeric value at all. *One must check such error flags!* In the present case, exceeding some hundred terms or so is sufficient to trigger error messages.

Needless to say, the infinite sum in Eq. (6.47) must also be truncated at some point. In Fig. 6.8, results are shown for the RCS for the sphere as the maximum number of terms is increased; this has been graphed on a semi-logarithmic scale, so that the variation is more easily seen. Plotting against ka is especially insightful, since it is clear that the number of terms required is approximately equal to this product. (This is not coincidental: these terms correspond to circumferentially varying modes, and modes with significantly more rapid variation than ka contribute primarily to the reactive near-field only.)

For electrically large spheres, Eq. (6.47) is clearly going to be problematic to evaluate directly, and one needs to use asymptotic forms to retain accuracy.

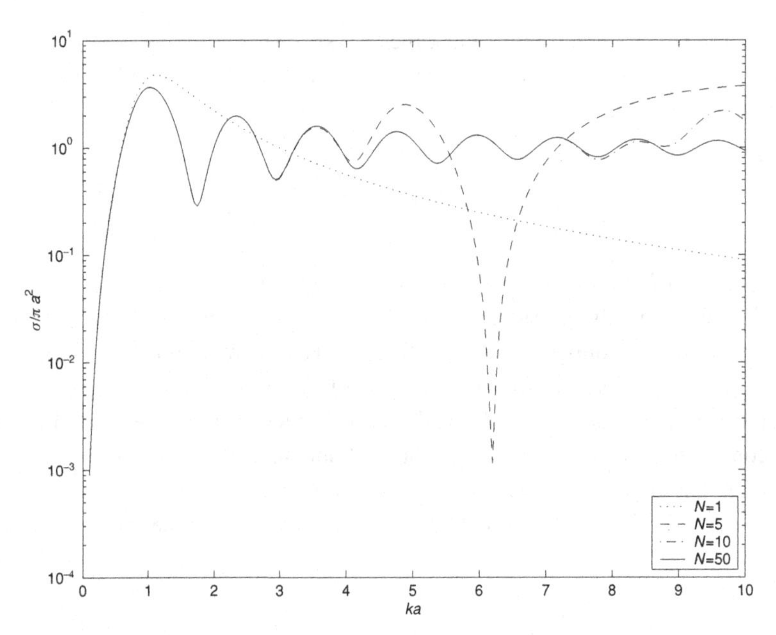

Figure 6.8 Convergence of the analytical solution for the RCS of a PEC sphere, as a function of the number of terms used.

A philosophical aside – on "exact" analytical solutions

The above discussion raises a number of interesting points about the nature of "exact" analytical solutions. Critics of our present-day reliance on numeric codes sometimes forget that even pristine analytical solutions are usually approximate in reality, when it comes to evaluating them; such solutions, derived from separation of variables and suitable special functions, usually involve infinite summations which must in practice be truncated. Furthermore, the evaluation of the special functions is almost always done computationally nowadays, and as we have commented, this process is by no means always reliable. (Even tables of functions are not always error free.) It is perhaps the ultimate irony that the author verified his MATLAB implementation of Eq. (6.47) by comparing the results to FEKO computations...

6.5 Modelling homogeneous material bodies using equivalent currents

In the preceding section, we discussed modelling structures consisting of PEC (perfect electric conductor) material.[7] The current which the MoM computes in this case is the real, physical current, and is what would be measured were one to probe the surface current using a loop, for instance. However, there is another interesting application of surface currents: modelling homogeneous material bodies, that is, dielectric (or magnetic) regions.

[7] The approach can be extended to work for highly conducting structures.

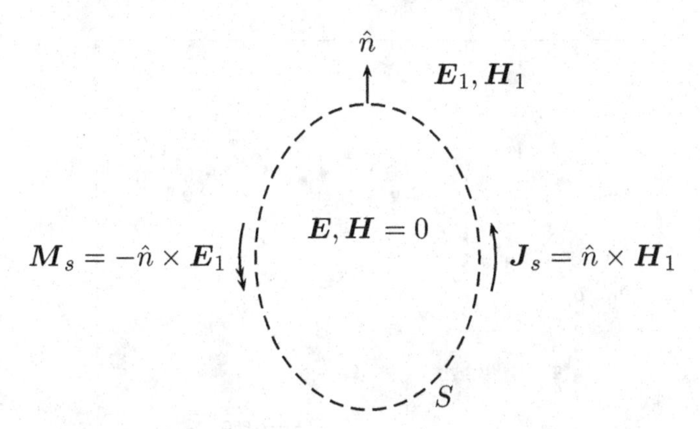

Figure 6.9 Love's form of the equivalence principle.

All of these rest on the application of the *surface equivalence theorem*, first introduced in 1936 by Schelkunoff. It states that the fields outside an imaginary closed surface can be obtained by placing, over the closed surface, suitable electric and magnetic current densities that satisfy the boundary conditions. Furthermore, the fields inside the surface can be chosen essentially arbitrarily, since the problem is only "equivalent" in the exterior region. When this imaginary surface coincides with a real surface, interesting physics emerges with specific choices of the internal fields. For PEC modelling, the form of the equivalence principle which is generally used is *Love's equivalence principle*, illustrated in Fig. 6.9 for a general surface. With this form, the fields inside the body are assumed zero; since the boundary condition at a PEC surface requires that the tangential total electric field be zero (and hence also the magnetic surface current), only the electric surface current is non-zero and since it is equal to $\hat{n} \times (\boldsymbol{H}_{\text{tot}} - 0)$, where $\boldsymbol{H}_{\text{tot}}$ is the total magnetic field just above the surface, and the 0 represents the internally zeroed fields, it is also the *actual* current. It is also very convenient because since the field has been chosen as zero in the internal region, the material in this region can be replaced arbitrarily; usually, it is chosen to have the same value as the exterior region, which is usually free space in antenna problems.[8] This is *very* important, since it permits the use of the free space Green function – we usually apply this without fully discussing the underlying justification.

In passing, note that there is another variant of this principle which one quite often encounters in the theoretical analysis of aperture antennas. In this case, instead of replacing the internal region with free space, one uses a PEC body. If this is a half-space, one can then use image theory and hence the Green function for free space (again) to solve the problem. There are yet other forms which are useful in specific circumstances.

When the material body is an homogeneous dielectric or magnetic structure, we can apply the same approach as with the PEC body; there are two differences, however. Firstly, the currents are now fictitious (in other words, one would not be able to measure them with some cleverly devised experiment), and secondly, *both* electric and magnetic

[8] Note that the whole argument also works in reverse for the interior region: in this case, it is the fields in the exterior region which are arbitrary. This is not very useful in antenna modelling, however.

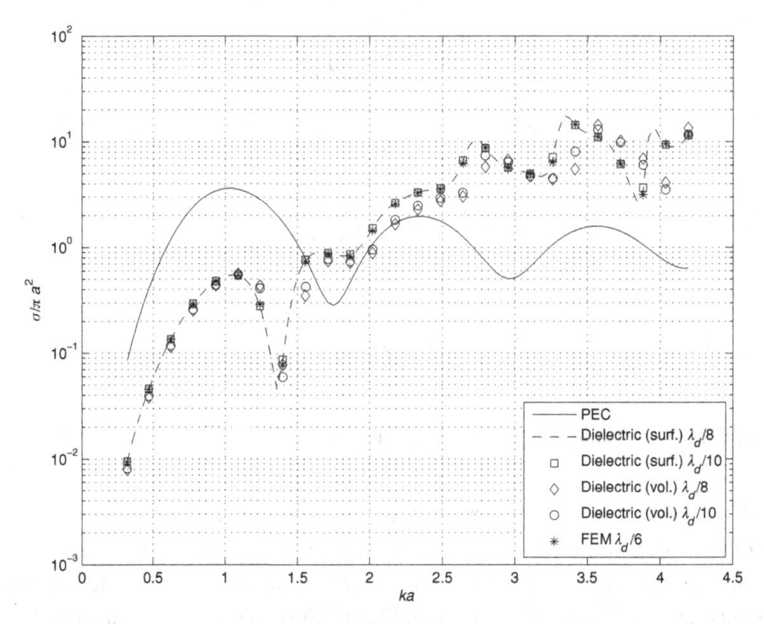

Figure 6.10 RCS of a dielectric sphere, radius a, $\epsilon_r = 4$, compared to a PEC sphere. All results are normalized by πa^2. k is the free-space wavenumber.

equivalent surface currents are required. Given that the fields in a region are uniquely specified by knowledge of *either* equivalent electric or magnetic currents, one might wonder why both are now needed – the reason is that the EFIE for homogeneous scatterers is actually a coupled EFIE,[9] connecting *two* equivalent problems. One of these considers a set of equivalent currents radiating in a homogenous space with the material properties of the exterior region (usually, but not necessarily, free space); the other has another set of equivalent currents radiating in a homogeneous space with the material properties of the scatterer. The continuity conditions on tangential electric and magnetic fields are used to couple the problems across the surface of the scatterer. For a detailed analysis, see [3, Section 1.9].

6.6 Scattering from a dielectric sphere

Having just discussed a PEC sphere, it is now an interesting exercise to repeat the analysis for a dielectric sphere. The model is very similar to the PEC sphere. Results are shown in Fig. 6.10. A moderate value, $\epsilon_r = 4$, has been chosen for the relative permittivity, otherwise the sphere has a very low signature. Results are normalized to the asymptotic limit for the PEC sphere, πa^2. It is interesting that the RCS of the dielectric sphere exceeds that of the PEC sphere for $ka > 1.5$.

To validate this computation, we can use another approach for modelling dielectrics available within FEKO, namely equivalent *volumetric* currents. In this case, the entire

[9] The same is true for the MFIE.

Table 6.1 Comparison of computational requirements for the PEC versus dielectric sphere

	PEC	Dielectric (surface)		Dielectric (volume)		FEM
		$\lambda_d/8$	$\lambda_d/10$	$\lambda_d/8$	$\lambda_d/10$	$\lambda_d/6$
N	663	$2 \times 663 =$ 1326	$2 \times 1008 =$ 2016	$3 \times 772 =$ 2316	$3 \times 1370 =$ 4110	16733 FEM, 507 MoM
Memory (Mbytes)	15.1	55.9	127	166	520	53.9
Relative runtime	1	3.13	6.99	6.90	22.0	3.51

N is the number of unknowns. Runtime is per frequency point. The edge lengths are given for $ka = \pi$.

volume is meshed using cubical cells – this permits the material properties to vary from cell to cell, but at much higher computational cost (we will discuss this shortly). Results computed using the volumetric approach using two meshes, as well as a surface current model using a slightly finer mesh, are also shown on Fig. 6.10. The agreement between the surface and volume formulations is good up to just above $ka = 2$, which is about the point at which the mesh density drops below $\lambda_d/10$, where λ_d is the wavelength in the dielectric. (Because we are effectively modelling fields in an electrically denser medium, it is the wavelength *in the dielectric* which concerns us.) Since the volume approach meshes the sphere with small rectangular cubes, as opposed to a conforming triangular surface mesh, one can expect the volume approach to be slightly less accurate geometrically, in particular at higher frequencies. This is confirmed by a calculation using a slightly smaller edge length for the surface mesh; the agreement between the two surface current meshes is good. (Refining the volumetric mesh affords little improvement, at greatly increased computational cost.) An additional calculation with yet another technique, a FEM/MoM hybrid (to be discussed subsequently), confirms the validity of the surface formulation results.

A note of caution here: such intracode validation is usually questionable, but in this case, we are using no less than three different techniques to compute the RCS, so we can place considerable faith in this result.

Although the equivalent surface current model is probably the most computationally efficient available for general problems,[10] the requirement to treat both the equivalent electric and magnetic surface currents doubles the number of unknowns, and hence quadruples the amount of memory, and increases the runtime by between four and eight, depending on the problem size, when compared to a PEC sphere. (Eight is the asymptotic limit, for problems with a very large number of unknowns where the matrix solution dominates the runtime – we discuss this shortly.) A summary of the computational requirements is given in Table 6.1.[11] The runtimes are given normalized to the PEC

[10] For the dielectric sphere, the Green function is known analytically, so for this special case only, one could develop a faster solver.

[11] Readers with the first edition will note that relative runtimes have changed slightly, reflecting code improvements, and that memory requirements have approximately halved. The latter is due to single precision being now the default in FEKO for the storage of the matrix elements; this is quite adequate for most problems, including the one to hand.

case. Note how execution time increases by a factor of about three for the dielectric sphere using surface equivalence (compared to the PEC, which is used as the baseline[12] on this table), and around six when using the volumetric mesh (which did not give a fully satisfactory answer). Also shown on this table is a FEM/MoM solution, computed using the hybrid scheme to be presented in Section 12.1. Although the mesh appears coarse, the FEM/MoM solution uses an LT/QN element (mixed second order) in the FEM region to improve accuracy, which results in a large number of FEM degrees of freedom. The FEM/MoM formulation is highly competitive with the surface equivalence solution, and additionally could handle an *in*homogeneous sphere with the same computational requirement. This FEM/MoM solution used the MoM closure *on* the surface of the scatterer; it is possible to trade-off FEM volume against MoM discretization by adding a thin "buffer" shell of free space, which permits a somewhat coarser MoM mesh to be used on the closure (as it is now in free space).

Also shown in this table is the effect of refining the discretization. Changing the edge length from $\lambda_d/8$ to $\lambda_d/10$, with the corresponding frequency in this case chosen as that corresponding to $ka = \pi$ (towards the upper end of the frequency band), results in large increase in computational requirements for especially the volumetric MoM.

6.7 Computational implications of surface and volume modelling with the MoM

As has just been seen with the analysis of the sphere, modelling surfaces is far more computationally expensive than modelling wires. As already discussed in Chapter 4, for a typical wire model the number of unknowns N is linearly related to the length of the wire. We will use the product kd to characterize this, with k the wavenumber and d the length of wire. There are two time-consuming operations required by an MoM code with N unknowns, viz. matrix filling and factoring. The former is of $\mathcal{O}(N^2)$, the latter $\mathcal{O}(N^3)$ when using direct solvers (iterative solvers will be discussed later). However, the constants associated with matrix filling can be quite large (that of the matrix solve is close to unity) and in practice one often finds that MoM codes are in the pre-asymptotic region as far as timing goes, spending more time filling than factoring the matrix. Since N is proportional to kd, we have an asymptotic cost of $\mathcal{O}([kd]^3)$. To store the interaction matrix $[Z]$, N^2 memory locations are required. Hence the amount of memory required is $\mathcal{O}([kd]^2)$ for wires. These properties are also known as the *frequency scaling* behavior of the algorithm.

For a surface, although triangles using the vector RWG basis functions are the approach generally used in practice, when doing a frequency scaling analysis it is easier to consider square patches. To model a surface of size $kd \times kd$, it is clear that the number of unknowns will now be $M = N \times N$; thus the asymptotic computational cost is clearly

[12] For reference, the results were computed on a Lenovo W500 ThinkPad laptop, using both cores of its Intel Core2 Duo 2.80 GHz processor. The actual wall-clock time for each frequency point for the PEC sphere was about 2.6 seconds.

$\mathcal{O}([kd]^6)$. (The asymptotic analysis neglects the fact that when modelling a *surface*, one needs to approximate the *two* components of current on each patch – so in practice surface modelling is costly. As for the one-dimensional case, however, the matrix fill tends to dominate the runtime for many problems, with a somewhat lower asymptotic behavior.) In terms of memory, the requirements are $\mathcal{O}([kd]^4)$ for surfaces.

To give a concrete example, consider doubling the size of ground plane in the helix example discussed in Chapter 5; equivalently, double the frequency – the product kd expresses this product of wavenumber and size succinctly. The runtime will increase by between $2^4 = 16$ and $2^6 = 64$, and the amount of memory required will increase by a factor of $2^4 = 16$. (This is approximate since the helix must also be modelled more finely, but as a wire structure, the frequency scaling is somewhat better; however, the requirements of the ground plane increasingly dominate the considerations.) A factor of 64 is almost precisely the difference between minutes and hours and one should appreciate that modelling surfaces may require powerful computers and take considerable time. Fortunately, there are some methods available to assist in this regard, which we will discuss shortly.

Modelling volumes is even more costly. To model a volume of size $kd \times kd \times kd$, it is clear that the number of unknowns will now be $M = N \times N \times N$; thus the asymptotic computational cost is clearly $\mathcal{O}([kd]^9)$. (Again, the asymptotic analysis neglects the fact that when modelling a volume, one needs to approximate the components of current on each cell – now three of them. On the other hand, once again the matrix fill tends to dominate the runtime for many problems.) In terms of memory, the requirements are $\mathcal{O}([kd]^6)$ for volumes. We saw these effects clearly at work in Table 6.1, where a slight change in edge length for the volumetric case meant a dramatic increase in both runtime and memory requirements.

Code tip – using symmetry

The results shown in Fig. 6.10, and the accompanying Table 6.1 giving computational requirements, were computed using *symmetry*. In codes such as FEKO there are generally three types of symmetry available: electric symmetry, magnetic symmetry and geometrical symmetry. A plane of electric (magnetic) symmetry is a plane which can be replaced by an ideal electrically (magnetically) conducting wall without changing the field distribution; put differently, there are normal but no tangential components of electric (magnetic) field. (It may be stating the obvious, but such a plane of field symmetry can be of only one or the other kind.) A plane of geometrical symmetry is a plane of symmetry in the *model*. Exploiting electric and/or magnetic symmetry reduces the number of unknowns, and hence matrix size. (The memory requirement approximately halves for each plane of symmetry used, and runtime also decreases very substantially.) Geometrical symmetry, on the other hand, does not reduce the number of unknowns, but does permit some calculations to be re-used (such as those related to the construction of the matrix elements). Although no savings in memory are made, runtime can still be reduced.

For the spherical scatterer above, the scatterer was located symmetrically about the origin. The incident field was specified as incident along the z-axis, and with E_x polarization. The $x = 0$ plane was then specified as a plane of electric symmetry; similarly, the $y = 0$ plane was a plane of magnetic symmetry. Finally, the $z = 0$ plane was set as a plane of geometric symmetry. In codes such as NEC-2, and earlier versions of FEKO, this would have required that only one-eighth of the structure be generated; in CADFEKO, symmetry can very conveniently be directly applied to the CAD model, which in this case was the entire sphere.

Another type of symmetry which sometimes be exploited is an even-odd decomposition of a problem with geometrical symmetry into two problems, one with electric and one with magnetic symmetry. Even though two problems must be solved, the overall computational burden is still reduced.

6.8 Hybrid MoM/asymptotic techniques for large problems

This section is based on a review paper originally published as [19].

6.8.1 Introduction

Any combination of CEM techniques can be termed a hybrid. Here it is convenient to distinguish between *exact* and *approximate* hybrid approaches. In the former, also known as MoM/Green's function hybrids, special Green's functions are used to take the effect of the scatterer in Fig. 6.11 into account *implicitly*. Although very powerful for appropriate problems, the restricted number of special Green's functions available limits the generality of this approach. In the latter case, high-frequency methods such as physical optics are used to describe approximately the interaction between parts of the structure far removed from one another.

Probably the best known of the exact hybrids is the Sommerfeld potential treatment for radiators near, on or within stratified media. This will be discussed subsequently in this book. For slotted waveguide array analyses, the appropriate waveguide Green function has been widely used in MoM formulations. Another special Green function that has been used is that for layered spheres [20].

Deriving such Green functions is a formidable task: [15] gives a good introduction to the process of deriving a Green function, but for more advanced purposes a detailed description of dyadic Green functions may be found in [21]. A review of this type of hybrid method may be found in [22].

We use the term "exact" hybrid method for this approach since the only approximations made involve the conventional MoM discretization of the current on the radiator/scatterer. There is some disagreement about the use of the term "hybrid" for the MoM/Green function method; we follow the nomenclature of [22] here. As regards the use of "exact," we have already commented that many special Green's functions involve theoretically infinite series expansions or, as we will see, pose challenging integration problems in the complex plane, as is the case with Sommerfeld potentials.

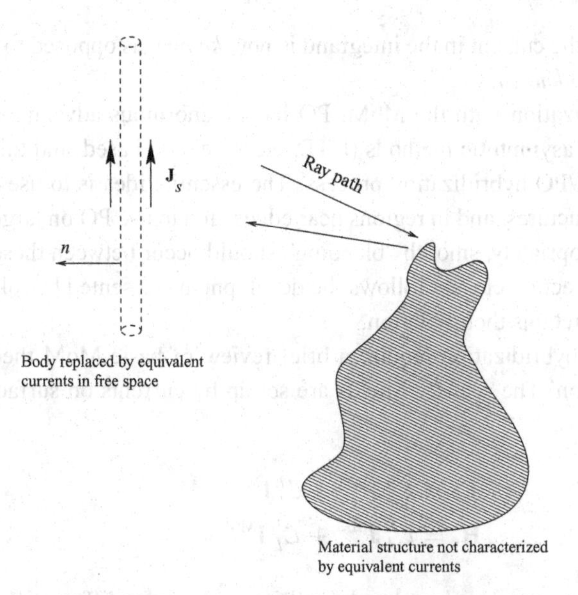

Figure 6.11 Wire radiator together with an electromagnetically large scatterer, whose effect is taken into account using a hybrid formulation. (After [19], ©1999 SAIEE.)

6.8.2 Moment method/asymptotic hybrids

Hybridizations of the MoM with various asymptotic techniques are approximate in the sense that in addition to the conventional MoM discretization, assumptions are made that are only exact in the high-frequency limit. (The MoM is sometimes described as a "numerically exact" formulation, in that the only approximations are those required to produce a linear system. This type of hybrid is no longer numerically exact – even if the equations could be solved exactly, without any errors introduced by discretization or numerical evaluation of integrals, the method is still approximate.) However, these methods are potentially more generally applicable than the MoM/Green function hybrids outlined above and we will now review physical optics for this purpose.

6.8.3 Physical optics and MoM hybridization

Physical optics (PO) is a well-established concept in electromagnetic theory [15, Section 7.10]. The essence is that the equivalent surface current on a smooth conducting surface is given by

$$\mathbf{J}_s = 2\hat{n} \times \mathbf{H}_i \tag{6.51}$$

We have already seen in Section 6.1 that this is an approximation of the MFIE. It may also be seen as an application of the equivalence principle, with the following approximations for a sufficiently large structure: firstly, $\mathbf{H}$ can be replaced by $2\mathbf{H}_i$ (this essentially assumes no end effects); secondly, currents can be "locally" imaged (hence the factor 2). Note that unlike a ray-based method, integration over the surface current is still required – but the current in the integrand is now *known*, as opposed to the MoM where the current is *unknown*.

In terms of hybridization with the MoM, PO has an enormous advantage in being *current based* – most asymptotic methods (UTD, etc.) are *field* based, and this leads to a rather natural MoM/PO hybridization process. The essential idea is to use the MoM on small, resonant structures, and in regions near edges, and to use PO on large, smooth areas. If applied appropriately, smooth "blending" should occur between these regions. The overview in this section closely follows the development presented by Jakobus and Landstorfer [23] and retains their notation.

The mechanics of hybridization require a brief review of basic MoM theory using linear operator notation. The scattered fields are set up by currents on surfaces ($\mathbf{J}^{\mathrm{MM}}$) and wires ($\mathbf{I}^{\mathrm{MM}}$):

$$\begin{aligned}\mathbf{E}_s &= \mathcal{L}_J^E \mathbf{J}^{\mathrm{MM}} + \mathcal{L}_I^E \mathbf{I}^{\mathrm{MM}} \\ \mathbf{H}_s &= \mathcal{L}_J^H \mathbf{J}^{\mathrm{MM}} + \mathcal{L}_I^H \mathbf{I}^{\mathrm{MM}}\end{aligned} \tag{6.52}$$

$\mathcal{L}_J^E$, $\mathcal{L}_J^E$, etc. are linear operator short-hand for the actual integrodifferential operators (for example, the EFIE and the MFIE as in Section 6.1). Standard MoM basis functions are used:

$$\begin{aligned}\mathbf{J}^{\mathrm{MM}} &= \sum_{n=1}^{N_J^{\mathrm{MM}}} \alpha_n \mathbf{f}_n \\ I^{\mathrm{MM}} &= \sum_{n=1}^{N_I^{\mathrm{MM}}} \beta_n g_n\end{aligned} \tag{6.53}$$

Jakobus and Landstorfer use piecewise linear basis functions for g_n and $\mathbf{f}_n$; the latter are the Rao–Wilton–Glisson triangular vector functions for surfaces as already discussed. For a PEC, the standard boundary condition $\mathbf{E}_{\tan} = 0$ is applied, resulting in:

$$-\mathbf{E}_{\tan}^i = \sum_{n=1}^{N_J^{\mathrm{MM}}} \alpha_n (\mathcal{L}_J^E \mathbf{f}_n)_{\tan} + \sum_{n=1}^{N_I^{\mathrm{MM}}} \beta_n (\mathcal{L}_I^E g_n)_{\tan} \tag{6.54}$$

Either collocation or weighted residuals can be used to solve for the unknown coefficients α_n and β_n (in total, $N_J^{\mathrm{MM}} + N_I^{\mathrm{MM}}$ of them).

Now, in the region of the scatterer *not* treated by the MoM, the PO surface current is approximated *using the same surface patch treatment as in the MoM region* as

$$\mathbf{J}^{\mathrm{PO}} = \sum_{n=N_J^{\mathrm{MM}}+1}^{N_J^{\mathrm{MM}}+N_J^{\mathrm{PO}}} \gamma_n \mathbf{f}_n \tag{6.55}$$

with $\mathbf{f}_n$ as before and γ_n coefficients of surface current in the PO region. It is *very important* to note that γ_n are known (in terms of the α_n and β_n coefficients) from the PO approximation, as shown later. Hence they are *not* obtained by the solution of a linear system – thus the matrix *size* remains $N_J^{\mathrm{MM}} + N_I^{\mathrm{MM}}$.

In the PO region, the PO current $\mathbf{J}^{\mathrm{PO}}$ is given by

$$\mathbf{J}(\mathbf{r})^{\mathrm{PO}} = 2\delta_i \cdot \hat{n} \times \mathbf{H}_i(\mathbf{r}) + \sum_{n=1}^{N_J^{\mathrm{MM}}} 2\alpha_n \delta_{J,n} \cdot \hat{n} \times \mathcal{L}_J^H \mathbf{f}_n + \sum_{n=1}^{N_I^{\mathrm{MM}}} 2\beta_n \delta_{I,n} \cdot \hat{n} \times \mathcal{L}_I^H g_n \tag{6.56}$$

$\delta_{J,n}$ and $\delta_{I,n}$ account for possible shadowing, with values of ± 1 or 0; the optical basis of the method will be recalled.

When currents in the PO region are included as well, the equation from the boundary condition *in the MoM region* becomes:

$$\mathcal{L}_J^E \mathbf{J}^{\mathrm{MM}} + \mathcal{L}_I^E \mathbf{I}^{\mathrm{MM}} + \mathcal{L}_J^E \mathbf{J}^{\mathrm{PO}} = -\mathbf{E}^{i,\tan} \tag{6.57}$$

Note that there are *two different* PO/MoM coupling mechanisms:

(1) The currents in the MoM region contribute to the PO currents via Eq. (6.56) (via the summation terms).
(2) The currents in the PO region in turn contribute to the fields in the MoM region and thus impact on the boundary condition represented by Eq. (6.57).

It might appear that this would require some iterative process for self-consistency, but the "feedback" effects can be taken into account in closed form. The PO currents can be found *in terms of the unknown MoM currents* as

$$\gamma_k = \tau_{i,k} + \sum_{n=1}^{N_J^{\mathrm{MM}}} \alpha_n \cdot \tau_{J,n,k} + \sum_{n=1}^{N_I^{\mathrm{MM}}} \beta_n \cdot \tau_{I,n,k} \tag{6.58}$$

with

$$\begin{aligned}
\tau_{i,k} &= (\hat{t}_k^+ + \hat{t}_k^-) \cdot (\delta_i \hat{n} \times \mathbf{H}_i) \\
\tau_{J,n,k} &= (\hat{t}_k^+ + \hat{t}_k^-) \cdot (\delta_{J,n} \hat{n} \times \mathcal{L}_J^H \mathbf{f}_n) \\
\tau_{I,n,k} &= (\hat{t}_k^+ + \hat{t}_k^-) \cdot (\delta_{I,n} \hat{n} \times \mathcal{L}_I^H g_n)
\end{aligned} \tag{6.59}$$

$\hat{t}_k^+$ and $\hat{t}_k^-$ are unit vectors associated with the kth triangle edge; see [23] for further details. It is important to note that *all* the terms in the above equation are *known*, being either derived from the geometry of the problem, the discretization or the chosen basis function.

The central idea here is that these PO currents *in terms of the MoM unknowns* can now be substituted into Eq. (6.57). The final result is the following:

$$\sum_{n=1}^{N_J^{\mathrm{MM}}} \alpha_n \cdot \left[(\mathcal{L}_J^E \mathbf{f}_n)_{\tan} + \sum_{k=N_J^{\mathrm{MM}}+1}^{N_J^{\mathrm{MM}}+N_J^{\mathrm{PO}}} \tau_{J,n,k} \cdot (\mathcal{L}_J^E \mathbf{f}_k)_{\tan} \right]$$

$$+ \sum_{n=1}^{N_I^{\mathrm{MM}}} \beta_n \cdot \left[(\mathcal{L}_I^E g_n)_{\tan} + \sum_{k=N_J^{\mathrm{MM}}+1}^{N_J^{\mathrm{MM}}+N_J^{\mathrm{PO}}} \tau_{I,n,k} \cdot (\mathcal{L}_J^E \mathbf{f}_k)_{\tan} \right]$$

$$= -\mathbf{E}_{i,\tan} - \sum_{k=N_J^{\mathrm{MM}}+1}^{N_J^{\mathrm{MM}}+N_J^{\mathrm{PO}}} \tau_{i,k} \cdot (\mathcal{L}_J^E \mathbf{f}_k)_{\tan} \tag{6.60}$$

Equation (6.60) summarizes the MoM/PO interaction: the effect of the PO is to *alter* the MoM matrix entries. Note that each MoM entry is modified by contributions from *all* the PO currents; this can become computationally expensive and can be neglected under certain conditions, usually when the PO and MoM regions are physically separated. (An example is a reflector antenna, where the feed is treated with the MoM and the reflector with the PO.) Note further that the boundary condition of zero tangential **E** is only rigorously enforced in the MoM region.

In the basic MoM/PO hybridization outlined above, edge effects are *not* taken into account by the PO. It is possible to use Fock theory to account for these effects; see for example [23]. The approach used is related to Umfitsev's physical theory of diffraction.

For very large structures, the integration over the entire structure can still become very time consuming – although the $\mathcal{O}(f^6)$ dependence of the MoM is reduced enormously, the PO asymptotic dependence is still $\mathcal{O}(f^2)$.

Hodges and Rahmat-Samii have shown that the MoM/PO hybrid can be seen as a special case of a more general EFIE/MFIE hybridization, with the MoM/PO as the first term in an iterative Neumann series technique [24]. They show good results for two monopoles mounted on opposite sides of a cylinder, and thus in each other's shadow region. However, the use of the MFIE restricts the method to smooth closed bodies.

6.8.4 A FEKO example using the MoM/PO hybrid

The above theory is available within FEKO, and we will now consider an example of its applications. For this example, one of the simplest (and also most effective) applications of the MoM/PO hybrid will be chosen. We will mount a $\lambda/2$ dipole antenna horizontally above a finite ground plane, of $1\lambda \times 1\lambda$ in size. From basic image theory, the "image" in the ground plane is out of phase, so the distance above the ground plane should be an odd multiple of $\lambda/4$ above the ground plane to produce constructive interference.

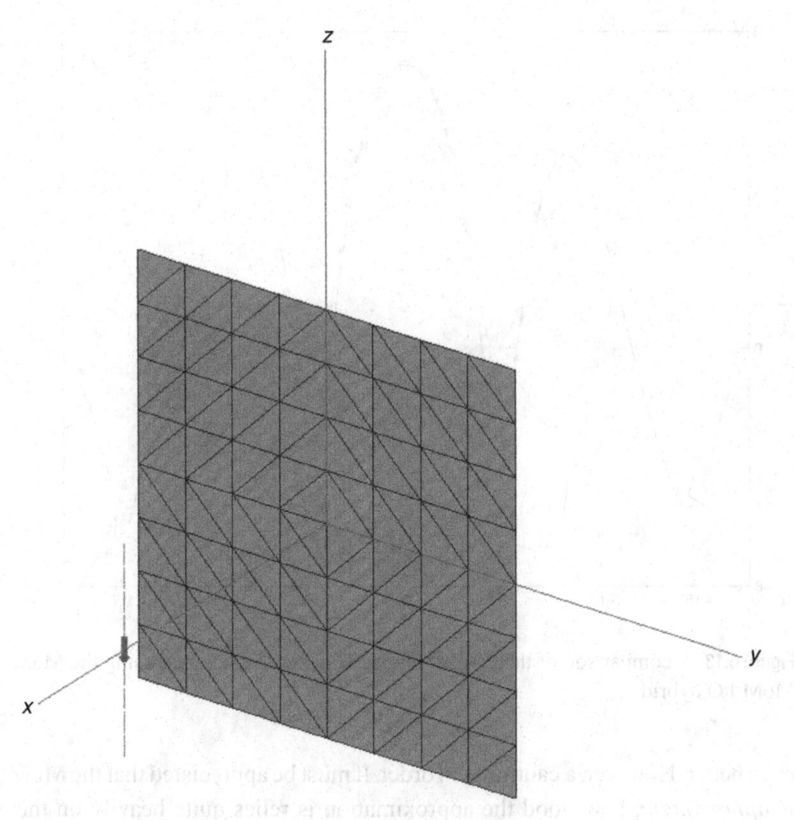

Figure 6.12 FEKO model of a dipole in front of a reflector.

Modelling hints – symmetry

Once again, symmetry can be exploited to build the model and improve the computational efficiency. By mounting and feeding the dipole symmetrically about the $y = 0$ and $z = 0$ planes, magnetic and electric symmetry can be used. Note that the quarter-ground plane is imaged first in the $y = 0$ plane before the half-dipole is added; one does not want to image a wire on top of itself! Following this, the half-ground plane and half-dipole are then imaged in the $z = 0$ plane to create the whole model, and the feed segment is then added.

Two approaches have been used to solve this problem: firstly, the MoM has been used for the entire problem; then, the MoM is applied to the dipole only, and the effect of the reflector is approximated using the PO. The FEKO models for both are shown in Fig. 6.12 – note that the models appear identical, since it is the mathematical approach, rather than the geometrical model, which differs. Results comparing the far-field H-plane ($z = 0$) radiation patterns computed using the two approaches are shown in Fig. 6.13.

The results shown in Fig. 6.13 compare favorably. Using some advanced methods within FEKO which correct the PO currents at the edge of the reflector, it is possible to do

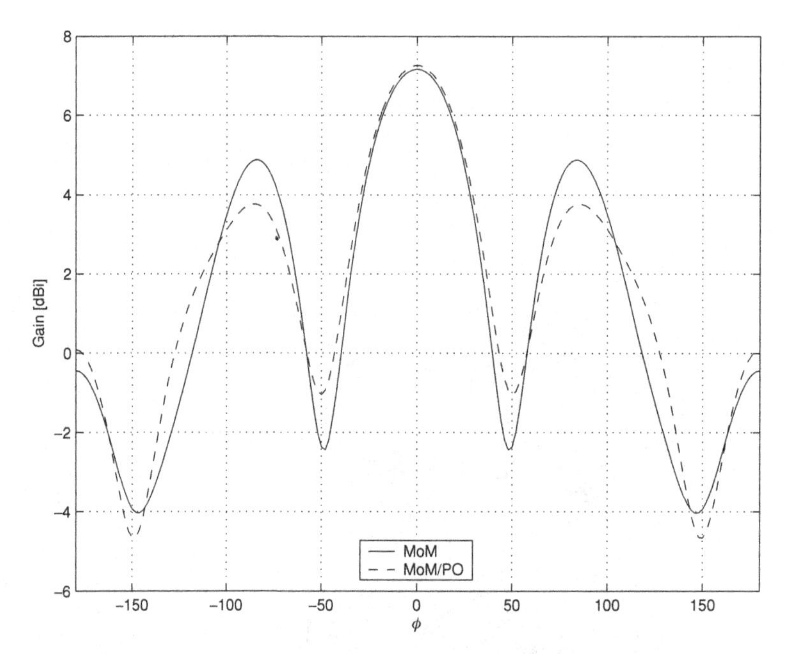

Figure 6.13 A comparison of the H-plane far-field patterns computed using the MoM and MoM/PO hybrid.

even better. However, a caution is in order. It must be appreciated that the MoM/PO hybrid is *approximate*; how good the approximation is relies quite heavily on the experience of the user. As such, it is useful to build confidence by initially comparing results using MoM/PO hybrids with full MoM solutions as far as possible. Efficient use of symmetry usually allows the solution of quite electrically large MoM problems, although these may of course take some time to compute. Once one is reasonably confident of the level of accuracy for a particular class of applications, one may then do production runs investigating changes to and optimization of the structure, etc. It would, however, be very unwise to base major design decisions on an MoM/PO hybrid solution which one has not carefully evaluated beforehand. (Of course, this is true in general of computed solutions, but even more so in this case.) Problems which generally lend themselves very well to the MoM/PO hybrid approach are reflector-type problems, since the radiating feed element is largely decoupled from the reflector, and ray-tracing issues are not problematic.

Modelling hints – using the MoM/PO hybrid within FEKO

FEKO is the only commercial code offering this functionality at the time of writing, so the following discussion only applies to FEKO. Physical optics is controlled using the `PO` card, which offers a number of parameters which require some brief discussion. The first parameter, requiring a label, is obvious; the PO is applied to the structures with this label. The second parameter controls *ray tracing*. Because the PO is an optics-based method, in general one needs to ray trace to determine whether a triangle is in the "lit" or "shadow" region relative to the source. In this case, it is clear that all triangles are illuminated, and ray tracing may be switched off to save time. The third parameter relates to the use of symmetry in ray tracing and is irrelevant here since ray tracing has been deactivated. The fourth parameter controls MoM–PO coupling, as described in the previous theoretical section; here, the full treatment is applied and the regions are fully coupled. The fifth parameter is another optics-based one; it determines the number of multiple reflections to be taken into account. In this case, none are required. The final parameter is for specialized use and the default should be used here.

FEKO offers additional functionality to improve PO modelling. The `KA` card permits one to define the boundary of the PO region, and "fringe wave" currents are then used in this region to improve the approximation. The `VS` card allows one to specify "visibility" information, to reduce the time required when multiple reflections are present. The `FO` card uses Fock theory to improve the PO surface current.

6.9 Other approaches for the solution of electromagnetically large problems

6.9.1 Background

By the late 1980s, research on the MoM was confronted with the basic problem of the high asymptotic cost of the method – $\mathcal{O}(N^3)$ in terms of number of unknowns, or $\mathcal{O}([kd]^6)$ for surfaces, as we have seen for direct solvers. Little can be done to improve this further, apart from the application of high-performance computing (of which more anon). Iterative solvers started attracting much attention in CEM the late 1980s – even though the basic algorithms, in particular the conjugate gradient (CG) algorithm, have been known since the 1940s – since the computational cost is $\mathcal{O}(N^2)$ per iteration, with overall cost $\mathcal{O}(N_{\text{iter}}N^2)$ for N_{iter} iterations. Clearly, if N_{iter} can be kept well below N, algorithms with better scaling properties are possible. It has to be said here that, unfortunately, the considerable experience accumulated by many researchers over the years has indicated that it is very difficult to predict N_{iter} for arbitrary problems; testing the algorithms on canonical problems, such as spheres, has frequently resulted in highly over-optimistic predictions. (The reason is the relatively simple eigenvalue structure of such problems; since the iterative methods usually used variants of the CG method, the rate of convergence is heavily determined by the eigenvalue spectrum.) So, using iterative techniques alone is not sufficient – and in any case, this does nothing to the $\mathcal{O}(N^2)$ memory requirements of the method, which is frequently as serious a problem as computational cost.

From a slightly different perspective, the integral equation formulations which we have worked with are essentially *convolutions* of the Green function with the currents. Familiarity with signal processing methods immediately suggests that convolution in one domain may be more easily implemented by multiplication in the Fourier transform domain; we will exploit this idea in Chapter 7, although for a slightly different purpose. But for now, the idea that one could use a Fourier transform immediately suggests the use of the *fast Fourier transform* (FFT), and, indeed, this was one of the first successful "fast" methods in electromagnetics. However, it was limited in terms of application to general structures with arbitrary meshes. An extension of this concept, the adaptive integral method, removes this restriction. However, it is an alternative approach, the fast multipole method (FMM), which provided the theoretical breakthrough in the early 1990s. In its most powerful multilevel form it reduced the asymptotic cost from $\mathcal{O}(N^2)$ to $\mathcal{O}(N \log N)$, and it is the most popular of the fast methods today. It was a breakthrough as significant as Berenger's PML absorber,[13] although the theory is rather more complex, and efficient implementation in particular is challenging. (By comparison, the PML is really quite straightforward to code.) Despite the complexity of the theory underlying the FMM, and the challenges of implementing it in parallel, it is offered by some commercial codes at the time of writing,[14] and hence an elementary introduction is certainly appropriate at this stage. Before looking at fast techniques, however, we will briefly discuss high-performance computing, which is also an important topic when the solution of large electromagnetic problems is considered.

6.9.2 High-performance computing

All the methods and technologies described in this section had their genesis in the late 1980s. One approach to the problem of high computational cost, and one which is still bearing fruit today, was exploitation of the emerging technology of parallel processing. Parallel processing – or indeed high-performance computing (HPC) in general – simply provides more computational power, it does not address the fundamental algorithmic issue of computational cost, but can significantly push the envelope of any particular computational technique. At its heart, there are only two ways of making a given computation faster: either increase the rate at which a computer can process information, or do more operations at the same time. The former of course has been the dominant technological drive through several decades, manifested by clock speeds which, for typical personal computers, have increased from some tens of MHz at the start of the 1990s to some GHz by the millennium, only a decade later. (In the early 2000s, however, clock speeds stagnated, and it appears that a few GHz is as fast as clock speeds are likely to go with current technology.) The latter has spawned a variety of methods; historically, parallel processing originally split into *pipelining* and *replication*.

Pipelining involves overlapping parts of operations in time and was the approach taken by the vector supercomputers, such as the early CRAY machines (the first of which was installed in 1976). Replication provides more than one functional unit (e.g. CPU), permitting operations to be performed simultaneously, and was the competing approach taken by large processing arrays. Another nomenclature encountered in the earlier literature (and still used today) was single instruction multiple data (SIMD) and

[13] Hopefully, this comparison will not cause confusion: the PML and FMM are entirely different methods, with quite different aims.

[14] FEKO appears to have been the first publicly available commercial code to incorporate the FMM; the frequently referenced Fast Illinois Solver Code (FISC) has numerous restrictions on its distribution, especially outside the USA, due to US military funding during its development.

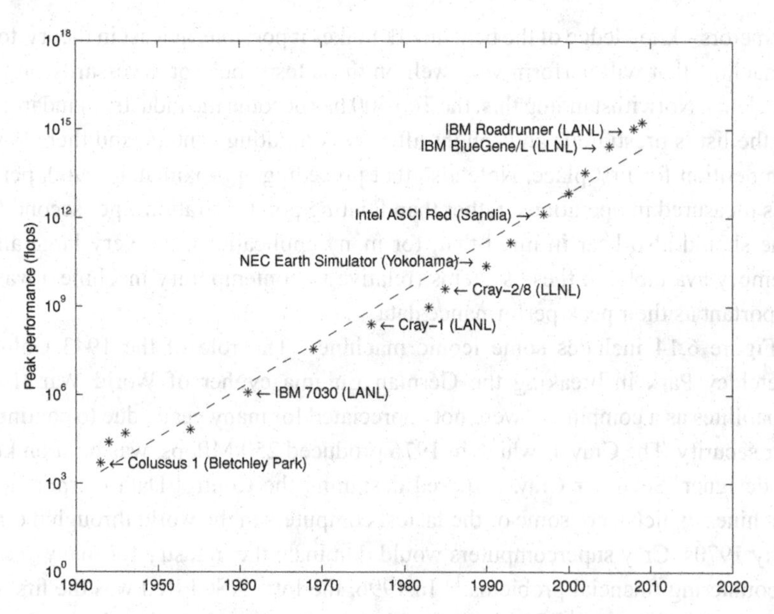

Figure 6.14 Historical development of supercomputer systems.

multiple instruction multiple data (MIMD) machines. This taxonomy was introduced by in the early 1970s [25]; a MIMD system described a computer consisting of a number of nodes, each with at least a processing element, operating independently on its own local instruction stream and data, whereas a SIMD system performed the same operation in lockstep to all data. Machines were also characterized in terms of how data were exchanged; many of the early experimental systems used were local memory, message passing systems. In these, all memory was divided up locally amongst the available processors, and a processor could only directly access its own memory. Access to the memory on other processors was done by explicit message passing, which was *much* slower than direct memory access. Message passing remains a core technology on modern clusters. However, the problem of memory contention that complicated the other main competing approach to memory allocation, namely global memory, was removed with this approach. Technological advances have however blurred many of the traditional distinctions. Even the ubiquitous CPUs encountered in personal computers contain significant elements of pipelining and replication – mutiple processing units (known as cores) are already commonplace at the time of writing – and increasingly sophisticated architectures have attempted to blur the global/local memory dichotomy with varying degrees of success, although memory access remains a major issue in high performance computing.

In Fig. 6.14, supercomputer development in terms of peak performance since the first electronic computers is highlighted by a selection of the fastest machines over time. For machines preceding 1993, the "fastest" may be disputed, as the figures are based on various historical claims, but from 1993 onwards, the Top 500 list, using a linear algebra benchmark, has provided uniformity. The Top 500 benchmark is not without detractors – knowledge of the benchmark makes it possible, at least in theory, to engineer a machine that will perform very well on these tests, but not necessarily on real-world problems. Notwithstanding this, the Top 500 has become the industry standard; inclusion on the list is prestigious and sought-after by computing centers, and there is very keen competition for first place. Note also that preceding approximately 1960, performance was measured in operations, rather than floating point operations, per second (or flops). One should also bear in mind that, for many applications, the very large amounts of memory available on these systems (relative to contemporary machines) was often as important as their peak performance data.

Figure 6.14 includes some iconic machines. The role of the 1943 Colussus 1 at Bletchley Park in breaking the German Enigma cypher of World War II – and its capabilites as a computer – were not appreciated for many years, due to continuing post-war security. The Cray 1, which in 1976 produced 250 Mflops, was built on knowledge its designer, Seymour Cray, acquired designing the Control Data Corporation (CDC) machines, which were some of the fastest computers in the world through the 1960s and early 1970s. Cray supercomputers would dominate the industry for many years, before encountering financial problems.[15] In 1996, the Intel ASCI/Red was the first computer to exceed 1 Tflop (10^{12} flops) on the Linpack benchmark, and maintained its position as the fastest machine in the world (number one on the Top 500 list) for three years, 1997–2000. In 2008, the IBM Roadrunner became the first to exceed 1 Pflop (10^{15} flops) on the benchmark.

A straight-line fit to the semi-logarithmic curve in Fig. 6.14 gives an increase per decade of just under 50 (actual 45.9) per decade. One should exercise caution with such analysis; this is not some law of nature being investigated, but the performance of engineered devices responding to needs in especially national security. (It is notable that Japan's Earth Simulator is one of the very few supercomputers which have been number one on the list which were not primarily designed for national security applications, and also one of very few outside the USA.) In the general context of computing, Moore's law (based on the 24-month doubling period of his 1975 paper [26]), predicts an increase in speed by a factor of 32 over a decade. The additional improvement is due to parallel processing. Early machines were largely single processors, but the Roadrunner (in the configuration which broke the 1 Pflop barrier) used 122 400 cores. It is also noticeable that, despite the general stagnation of clock speeds from the early 2000s already noted, the machines charted in this decade are above the historical trend, due to other engineering advances.

With the Pflop barrier now exceeded, the next barrier is the Exaflop (10^{18} flops). A simplistic extrapolation of the historical curve in Fig. 6.14 would imply that this should be reached some time in the late 2020s. If the trend through the decade of the 2000s is continued, it could be somewhat sooner, perhaps the early 2020s. However, predictions

[15] The name Cray has been associated with a number of business entities which derive in some way from the original; at the time of writing, early 2010, a Cray XT5 machine, the Jaguar, was back in number one position on the Top 500.

of this type ignore all manner of practical issues which will require attention. Thermal dissipation of the heat generated by tens or hundreds of thousands of processors (cores) is already a major problem in contemporary supercomputers, and memory speeds have not remotely kept up with processor speeds. (As a rule of thumb, there has been about 1 byte of RAM for each flop of performance for many years, but this may not be sustainable, and, in the future, relatively less RAM may be available.) There is also no new technology at present which holds the promise of alleviating these issues. Compared to the avant-garde technology used in earlier supercomputers (such as the GaAs semiconductors used in some Cray designs), today's supercomputers are without exception based on commodity silicon processors, but with very fast interconnection technologies. Nonetheless, work can be expected towards Exaflop machines, and, in the past, the ingenuity of computer engineers has overcome barriers to performance improvement which many claimed were insurmountable at the time.

In practice, it is useful to distinguish between capability computing, which is the use of a machine at, or close to, its limit to solve one very large problem (which is what the Top 500 list monitors), and capacity computing, which is the efficient use of cost-effective computing power to solve reasonably large problems, or many small problems. The latter is by far the most common use of most computing centers. Also, it is interesting to observe that there is typically some three orders of magnitude difference in peak performance between the number 1 and number 500 machines on the list (corresponding at present to Pflops vs Tflops), and that this repeats again with typical desktop computers (presently in the Gflop regime).

Moore's law

This "law" has been evolved somewhat in time. In Moore's original 1965 paper (in a now very hard to find magazine[a]), he commented that "the complexity for minimum component costs has increased at a rate of roughly a factor of two per year" and predicted that the rate would be sustained for at least another ten years. In 1975, he altered his projection to a doubling every two years [26]. There have been another widely quoted variant of the law giving a doubling period of 18 months; its attribution is disputed.

[a] *Electronics Magazine*, 19 April 1965.

The basic concept of parallel processing was, and still remains, to provide P processors or processing elements, and by splitting the computational load, reduce the overall runtime by a factor as close to P as possible. Several methods have been proposed to characterize parallel computers, but the most widely used are speed-up and efficiency. Speed-up, S, is the ratio of time taken by an equivalent serial algorithm running on one processor, T_s, to the time taken by the parallel algorithm using P processors, T_P.

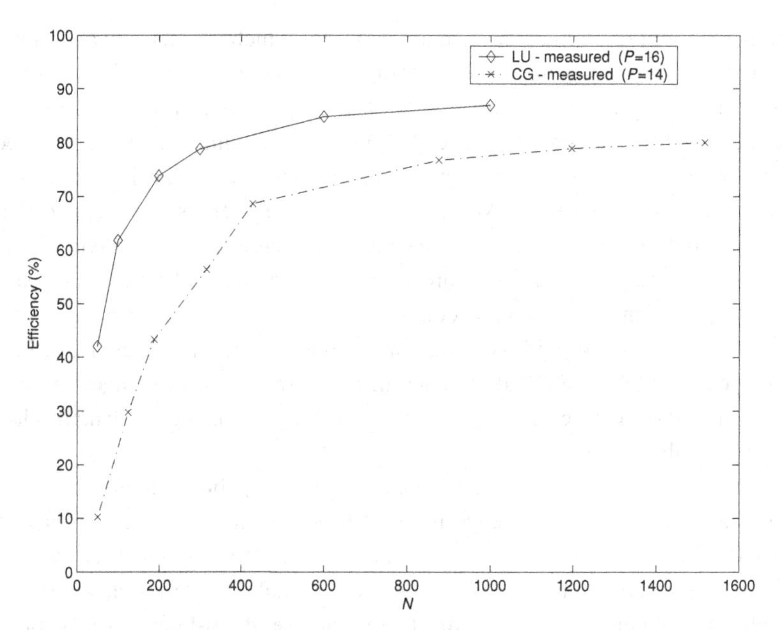

Figure 6.15 Comparison of measured efficiencies of parallel CG and LU algorithms on a transputer array, for an MoM problem with a total of N unknowns running on P processors. (Adapted from [27, Fig. 18] and [28, Fig. 12].)

Efficiency, ϵ, is the speed-up normalized by the number of processors. Formally,

$$S = \frac{T_s}{T_p} \tag{6.61}$$

$$\epsilon = \frac{S}{P} \tag{6.62}$$

S is usually bounded from above[16] by P, and ϵ is hence usually bounded from above by 1. As already noted, for many applications, the large amounts of memory often available on parallel systems can be as significant as their processing power.

Some algorithms can be parallelized very easily and efficiently: examples are the FDTD and iterative methods. Some, such as LU decomposition, are rather less obvious, but can nonetheless be very efficiently parallelized with some clever data decomposition techniques. All the major algorithms in CEM have been parallelized with varying degrees of success over the last decade; perhaps the most problematic one has been the FEM, due to the large, unstructured, but highly sparse matrix characterizing the method. Examples of measured efficiencies on a transputer array are shown in Fig. 6.15. (The results are shown for slightly different numbers of processors; this was due to different interconnection topologies used for the algorithms.) These data were measured in the early 1990s on a transputer array, hence the problem sizes are small by contemporary standards, but, nonetheless, establish the principle. Contemporary results are shown in

[16] Sometimes, architectural quirks have resulted in "superlinear" improvement on specific problems, i.e. a speed-up in excess of P; usually, this was a result of the cache design.

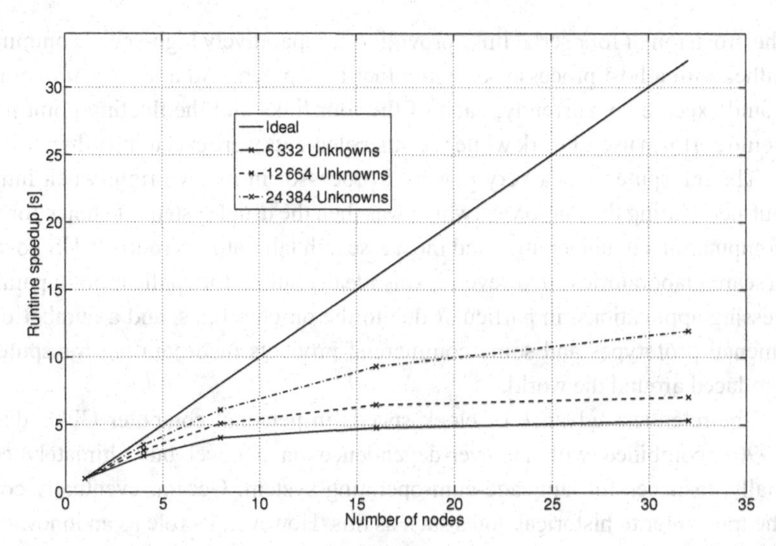

Figure 6.16 Speedup vs. the number of nodes on an IBM e1350 cluster, using the Infiniband interconnect. (Figure courtesy D. Ludick [5].)

Fig. 6.16; these results were measured in 2009 using a parallel version of FEKO 5.3, running on the IBM e1350 cluster installed at the Centre for High Performance Computing (CHPC), Cape Town, South Africa.[17] This cluster had 160 nodes, each with two dual-core AMD Opteron 2.6 GHz processors, and 16 GB of RAM, giving a total of 640 processors with a peak processing power of around 2.5 Tflops and 2.56 TB of RAM. The system has both 1 Gbit/s Ethernet and 10 Gbit/s Infiniband interconnects. (Although faster, the latter is not as widely supported in application programs.)

An historical aside – the transputer

In the late 1980s, PCs were limited by the 640 kB limitation on RAM imposed by the then dominant operating system, DOS, and clock speeds were low. Supercomputers were (and for that matter still are) extremely expensive. A British company, INMOS, introduced the transputer, one of the first "computers on a chip," incorporating a CPU, floating point unit, memory and communication links. (This was to become quite standard later, but at the time was revolutionary.) The transputer came in several different variants – the T800 model was the one widely used in parallel processing.

The transputer was a 32-bit RISC[a] design, capable of internal operation at up to 30 MHz – again, this must be seen from the viewpoint of the technology of the time! One T800 transputer was able to produce a peak floating point throughput of 1.5 Mflops. A novel feature, still not widely seen on other systems to this day, was

[17] This center, which opened in 2008, was able to learn many lessons from international centers and epitomizes a modern HPC center. At the time of writing, early 2010, the CHPC has an IBM P690 SMP, the e1350 cluster, an IBM BlueGene/P and a SunBlade cluster; the last appeared at number 312 on the Nov 2009 Top 500 list, with a peak measured performance of 25.440 Tflops.

the provision of four serial links providing comparatively high-speed communication either with a host processor or with other transputers. Additionally, *all* components could execute concurrently; each of the four links and the floating point processor could perform useful work while the other elements were executing other instructions.

The transputer was a very powerful processor in its own right when introduced, out-performing the microVax, which was then the usual system of choice for numeric computations in universities and most research laboratories (outside US government research laboratories). However, it was ideally suited for application in parallel processing applications, in particular due to the on-chip links, and a number of experimental prototypes and some commercial products incorporating transputers were produced around the world.

The relentless advance of clock speeds in personal computer CPUs during the 1990s, combined with an over-dependence on a novel but ultimately commercially unsuccessful language-cum-operating system, Occam, eventually consigned the transputer to historical notes such as this. However, its role as an innovative catalyst in affordable parallel processing should not be underestimated; its do-it-yourself bargain-basement philosophy, if not technology, inspired a generation of computational scientists working at institutions unable to afford the extremely expensive supercomputers of the time, and still resonates today in current systems using Linux clusters. The idea has recently resurfaced again with the use of graphical processing units (GPUs) for general-purpose computing.

[a] Reduced instruction set computer.

In this context, it is necessary to mention Amdahl's "law,"[18] which states that if an algorithm contains both a serial and a parallel part, the relative time taken by the serial part increases as parallelization reduces that of the parallel part, and a law of diminishing returns holds: further parallelization has increasingly little influence on runtime. While this observation is perfectly true, for many problems the ultimate aim is to increase the problem size that can be handled. Thus as more parallelization is made available, larger problems are tackled and the overall serial/parallel split remains fairly constant. In particular, the efficiency of many parallel algorithms is a function of grain size – the number of unknowns per processor, N/P. An example of this is shown in Fig. 6.17, which indicates that for a particular grain size, the algorithm has approximately constant efficiency. It is very interesting to compare this with Fig. 6.18, measured in 2009 on the contemporary e1350 cluster at CHPC discussed earlier, which shows very much the same trends, although of course on far larger datasets, reflecting twenty years of development.

When HPC first came to the attention of the CEM community, it was often accompanied by highly specialized hardware, frequently purpose built, such as the transputer-based arrays mentioned here. Throughout the 1990s, relatively mainstream environments

[18] As with Moore's law, this is really an observation rather than a law in the sense as used in physics.

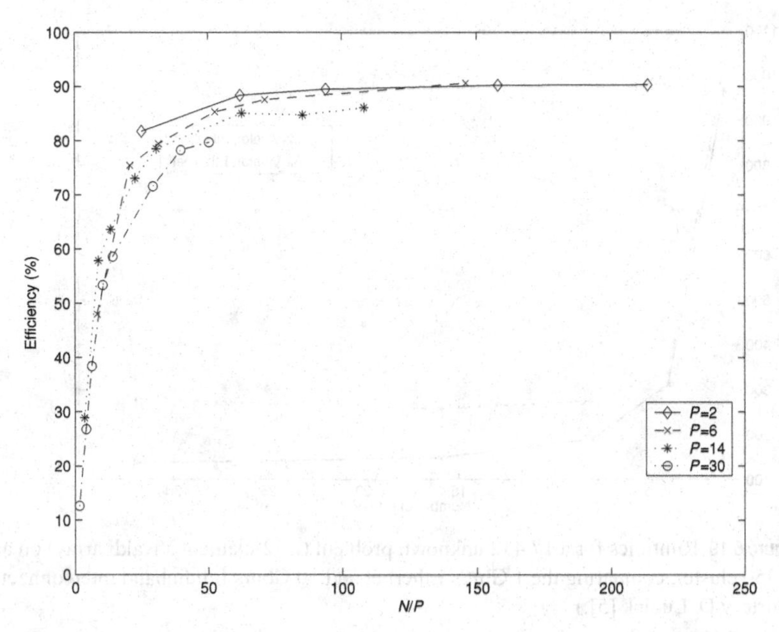

Figure 6.17 Measured efficiency of a parallel CG algorithm on a transputer array, for an MoM problem with a total of N unknowns running on P processors. (After [27, Fig. 7], ©1993 ACES, reprinted with permission.)

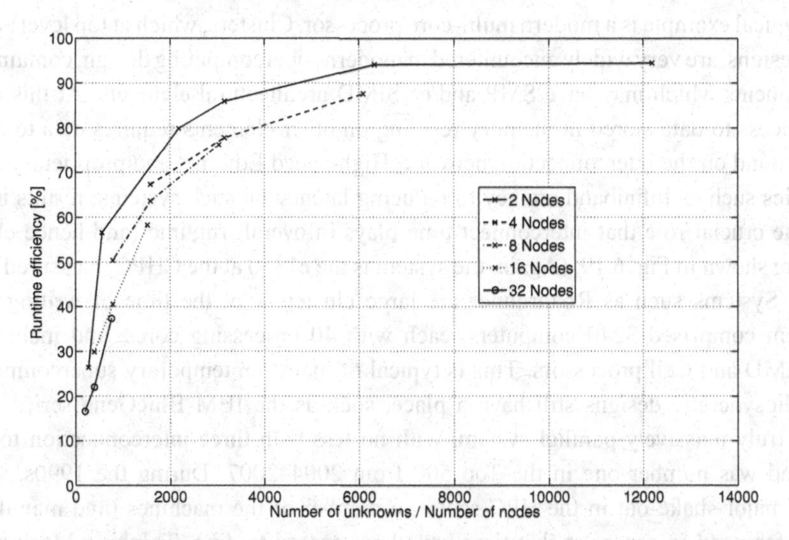

Figure 6.18 Efficiency vs. grain size on an IBM e1350 cluster, using the Infiniband interconnect. (Figure courtesy D. Ludick [5].)

increasingly became the norm, apparently reflecting a degree of maturity in the field, but at the time of writing, special-purpose hardware, most notably the use of graphical processing units for general-purpose computing, was again a topic of very active research (see, for instance, [29]). Although the old SIMD–MIMD classification is not used as much nowadays, it is still a useful tool for distinguishing elements

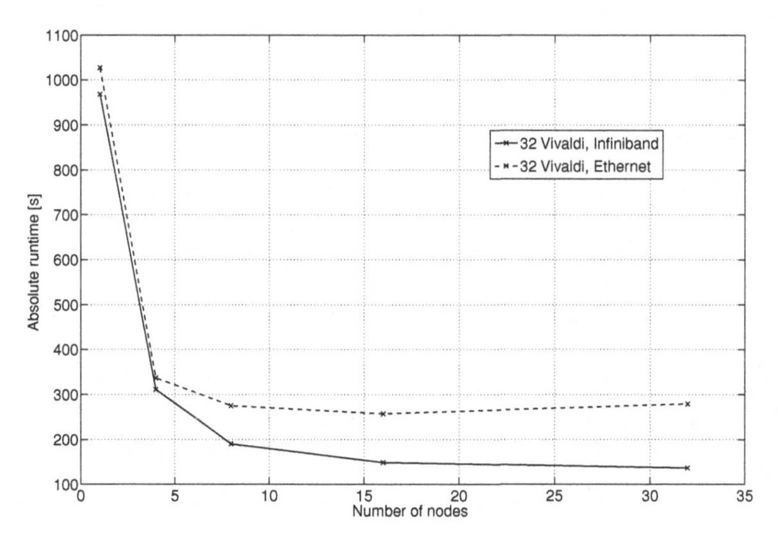

Figure 6.19 Runtimes for a 17 472 unknown problem (a 32-element Vivaldi array) on an IBM e1350 cluster, comparing the 1 Gbit/s Ethernet and 10 Gbit/s Ininfinband interconnects. (Figure courtesy D. Ludick [5].)

of high-performance computers, which increasingly hybridize such approaches. SMP (symmetric multi-processors) are machines which in essence have shared memory; a typical example is a modern multi-core processor. Clusters, which at top level are MIMD designs, are very widely encountered in modern supercomputing design, containing components which may have SMP and/or SIMD architectural elements. At this top level, access to data stored in memory residing on other elements requires data to be moved around on the interconnection network. High-speed Ethernet and proprietary technologies such as Infiniband are key to reducing latency on such systems; results indicating the crucial role that interconnect time plays in overall runtime (and hence efficiency) are shown in Fig. 6.19. (Again, the system is the e1350 at the CHPC, measured in 2009.)

Systems such as Roadrunner are large clusters – at the time of writing, the system comprised 3240 computers, each with 40 processing cores, and including both AMD and Cell processors. This is typical of many contemporary supercomputers, but idiosyncratic designs still have a place, such as the IBM BlueGene series, which is a truly massively parallel system, with no less than three interconnection topologies, and was number one in the Top 500 from 2004–2007. During the 1990s, there was a major shake-out in the HPC sector; a number of the machines (and manufacturers) referenced in papers at that time have long ceased trading. Thinking Machines Corp. and their Connection Machines (CM-2 and CM-5), which at the time were some of the few truly deserving the *massively* parallel tag, with thousands of SIMD processors, are gone. Kendall Square Research, whose machines had some innovative features, not least a physically distributed memory which was accessed as shared memory by application programs, using a system of multilevel caches, has also long ceased to function commercially. The past decade of the 2000s saw a continued consolidation in this specialized industry. At the time of writing, IBM and HP dominate the Top 500 list, but Cray (with the number one machine, an XT5 known as the Jaquar, with a peak measured performance speed of 1.75 Pflops and almost a quarter of a million cores), SGI, Dell and Sun continue to be very influential vendors in this field.

It is perhaps ironic that the most efficient languages for programming these systems are some of the oldest, namely FORTRAN and C. These languages have not abstracted away from the hardware to the extent of more modern languages, permitting more efficient use of the hardware. MPI (message passing interface) is the de-facto standard library for parallel programming, permitting programmers to move data around the machine with high-level library calls.

Due to the highly connected nature of CEM algorithm, new paradigms such as "grid computing" and "cloud computing" have had little impact on large-scale EM analysis, although "embarrassingly parallel" applications (such as repeating an analysis over a frequency loop) may well be able to exploit these concepts.

A noteworthy aspect of the work reported in the literature on parallel processing is that no new specifically parallel algorithms have arisen in computational electromagnetics. Two decades back, when parallel computing first attracted serious interest, there was speculation in some quarters that the rise of massively parallel computers would trigger entirely new algorithms that were only feasible in massively parallel computing environments. With hindsight, such claims appear as primarily marketing "hype." The usability of current high-performance computing platforms is improving with time, but parallel programming remains challenging, easy-to-use parallel debuggers are still scarce, and remote access, whilst greatly improved with high-speed internet, is still not as convenient as on-site access.

Nonetheless, despite implementation issues which remain challenging, parallel processing has emerged as a very useful enabling technology; several commercial codes (such as FEKO) are available in parallelized versions for various platforms. Whilst one does not always appreciate the impact of incremental increases in performance, when compounded over decades the results are deeply impressive. In Fig. 6.20, the time required for direct matrix solution (LU decomposition) on systems capable of sustaining 1 Mflop, 1 Gflop, 1 Tflop and 1 Pflop respectively are compared.[19] Comparing a 1 Mflop (typical of the late 1980s) and a 1 Gflop machine (typical of current systems), one sees that for a problem with around 1000 unknowns, the time has dropped from around an hour to a few seconds. A similar improvement is noted for a 10 000 unknown problem when comparing a 1 Gflop and a 1 Tflop machine, the latter typical of entries towards the bottom of the Top 500 list at the time of writing. Since the first edition of this text, a Pflop machine has also been added to the figure, noting the accomplishments of Roadrunner and Jaquar.

In conclusion, perhaps of most general significance in the CEM community, as multiple cores become standard on desktops and laptops, is that simulation packages supporting parallel processing will increasingly – and invisibly – benefit routine usage.

[19] The operation count for LU decomposition for a matrix of dimension N with complex valued entries is approximately $8/3N^3$ floating point operations.

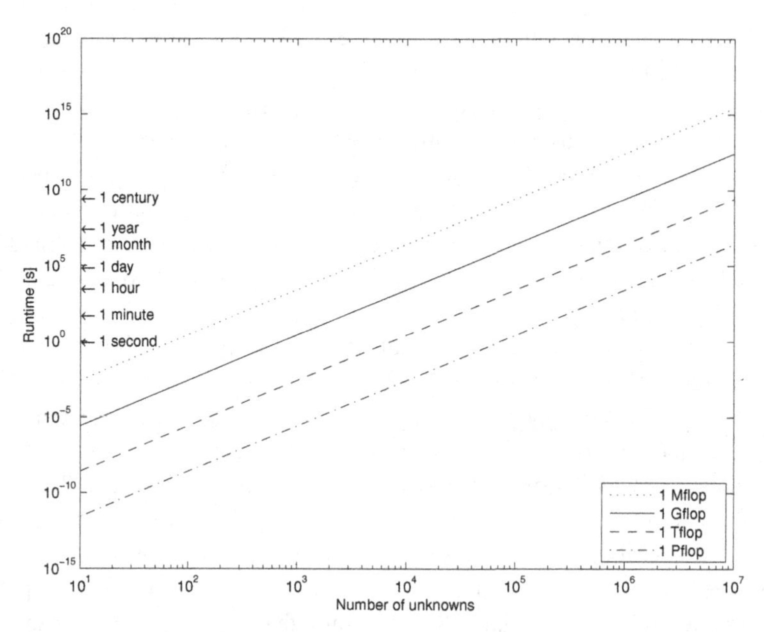

Figure 6.20 Runtimes for LU decomposition, compared for systems capable of sustaining 1 Mflop, 1 Gflop, 1 Tflop and 1 Pflop.

6.9.3 FFT-based methods

If we refer back to the very simple introductory thin-wire example of Chapter 4, specifically to Eqs. (4.15) and (4.16), we note that Z_{mn} is a function of only $m - n$ and Δ. With a uniform discretization, as used there, the latter is constant, and hence we actually only need to compute one row of the matrix. This is known as *Toeplitz* (or translational) symmetry. The reason that this observation is important is that in this case, the product of this matrix with a vector can be implemented as a discrete convolution.

In general, a discrete convolution is an operation of the form

$$e_m = \sum_{n=0}^{N-1} j_n g_{m-n} \tag{6.63}$$

or in matrix form:

$$\begin{bmatrix} g_0 & g_{-1} & g_{-2} & \cdots & g_{1-N} \\ g_1 & g_0 & g_{-1} & \ddots & \\ g_2 & g_1 & g_0 & \ddots & \\ \vdots & & & & \\ g_{N-1} & g_{N-2} & g_{N-3} & \cdots & g_0 \end{bmatrix} \begin{bmatrix} j_0 \\ j_1 \\ \vdots \\ j_{N-1} \end{bmatrix} = \begin{bmatrix} e_0 \\ e_1 \\ \vdots \\ e_{N-1} \end{bmatrix} \tag{6.64}$$

The $N \times N$ matrix in the above is a general *Toeplitz* matrix; all the elements of this matrix are described by the $2N - 1$ entries in the first row and column. If the elements repeat with period N, so that

$$g_{n-N} = g_n, \qquad n = 1, 2, \ldots, N - 1 \tag{6.65}$$

then the operation is known as a *circular* discrete convolution, and the $N \times N$ matrix above is *circulant*. Otherwise, the operation is a *linear* discrete convolution. Any linear discrete convolution of length N can be embedded into a circular discrete convolution of length $2N - 1$ by extending the original sequence g to repeat with period $2N - 1$, zero padding the sequence to length $2N - 1$ and changing the upper limit of summation in Eq. (6.63) to $2N - 2$.

The discrete convolution theorem states that if Eq. (6.63) is a circular discrete convolution, it is equivalent to

$$\tilde{e}_n = \tilde{j}_n \tilde{g}_n, \qquad n = 0, 1, \ldots, N - 1 \tag{6.66}$$

where the $\tilde{e}$ is the N-point discrete Fourier transform (DFT) of e, and similarly $\tilde{j}_n$ and $\tilde{g}_n$. The DFT will of course be implemented using the FFT algorithm. If Eq. (6.63) is a linear discrete convolution, then embedding as described above is used.

Hence, the matrix-vector product of Eq. (6.64) can be efficiently implemented as

$$e = \mathrm{FFT}_N^{-1}\{\mathrm{FFT}_N(j)\mathrm{FFT}_N(g)\} \tag{6.67}$$

In the MoM context, with a Toeplitz matrix, the matrix-vector product is thus expressed as

$$\sum_{i=1}^{n} Z_{mn} I_n = Z_m \otimes I_m \tag{6.68}$$

where $Z_m = Z_{m1}$ and $\otimes$ indicates cyclic convolution, evaluated as

$$[Z]\{I\} = \mathrm{FFT}_N^{-1}\{\mathrm{FFT}_N(Z_m)\mathrm{FFT}_N(I)\} \tag{6.69}$$

Usually, $\{I\}$ is an approximation of the current, typically $\{I\}_k$ at the kth iteration of an iterative solver.

Note that the convolution has become the Hadamard, or outer, product (i.e. element-by-element) and hence for an iterative algorithm, the $\mathcal{O}(N^2)$ cost of the matrix-vector product (usually required once or twice per iteration) has been reduced to $\mathcal{O}(N \log N)$. Also very importantly, the memory requirement is reduced from $\mathcal{O}(N^2)$ to $\mathcal{O}(N)$.

This can of course be extended to two and three dimensions, using two- and three-dimensional FFTs as appropriate; the requirement remains that the grid should be a regular Cartesian one. Indeed, three-dimensional FFT-based methods provide quite efficient ways of dealing with the volume integral MoM discretizations.

The adaptive integral method is an extension of this idea to triangular surface grids. In this case, the triangular subdomain basis functions are projected onto a rectangular grid so that the FFT can be applied for the matrix-vector product.

A mathematical aside – what makes the fast Fourier transform (FFT) fast?

The FFT must rate as one of the top numerical algorithms of the twentieth century. Although first popularized by J. W. Cooley and J. W. Turkey in the mid 1960s, perhaps as many as a dozen individuals had independently discovered, and in some cases implemented, efficient methods for evaluating the discrete Fourier transform (DFT), starting with no less a figure than Gauss in 1805. As usual, the treatment in [30] is both highly entertaining and informative, and the following is a summary thereof.

Firstly, until the mid 1960s, the standard method for evaluating an N-point DFT of the discrete function h_k,

$$H_n \equiv \sum_{k=0}^{N-1} h_k \, e^{2\pi i k n/N} \tag{6.70}$$

was to define the complex number W as (note that $i = \sqrt{-1}$, the unit imaginary number, not a counter!)

$$W \equiv e^{2\pi i/N} \tag{6.71}$$

and then the DFT can be written as

$$H_n = \sum_{k=0}^{N-1} W^{nk} h_k, \qquad n = 0, 1, \ldots, N-1 \tag{6.72}$$

Clearly, for each n, this is the product of a matrix of size $N \times N$ (whose (n, k)th entry is W to the power of $n \times k$) times a vector of length N; this must be done N times (for each value of n) yielding an $\mathcal{O}(N^2)$ algorithm.

One of the "rediscoveries" of the algorithm which provides one of the clearest derivations of the FFT is that of Danielson and Lanczos in 1942. The DFT is written as the sum of two DFTs, each of length $N/2$. One is formed from the even-numbered points, one from the odd-numbered points. Mathematically,

$$\begin{aligned} F_k &= \sum_{j=0}^{N-1} e^{2\pi i j k/N} f_j \\ &= \sum_{j=0}^{N/2-1} e^{2\pi i k(2j)/N} f_{2j} + \sum_{j=0}^{N/2-1} e^{2\pi i k(2j+1)/N} f_{2j+1} \\ &= \sum_{j=0}^{N/2-1} e^{2\pi i k j/(N/2)} f_{2j} + W^k \sum_{j=0}^{N/2-1} e^{2\pi i k j/(N/2)} f_{2j+1} \\ &= F_k^e + W^k F_k^o \end{aligned} \tag{6.73}$$

F_k^e is the kth component of the Fourier transform of length $N/2$ formed from the even components of the original f_j, and similarly F_k^o is the corresponding transform formed from the odd components. Although in the last line of Eq. (6.73), k varies from 0 to $N-1$, not just $N/2-1$, the transforms F_k^e and F_k^o are periodic in k with length $N/2$, so each is simply repeated through two cycles.[a]

The neat point about this algorithm is that is can be applied *recursively*. For instance, F_k^e can now be subdivided in F_k^{ee} and F_k^{eo}. For N a power of two, this can be continued down to the point where one is left with the transform of length one – which simply copies the input to the output. There are $\log_2 N$ such recursions. These one-point transforms are then combined appropriately. Each such combination takes of order N operations, there are $\log_2 N$ such combinations, hence we have the $\mathcal{O}(N \log_2 N)$ operation count of the FFT.

The above is not a complete description of the algorithm; one still needs to perform some book-keeping to keep track of which one-point transform corresponds to which combination of even–odd subdivisions, e.g. F^{eoe} for an eight-point transform. By *bit-reversing* the binary representation of each index of the input vector, it turns out that this can be done very efficiently. The interested reader can refer to [30, Section 12.2] for the details.

[a] Another way of looking at this is that taking even-numbered points is equivalent to halving the sampling density, hence the aliasing frequency also halves.

6.9.4 The fast multipole method

A two-dimensional FMM prototype

Whereas the FFT-based methods rely on the algebraic properties of the DFT, the fast multipole method (FMM) is based on the analytical properties of the Green function. Before we briefly introduce the full FMM, it is worth discussing a two-dimensional example originally developed by Lu and Chew, which captures the essence of the algorithm in a far more readily accessible form; it is presented in the following form in [3, Section 4.13].

Assume a TM_z PEC scattering problem. In this case, the EFIE is (see Section 4.6.2)

$$E_z^{\text{inc}}(t) = jk\eta A_z(t) \tag{6.74}$$

where

$$A_z(t) = \frac{1}{4j} \int J_z(t') H_0^{(2)}(kR)\, dt'$$
$$R = \sqrt{[x(t) - x(t')]^2 + [y(t) - y(t')]^2} \tag{6.75}$$

with t a parametric variable describing position around the contour of the cylinder surface, and the unprimed and primed coordinates indicate source and field points as usual. Using subsectional pulse basis functions, as in Section 4.6.2, one obtains the usual MoM matrix equation, with impedance matrix entries which, for segments small

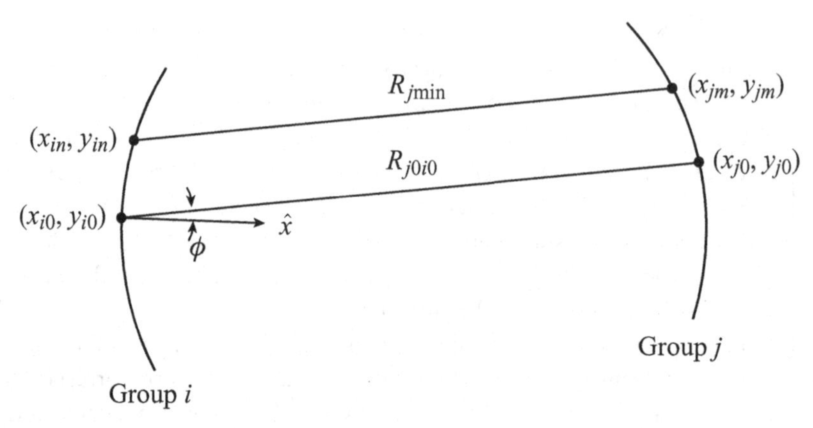

Figure 6.21 Groups i and j, with segments in and jm.

compared to a wavelength, may be approximated by (see Eq. (4.74))

$$Z_{mn} \approx \frac{k\eta}{4} w_n H_0^{(2)}(kR_{mn})\, dt', \qquad \forall\, m \neq n \tag{6.76}$$

with w_n the width of cell n and

$$R_{mn} = \sqrt{(x_m - x_n)^2 + (y_m - y_n)^2} \tag{6.77}$$

This, then, is the conventional MoM solution of this problem. We will assume that there are no geometrical properties of the shape of the circumference that we can exploit. (For instance, if it is a right circular cylinder, and the discretization is uniform, we have a Toeplitz matrix and we can apply the FFT approach to reduce the cost.) If we seek the solution of $[Z]\{I\} = \{V\}$ using a conventional iterative solver, the cost per iteration will be $\mathcal{O}(N^2)$. The memory requirement is also $\mathcal{O}(N^2)$.

Now, consider a fast approach for computing the product of the matrix-vector product. As usual, the circumference of the cylinder will be divided into N segments (which need *not* be equal in size in this approach). Now, the new idea: we collect these segments into p *groups* [20] of roughly equal size and number of unknowns. We index the groups as $i = 1, 2, \ldots, p$; there are now N/p segments per group, indexed as $n = 0, 1, \ldots, N/p - 1$ in each group. One segment per group will be centered at a local origin (x_{i0}, y_{i0}), whilst the other segment centroids are denoted by (x_{in}, y_{in}). For source and field cells closely located, the "near-zone," the calculation proceeds as usual. However, for other segments, sufficiently far separated that they are in the "far-zone," an approximation will be used as follows.

Consider the calculation of the field at (x_{jm}, y_{jm}) due to sources on group i (see Fig. 6.21):

$$E_z^{\text{scat}}(x_{jm}, y_{jm}) = -\frac{\omega\mu}{4} \sum_{n=0}^{N/p-1} j_n w_n H_0^{(2)}(kR_{j\min}) \tag{6.78}$$

[20] In the presentation of [3, Section 4.13], the terms "cells" and "segments" are used respectively. The latter is rather confusing, since a segment in an MoM formulation is usually the sub-domain spanned by one (or sometimes a few) basis functions. The nomenclature used in this section corresponds to typical FMM usage.

The distance function $R_{j\min}$ is approximated as

$$R_{j\min} \approx R_{j0i0} + R_{jm} - R_{in} \tag{6.79}$$

where

$$R_{j0i0} = \sqrt{(x_{j0} - x_{i0})^2 + (y_{j0} - y_{i0})^2} \tag{6.80}$$

$$R_{jm} = (x_{jm} - x_{j0})\cos\phi + (y_{jm} - y_{j0})\sin\phi \tag{6.81}$$

$$R_{in} = (x_{in} - x_{i0})\cos\phi + (y_{in} - y_{i0})\sin\phi \tag{6.82}$$

The angle ϕ denotes the orientation of R_{j0i0} with respect to the x-axis. (This is just the usual far-field approximation used in the derivation of the potential of a two-dimensional dipole.)

Now, the asymptotic form of the Hankel function for large arguments,

$$H_0^{(2)}(k\rho) = \sqrt{\frac{2j}{\pi k\rho}}e^{-jk\rho} \tag{6.83}$$

is applied, yielding

$$H_0^{(2)}(kR_{j\min}) \approx H_0^{(2)}(kR_{j0j0})e^{-jkR_{jm}}e^{+jkR_{in}} \tag{6.84}$$

and thus Eq. (6.78) can be replaced by

$$E_z^{\text{scat}}(x_{jm}, y_{jm}) \approx -\frac{\omega\mu}{4}H_0^{(2)}(kR_{j0i0})e^{-jkR_{jm}}\sum_{n=0}^{N/p-1} j_n w_n e^{+jkR_{in}} \tag{6.85}$$

Hence, *all* interactions between the cells in groups i and j can be obtained from a *single* summation over the coefficients j_n, and one Hankel function calculation. This involves $\mathcal{O}(N/p)$ operations. There are approximately p^2 combinations of far-zone groups, so the overall complexity grows as $\mathcal{O}(Np)$. It can be shown that the optimal grouping is $p = \sqrt{N}$, in which case the complexity is $\mathcal{O}(N^{3/2})$.

It is useful to separate the operations contained in Eq. (6.85). First, the sources on group i are *aggregated* together via the summation

$$S^i \approx \sum_{n=0}^{N/p-1} j_n w_n e^{+jkR_{in}} \tag{6.86}$$

Then, *translation* uses the Hankel function

$$E_z^{\text{scat}}(x_{j0}, y_{j0}) \approx -\frac{\omega\mu}{4}H_0^{(2)}(kR_{j0i0})S_i \tag{6.87}$$

to shift the field to the center of group j. Finally, the scattered field is *disaggregated* throughout group j by a multiplication with the phase correction

$$E_z^{\text{scat}}(x_{jm}, y_{jn}) \approx e^{-jkR_{jm}}E_z^{\text{scat}}(x_{j0}, y_{j0}) \tag{6.88}$$

We find analogous steps in the full FMM.

The full three-dimensional FMM

The FMM rests on two identities. The first, a form of Gegenbauer's addition theorem, states that

$$\frac{e^{-jk_0|\boldsymbol{r}+\boldsymbol{d}|}}{|\boldsymbol{r}+\boldsymbol{d}|} = -jk_0 \sum_{l=0}^{\infty} (-1)^l (2l+1) j_l(k_0 d) h_l^{(2)}(k_0 r) P_l(\boldsymbol{d} \cdot \boldsymbol{r}) \tag{6.89}$$

where $j_l(x)$ is a spherical Bessel function of the first kind, $h_l^{(2)}(x)$ is a spherical Hankel function of the second kind, $P_l(x)$ is a Legendre polynomial, and $d < r$. All the special functions are as defined in standard texts, e.g. [31]. The second identity is a *spectral decomposition* of the product of the Bessel function and the Legendre polynomial, into propagating plane waves:

$$4\pi(-j)^l j_l(k_0 d) P_l(\hat{d} \cdot \hat{d}) = \oint_S e^{-j\boldsymbol{k}\cdot\boldsymbol{d}} P_l(\hat{d} \cdot \hat{r})\, d^2\hat{k} \tag{6.90}$$

where the integral is over a unit sphere S and $\boldsymbol{k} = k_0\hat{k}$. Substituting Eq. (6.90) into Eq. (6.89), and interchanging the order of addition and summation (which has been described as "illegitimate but expedient" [32]), we obtain the approximation

$$\frac{e^{-jk_0|\boldsymbol{r}+\boldsymbol{d}|}}{|\boldsymbol{r}+\boldsymbol{d}|} = -\frac{jk_0}{4\pi} \oint_S e^{-j\boldsymbol{k}\cdot\boldsymbol{d}} T_L(\hat{k} \cdot \hat{r})\, d^2\hat{k} \tag{6.91}$$

with

$$T_L(\hat{k} \cdot \hat{r}) = \sum_{l=0}^{L} (-j)^l (2l+1) h_l^{(2)}(k_0 r) P_l(\hat{k} \cdot \hat{r}) \tag{6.92}$$

The first key point in the FMM is the function $T_L(\hat{k} \cdot \hat{r}) = T_L(\kappa, \theta)$ with $\kappa = k_0 r$ *precomputed* for various values of distance κ and various angles θ. This is a truncated multipole expansion, hence the name: it has been shown semi-empirically that the number of multipoles is approximately $k_0 D + 6(k_0 D)^{1/3}$ (with D the maximum dimension applicable) for an accuracy of 10^{-6}.

The second key point of the FMM is that the interaction matrix is divided into *near* and *far* parts. *Near* interactions are computed as usual with the MoM, and the FMM does not change these at all (by contrast, FFT methods evaluate *all* matrix elements). *Far* interactions are evaluated *approximately*, using the above function T_L. Basis functions in the *far* region are grouped into M localized groups – it has been shown that the optimal value of this is $\sqrt{N}$, with N the number of basis functions.

The third key point in the FMM is that the (approximate) matrix-vector product may be done in $\mathcal{O}(N^{3/2})$ operations. This is done by first computing the far fields of each *group*, then computing the Fourier components of the field in the neighborhood of each group generated by non-near sources, and finally adding the effects of the near- and far-group interactions. These steps are also known as *aggregation*, lumping the fields radiated by a group to the group center, *translation and summation*, which sends the fields from one group to another and then sums them, and finally *disaggregation*, which distributes the received field to each point within the receiving group.

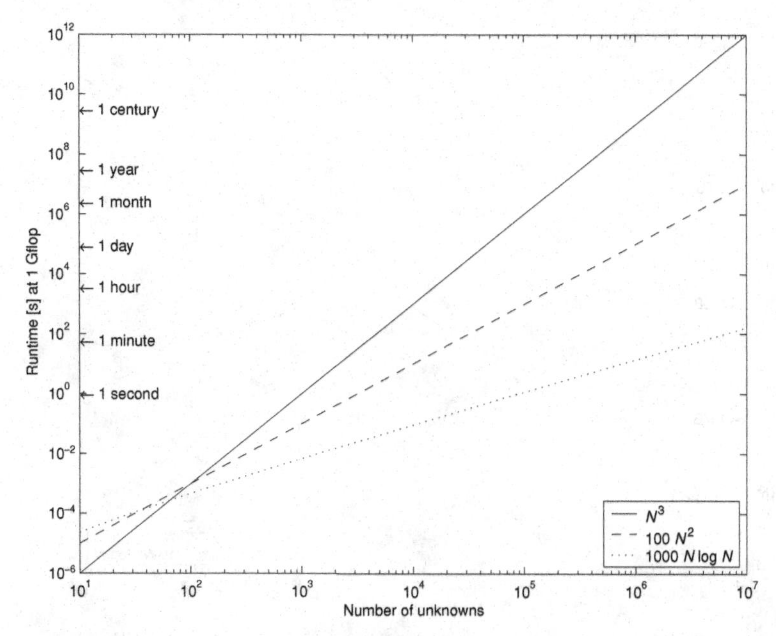

Figure 6.22 Runtimes for typical LU decomposition, a very rapidly converging iterative solution and a well-optimized FMM solution on a 1 Gflop system. (Adapted from [33, Fig. 14.8].)

By introducing a recursive hierarchy of groups, the operation count can be further reduced to $\mathcal{O}(N \log N)$; this is known as the multilevel fast multipole algorithm (MLFMA).

The above description is very cursory, and the interested reader is referred to Section 6.10 for references which provide far more detail. We should caution that the constants in the operation counts can be very large, easily on the order of many thousands or more (by contrast, for direct methods or matrix-vector multiplication, the constants are usually on the order of unity) so the FMM and MLFMA are only *asymptotically* "fast"; indeed for small to medium size problems, the FMM will probably be slower than the MoM. Furthermore, for large problems, highly efficient implementation is essential, otherwise the benefits are lost, so an FMM implementation is emphatically *not* a project for beginners.

The impact of a reduction in asymptotic cost is not always immediately apparent. To illustrate this, Fig. 6.22 compares the runtime on a system capable of sustaining 1 Gflop for N^3, $100N^2$ (as one might hope to obtain with a very rapidly converging iterative solver) and $1000N \log N$, as one might obtain with a very well optimized FMM code, as suggested by [33, Fig. 14.8]. Clearly, the impact of reducing this asymptotic cost is enormously significant for large problems; the difference with the assumed operation counts for 1 million unknowns is that of minutes versus decades! (In reality, the FMM code is likely to run for many hours at least, but the point remains valid.) It must be commented that the constants assumed in both the iterative and FMM cases above may well be extremely optimistic.

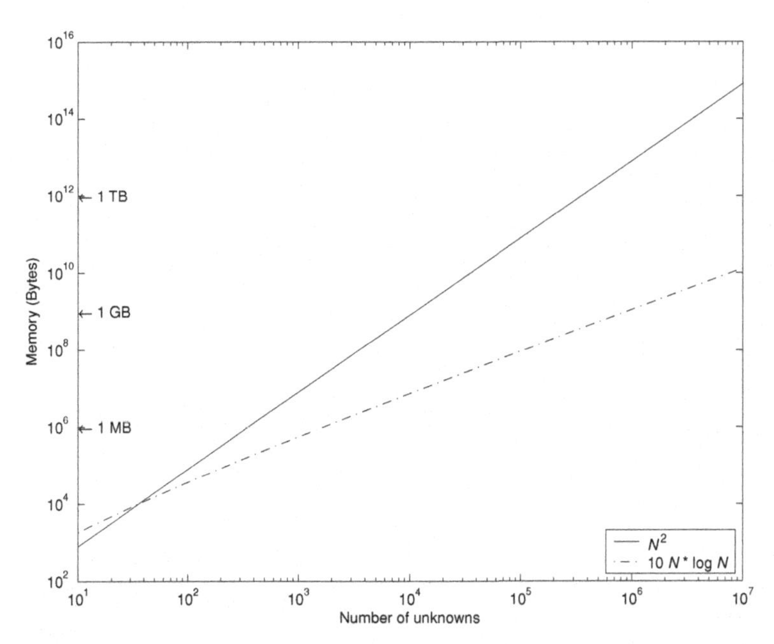

Figure 6.23 Memory required for LU decomposition and a proposed FMM implementation. (Adapted from [33, Fig. 14.8].)

The impact on memory is also highly significant; Fig. 6.23 compares the memory required to store the full MoM matrix compared to the storage requirements of a proposed FMM implementation, as suggested by [33, Fig. 14.9]. (Note that each complex word requires 8 bytes to store in single precision on typical systems.) Again, one should note that a real FMM implementation is unlikely to be this memory efficient.

FEKO incorporates an MLFMM implementation, and in Fig. 6.24 results for the computation of the bistatic RCS of a PEC sphere are presented. The sphere diameter was 10.264λ, and the problem had 100 005 unknowns. In terms of memory, the MLFMM solution required 405.89 Mbytes of RAM; and the MoM solution would have required 149.02 Gbytes. The runtime on a contemporary desktop (with an Intel Core2 E8400 processor, clock speed 3.00 GHz) was 276.85 seconds. (This was a sequential run, using a single thread only.) For obvious reasons, the MoM could not be run on such a system. It is interesting to note that by comparison, in 2003, when the first edition of this text was under preparation, this run required 14 hours, and 1.45 Gbytes of RAM, on a Pentium P4 1.8 GHz processor. The improved processor makes only a small contribution to this great improvement in runtime, which is almost entirely due to algorithmic improvements (e.g fast far-field calculations, single precision storage for some arrays, pre-conditioner improvements, and support of the CFIE for closed PEC bodies). A 100 000 unknown problem is no longer a large problem; at the time of writing, MLFMM problems with many millions of unknowns were being run by various groups around the world using parallel implementations of the algorithm. Figure 6.25 provides data for the actual memory usage of the MLFMM impelementation in FEKO, which may be compared to the predictions of Fig. 6.23. (Note that the x-axis in Fig. 6.25 is not logarithmic.)

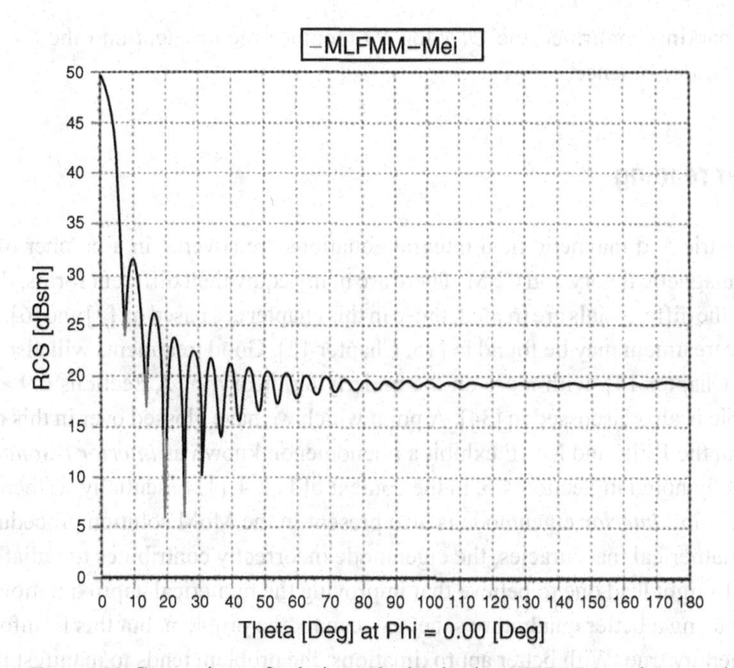

Figure 6.24 Comparison of solutions obtained using MLFMM on the canonical problem of scattering from a sphere. The curves lie on top of one another. (Courtesy EM Software and Systems-SA.)

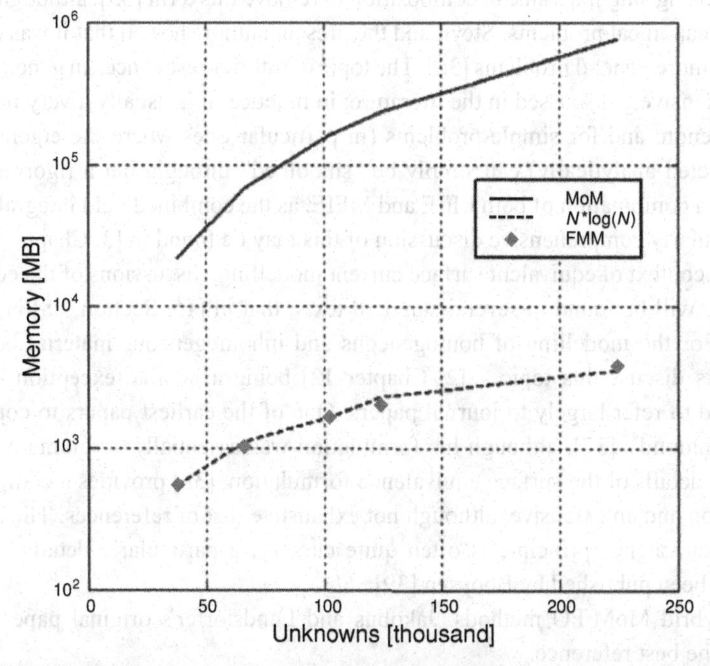

Figure 6.25 Memory required for LU decomposition and an actual commercial FMM implementation. (Courtesy EM Software and Systems-SA.)

Benchmarking confirmed the $\mathcal{O}(N \log N)$ memory requirement and the $\mathcal{O}(N \log^2 N)$ CPU time dependence.

6.10 Further reading

The electric and magnetic field integral equations are covered in a number of texts on electromagnetic theory and CEM. There are many equivalent different forms, depending on how the differentials are treated; those in this chapter are based on [1] and [6]. An introductory treatment may be found in [15, Chapter 12]. Good treatments will also be found in [33, Chapter 14], with more on the underlying theory in [2, Sections 6.9 and 12.3]. The topic is also discussed in [34]. A point which we have glossed over in this chapter is that both the EFIE and MFIE exhibit a phenomenon known as *interior resonance*. This was briefly noted in Section 4.6, in the context of Fig. 4.11. Essentially, a (theoretically) non-radiating *interior* eigenmode is also present in the MoM solution procedure,[21] and due to numerical inaccuracies, the eigenmode incorrectly contributes to radiation. (Figure 4.11 might lead one to believe that improving the numerical approximation – in that case by using a better quadrature scheme – can fix the problem, but this is unfortunately not generally true. With better approximations, the problem tends to manifest itself over a smaller frequency range, but is still present.) Canning showed that there is a component of the field equations which should annihilate this term, but that this term is slightly "off" in frequency in the discrete MoM solution, hence the problem. He proposed a method using singular value decomposition to remove this term [35]; although it worked well for canonical problems, Steyn and the present author showed that it was difficult to apply to more general problems [36]. The topic of interior resonances in general has been quite extensively discussed in the literature; in practice, it is usually a very narrowband phenomenon, and for simple problems (in particular ones where the eigenvalues can be predicted analytically) can simply be "smoothed" through, but a rigorous solution requires a combination of both EFIE and MFIE, as the combined field integral equation. A particularly comprehensive discussion of this may be found in [3, Chapter 6].

In the context of equivalent surface current modelling, discussions of the equivalence principle will be found in several standard texts; that in [15, Section 7.8] is especially useful. For the modelling of homogeneous and inhomogeneous material bodies, few textbooks discuss this topic – [2, Chapter 12] being a notable exception – and one will need to refer largely to journal papers. One of the earliest papers to consider this was Richmond's [37], although his formulation was essentially a volume equivalence one. For details of the surface equivalence formulation, [38] provides a comprehensive discussion and an extensive, although not exhaustive, list of references. The discussion of the equivalence principle is often quite cursory; a particularly detailed study has recently been published by Booysen [39].

On hybrid MoM/PO methods, Jakobus and Landstorfer's original papers [23, 40] remain the best reference.

[21] We assume here the usual exterior field problem.

Regarding parallel processing, the present author made some of the earlier contributions in this regard [27, 28, 41]; other early work may be found in [42]. With Cwik, the present author summarized much of the state-of-the-art then [43]; this special issue contains papers by many of the researchers active in the field in the mid to late 1990s.

There is now a large body of literature dealing with fast techniques in CEM. A very readable introductory treatment will be found in [3, Chapter 4]. Jin provides a detailed, up-to-date and yet succinct overview of fast methods in general in [33, Chapter 14], and this would serve well as a first reference for more detailed study; a fairly extensive list of references complements the technical descriptions. On a historical note, Bojarski is credited with the first use of the FFT method in electromagnetics for this purpose,[22] in a US Air Force technical report of 1971, although the work was only published in the archival open literature a decade later [44]. The application of the FFT to surface and volumetric scattering is well illustrated by the work of Zwamborn and van den Berg, of which [45] is a good example, and also by Borup and Gandhi [46]. For some of the early work on iterative methods, the papers by Sarkar contain useful descriptions of the iterative algorithms ([47] is typical), but it should be noted that there are misconceptions in this and other papers about the nature of discrete operators. This led to a lengthy debate in the literature (see [48], for instance, as well as comments in [34, Chapter 1]); this was finally settled by Ray and Peterson [49]. Their closing comment is conclusive:

> While direct iterative methods may be very efficient for some problems, they are no more accurate than their moment-method analogs.

On the FMM, the paper by Coifman *et al.* [32] remains a classic; the paper belies its title, providing the essential ideas and outlining the implementation in only six pages. (Note that they use the $e^{-i\omega t}$ convention widely used in physics, so the signs of i are reversed relative to the discussion in Section 6.9.4, and the spherical Hankel function is of the first kind.) Chew and colleagues at Illinois have been prolific users of the method; their book provides a detailed discussion of the many applications [50], and their review paper provides a succinct overview of the field [51]. On the question of error control, the paper by Botha and the present author presents a detailed discussion [52].

There has been recent work on what are generically called "macro" basis function formulations. Whilst the FMM is a very powerful and fairly general technique for electromagnetically large structures, knowledge of the electromagnetics of the device being simulated can admit specialized but effective modelling methods. The idea of these is to build macro basis functions from the usual RWG ones, and one of the best-known methods is the characteristic basic function method, CFBM. These macro basis functions exploit a physics-based domain decomposition of the structure. A simple example is scattering from a number of identical, unconnected plates, where the method starts by essentially decoupling the problem and then progressively improving it with higher-order macro basis functions, taking more and more of the complete coupling into account. (The primary ones ignore mutual coupling between the plates entirely, for instance.) A more complex example is an array of Vivaldi antennas; here, the coupling between

[22] He used the term "k-space" in his work rather than CGFFT.

the macro-basis functions (which can be one or several array elements) is important and higher-order terms are needed. See, for instance [53] for a typical application. It is inherently amenable to parallel processing; a discussion of this, with measured performance results, may be found in [5]. The method was originally developed for application to the MoM, but has since seen application to other methods, especially FEM.

6.11 Concluding comments

In this chapter, we have studied methods of solving currents on surfaces using the MoM, starting with the electric and magnetic field integral equations. These may be real currents, in the case of a PEC, or fictitious ones, in the case of an homogeneous dielectric (or magnetic) body. Some theoretical background on the RWG surface basis functions has also been provided, since these are widely used in commercial codes. A Galerkin MoM formulation based on the mixed potential electric field integral equation, using the RWG basis functions, was developed in detail and results shown. The ability to model homogeneous material bodies using fictitious equivalent surface currents is very useful indeed; some MoM codes, such as FEKO, can also handle *inhomogeneous* material bodies, using an equivalent volume current method, but the computational cost associated with this is extremely high, as we have seen (unless FFT-based methods are used).

The much larger computational requirements of surface modelling as opposed to thin-wire modelling have been discussed comprehensively. A hybrid MoM/PO formulation has been outlined. Although inherently approximate, this permits large structures to be modelled with good accuracy provided caution is exercised; it is particularly useful for what is often called "installed antenna performance modelling," which frequently involves electrically small antennas mounted on electrically large vehicles (used here in the general sense to include aircraft, spacecraft and ships). A commercial implementation of this theory is available and we have shown an example of its use. High-performance computing has also been discussed; this continues to be an important enabling technology driving very large applications of the method. Finally, "fast" methods have been considered, including the original FFT-based methods, extensions in the form of the adaptive integral method, and of course the fast multipole method. The last in particular rejuvenated the method of moments in the early 1990s and has proven one of the most important theoretical advances in the MoM over the last two decades.

References

[1] S. M. Rao, D. R. Wilton and A. W. Glisson, "Electromagnetic scattering by surfaces of arbitrary shape," *IEEE Trans. Antennas Propagation*, **30**, 409–418, May 1982.

[2] A. Ishimaru, *Electromagnetic Wave Propagation, Radiation and Scattering*. Engelwood Cliffs, NJ: Prentice-Hall, 1991.

[3] A. F. Peterson, S. L. Ray and R. Mittra, *Computational Methods for Electromagnetics*. Oxford & New York: Oxford University Press and IEEE Press, 1998.

[4] S. Makarov, "MoM antenna simulations with Matlab: RWG basis functions," *IEEE Antennas Propagat. Mag.*, **43**, 100–107, October 2001.

[5] D. Ludick, "Efficient numerical analysis of focal plane antennas for the SKA and MeerKAT." Master's thesis, Dept. Electrical & Electronic Engineering, University of Stellenbosch, March 2010.

[6] P. P. Silvester and R. L. Ferrari, *Finite Elements for Electrical Engineers*. Cambridge: Cambridge University Press, 3rd edn., 1996.

[7] M. A. Khayat and D. R. Wilton, "Numerical evalution of singular and near-singular potential integrals," *IEEE Trans. Antennas Propagat.*, **53**, 3180–3190, October 2005.

[8] M. Khayat, D. Wilton and P. Fink, "An improved transformation and optimized sampling scheme for the numerical evaluation of singular and near-singular potentials," *IEEE Antennas Wireless Propagat. Lett.*, **7**, 377–380, 2008.

[9] R. Graglia and G. Lombardi, "Machine precision evaluation of singular and nearly singular potential integrals by use of Gauss quadrature formulas for rational functions," *IEEE Trans. Antennas Propagat.*, **56**, 981–998, April 2008.

[10] G. Vecchi, "Loop-star decomposition of basis functions in the discretization of the EFIE," *IEEE Trans. Antennas Propagat.*, **47**, 339–346, February 1999.

[11] J.-F. Lee, R. Lee and R. Burkholder, "Loop star basis functions and a robust preconditioner for EFIE scattering problems," *IEEE Trans. Antennas Propagat.*, **51**, 1855–1863, August 2003.

[12] F. Andriulli, K. Cools, H. Bagci, F. Olyslager, A. Buffa, S. Christiansen and E. Michielssen, "A multiplicative Calderon preconditioner for the electric field integral equation," *IEEE Trans. Antennas Propagat.*, **56**, 2398–2412, August 2008.

[13] G. S. Smith, *An Introduction to Classical Electromagnetic Radiation*. Cambridge: Cambridge University Press, 1997.

[14] H. A. Haus and J. R. Melcher, *Electromagnetic Fields and Energy*. Englewood Cliffs, NJ: Prentice-Hall, 1989.

[15] C. A. Balanis, *Advanced Engineering Electromagnetics*. New York: Wiley, 1989.

[16] J. A. Stratton, *Electromagnetic Theory*. New York: McGraw-Hill, 1941. Reprinted by IEEE, 2007.

[17] R. F. Harrington, *Time-Harmonic Electromagnetic Fields*. New York: McGraw-Hill, 1961.

[18] W. H. Press, S. A. Teukolsky, W. Vettering and B. R. Flannery, *Numerical Recipes: the Art of Scientific Computing*. Cambridge: Cambridge University Press, 3rd edn., 2007.

[19] D. Davidson and S. Keunecke, "Hybrid techniques using the MOM and PO/UTD: a tutorial overview," *Trans. SAIEE*, **90**(2), 69–82, 1999.

[20] U. Jakobus, I. Sulzer and F. M. Landstorfer, "Parallel implementation of the hybrid MoM/Green's function technique on a cluster of workstations," in *10th International Conference on Antennas and Propagation*, vol. 1, pp. 1.182–1.185, 1996. IEE Conference Publication No. 436.

[21] C. T. Tai, *Dyadic Green's Functions in Electromagnetic Theory*. New York: IEEE Press, 2nd edn., 1994.

[22] E. H. Newman, "An overview of the hybrid MM/Green's function method in electromagnetics," in *Moment Methods in Antennas and Scattering* (R. C. Hansen, ed.), pp. 449–461. Norwood, MA: Artech House, 1990.

[23] U. Jakobus and F. M. Landstorfer, "Improved PO-MM hybrid formulation for scattering from three-dimensional perfectly conducting bodies of abritrary shape," *IEEE Trans. Antennas Propagat.*, **43**, 162–169, February 1995.

[24] R. E. Hodges and Y. Rahmat-Samii, "An iterative current-based hybrid method for complex structures," *IEEE Trans. Antennas Propagat.*, **45**, 265–276, February 1997.

[25] M. J. Flynn, "Some computer organizations and their effectiveness," *IEEE Transactions on Computers*, **C-21**, 948–960, September 1972.

[26] G. E. Moore, "Progress in digital integrated electronics," in *International Electron Devices Meeting*, **21**, 11–13, 1975.

[27] D. B. Davidson, "Parallel matrix solvers for moment method codes for MIMD computers," *Appl. Comput. Electromagnetics Soc. J.*, **8**(2), 144–175, 1993.

[28] D. B. Davidson, "Parallel processing revisited: a second tutorial," *IEEE Antennas Propagat. Mag.*, **34**, 9–21, October 1992.

[29] E. Lezar and D. B. Davidson, "GPU-based Arnoldi factorisation for accelerating finite element eigenanalysis," in *International Conference on Electromagnetics in Advanced Applications (ICEAA '09)*, pp. 380–383, September 2009.

[30] W. H. Press, S. A. Teukolsky, W. Vettering and B. R. Flannery, *Numerical Recipes in Fortran: the Art of Scientific Computing*. Cambridge: Cambridge University Press, 2nd edn., 1992.

[31] M. Abramowitz and I. A. Stegun, eds., *Handbook of Mathematical Functions*. New York: Dover, 1972.

[32] R. Coifman, V. Rohklin and S. Wandzura, "The fast multipole method for the wave equation: a pedestrian prescription," *IEEE Antennas Propagat. Mag.*, **35**, 7–12, June 1993.

[33] J.-M. Jin, *The Finite Element Method in Electromagnetics*. New York: Wiley, 2nd edn., 2002.

[34] J. J. H. Wang, *Generalized Moment Methods in Electromagnetics*. New York: Wiley, 1991.

[35] F. X. Canning, "Singular value decomposition of integral equations of EM and applications to the cavity resonance problem," *IEEE Trans. Antennas Propagat.*, **37**, 1156–1163, September 1989.

[36] P. Steyn and D. B. Davidson, "A technique for avoiding the EFIE 'interior resonance' problem applied to an MM solution of electromagnetic radiation from bodies of revolution," *Appl. Comput. Electromag. Soc. J.*, **10**(3), 116–128, 1995.

[37] J. H. Richmond, "Scattering by a dielectric cylinder of arbitrary cross section shape," *IEEE Trans. Antennas Propagat.*, **13**, 334–341, May 1965.

[38] K. Umashankar, A. Taflove and S. M. Rao, "Electromagnetic scattering by abritrary shaped three-dimensional homogeneous lossy dielectric objects," *IEEE Trans. Antennas Propagat.*, **34**, 758–766, June 1986.

[39] R. Booysen, "Aperture theory and the equivalence principle," *IEEE Antennas Propagat. Mag.*, **45**, 29–40, June 2003.

[40] U. Jakobus and F. M. Landstorfer, "Improvement of the PO-MM hybrid method by accounting for effects of perfectly conducting wedges," *IEEE Trans. Antennas Propagat.*, **43**, 1123–1129, October 1995.

[41] D. B. Davidson, "A parallel processing tutorial," *IEEE Antennas Propagat. Mag.*, **32**, 6–19, April 1990.

[42] T. Cwik and J. Patterson, eds., *Computational Electromagnetics and Supercomputer Architecture*. PIER7 Progress in Elecromagnetics Research, Cambridge, MA: EMW Publishing, 1993.

[43] D. B. Davidson and T. Cwik, Guest eds., "Special issue on computational electromagnetics and high-performance computing," *Appl. Comput. Electromag. Soc. J.*, **13**, 87–225, July 1998.

[44] N. N. Bojarski, "The k-space formulation of the scattering problem in the time domain," *J. Acoust. Soc. Am.*, **72**, 570–584, August 1982.

[45] A. P. M. Zwamborn, P. M. van den Berg, J. Mooibroek and F. T. C. Koenis, "Computation of three-dimensional electromagnetic-fields distributions in a human body using the weak form of the CGFFT method," *Appl. Comput. Electromag. Soc. J.*, **7**, 26–42, 1992.

[46] D. T. Borup and O. P. Gandhi, "Fast-fourier transform method for calculation of SAR distributions in finely discretized inhomogeneous models of biological bodies," *IEEE Trans. Microwave Theory Tech.*, **32**, 355–360, April 1984.

[47] T. K. Sarkar, "The conjugate gradient method as applied to electromagnetic field problems," *IEEE Antennas Propagat. Soc. Newsletter*, **28**, 5–14, August 1986.

[48] T. K. Sarkar, "Comments on 'Comparison of the FFT conjugate gradient method and the finite-difference time domain method for the 2-D absorption problem' and reply by D. T. Borup and O. P. Gandhi," *IEEE Trans. Microwave Theory Tech.*, **36**, 166–170, January 1988.

[49] S. L. Ray and A. F. Peterson, "Error and convergence in numerical implementation of the conjugate gradient method," *IEEE Trans. Antennas Propagat.*, **36**, 1824–1827, December 1988.

[50] W. C. Chew, J. Jin, E. Michielssen and J. Song, *Fast and Efficient Algorithms in Computational Electroamagentics*. Boston, MA: Artech House, 2001.

[51] W. C. Chew, J. Jin, C. Lu, E. Michielssen and J. M. Song, "Fast solution methods in electromagnetics," *IEEE Trans. Antennas Propagat.*, **45**, 533–543, March 1997.

[52] M. M. Botha and D. B. Davidson, "Application of the fast multipole method to the FE-BI analysis of cavity backed structures with comprehensive FMM error control," *Electromagnetics*, **22**, 405–417, July 2002.

[53] R. Maaskant, R. Mittra and A. Tijhuis, "Fast analysis of large antenna arrays using the characteristic basis function method and the adaptive cross approximation algorithm," *IEEE Trans. Antennas Propagat.*, **56**, 3440–3451, November 2008.

Problem

P6.1 Consider the basis function defined in Eq. (6.8). With reference to Fig. 10.8 in Chapter 10, consider edge 2. Show that requiring that the normal component vanishes on edge 1 and 3 results in $D = -Ex$ and $A = -Cy$, i.e. their removal from the basis function. Then show that requiring that the normal component on edge 2 is unity at nodes 1 and 3 produces a basis function which is within a constant of $\hat{z} \times \boldsymbol{N}_2$, with $\boldsymbol{N}_2$ as in Eq. (10.108).

7 The method of moments and stratified media: theory

7.1 Introduction

Modelling *stratified media* is an important application of the MoM. A stratified medium is one consisting of homogeneous layers of material, each layer having different electromagnetic properties. This includes the general category of *printed antennas*, of which microstrip is the best known. (Microstrip technology is discussed in more detail in the next chapter.) It also brings with it the problem of dealing with dielectric materials. Central to this is the issue of the *Green function*[1] for the problem. The MoM relies on an appropriate Green function as the "field propagator." Due to its perceived complexity, the topic of stratified media is generally regarded as an advanced one, and the coverage tends to be highly theoretical, and frequently impenetrable without lengthy study. One reason for this is that, historically, analysis focussed on the problem of a dipole above a dielectric half-space. There are a number of complex issues which this raises, requiring quite sophisticated analytical techniques to understand, in particular for the asymptotic cases where interesting radiation physics can be extracted. However, the analysis of a very important special case, namely the grounded single-layer microstrip line (or patch antenna), can be undertaken without undue complexity, at least for most practical cases where the substrate is relatively thin.

In this chapter, a static analysis of a microstrip transmission line is first undertaken, to demonstrate the basic principles of the spectral domain and the derivation of the Green function. Following this, the dynamic analysis is introduced, and the Sommerfeld potentials derived from first principles. Although the work in this chapter is certainly not original, being based on a synthesis of the literature – in particular [1] – the presentation in the present format does not appear to have been thus undertaken in other works to date.

7.2 Dyadic Green functions: some introductory notes

The main reason for efficiency of the MoM formulation already discussed is the existence of suitable Green functions. The Green function $G(\boldsymbol{r})$ is equivalent to the *impulse*

[1] Contemporary usage is "Green function" rather than "Green's function," in line with "Dirac delta function," "Heaviside step function," etc.

response $h(t)$ of system theory. Just as $h(t)$ gives the response (in time) to a *temporally* impulsive source, so $G(\boldsymbol{r})$ gives the response (in space) to a *spatially* impulsive (current) source. The response to a spatially distributed source is obtained by integration, and plays the same role in space that *convolution* in system theory does in time:

$$y(t) = h(t) * x(t) \iff \boldsymbol{E}(\boldsymbol{r}) = \bar{\bar{G}}(\boldsymbol{r}, \boldsymbol{r}') * \boldsymbol{J}(\boldsymbol{r}') \tag{7.1}$$

We have already encountered the *free-space Green function* in our work in Chapter 4, although we made only passing reference to it then. In free space, the function is (moderately) simple:

$$\bar{\bar{G}}(\boldsymbol{r}, \boldsymbol{r}') = \left(k^2 \bar{\bar{I}} + \nabla\nabla\right) g(\boldsymbol{r}, \boldsymbol{r}'), \qquad g(\boldsymbol{r}, \boldsymbol{r}') = \frac{e^{-jkR}}{4\pi R} \tag{7.2}$$

where $R = |\boldsymbol{r} - \boldsymbol{r}'|$ is the distance from source to field point. Green functions can be obtained for either fields or potentials, and in the above, $\bar{\bar{G}}(\boldsymbol{r}, \boldsymbol{r}')$ is the electric field Green function for free space, and $g(\boldsymbol{r}, \boldsymbol{r}')$ is the potential Green function for free space. We will primarily use Green functions for potentials in this chapter. It is worth highlighting that the Green function for free space is given in *closed form* and is trivial to compute (although the singularities which accompany it make an accurate MoM implementation anything but!).

Some new notation has been introduced in the above. The double-overbar notation indicates a *dyad*; this is a mathematical device which after multiplication by a vector, yields a vector. A dyad typically consists of the following terms, when written as a matrix:

$$\bar{\bar{G}} = \begin{bmatrix} G^{xx} & G^{xy} & G^{xz} \\ G^{yx} & G^{yy} & G^{yz} \\ G^{zx} & G^{zy} & G^{zz} \end{bmatrix} \tag{7.3}$$

It is also frequently written out in its component form:

$$\bar{\bar{G}} = G^{xx}\hat{x}\hat{x} + G^{xy}\hat{x}\hat{y} + G^{xz}\hat{x}\hat{z} + G^{yx}\hat{y}\hat{x} + G^{yy}\hat{y}\hat{y} + G^{yz}\hat{y}\hat{z} + G^{zx}\hat{z}\hat{x} + G^{zy}\hat{z}\hat{y} + G^{zz}\hat{z}\hat{z} \tag{7.4}$$

The product of a dyad and vector is then computed using normal matrix theory or the usual vector dot-products. $\bar{\bar{I}}$ is the identity dyad. Note that although both operations $\hat{s} \cdot \bar{\bar{G}}(\boldsymbol{r}, \boldsymbol{r}')$ and $\bar{\bar{G}}(\boldsymbol{r}, \boldsymbol{r}') \cdot \hat{s}$ with $\hat{s}$ a unit vector (i.e. $\hat{x}$, $\hat{y}$ or $\hat{z}$) are defined, only the latter has physical meaning as the potential due to an $\hat{s}$-oriented source.

However, for many applications (such as printed antennas, antennas above or buried in a real earth) radiation occurs in a *stratified media* environment, not free space. The presence of the stratified media greatly complicates the analysis. The Green function for an elementary dipole radiating in the vicinity of the stratified medium needs to be worked out. This was done many years ago by Arnold Sommerfeld – in 1909, he determined the field radiated by a short vertical electrical dipole above a dielectric interface. However, the passage of time has not made the theory any easier. In particular, the required integration in the complex plane brings with it a number of complex issues. Finally, the Green functions obtained are *not* given in closed form, and are computationally

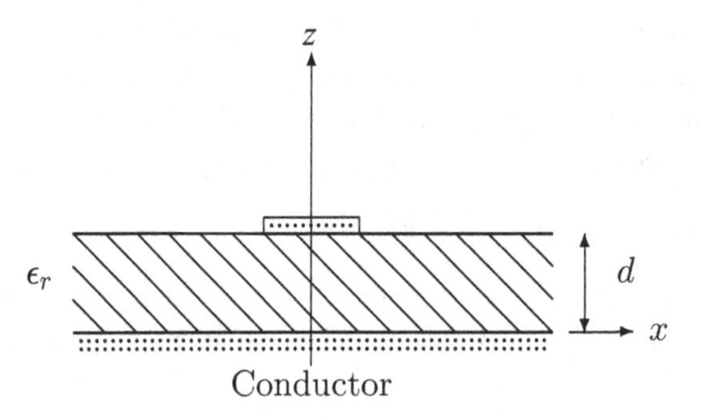

Figure 7.1 Typical microstrip structure.

expensive to compute, so even implementations of seemingly simple problems require some thought.

Before concluding this introductory section, it should be commented that there are a number of MoM formulations for stratified media. This chapter uses the mixed potential integral equation formulation (MPIE), introduced by Mosig and Gardiol [2] and used with great success for MoM formulations by a number of workers. However, before we outline this approach, we will consider a much simpler problem, which illustrates many of the issues: deriving the Green function for stratified media for electrostatics from first principles.

7.3 A static example of a stratified medium problem: the grounded dielectric slab

Central to stratified media formulations is the spectral domain transform. The Fourier transform is used to simplify the problem by transforming the partial differential equation(s) of electromagnetics in the spatial domain into an ordinary differential equation in the spectral domain. (Once again, the analogy with linear systems theory is strong.) To illustrate the basic concepts, we will derive the static spectral domain Green function for a microstrip structure, as shown in Fig. 7.1. This does *not* include radiation effects, which requires the full-wave solution of the problem, the topic of later parts of this chapter. This is still quite useful, nonetheless: the quasi-TEM approach often used for transmission-line analysis renders the problem (quasi-)static. A solution can be used to compute the characteristic impedance and phase constant of the transmission line by making the calculation twice – once with the dielectric present, and once with the dielectric replaced by free space [3, p. 166]. (See also Section 10.2.3.) Note that the structure is assumed to be of infinite length, thus there is no variation in y.

This formulation appears to have been originally presented in the engineering literature by Yamashita and Mittra [4]. They did not actually derive the Green function; they were formulating a variational expression for the unknown charge distribution on the strip, but the extension is straightforward. Their notation is largely followed here, except that k_x is used as the Fourier transform variable instead of β, and d instead of h for the substrate

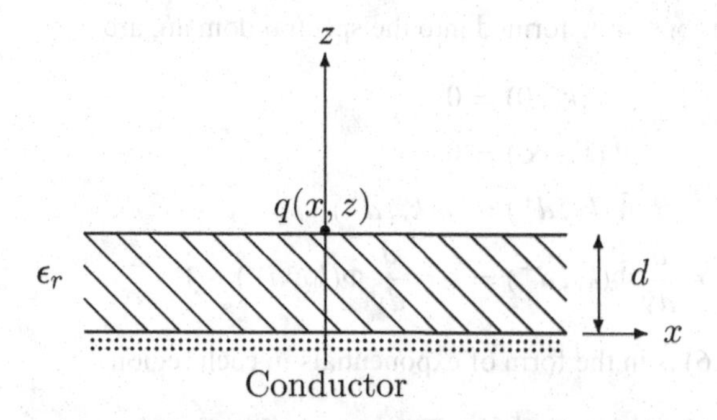

Figure 7.2 Stratified medium equivalent with impulsive source $q(x, z) = \delta(x)\delta(z - d)$.

thickness. Booton provides a similar derivation [5, Section 10.3]. It is interesting to note that an almost identical derivation may be found in Schwinger's lecture notes [6, Chapter 14]; although only recently published, these lectures were originally given in 1976.

To derive the Green function, the Poisson equation for a spatially impulsive source of unit magnitude located at $x = 0$, $z = d$ must be solved (subsequently, the case $x \neq 0$ is also considered); see Figs. 7.1 and 7.2. Thus the partial differential equation to solve is

$$\nabla^2 \Phi(x, z) = -\frac{1}{\epsilon}\delta(x)\delta(z - d) \tag{7.5}$$

The equation is transformed into the spectral domain; using the linearity of the Fourier transform, the $\frac{\partial}{\partial x} \Longleftrightarrow jk_x$ transform property, and the Fourier transform of the Dirac delta function, one obtains

$$\left[-k_x^2 + \frac{d^2}{dz^2}\right]\tilde{\Phi}(k_x, z) = -\frac{1}{\epsilon}\delta(z - d) \tag{7.6}$$

with $\tilde{\Phi}(k_x, z)$ the Fourier transform of the potential, also known as the spectral domain representation:

$$\tilde{\Phi}(k_x, z) = \int_{-\infty}^{\infty} \Phi(x, z) e^{-jk_x x}\, dx \tag{7.7}$$

Note that this is now an *ordinary* differential equation in $\tilde{\Phi}(k_x, z)$. The homogeneous differential equation (with the inhomogeneous source taken into account via a Neumann boundary condition) is now solved:

$$\left[-k_x^2 + \frac{d^2}{dz^2}\right]\tilde{\Phi}(k_x, z) = 0, \qquad \forall z \neq d \tag{7.8}$$

The boundary conditions are: zero potential at $z = 0$ and $z \to \infty$; continuous potential at the material interface at $z = d$; and flux discontinuous by the source singularity at $z = d$. These boundary conditions transform in a straightforward fashion to the spectral domain. The solution to Eq. (7.8) must be written in the two regions demarcated by the material interface. Note that even if $\epsilon_r = 1$, this two-region approach is still necessary, so that the jump discontinuity can be enforced.

The boundary conditions, transformed into the spectral domain, are:

$$\tilde{\Phi}(k_x, 0) = 0 \tag{7.9}$$

$$\tilde{\Phi}(k_x, \infty) = 0 \tag{7.10}$$

$$\tilde{\Phi}(k_x, d^+) = \tilde{\Phi}(k_x, d^-) \tag{7.11}$$

$$\epsilon_0 \frac{d}{dy} \tilde{\Phi}(k_x, d^+) = \epsilon_0 \epsilon_r \frac{d}{dy} \tilde{\Phi}(k_x, d^-) - 1 \tag{7.12}$$

The solution of Eq. (7.6) is in the form of exponentials in each region:

$$\tilde{\Phi}_1(k_x, z) = A\, e^{-k_x z} + B\, e^{k_x z}, \qquad \forall 0 \leq z < d \tag{7.13}$$

$$\tilde{\Phi}_2(k_x, z) = C\, e^{-|k_x| z} + D, \qquad \forall z \geq d \tag{7.14}$$

Equation (7.10) immediately yields $D = 0$, and Eq. (7.9) yields $A = -B$. Thus

$$\tilde{\Phi}_1(k_x, z) = -2A \sinh k_x z \tag{7.15}$$

Applying Eq. (7.11) in the limit $d^\pm \to d$ one obtains

$$A = -C \frac{e^{-|k_x| d^+}}{2 \sinh k_x d^-} \tag{7.16}$$

The d/dz terms in Eq. (7.12), again in the limit $d^\pm \to d$, are thus:

$$\left. \frac{d\tilde{\Phi}_1(k_x, z)}{dz} \right|_{z=d^-} = +C k_x\, e^{-|k_x| d} \coth k_x d^+$$

$$\left. \frac{d\tilde{\Phi}_2(k_x, z)}{dz} \right|_{z=d^+} = -C |k_x|\, e^{-|k_x| d^-}$$

Equation (7.12) yields:

$$C = \frac{e^{|k_x| d}}{\epsilon_0 |k_x| [1 + \epsilon_r \coth |k_x| d]} \tag{7.17}$$

where the even property of the product of $\coth(k_x d)$ and $k_x d$ has been used to make the required simplification $k_x \coth k_x d = |k_x| \coth |k_x| d$ (assuming $d \geq 0$). The solution for $\tilde{\Phi}_2(k_x, z)$, valid in the limit $d^+ \to d$ for $z \geq d$ is thus:

$$\tilde{\Phi}(k_x, z) = \frac{e^{|k_x|(d-z)}}{\epsilon_0 |k_x| [1 + \epsilon_r \coth |k_x| d]} \tag{7.18}$$

We have dropped the subscript 2 since we are now on the interface. Note that, for $z = d$, this reduces to

$$\tilde{\Phi}(k_x, d) = \frac{1}{\epsilon_0 |k_x| [1 + \epsilon_r \coth |k_x| d]} \tag{7.19}$$

This can also be written as

$$\tilde{\Phi}(k_x, d) = \frac{\sinh |k_x| d}{\epsilon_0 |k_x| \{\sinh |k_x| d + \epsilon_r \cosh |k_x| d\}} \tag{7.20}$$

(An interesting special case can be identified, viz. $\epsilon_r = 1$. For this case, by expanding the hyperbolic terms in the denominator, Eq. (7.20) reduces to

$$\tilde{\Phi}(k_x, d) = \frac{1}{\epsilon_0} \frac{e^{-|k_x| d} \sinh |k_x| d}{|k_x|} \tag{7.21}$$

This can be useful in asymptotic analysis, where the Green function for a homogeneous dielectric is used.)

Equation (7.20) is the spectral domain Green function for a source located on the z-axis. The Green function is then the inverse Fourier transform of this:

$$G(x, 0) = \frac{1}{2\pi} \int_{-\infty}^{\infty} \tilde{\Phi}(k_x, d)\, e^{jk_x x}\, dk_x \tag{7.22}$$

and for the general case of the source located at x', this becomes

$$G(x, x') = \frac{1}{2\pi} \int_{-\infty}^{\infty} \tilde{\Phi}(k_x, d)\, e^{jk_x (x-x')}\, dk_x \tag{7.23}$$

The required integral equation for the potential in terms of the charge distribution $\rho(x, d)$ is thus

$$\Phi(x, d) = \int_{-\infty}^{\infty} G(x, x')\rho(x', d)\, dx' \tag{7.24}$$

This, then, is the spectral domain static Green function for a grounded dielectric slab. Unfortunately, we note that it must first be inverse Fourier transformed to the spatial domain, and doing this for each possible value of the argument $x - x'$ is very time consuming, since numerical integration is required. Interpolation tables are often used to accelerate the evaluation of the functions. Another approach is to formulate the entire MoM problem in the spectral domain, by using basis functions which have analytical Fourier transforms. This is described in detail for the quasi-static microstrip analysis problem in [7]. However, we will not pursue this further here. Instead, we turn our attention to the full-wave case, after first revising some concepts from electromagnetic theory regarding scalar and vector potential representations.

7.4 The Sommerfeld potentials

7.4.1 A brief revision of potential theory

Before confronting the full-wave stratified medium problem, we will briefly revise some basic electromagnetic theory, in particular, potential theory. It is often useful to represent fields in terms of potentials. Classic elementary electrostatics uses $\boldsymbol{E} = -\nabla\Phi$. For high-frequency electromagnetics the electrostatic potential is of course incomplete, and a very widely used set of potentials is

$$\boldsymbol{E} = -\nabla\Phi - \frac{\partial \boldsymbol{A}}{\partial t} \tag{7.25}$$

$$\boldsymbol{B} = \nabla \times \boldsymbol{A} \tag{7.26}$$

It will be recalled that there is considerable arbitrariness surrounding the choice of potential (as is well known, a potential $A' = A + \nabla\phi$ with ϕ any suitable scalar function results in the same set of fields); this is usually resolved via a *gauging* process. The most widely used in RF engineering is the "Lorenz gauge," with

$$\nabla \cdot \boldsymbol{A} = -(1/c^2)\frac{\partial \Phi}{\partial t} \tag{7.27}$$

A and ϕ must be worked out from

$$\nabla^2\phi - \frac{1}{c^2}\frac{\partial^2 \Phi}{\partial t^2} = -\frac{\rho}{\epsilon} \tag{7.28}$$

$$\nabla^2\boldsymbol{A} - \frac{1}{c^2}\frac{\partial^2 \boldsymbol{A}}{\partial t^2} = -\mu\boldsymbol{J} \tag{7.29}$$

In the frequency domain, these become

$$\left(\nabla^2 + k^2\right)\Phi = -\frac{\rho}{\epsilon} \tag{7.30}$$

$$\left(\nabla^2 + k^2\right)\boldsymbol{A} = -\mu\boldsymbol{J} \tag{7.31}$$

and these solutions – for differential current elements $d\rho$ and $d\boldsymbol{J}$ – are the potential Green functions.

We have already commented that within one potential representation, the potentials are not unique. There is also more than one possible potential representation. Another set involving *only* electric and magnetic *vector* potentials may be used; this was originally introduced by Hertz. In this case, the potentials satisfy the following Helmholtz equations:

$$\left(\nabla^2 + k^2\right)\boldsymbol{A} = -\mu\boldsymbol{J} \tag{7.32}$$

$$\left(\nabla^2 + k^2\right)\boldsymbol{F} = -\epsilon\boldsymbol{M} \tag{7.33}$$

where $\boldsymbol{M}$ is the (fictitious) magnetic current. These are also sometimes written as $\boldsymbol{\Pi}^e = \boldsymbol{A}/j\omega\mu\epsilon$ and $\boldsymbol{\Pi}^h = \boldsymbol{F}/j\omega\mu\epsilon$. For the Hertz potentials, the fields in the spatial domain are given as

$$j\omega\mu\epsilon\boldsymbol{E} = k^2\boldsymbol{A} + \nabla\cdot\nabla\boldsymbol{A} - j\omega\mu\nabla\times\boldsymbol{F} \tag{7.34}$$

$$j\omega\mu\epsilon\boldsymbol{H} = k^2\boldsymbol{F} + \nabla\cdot\nabla\boldsymbol{F} + j\omega\mu\nabla\times\boldsymbol{A} \tag{7.35}$$

7.4.2 The Sommerfeld potentials

Preliminaries

In the stratified medium case, at least two approaches using potentials have been used. The former uses the field components normal to the interface as potentials. We will retain the convention of the preceding sections that the interfaces are in planes of constant z; hence, in this case, the potentials would be E_z and H_z. Another possibility is the use of the (Hertz) potentials, of both electric ($\boldsymbol{A}$) and magnetic ($\boldsymbol{F}$) type. If only z-directed components A_z and F_z are retained, this choice is traditionally called the Hertz–Debye potentials. The final possibility, and the one we will investigate since it is the most popular, is the Sommerfeld potentials.

The Sommerfeld potentials, in the absence of magnetic currents, assume $\boldsymbol{F} = 0$. A vertical electric dipole (VED), i.e. z-directed in our convention, needs only the A_z component. A horizontal electric dipole (HED) (i.e. parallel to the xy-plane) will require a component parallel to the source. Hence, the dyadic in this approach will have only five non-zero terms:

$$\bar{\bar{G}}_A = (\hat{x}G_A^{xx} + \hat{z}G_A^{zx})\hat{x} + (\hat{y}G_A^{yy} + \hat{z}G_A^{zy})\hat{y} + \hat{z}G_A^{zz}\hat{z} \tag{7.36}$$

In order to find these terms, we first need some additional background on the spectral domain.

The spectral domain transform

In the static case discussed previously, no $\hat{y}$ variation was assumed, and the Fourier transform was the usual one-dimensional one. For a general structure, we cannot make this assumption, and the transform (and inverse) becomes two dimensional:

$$\tilde{f}(k_x, k_y) = \frac{1}{2\pi}\int\int_{-\infty}^{\infty} f(x, y)\, e^{-jk_x x}\, e^{-jk_y y}\, dx\, dy \tag{7.37}$$

$$f(x, y) = \frac{1}{2\pi}\int\int_{-\infty}^{\infty} \tilde{f}(k_x, k_y)\, e^{jk_x x}\, e^{jk_y y}\, dk_x\, dk_y \tag{7.38}$$

It is useful to introduce the polar vector $\rho = x\hat{x} + y\hat{y}$ (this is simply the usual radius vector in cylindrical coordinates, $|\boldsymbol{\rho}| = \sqrt{x^2 + y^2}$) and the radial spectral variable $\boldsymbol{k}_\rho = k_x\hat{x} + k_y\hat{y}$. This permits the "del" operator ∇ to be split into its transverse and normal parts as $\nabla = \nabla_t + \frac{\partial}{\partial z}\hat{z}$. In the spectral domain, this becomes

$$\tilde{\nabla} = j\boldsymbol{k}_p + \frac{\partial}{\partial z}\hat{z} \tag{7.39}$$

Since the only spatial derivative remaining in the spectral domain is with respect to z, the shorter dot notation for derivatives will frequently be used in the following, for example $\partial\tilde{\Psi}/\partial z = \dot{\tilde{\Psi}}$. Using the Bessel function J_0, the above transforms may be written as

$$\tilde{f}(k_\rho) = \int_0^{\infty} J_0(k_\rho\, \rho) f(\rho)\, \rho\, d\rho \tag{7.40}$$

$$f(\rho) = \int_0^{\infty} J_0(k_\rho\, \rho)\tilde{f}(k_\rho)\, k_\rho\, dk_\rho \tag{7.41}$$

This is known as the Fourier–Bessel or Hankel integral transform pair. These are best known amongst RF and microwave engineers as Sommerfeld integrals.

As in the two-dimensional static case, the introduction of these transforms permits the spatial domain differential equation (the Helmholtz, rather than the Laplace of the static case)

$$(\nabla^2 + k^2)\,\Psi = 0 \tag{7.42}$$

to be written in the spectral domain as the solution of an ordinary differential equation

$$\left(\frac{\partial^2}{\partial z^2} - u^2\right)\tilde{\Psi} = 0 \tag{7.43}$$

where the parameter u in the traditional notation of Sommerfeld is given by

$$u^2 = -k_z^2 = k_x^2 + k_y^2 - k^2 = k_\rho^2 - k^2 \tag{7.44}$$

The spectral variable k_ρ is complex valued, and by convention written as $k_\rho = \lambda + j\nu$. λ in this context is the real part of k_ρ, and should not be confused with wavelength.

Normal component representation

One possibility for stratified media is the use of the normal fields E_z and H_z as potentials. The normal components satisfy Eq. (7.42) or (7.43) in the spatial or spectral domain respectively. In the spectral domain, the transverse components are given by

$$k_\rho^2 \tilde{E}_x = jk_x \dot{\tilde{E}}_z - \omega\mu k_y \tilde{H}_z \tag{7.45}$$

$$k_\rho^2 \tilde{E}_y = jk_y \dot{\tilde{E}}_z + \omega\mu k_x \tilde{H}_z \tag{7.46}$$

$$k_\rho^2 \tilde{H}_x = jk_x \dot{\tilde{H}}_z + \omega\epsilon k_y \tilde{E}_z \tag{7.47}$$

$$k_\rho^2 \tilde{H}_y = jk_y \dot{\tilde{H}}_z - \omega\epsilon k_x \tilde{E}_z \tag{7.48}$$

As in the static case, the boundary conditions transform in a straightforward fashion to the spectral domain. Hence, tangential field continuity across the layers is satisfied if $\epsilon\tilde{E}_z$, $\dot{\tilde{E}}_z$, $\mu\tilde{H}_z$ and $\dot{\tilde{H}}_z$ are continuous. Rather importantly, this means that the boundary conditions do *not* introduce coupled equations in $\tilde{E}_z$ and $\tilde{H}_z$. From the viewpoint of the Green functions, the potentials are the normal components, but we will not pursue this further now. The Sommerfeld potentials make use of some normal components, hence the discussion here.

Sommerfeld potentials

In the absence of magnetic currents,[2] the Sommerfeld approach assumes $\boldsymbol{F} = 0$. A VED requires only the A_z component, obtained from the spectral domain relationship

$$j\omega\mu\epsilon\tilde{E}_z = k_\rho^2 \tilde{A}_z \tag{7.49}$$

This is obtained from the spectral domain equivalent of Eq. (7.34). $\tilde{E}_z$ is obtained as above. The other components may be computed from the spectral domain equivalents of Eqs. (7.34) and (7.35). It may be shown that one obtains the following in terms of the

[2] As an aside, it should be noted that it is possible to have non-zero $\boldsymbol{F}$ even with zero magnetic current $\boldsymbol{M}$, due to the amount of arbitrariness in the potentials.

normal component representation:

$$\tilde{G}_A^{xx} = -\frac{\mu \tilde{G}_H^{zx}}{jk_y} \tag{7.50}$$

$$k_\rho \tilde{G}_A^{zx} = j\omega\mu\epsilon \tilde{G}_E^{zx} + \frac{k_x \mu \dot{\tilde{G}}_H^{zx}}{k_y} \tag{7.51}$$

$$\tilde{G}_A^{yy} = \frac{\mu \tilde{G}_H^{zy}}{jk_x} \tag{7.52}$$

$$k_\rho \tilde{G}_A^{zy} = j\omega\mu\epsilon \tilde{G}_E^{zy} - \frac{k_y \mu \dot{\tilde{G}}_H^{zy}}{k_x} \tag{7.53}$$

$$k_\rho \tilde{G}_A^{zz} = j\omega\mu\epsilon \tilde{G}_E^{zz} \tag{7.54}$$

Regarding boundary conditions at the interface, it may be shown – from Eqs. (7.34) and (7.35) – using these Sommerfeld potentials, that transverse field continuity implies that $\tilde{A}_z$ and $\tilde{A}_z/\epsilon$ must be continuous for a VED. For an x-directed HED, $\tilde{A}_x$, $\dot{\tilde{A}}_x$, $\tilde{A}_z$ and $\nabla \cdot \tilde{A}/\epsilon$ must be continuous, and a similar expression holds for a y-directed HED. The last condition couples normal and transverse components of the Green function, which hence cannot be independently computed. For this reason, it is usually easier to work with the normal field components, as will be done shortly.

Symmetry also results in the following expressions, which we note although we will not use them further:

$$\tilde{G}_A^{xx} = \tilde{G}_A^{yy} \tag{7.55}$$

$$\frac{\tilde{G}_A^{zx}}{jk_x} = \frac{\tilde{G}_A^{zy}}{jk_y} \tag{7.56}$$

7.4.3 An example: derivation of G_A^{xx} for single-layer microstrip

General multi-layered substrates are best handled using a matrix formulation. Within each substrate, the normal field components are computed for a unit Hertz dipole embedded in the layered medium. The boundary conditions are handled using "chain" matrices. A particularly complete description may be found in [1]. However, for the simple but very important case of a single-layer microstrip, we can directly compute the potentials in a fashion very similar to that described in Section 7.3. Once again, Fig. 7.2 is relevant, although now the impulsive source is a horizontal Hertzian dipole, and for convenience the air–dielectric interface, rather than the ground plane, is at $z = 0$ (and hence the ground plane is located at $z = -d$). In general, the derivation must be repeated for the five non-zero components of the Green function, viz. Eq. (7.36), but we will only derive one of these here – the x-directed magnetic Green function. We also restrict the derivation to non-magnetic lossy dielectric substrates, i.e. $\mu_1 = \mu_0$ and $\epsilon_1 = \epsilon_0\epsilon_r'(1 - j\tan\delta)$. We will use $\epsilon_r = \epsilon_r'(1 - j\tan\delta)$ to represent the complex relative permittivity in the following; it is useful to be able to distinguish between ϵ_r and ϵ_r'.

Table 7.1 Values of the amplitude coefficients U_i and L_i associated with the upper and lower parts of the layer containing the source (after [1, Table 1, p. 150])

	G_H^{zx}	G_H^{zy}	G_E^{zx}	G_E^{zy}	G_E^{zz}
U_i	$-jk_y/4\pi u_0$	$jk_x/4\pi u_0$	$-jk_x/4\pi j\omega\epsilon$	$-jk_y/4\pi j\omega\epsilon$	$k_\rho^2/4\pi j\omega\epsilon u_0$
L_i	U_i	U_i	$-U_i$	$-U_i$	U_i

The source-free ODE to be solved for the normal magnetic field in the spectral domain is of the form of Eq. (7.43), repeated here for the H_z case:

$$\left(\frac{\partial^2}{\partial z^2} - u^2\right)\tilde{H}_z = 0 \tag{7.57}$$

The solution in each region may either be written as the sum of exponentials, as in Section 7.3, or as hyperbolic functions. In the upper region $z \geq 0$, the solution is of the form

$$\tilde{H}_z = a_0\, e^{-u_0 z}. \tag{7.58}$$

which already incorporates the boundary condition at infinity. In the dielectric region, the solution is of the form

$$\tilde{H}_z = a_1 \cosh u_1(z+d) + b_1 \sinh u_1(z+d) \tag{7.59}$$

The remaining boundary conditions on $\tilde{H}_z$ are:

$$\mu_0\mu_r \tilde{H}_z|_{z=0-} = \mu_0 \tilde{H}_z|_{z=0+} \tag{7.60}$$

$$\dot{\tilde{H}}_z|_{z=0-} = \dot{\tilde{H}}_z|_{z=0+} \tag{7.61}$$

$$\dot{\tilde{H}}_z|_{z=-d} = 0 \tag{7.62}$$

The last boundary condition may not be immediately apparent. The perfect electric conductor at $z = -d$ imposes a zero tangential *electric* field condition, implying zero *normal* derivative of magnetic field.

The above are for the source-free case. In Section 7.3, the effect of the source was introduced via a boundary condition. Here, we will introduce another method of dealing with this. For a layer with a source inside it, this can be taken into account by adding a solution ψ^∞, which is the particular solution corresponding to the source embedded in an unbounded homogeneous medium. In the spectral domain, the solution can be written as

$$\psi_i^\infty = \begin{cases} U_i\, e^{-u_i(z_i - D)} & D \leq z_i \leq d_i \\ L_i\, e^{+u_i(z_i - D)} & 0 \leq z_i < D \end{cases} \tag{7.63}$$

for a source at $z_i = D$, with $z_i = z + d_i$ the local normal coordinate in each layer. The amplitude coefficients U_i and L_i depend on the physical quantity represented by ψ, and are tabulated in Table 7.1. (In the spectral domain, the transform of an HED of unit magnitude, $\delta(x)\delta(z = -D)$, is $1/2\pi$. The table takes this and other factors into account.) In the present case, this source will be located in the upper medium (free space) at $D > 0$; the limit case $D \to 0$ will be considered subsequently.

In the free-space region then, the solution is

$$\tilde{H}_z = a_0\, e^{-u_0 z} - \frac{jk_y}{4\pi u_0} e^{+u_i(z_i - D)}, \qquad \forall\, d \leq z < D \tag{7.64}$$

in the region just above the interface, and for the rest of the region

$$\tilde{H}_z = a_0\, e^{-u_0 z} - \frac{jk_y}{4\pi u_0} e^{-u_i(z_i - D)}, \qquad \forall\, z \geq D \tag{7.65}$$

It is tempting to set D to zero and use this latter equation immediately, but it yields the incorrect solution.

We now apply the boundary conditions and eliminate the three unknown coefficients, a_0, a_1 and b_1. Application of Eq. (7.62) immediately yields $a_1 = 0$. Applying Eq. (7.60) for the non-magnetic substrate case ($\mu_r = 1$) in the limit $D \to 0$ yields

$$b_1 = \frac{a_0 - \frac{jk_y}{4\pi u_0}}{\sinh u_1 d} \tag{7.66}$$

Application of Eq. (7.61), again in the limiting case, gives

$$a_0 = \frac{-\frac{jk_y}{4\pi} + \frac{jk_y}{4\pi u_0} u_1 \coth u_1 d}{D_{\mathrm{TE}}} \tag{7.67}$$

where

$$D_{\mathrm{TE}} = u_0 + u_1 \coth u_1 d \tag{7.68}$$

The D_{TE} term (and a similar D_{TM} term, to be defined shortly) are written in this specific notation because they are linked to *surface waves*. These can be important as a mechanism both for loss, and for increasing coupling between elements in a microstrip patch array. Neither is usually desirable. We will return to this later.

The last coefficient, b_1, may now be obtained, and we find for the fields in the dielectric that

$$\tilde{H}_z = -\frac{jk_y}{2\pi \sinh u_1 d} \frac{1}{D_{\mathrm{TE}}} \sinh u_1(z + d) \tag{7.69}$$

For the case where both source and observer lie on the air–dielectric interface, $z \to 0$ and this reduces to

$$\tilde{H}_z = -\frac{jk_y}{2\pi} \frac{1}{D_{\mathrm{TE}}} \tag{7.70}$$

What has now been computed is the spectral domain normal magnetic field due to an elementary x-directed dipole, i.e. $\tilde{G}_H^{zx}$. From Eq. (7.50), we find that

$$\tilde{G}_A^{xx} = -\frac{\mu \tilde{G}_H^{zx}}{jk_y} = \frac{\mu_0}{2\pi} \frac{1}{D_{\mathrm{TE}}} \tag{7.71}$$

The other components required for a HED may be derived in a similar fashion. The results are given in Table 7.2. Here, the subscript 1 has been dropped on u, since it

Table 7.2 Spectral domain Green functions for a single-layer grounded microstrip structure

Sommerfeld potentials
$\frac{2\pi \tilde{G}_A^{xx}}{\mu_0} = \frac{1}{D_{\text{TE}}}$
$\frac{2\pi \tilde{G}_A^{zx}}{\mu_0} = \frac{jk_x(\epsilon_r - 1)}{D_{\text{TE}} D_{\text{TM}}}$
$2\pi \epsilon_0 \tilde{G}_V = \frac{u_0 + u \tanh ud}{D_{\text{TE}} D_{\text{TM}}}$
$D_{\text{TE}} = u_0 + u \coth ud,\ D_{\text{TM}} = \epsilon_r u_0 + u \tanh ud$
$u^2 = k_\rho^2 - k^2,\ u_0^2 = k_\rho^2 - k_0^2$

Both source and observer are on the air–dielectric interface (after [1, Table 2, p. 153]). k_0 is the wavenumber in free space, and k is the wavenumber in the dielectric.

clearly refers to the substrate. For convenience, the spectral domain parameters u and u_0 are also listed.

7.4.4 The scalar potential and the mixed potential integral equation

The third entry in Table 7.2 lists a term which requires a brief comment, viz. $\tilde{G}_V$. In Section 7.4.1, the usual "mixed potential" formulation, Eq. (7.25) (which is valid for $F = 0$) was presented. It is actually by no means obvious that the usual scalar potential,

$$V(\boldsymbol{r}) = \int_S G_V(\boldsymbol{r}, \boldsymbol{r}') q_s(\boldsymbol{r}')\, dS' \tag{7.72}$$

can be extended to a layered medium under dynamic conditions. Fortunately, in the case of horizontal conducting surfaces, it can be shown that this is indeed valid, and further that the required scalar Green function is given in the spectral domain by [1, Section 3.3]

$$\tilde{G}_V = \frac{j\omega}{k_\rho^2} \left(\frac{\tilde{G}_E^{zx}}{jk_x} \right) - \left(\frac{k}{k_\rho} \right)^2 \left(\frac{\tilde{G}_H^{zx}}{jk_y \epsilon} \right) \tag{7.73}$$

for the Sommerfeld potentials.

Once the potentials are known, the fields can be computed from the potentials, as in Section 7.4.1. Before proceeding, it is worthwhile reminding the reader that the Green functions we have obtained are *spectral domain* representations; the spatial domain equivalents are of course defined by

$$G_A^{xx}(\boldsymbol{\rho} \mid \boldsymbol{\rho}' = 0) \equiv A_x(\boldsymbol{\rho}) = \frac{\mu_0}{2\pi} \int_0^\infty J_0(k_\rho \rho) \frac{k_\rho}{D_{\text{TE}}} dk_\rho \tag{7.74}$$

$$G_V(\boldsymbol{\rho} \mid \boldsymbol{\rho}' = 0) \equiv V(\boldsymbol{\rho}) = \frac{1}{2\pi \epsilon_0} \int_0^\infty J_0(k_\rho \rho) k_\rho \frac{u_0 + u \tanh ud}{D_{\text{TE}} D_{\text{TM}}} dk_\rho \tag{7.75}$$

and these are the functions we require. Again, as a reminder, ρ is radial distance on the patch surface, $\sqrt{x^2 + y^2}$; k_ρ is the integration variable; by convention, $z = 0$ is the air–dielectric interface; and $J_0(x)$ is the Bessel function of the first kind of order zero:

$$J_0(x) \equiv \frac{1}{\pi} \int_0^{\pi} \cos(x \sin \psi) d\psi \tag{7.76}$$

Note also that these are the Green functions for a source located at $\rho' = 0$; due to the translation symmetry, for sources located at a point other than the origin, all we need do is interpret the radial parameter as the distance from the observer to the source, i.e. $\rho = \sqrt{(x - x')^2 + (y - y')^2}$. This is also sometimes expressed as

$$G(x, y|x', y') = G(x - x', y - y'|0, 0) \tag{7.77}$$

Equipped with these Sommerfeld potentials, we can now write the mixed potential integral equation (MPIE) for the x-directed HED:

$$\mathbf{z} \times \mathbf{E}^{\text{inc}} = \mathbf{z} \times \left[j\omega \int_S \bar{\bar{G}}_A \cdot \mathbf{J}_S \, dS' + \nabla \int_S G_V q_S \, dS' + Z_S \mathbf{J}_S \right] \tag{7.78}$$

The vector potential $\bar{\bar{G}}_A$ and scalar potential G_V are as in the preceding section and are of course known, even if difficult to compute, as is the excitation $\mathbf{E}^{\text{inc}}$.

7.4.5 Surface waves

We commented earlier that the D_{TE} and D_{TM} terms are written in this specific form since they can be interpreted as *surface waves*. It can be shown that these expressions are the characteristic equations for the surface waves of, respectively, TE and TM waves propagating in a dielectric layer backed by a perfect conductor [1, Section 6]. Surface waves can decay as slowly as $1/\sqrt{\rho}$, and hence can be an important coupling mechanism between patches in a microstrip patch array. In the integrals required to compute the spatial domain Sommerfeld potentials, Eqs. (7.74) and (7.75), these enter in the denominator of the integrand, and zeros in D_{TE} and D_{TM} hence represent poles in the kernel, complicating the integration process. Fortunately, if $k_0 d \sqrt{\epsilon_r' - 1} < \pi/2$, then D_{TE} has no zeros and D_{TM} has only one, corresponding to the dominant zero-cutoff TM surface wave. This condition is equivalent to the following restriction:

$$f[\text{GHz}] \leq \frac{75}{d[\text{mm}]\sqrt{\epsilon_r' - 1}} \tag{7.79}$$

For practical substrates, this condition is generally satisfied over most of the microwave band. Only in the case of a thick substrate of high dielectric constant need one be concerned with this requirement.

The position of the pole is also required for the integration process. For lossless substrates, the pole is real ($k_\rho = \lambda_{p0}$) and lies inside the segment of the real axis $1 < \lambda_{p0}/k_0 < \sqrt{\epsilon_r}$. For thin substrates, an approximation of its position is [1, Section 6]

$$\lambda_{p0}/k_0 \approx 1 + (k_0 d)^2 \frac{(\epsilon_r - 1)^2}{2\epsilon_r^2} \tag{7.80}$$

This expression also holds for low-loss substrates, although the pole then migrates below the real axis, as in Fig. 7.3:

$$\lambda_p \approx \lambda_{p0}$$
$$\nu_p \approx (\epsilon_r' - 1)\tan\delta \left(\frac{k_0 d}{\epsilon_r'}\right)^2 \tag{7.81}$$

7.5 Evaluating the Sommerfeld integrals

7.5.1 Approximate evaluation of the Sommerfeld integrals

In general, the semi-infinite integrals in the spatial domain Sommerfeld potentials, Eqs. (7.74) and (7.75), have no closed-form solution and numerical evaluation, the topic of this section, is required. In certain cases, however, approximate solutions can be used, and one useful one in the present context is for the magnetic vector potential A_x for the HED case. Equation (7.74) does not contain the TM pole, with the result that the vector potential can be approximated by the vector potential for the homogeneous region $\epsilon_r = 1$. (Physically, the argument is that this is the *magnetic* vector potential, which should not be much affected by thin dielectric sheets.) In this case, the approximation is

$$\frac{4\pi}{\mu} A_x = \frac{e^{-jk_0 R_0}}{R_0} - \frac{e^{-jk_0 R_1}}{R_1} \tag{7.82}$$

with $R_0^2 = \rho^2$ and $R_1^2 = \rho^2 + (2d)^2$. The latter is of course the distance from the image of the HED in the ground plane, and we recognize this expression as that of a dipole and its (reversed) image. Although not generally valid, this is a useful approximation, especially for thin substrates of moderate dielectric constant. Although an approximation of the scalar potential is also available [1, Section 7.2], it turns out to be far less useful in this case and will not be discussed here.

Before proceeding further, the very important point must be made that the techniques to be discussed here emphasize simplicity, frequently exploiting knowledge of the specific problem: for instance, we restrict the analysis to the case of a single pole, and concentrate largely on the lossless substrate case. General-purpose programs using the Sommerfeld potentials have to handle potentially far more complex problems, and research still continues on efficient and robust implementations.

A mathematical aside – integration on the complex plane

The Sommerfeld integrals involve integration on the complex plane, $k_\rho = \lambda + j\nu$ in the present context, or more usually $z = x + iy$ in mathematical notation which we will use in this brief note. A few refreshers might be useful here. Firstly, a function $f(z)$ is *analytic* (or *regular*) in a region of the complex plane if it has a unique derivative at every point of the region. This is a far stronger condition in the complex plane than on the real line, since an analytic function has derivatives of *all* orders. (Many real functions have only derivatives to a certain order.) The *Cauchy–Riemann* conditions can be used to test whether a function is analytic in a region. A singularity is a point where $f(z)$ is not analytic; in the present context, it usually corresponds to an infinite value of the function.

Cauchy's theorem, and the resulting integral formula, are crucial: the theorem states that on a closed contour C^a:

$$\oint_C f(z)\,dz = 0$$

provided that the function is analytic on and inside C.

A very important consequence of this is that if $C = C_1 + C_2$, then $\int_{C_1} f(z)\,dz = \int_{C_2} f(z)\,dz$. This is so important in the context of the Sommerfeld potentials that it is worth reiterating: provided that the function is analytic, *different integration paths between two points in the complex plane yield the same result.*

Cauchy's integral formula states that under the same limitations as above, the value of $f(z)$ at $z = a$, a inside C, is given by

$$f(a) = \frac{1}{2\pi i}\oint_C \frac{f(z)}{z-a}\,dz$$

We usually apply this in reverse: for a function analytic except for a simple pole at $z = a$, the above theorem permits us to evaluate the integral. Combined with Laurent's theorem, this produces the residue theorem, which states that for isolated singularities within C,

$$\oint_C f(z)\,dz = 2\pi i \Sigma_k R_k$$

where R_k are the residues of $f(z)$ inside C. We will discuss finding the residues subsequently.

[a] There are some limitations on the form of C – it must not cross itself, and only a finite number of corners are permitted.

7.5.2 Numerical integration in the spectral domain

The spatial domain Sommerfeld potentials, Eqs. (7.74) and (7.75), require integration over the real positive axis λ.[3] We also note that since the integration is in the complex plane, the theory of complex functions permits deformation of the integration path, and a number of approaches avoid the pole(s), deforming the integral into the first quadrant. (The reason that the deformation takes this route is as follows. As already noted, for a lossy dielectric, the pole lies below the real axis, and the integration (along the real axis) lies above it. In the limit, as the loss tends to zero, the integration path must remain above the pole.) However, the most straightforward approach for the case of a simple

[3] Once again, readers are reminded that in this context, $\lambda = \text{Re}[k_\rho]$. Since we will continue to use λ_0 as the free-space wavelength, the potential for confusion is present, but we follow the notation of the literature in this context.

pole is to integrate along the real positive axis and this is the approach discussed here. There are, however, two points along the axis that require special care – the branch cut and the pole – and an asymptotic case needing caution.

Firstly, at $k = k_0$, the function $u_0 = \sqrt{k_\rho^2 + k_0^2}$ introduces a *branch point*. This is due to the multi-valued nature of the complex valued square root function. Which value to choose is mathematically described as the process of selecting the correct *Riemann sheet*. Fortunately, all we need note here is that we should choose $\mathrm{Re}[u_0] \geq 0$; since the integrand remains bounded at this point, we can integrate straight through the branch point.

A mathematical aside – branch points and branch cuts

Branch points and cuts arise due to multi-valued functions in the complex plane. The branch cut is used to demarcate "Riemann sheets," which resolve the ambiguities. As a simpler example, consider $f(z) = z^{1/2}$. Obviously, with $z = Ce^{i\theta}$, $f(z) = \sqrt{C}e^{i\theta/2}$. This is periodic, but with period 4π, and this is where the problems arise. For instance, consider $\theta = 3\pi/2$ and $\theta = -\pi/2$, the *same* point on the complex plane. Now, the two solutions for $f(z)$ are $\sqrt{C}e^{i3\pi/4}$ and $\sqrt{C}e^{i\pi/4}$, clearly *not* the same point anymore.

Riemann sheets adopt some convention to resolve this ambiguity. In this case, $f(z)$ for $-\pi < \theta < \pi$ is associated with the "top" Riemann sheet, and $f(z)$ for $\pi < \theta < 3\pi$ with the "bottom" Riemann sheet. This is best illustrated as below:

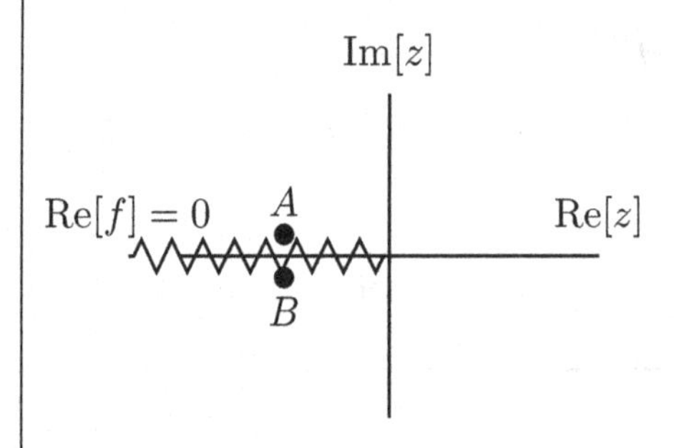

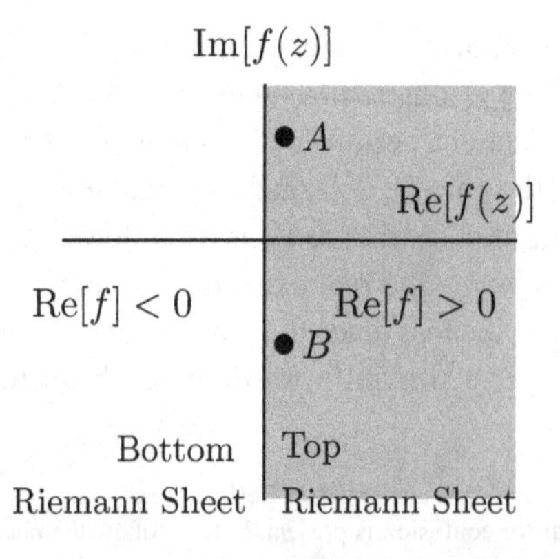

The negative real axis forms the *branch cut* in the z-plane, which opens up to define the boundary between the Riemann sheets in the $f(z)$-plane. By alternating between Riemann sheets, the function $f(z)$ can be made continuous. For instance, as one moves from $\theta = \pi^-$ (on the top Riemann sheet) to $\theta = \pi^+$, one must move onto the *bottom* Riemann sheet, which effectively resolves the ambiguity of which value of $\sqrt{-1}$ to choose, since we now know we must use π^+ and not $-\pi^-$ when evaluating the function with this convention. In this case, there were only two Riemann sheets. Other multi-valued functions, such as $\ln z$, can have infinitely many values and require an infinite number of Riemann sheets.

Which Riemann sheet one must work in the present context of Sommerfeld integrals often requires physical arguments, such as the radiation condition. This, and related issues, have caused many problems in the history of Sommerfeld potentials, with incorrect choices having led to unphysical artifacts and much debate in the literature. An extended discussion may be found in [8, 2.2].

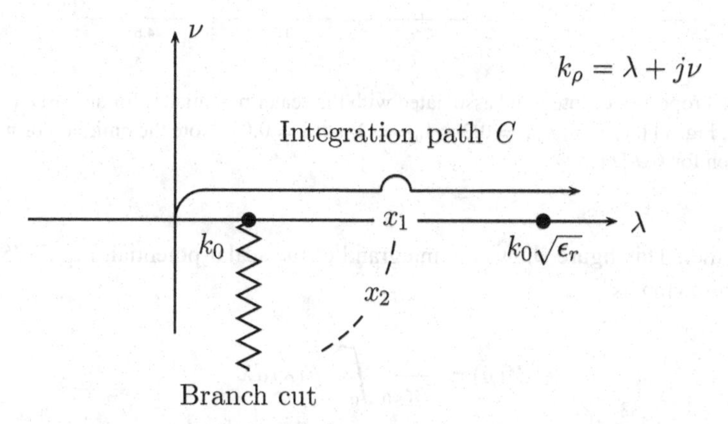

Figure 7.3 Topology of the complex plane for a thin grounded substrate, showing the branch cut, pole positions and the integration path C. For a lossless dielectric, the pole is on the real axis (x_1); when loss is present, it migrates into the fourth quadrant (x_2). (Adapted from [1, Fig. 5].)

The second point requiring attention is the pole, due to the TM surface wave. This introduces a rapidly varying integrand. Here, we follow [1, Section 8] and integrate through the pole (which lies on the real positive axis in the case of a lossless substrate), using a special method to extract the singularity which we will describe shortly. Note that for the HED, and assuming that the inequality of Eq. (7.79) holds (i.e. only the TM pole is present) it is *only* the scalar potential V which is thus affected.

The final point which one must bear in mind is that the oscillating integrands have an envelope which converges very slowly in the asymptotic case $\lambda \to \infty$. All these issues are summarized in Fig. 7.3.

In Fig. 7.4, the general properties of the function to be integrated are shown for a rather thick substrate with relatively large dielectric contrast; this has been done for clarity, to separate clearly the pole and the branch point, which in many practical cases lie close to

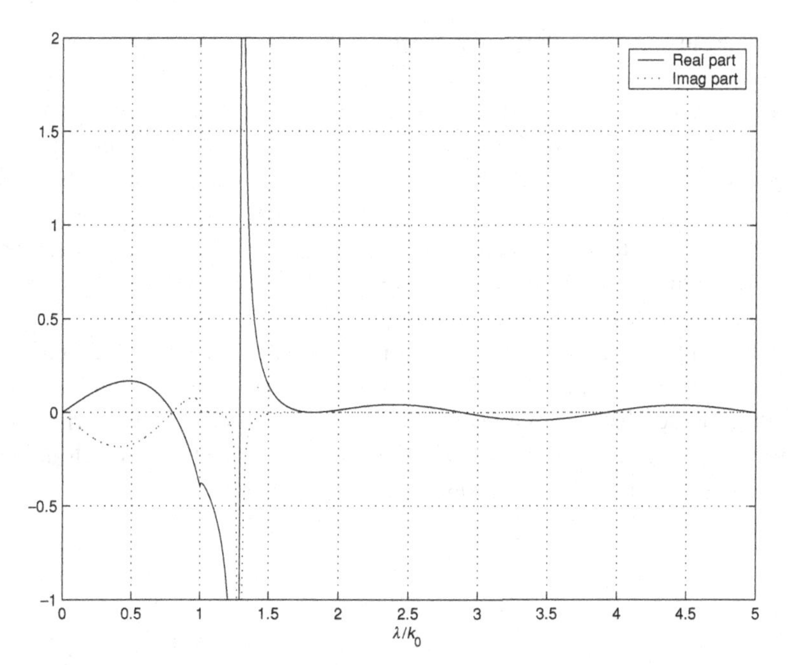

Figure 7.4 Properties of integrand associated with the scalar potential V for an HED. Parameters as for [1, Fig. 11]: $\epsilon_r' = 5$; $k_0 d = 0.2\pi$; $k_0 \rho = 3$; $\tan\delta = 0.01$. Note the omission of π in the expression for $k_0 d$ [1].

one another. This figure shows the integrand of the scalar potential, Eq. (7.75), written in the following as

$$V(\rho) = \frac{1}{2\pi\epsilon_0}\int_0^{\infty} F(\lambda)\,d\lambda \tag{7.83}$$

$$\begin{aligned} F(\lambda) &= J_0(\lambda\rho)\,\lambda\,\frac{u_0 + u\tanh ud}{D_{\mathrm{TE}}D_{\mathrm{TM}}} \\ &= J_0(\lambda\rho) f(\lambda) \end{aligned} \tag{7.84}$$

where we have used $k_\rho = \lambda + j\nu$ since the integration is on the real axis.

It has been proposed [1, Section 8] that the real axis be split into three subintervals, namely $[0, k_0]$, $[k_0, k_0\sqrt{\epsilon_r}]$ and $[k_0\sqrt{\epsilon_r}, \infty]$, and we will follow this approach here. We will investigate only the scalar potential V, since as mentioned above, the vector potential does not contain the TM pole and can be approximated using Eq. (7.82) for the case we will study. In each region, we proceed as follows.

Region 1 $[0, k_0]$

No special care is needed in this region, since the function is well behaved, apart from an infinite derivative at $\lambda = k_0$. A change of variables $\lambda = k_0 \cos t$ suffices to make the function very smooth and easily integrated using standard procedures. Hence, in

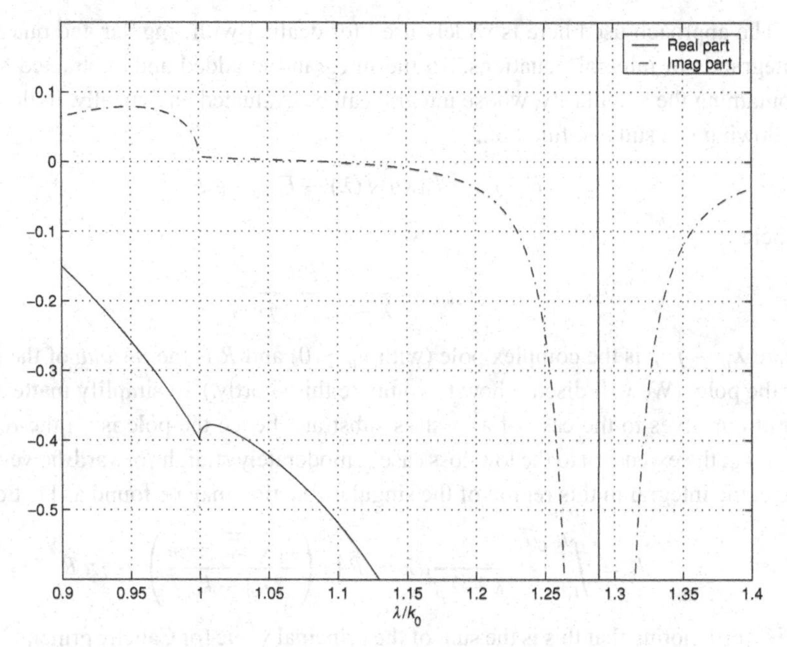

Figure 7.5 Detail of Fig. 7.4 in the region $k_0 \in [0.9k_0, 1.4k_0]$.

region 1, the integral to evaluate numerically is

$$\int_0^{\pi/2} F(k_0 \cos t) k_0 \sin t \, dt \tag{7.85}$$

Note that the minus sign present in the differential $d\lambda = -k_0 \sin dt$ is cancelled by the interchange of the lower and upper limits of integration required.

The numerical integration in this and all the remaining regions can be performed in MATLAB using the `quad` function, which implements adaptive Simpson quadrature. (Simpson quadrature, the classic numerical integration routine, fits a quadratic polynomial to the data points to be integrated; due to symmetry, it is exact to third order. The adaptive variant recursively divides the intervals until the difference between successive evaluations is less than some specified tolerance.) Many other types of numerical integration are available and can be applied; see, for instance, [9, Chapter 4] for an especially entertaining discussion.

Region 2 $[k_0, k_0\sqrt{\epsilon_r}]$

In this region, enlarged in Fig. 7.5, the singularity caused by the pole is clearly present. Strictly speaking, with finite loss this is a numerical singularity (or a quasi-singularity), since the pole is now slightly below the real axis and the value of the function is not truly infinite at the pole; however, for practical situations with low-loss substrates, the values are numerically so large that the effect is that of a singularity; furthermore, for a lossless substrate, this is a true mathematical singularity.

The approach used here is widely used for dealing with singular and quasi-singular integrands in integral equations. To the integrand is added and subtracted a function containing the singularity, whose integral can be evaluated analytically. In this case, the following is a suitable function:

$$F(\lambda) = \left[J_0(\lambda\rho)f(\lambda) - F_{\text{sing}}\right] + F_{\text{sing}} \tag{7.86}$$

where

$$F_{\text{sing}} = \frac{R}{\lambda - (\lambda_p - j\nu_p)} \tag{7.87}$$

Here $\lambda_p - j\nu_p$ is the complex pole (with $\nu_p > 0$) and R is the *residue* of the integrand at the pole. (We will discuss how to compute this shortly.) To simplify matters, we will limit ourselves to the case of a lossless substrate, hence the pole is on the real axis at $\lambda = \lambda_p$; the extension to the low-loss case is moderately straightforward, however. In this case, the integral in this region of the singular function may be found as [1, Eq. (110)][4]

$$I_s = \int_{k_0}^{k_0\sqrt{\epsilon_r'}} \frac{R}{\lambda - \lambda_p} d\lambda = R \ln\left(\frac{k_0\sqrt{\epsilon_r'} - \lambda_p}{\lambda_p - k_0}\right) - j\pi R \tag{7.88}$$

It is worth noting that this is the sum of the principal value (or Cauchy principal value) of the integral, and the contribution of the pole. (The principle value of a singular integral avoids the singularity.) The result for lossy materials is useful [1, Eq. (109)]:

$$I_s = \frac{R}{2}\ln\left(\frac{\upsilon_p^2 + (k_0\sqrt{\epsilon_r'} - \lambda_p)^2}{\upsilon_p^2 + (k_0 + \lambda_p)^2}\right) + jR\arctan\frac{k_0\sqrt{\epsilon_r'} - \lambda_p}{\nu_p} + jR\arctan\frac{\lambda_p - k_0}{\nu_p} \tag{7.89}$$

In Fig. 7.6, the original function F, the singular function F_{sing} and the difference function have been plotted. The last is clearly smooth and readily integrated numerically. The smoothness has been enhanced by the change of variables $\lambda = k_0 \cosh t$. The integral in this region is the sum of I_s, the analytically integrated singular function as above, and I_d, the numerically integrated difference function:[5]

$$\begin{aligned} I_d &= \int_{k_0}^{k_0\sqrt{\epsilon_r'}} \left[F(\lambda) - F_{\text{sing}}\right] d\lambda \\ &= \int_0^{\operatorname{arccosh}\sqrt{\epsilon_r'}} \left[F(k_0\cosh t) - F_{\text{sing}}(k_0\cosh t)\right] k_0 \sinh t\, dt \end{aligned} \tag{7.90}$$

One point that should be mentioned here is that for $\lambda = k_0\sqrt{\epsilon_r'}$, $u = 0$, and the $\coth ud$ term in D_{TE}, in the denominator of the integrand, results in a zero at this point. Attempting to evaluate this numerically is inadvisable, and the upper integration limit should be set fractionally below this value. (Since this is a zero and not a pole, this simple remedy suffices.)

[4] Note that this reference incorrectly includes the $j\pi R$ term, $j\pi P$ in their notation, on the left-hand side as well. Alternatively, the integral on the left-hand side should be a principal value integral.

[5] When performing the change of variables, recall that the derivative of $\cosh t$ is $+\sinh t$!

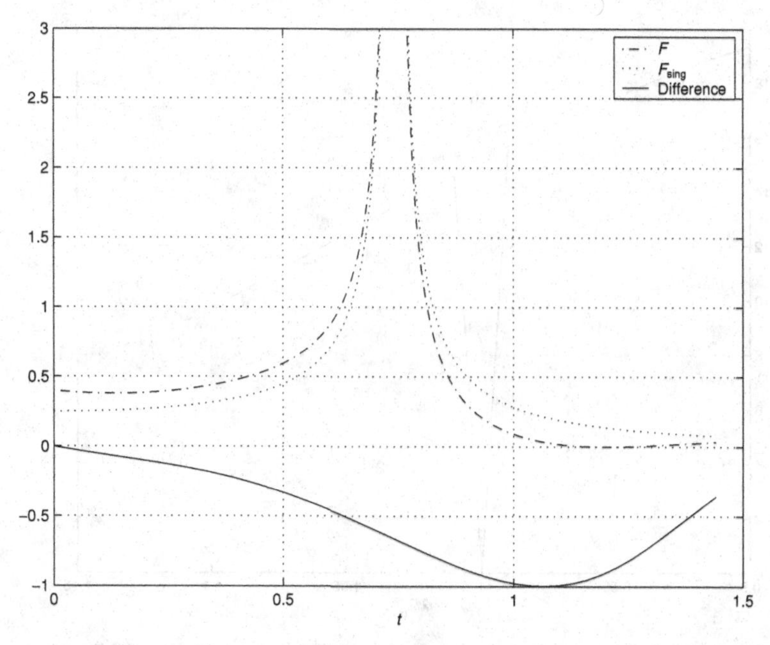

Figure 7.6 The original function (F), the singular function F_{sing} and the difference function, with the change of variables $\lambda = k_0 \cosh t$. All parameters as in Fig. 7.4, except that $\tan\delta = 0$.

One final point requiring discussion is the evaluation of the residue. For a function of a complex variable z, with simple pole at $z = z_p$, which is the case we have here, the residue can be computed by multiplying the function by $z - z_p$ and evaluating the result at $z = z_p$. It is instructive to attempt this numerically, as shown in Fig. 7.7. The theoretical value is $R = 15.1107$; if the numerical result is interpolated through the pole, one will obtain a value very close to this. The reason that the curve in Fig. 7.7 exhibits a linear decay to zero in a small region around the pole is no doubt due to numerical approximations made (by MATLAB, in this case) when evaluating extremely large-valued functions.

The residue may be found rigorously noting that the integrand is of the form $g(z)/h(z)$, with $h(z_p) = 0$, but $h'(z_p) \neq 0$ and $g(z_p) \neq 0$. In this case, the residue may be computed from

$$R(z_p) = \frac{g(z_p)}{h'(z_p)} \tag{7.91}$$

For the TM pole, the result is

$$R(\lambda_p) = \frac{J_0(\lambda_p \rho)\lambda_p(u_0 + u \tanh ud)}{D_{\text{TM}} \frac{d}{d\lambda} D_{\text{TE}} + D_{\text{TE}} \frac{d}{d\lambda} D_{\text{TM}}} \tag{7.92}$$

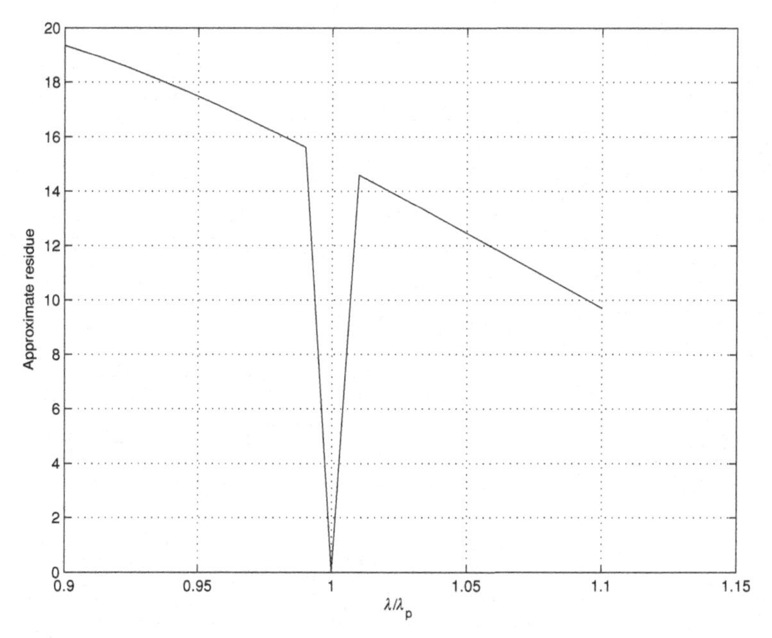

Figure 7.7 Result of attempting to evaluate the residue at the pole numerically. Parameters as in Fig. 7.6.

with

$$\frac{d}{d\lambda} D_{\mathrm{TE}} = \frac{\lambda}{u_0} + \frac{\lambda}{u} \coth ud - \lambda\, d \,\mathrm{csch}^2 ud \tag{7.93}$$

$$\frac{d}{d\lambda} D_{\mathrm{TM}} = \epsilon_r \frac{\lambda}{u_0} + \frac{\lambda}{u} \tanh ud + \lambda\, d \,\mathrm{sech}^2 ud \tag{7.94}$$

In deriving this result, note that $du/d\lambda = \lambda/u$ and $du_0/d\lambda = \lambda/u_0$.

Region 3 $[k_0\sqrt{\epsilon_r}, \infty]$

In this region, the function has no singularities or branch points, but contains a slowly converging integrand, as shown in Fig. 7.8. To accelerate the convergence, the static term

$$\frac{J_0(\lambda\rho)}{1+\epsilon_r}$$

is extracted. Beyond a certain point $\lambda > \lambda'$, the resulting integral is negligible. Using the standard result (for example, [10, Eq. 24.92])

$$\int_0^\infty J_0(\lambda\rho)\, d\lambda = \frac{1}{\rho}$$

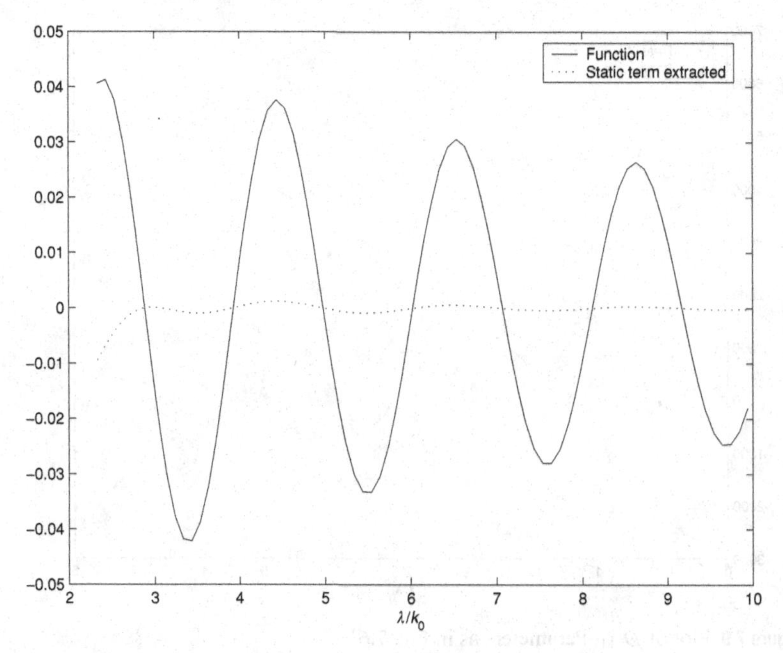

Figure 7.8 The integrand in region 3, before and after subtraction of the static term. Parameters as in Fig. 7.6.

one obtains

$$\int_{\sqrt{\epsilon_r'}k_0}^{\infty} F(\lambda)\,d\lambda \approx \int_{\sqrt{\epsilon_r'}k_0}^{\lambda'} \left[F(\lambda) - \frac{J_0(\lambda\rho)}{1+\epsilon_r}\right] d\lambda + \frac{1}{\rho(1+\epsilon_r)} - \frac{1}{1+\epsilon_r}\int_0^{\sqrt{\epsilon_r'}k_0} J_0(\lambda\rho)\,d\lambda \tag{7.95}$$

The question of how large to set λ' can be determined iteratively. The results to be shown started with $\lambda' = 10k_0$; the resulting integral was evaluated, as well as the integral with $\lambda' = 20k_0$. The difference, normalized by the integral in region 2, between the integrals was then compared, and if too large, the procedure was repeated with the upper limits doubled. (The integral in region 2 is usually the largest contributor to the integral, since it includes the contribution of the pole, and hence was used to normalize this result.) This process is not especially robust, and more sophisticated procedures are available [1, Section 8.2].

7.5.3 Locating the pole

The position of the pole must of course be found with considerable accuracy for the above process to work properly, in particular in region 2. An approximation of its position has already been given in Eq. (7.80), but this is not sufficient for the singularity extraction procedure. Finding the pole is equivalent to locating the roots of D_{TM}. In general, finding the roots of a non-linear function is a very challenging problem, but in the case under

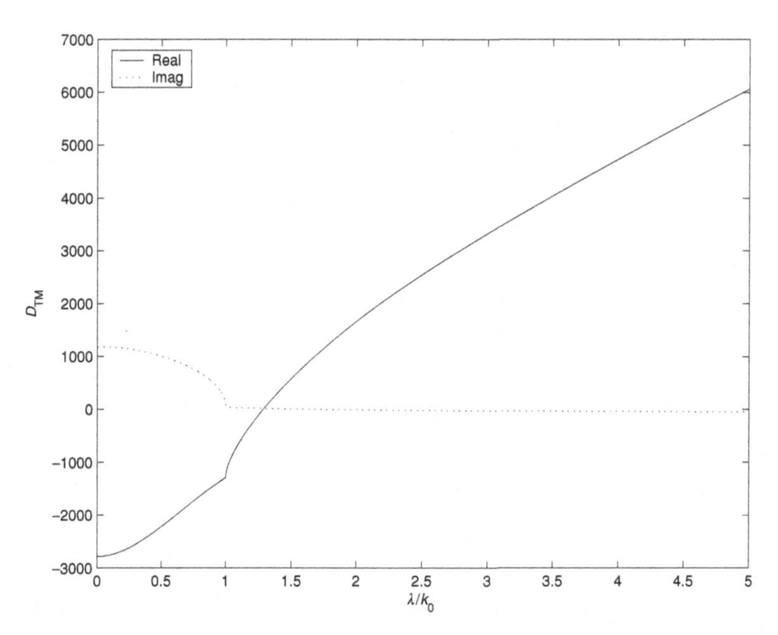

Figure 7.9 Plot of $D_{\rm TM}$. Parameters as in Fig. 7.6.

consideration, the pole is known to be single, and located on the real axis in the interval $[k_0, \sqrt{\epsilon_r'}k_0]$. Furthermore, as Fig. 7.9 shows, the function is purely real valued for $\lambda > k_0$ (the branch point) and changes sign in this interval $[k_0, \sqrt{\epsilon_r'}k_0]$. A very simple algorithm, such as interval bisection, yields the root easily. Interval bisection starts with an interval containing a root, with the function having opposite signs at the interval limits. The function is then evaluated at the mid-point of the interval, which then replaces whichever limit has the same sign. This proceeds until the root is found with satisfactory precision. Despite its simplicity, the algorithm is failsafe in the present case – since it will always find at least one root, and there is only one. The method also converges linearly which is more than sufficient. The algorithm is so simple as not to require listing; details can be found in any book on numerical analysis, such as [9].

Slightly lossy materials can also be accommodated, although the root finder must now work with complex values; fortunately, although $D_{\rm TM}$ is now complex-valued in the search interval, the overall shape of the function remains very similar to that of Fig. 7.9, and the imaginary part is small in the search interval.

Here we should comment that all the above holds *only* for the case of the single pole. As soon as more than one pole is present, the pole finding becomes far more complex. It is this type of complexity which makes robust, general-purpose codes so time-consuming to develop. Further details may be found in [1].

7.5.4 General source locations

The above potentials all assume that the source is located at $(x' = 0; y' = 0)$, i.e. $\rho' = 0$. For sources at other locations, all that is required is to substitute $\rho = \sqrt{(x-x')^2 + (y-y')^2}$. This is sometimes written as $V(\rho|\rho')$.

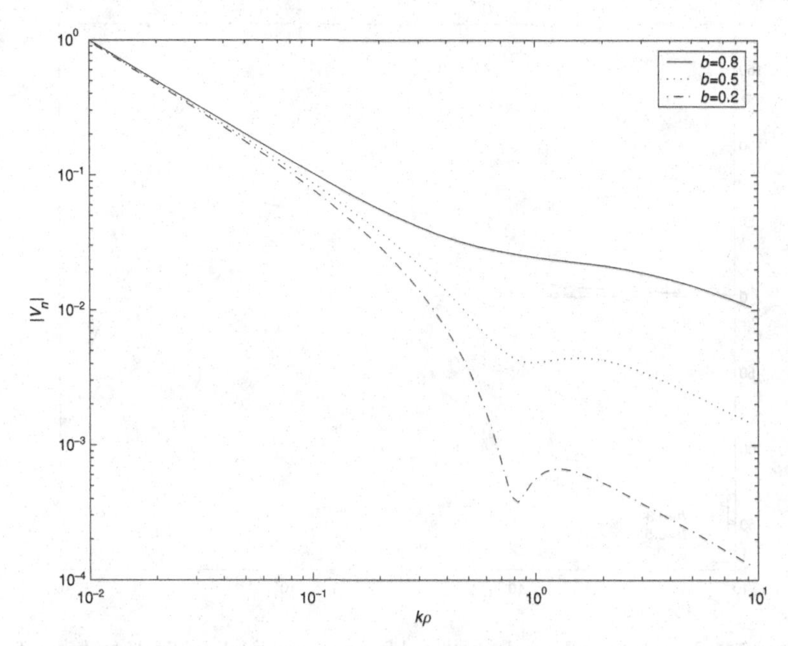

Figure 7.10 The modulus of the normalized scalar potential for various normalized thicknesses as a function of normalized distance. $\epsilon_r = 10$; $b = 2k_0 d\sqrt{\epsilon_r - 1}/\pi$.

7.5.5 Some results for the Sommerfeld potentials

Now that the question of the integration of the potentials has been addressed, we can turn our attention to the potentials themselves. Results are shown in Figs. 7.10 and 7.11, which illustrate the variation due to different substrate thicknesses; note that the results all converge for very small distances; this is the quasi-static limit. Figure 7.12 shows the effect of various dielectric constants. It is interesting to note the "knee" which sets in at progressively smaller distances as the dielectric constant increases; this corresponds to the transition from static to surface wave behavior. It will be noted that the potential decays at a slower rate once the surface wave sets in. It will also be noted that the surface wave is absent in the case of $\epsilon_r = 1.01$; this is essentially free space, which does not support a surface wave. (To avoid problems in the routines used, a value slightly larger than unity was used.) The effect of increasing dielectric constant has already been noted; for practical antenna design, this means that high-ϵ_r substrates are likely to have more problems with mutual coupling between array elements. The same effect is also present as the substrate thickness is increased.

These results are very similar to [1, Figs. 19–21] and serve to validate the implementation thus far.

7.6 MoM solution using the Sommerfeld potentials

Now that the potentials are available, the MoM discretization of the MPIE, Eq. (7.78), can be undertaken. Before we do this, it is useful to identify a suitable problem. Although

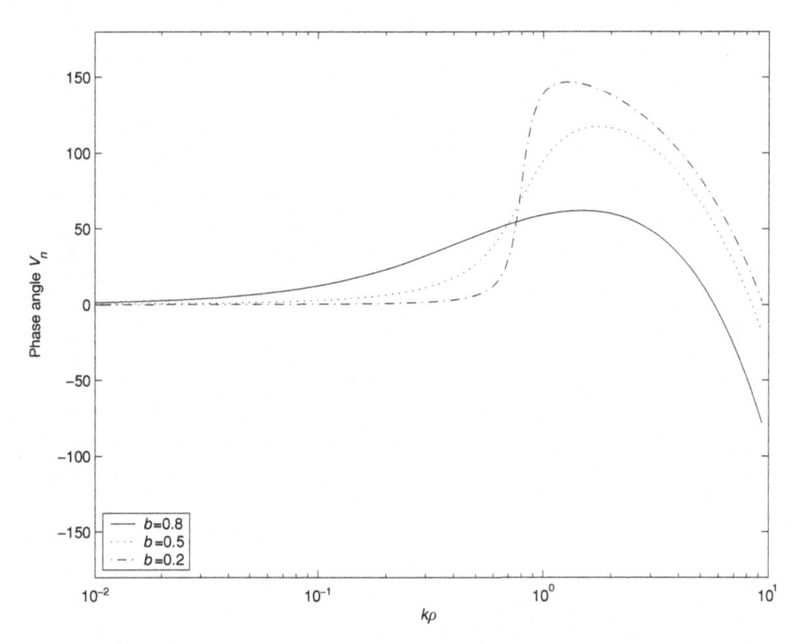

Figure 7.11 The phase of the scalar potential for various normalized thicknesses as a function of normalized distance. $\epsilon_r = 10$; $b = 2k_0 d\sqrt{\epsilon_r - 1}/\pi$.

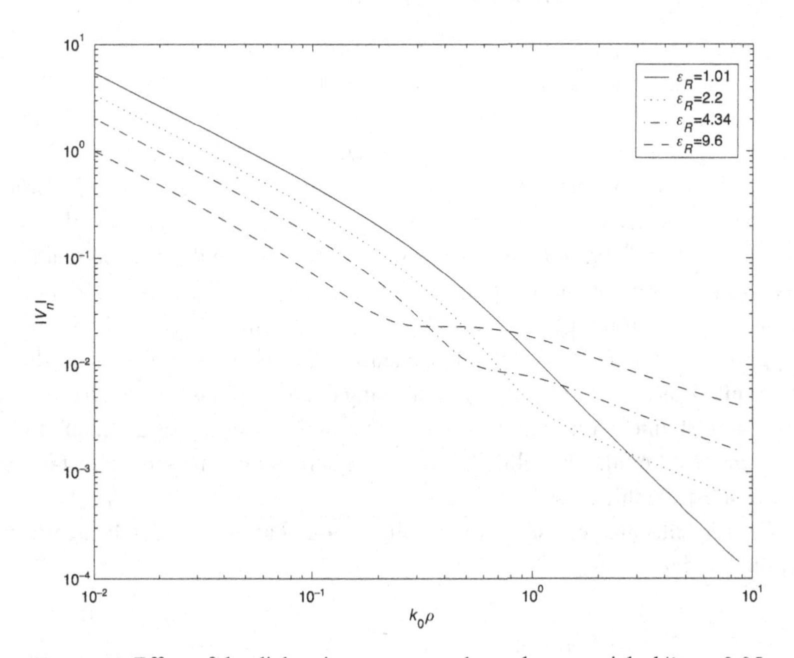

Figure 7.12 Effect of the dielectric constant on the scalar potential. $d/\lambda_0 = 0.05$.

microstrip patch antennas[6] are the dominant application of this theory at present, they require a surface discretization, supporting vector currents (that is, the basis function must be able to support both $\hat{x}$- and $\hat{y}$-directed currents). A printed dipole is a rather easier problem, since so long as the dipole is relatively thin, the current flows essentially along the axis of the structure, much as for the thin dipole in free space that we have already studied in Chapter 4. A printed dipole is also easily simulated using a commercial code, such as FEKO.

With a suitable problem identified, various possibilities arise with the MoM. Perhaps the most popular, especially for "do-it-yourself" research codes, have been "rooftop" basis functions, defined on rectangular elements.[7] More sophisticated codes generally use the Rao–Wilton–Glisson element. For testing functions, Galerkin procedures have been widely used; another popular option has been a pulse-doublet testing function. Collocation techniques have also been used. We will take the opportunity to do something a little different (although also used in the literature), namely utilize *entire domain* basis functions. A very obvious one here is a Fourier series expansion; for a symmetrically excited dipole (e.g. center fed) only a cosine series is needed, and only the odd numbered terms.

It is useful to develop the MoM equations from basic principles as another example of the application of the method. Referring back to the mixed potential integral equation, Eq. (7.78), repeated here, but with the last term dropped:

$$\boldsymbol{z} \times \boldsymbol{E}^{\text{inc}} = \boldsymbol{z} \times \left[j\omega \int_S \bar{\bar{G}}_A \cdot \boldsymbol{J}_S \, dS' + \nabla \int_S G_V q_S \, dS' \right] \tag{7.96}$$

the current is expanded as

$$\boldsymbol{J} \approx \sum_{m=1}^{N} \alpha_m \boldsymbol{F}_m \tag{7.97}$$

From the continuity equation, the charge is therefore expanded as

$$q_S \approx \sum_{m=1}^{N} \alpha_m \frac{-\nabla \cdot \boldsymbol{F}_m}{j\omega} \tag{7.98}$$

Note that the basis functions are effectively scalar in this case.

Introducing testing functions W_n and carrying out the weighted residual process as usual, we obtain:

$$\begin{aligned} \boldsymbol{z} \times \int_S \boldsymbol{W}_n \cdot \boldsymbol{E}^{\text{inc}} dS = \boldsymbol{z} \times \sum_{m=1}^{N} \alpha_m \bigg[& j\omega \int_S \boldsymbol{W}_n \int_S \bar{\bar{G}}_A \cdot \boldsymbol{F}_m \, dS' \, dS \\ & - \frac{1}{j\omega} \int_S \boldsymbol{W}_n \cdot \nabla \int_S G_V \nabla' \cdot \boldsymbol{F}_m \, dS' \, dS \bigg] \end{aligned} \tag{7.99}$$

[6] Readers not familiar with this technology should note that some more background on these antennas is presented in Chapter 8.

[7] The term *patch* instead of *element* is frequently encountered in the literature; the potential for confusion with the patch antenna is obvious and hence element is used here.

One subtlety worth commenting on here is the manipulation of the second surface integral on the right-hand side of the above equation. Using the vector identity $\nabla(a\boldsymbol{b}) = a\nabla\cdot\boldsymbol{b} + \boldsymbol{b}\nabla a$, and identifying $\boldsymbol{b} = \boldsymbol{W}$ and a as the inner integral, one obtains

$$z\times\int_S \boldsymbol{W}_n\cdot\nabla\int_S G_V\nabla\cdot\boldsymbol{F}_m\,dS'dS = z\times\int_S \nabla\left[\boldsymbol{W}_n\int_S G_V\nabla'\cdot\boldsymbol{F}_m\,dS'\right]dS - z\times\int_S \nabla\cdot\boldsymbol{W}_n\int_S G_V\nabla'\cdot\boldsymbol{F}_m\,dS'dS \tag{7.100}$$

The first term on the right-hand side in the above may be eliminated by applying a variant of the divergence theorem, known as the *surface divergence theorem*. For an open surface S bounded by contour C, this states that for a vector function $\boldsymbol{f}$,

$$\int_S \nabla_s\cdot\boldsymbol{f} = \oint_C \hat{m}\cdot\boldsymbol{f}\,dC \tag{7.101}$$

with $\hat{m}$ the unit vector normal to contour C, but tangential to surface S [11, p. 712]. $\nabla_s\cdot$ is the divergence operator in the surface, and this is precisely what $z\times$ selects.[8] Note that, *unlike* Stoke's theorem, the contour integral in this case evaluates *normal* fields on the boundary. Hence this term can be written in terms of a contour integral of a quantity related to current, normal to the bounding contour. Since normally directed current should go to zero at the edge of the dipole, this term is zero. Strangely, few references on this topic explain this point.

The MPIE thus results in the standard MoM matrix equation $[Z]\{I\} = \{V\}$. For convenience, it is useful to split the impedance matrix in two:

$$Z_{mn} = a_{mn} + v_{mn} \tag{7.102}$$

with matrix and vector entries as follows:

$$\begin{aligned} a_{mn} &= j\omega\int_S \boldsymbol{F}_m(\rho)\cdot\int_{S'}\bar{\bar{G}}_A\cdot\boldsymbol{F}_n\,dS'\,dS \\ v_{mn} &= \frac{1}{j\omega}\int_S \nabla\cdot\boldsymbol{F}_m(\rho)\cdot\int_{S'} G_V\nabla'\cdot\boldsymbol{F}_n\,dS'\,dS \\ b_m &= \int_S F_m(\rho)\cdot\boldsymbol{E}^{\text{inc}}\,dS \end{aligned} \tag{7.103}$$

For the case of a thin printed dipole, we will make a number of assumptions similar to those of our earlier work on the thin-wire dipole. It will be assumed that the current flows only in the $\hat{x}$-direction, and that the surface integrals can be approximated as line integrals. In this case, the integral in the transverse direction, $\hat{y}$, simply results in a constant W, present in both $[Z]$ and $[V]$, and thus cancelling. Further, the equations (7.103)

[8] The surface divergence operator can be defined in terms of general curvilinear coordinates for curved surfaces, but in the present case it is unnecessary.

can be rewritten in scalar form. The result is the following:

$$a_{mn} = j\omega \int_{\ell} F_m(x) \int_{\ell'} A_x(|x - x'|) F_n(x')\, dx'\, dx$$

$$v_{mn} = \frac{1}{j\omega} \int_{\ell} \frac{\partial}{\partial x'} F_n(x') \int_{\ell'} V(|x - x'|) \frac{\partial}{\partial x} F_n(x)\, dx'\, dx$$

$$b_m = \int_{\ell} F_m(x) E_x^{\text{inc}}\, dx \tag{7.104}$$

As already mentioned, we intend using entire domain basis functions. In this case, the source (primed coordinates) and field integrals are over the same domain, namely the length of the wire. Assuming that we center the wire at the origin, suitable entire domain basis functions are:

$$F_m = \cos\left(\frac{m\pi x}{L}\right), \qquad m = 1, 3, \ldots \tag{7.105}$$

Note that by this choice, the current goes to zero at the ends of the wire ($x = \pm L/2$) as required – for *all* the basis functions and hence also for their sum. With the above geometrical assumptions and these basis functions, and noting that the domain of integration is the same for both source and field points (the length of the wire), the matrix entries become:

$$a_{mn} = j\omega \int_{-L/2}^{L/2} \cos\left(\frac{m\pi x}{L}\right) \int_{-L/2}^{L/2} A_x(|x - x'|) \cos\left(\frac{n\pi x'}{L}\right) dx'\, dx$$

$$v_{mn} = \frac{1}{j\omega} \frac{mn\pi^2}{L^2} \int_{-L/2}^{L/2} \sin\left(\frac{m\pi x}{L}\right) \int_{-L/2}^{L/2} V(|x - x'|) \sin\left(\frac{n\pi x'}{L}\right) dx'\, dx$$

$$b_m = \int_{-L/2}^{L/2} \cos\left(\frac{m\pi x}{L}\right) E_x^{\text{inc}}\, dx \tag{7.106}$$

For the source, we will assume a very short feed section, of length Δs. The incident (impressed) electric field is thus V_s/Δ_s, where V_s is the source voltage. The result is that

$$b_m \approx V_s \tag{7.107}$$

It is interesting to note that the same result is obtained by assuming an infinitely thin Dirac delta source, with $E_x^{\text{inc}} = V_s \delta(x)$.

The code can now be developed. The integration required must be performed numerically. In this case, a simple trapezoidal scheme will suffice (implemented in MATLAB as `trapz`). An issue which requires a little care is that of singularities; both the vector and scalar potentials exhibit singularities at the origin. Fortunately, the singularities are of low order – this is one of the appealing features of the MPIE. The rigorous method for handling this extracts the singular component (which in both cases is the static limit), integrates this analytically and the remaining part is integrated numerically, in a fashion already applied in region 2 when evaluating the scalar Sommerfeld potential. This works very well for subdomain MoM methods and is relatively easy to implement, since it need only be applied to the "self" term; unfortunately, with entire domain basis and testing functions, it is rather more difficult to use. Because the singularity is of relatively low order, it can be side-stepped numerically, by using integration points for the field and source point integrals which are slightly offset from one another. If there are N equally spaced integration points $\Delta = L/N$ apart, instead of sampling at

$$x_j = -L/2 + \Delta/2 + \Delta(j-1) \tag{7.108}$$

(and similarly for x'_k) one can use, for instance,

$$x_j = -L/2 + \Delta/3 + \Delta(j-1) \tag{7.109}$$

and

$$x'_k = -L/2 + 2\Delta/3 + \Delta(j-1) \tag{7.110}$$

which offsets the points by $\Delta/3$.

To keep things simple, it will also be assumed that the substrate is thin enough that the low-frequency approximation of the magnetic vector potential may be used, namely Eq. (7.82).

One other issue which requires attention is computational efficiency. Usually, the first implementation of a new method can be done with little regard for this. However, the Sommerfeld potentials are sufficiently time consuming to evaluate that if some thought is not given to this, even simple problems take far too long to solve. Because of the dependence on wavenumber, the potentials are frequency dependent, and nothing can be done about this. However, for a particular antenna geometry at a specific frequency, the potentials are only a function of radial distance ρ (and in this one-dimensional case, $|x - x'|$) and a widely used approach is to pre-compute the potentials and use interpolation when constructing the MoM matrices. This significantly reduces the time required to fill the impedance matrix.

Results for a MATLAB implementation are shown in Fig. 7.13. The printed dipole has length $L = 0.39\lambda_0$ and width $W = 0.002\lambda_0$, with relative permittivity $\epsilon_r = 2.55$, as in [12]. This dipole was designed as an element in a very large array, with λ_0 the free-space wavelength corresponding to the center frequency. For this simulation, this was chosen as 10 GHz, well into the microwave band and a typical frequency where microstrip is an attractive technology. (Because this is a single element, one can expect the actual center frequency to differ from this value; it turns out to be around 0.9 of the design value.) The substrate used in [12] is very thick (although only the TM mode propagates), and the approximation of the magnetic vector potential with its static value is insufficiently accurate, so the simulation here used a thinner substrate, $h = 0.12\lambda_0$ thick.

Figure 7.13 shows three results: one computed using FEKO ($h/\lambda = 50$ discretization), and two computed with a MATLAB code based on the formulation developed here. The "coarse" result was computed using only 1 mode, with 32 integration points; the "fine" result used 5 modes and 128 integration points. The reflection coefficient is computed in a $Z_0 = 50\,\Omega$ system. (It should be commented that this antenna is not very well matched: the reason is that the substrate is not thick enough to provide sufficient spacing between the antenna and its image in the ground plane.) Improving the MoM model (the "fine" result) produces a value for minimum S_{11} very similar to the FEKO result, although at a frequency some 8% higher.

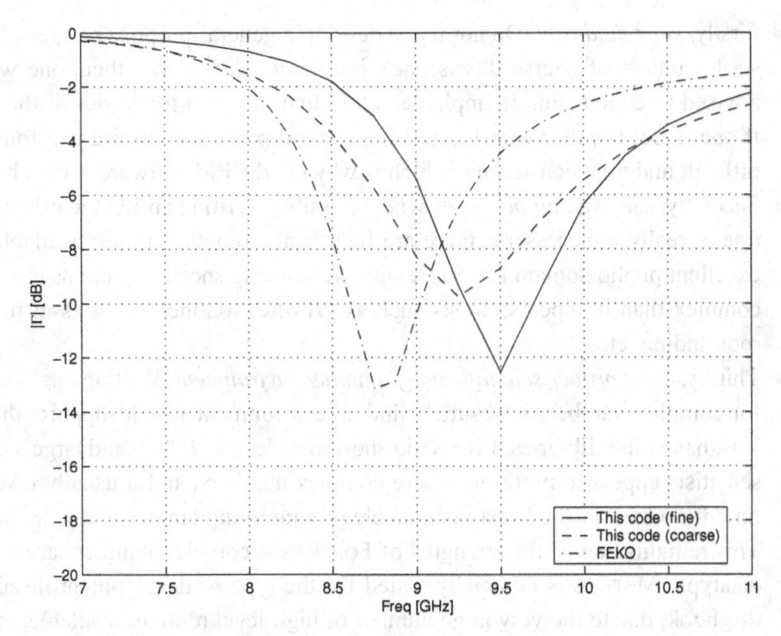

Figure 7.13 Reflection coefficient of a thin printed dipole.

This is not very accurate – certainly not sufficient for engineering design purposes – but provides verification of our formulation and implementation. The aim of this section has not been to develop an accurate engineering tool per se, but rather to demonstrate the basic operation of the Sommerfeld approach and this has been achieved. Nonetheless, there are various things one could to improve this scheme. Firstly, the magnetic vector potential should be implemented as a full Sommerfeld integral, rather than approximated by its low-frequency value as at present. Secondly, the integration scheme used in region 3 of the Sommerfeld integral would benefit from some refinement. Thirdly, the singularities in the MoM impedance matrices should be properly addressed; a subdomain MoM scheme might make this easier. We will not, however, pursue this here. There are sufficient problems remaining in this field that entire books can (and have) been written on this topic. Instead, we turn our attention in the next chapter to the use of a commercial package which has a very comprehensive implementation of this theory [13].

Coding hints – coping with complexity

The implementation of the Sommerfeld formulation discussed here is one of the most complex coding tasks in this book. The author's implementation used one main MATLAB m-file and some nine or ten functions, each of course in its own file. The total code ran to around 300 lines of MATLAB. This sounds quite modest, but MATLAB is particularly terse due to the vector nature of much of the code, implicit typing and high-level functions available (e.g. matrix solution, Bessel functions), so this would probably run to several thousand lines of code in languages such as FORTRAN, C, C++ or Java. How does one cope with the complexity that this brings? Here are some tips gleaned from twenty years of coding:

- Firstly, *start modestly*. Do not try to develop a general-purpose program from the start – unless of course this is one's job description. (Even then, one would be advised to code a simple implementation first, to learn the basics of the method if one is not familiar with it.) Writing general-purpose software is astonishingly difficult and time consuming, which is why good CEM software is not cheap.
- Secondly, *use existing packages* where possible. Writing an LU factorization routine is really unnecessary: there are industrial strength routines available in the excellent public domain LAPACK suite. Evaluating special functions is also more complex than it appears; books such as [9] offer routines[a] for Bessel functions, root finding, etc.
- Thirdly, *use a proper scientific programming environment*. Writing one's own code for complex numbers is absurd – find an environment which supports this, or at least has proper libraries. Life is too short to code $a + jb$! By and large, computer scientists appear to prefer to ignore complex numbers, and it usually takes some time for whatever the latest fashionable programming language is to include this. This remains one of the strengths of FORTRAN – complex numbers are a built-in datatype. MATLAB is especially suited for the type of development discussed in this book, due to the very large number of high-level routines available, excellent support for complex numbers, and ease of graphing. For CEM coding, systematic, disciplined and modular work usually leads to far better code than supposedly state-of-the-art advances in languages.
- Fourthly, *modularity* is a key to successful code development. Whilst languages such as C++ have taken this concept much further with object orientation, the basic idea of this is common sense: test sections of the code independently as far as possible. It is much easier to locate the problem in the evaluation of, for instance, the scalar Green function in region 2 when this is implemented and tested separately, than to track this down as part of a complex code.
- Fifthly, *debug intelligently*. This is best discussed by anecdote. The key question is: what are the symptoms of the bug? Two which caused problems in the present case were: evaluation of a special function at a singular point (fortunately, MATLAB warns of this, and then it was just a question of locating the specific call and value of the argument); and overlooking the factor 2 in $R_1^2 = \rho^2 + (2d)^2$ in Eq. (7.82). The symptoms in the latter case were an incorrect reactance, which was traced to incorrect $[Z]$ elements; since the contributions from the scalar potential had already been validated, the error probably lay in the vector potential, and thus the bug was located. This latter case is an example of a strange phenomenon of bugs: they are frequently located in some part of the code which should be very simple. Perhaps it is human nature to concentrate on the hard tasks and pay insufficient attention to the straightforward ones?
- Finally, *validate* your code carefully. This is very important, and we have emphasized this on several occasions.

[a] Be warned that these are *not* public domain codes.

7.7 Further reading

The development in this chapter is largely based on that of Mosig [1]. A similar, although not quite as comprehensive, treatment may be found in [14], and most of the key equations are also available in this source. Both of these contain quite extensive lists of references for further reading. For the specific development of an MoM code for microstrip antennas using the Sommerfeld potentials, these are the key references, containing a wealth of detail of implementation issues. Another contemporary publication was the monograph by Hansen [15]; this is somewhat more general in scope, addressing not just microstrip structures, but also computational issues in detail.

The formulation as discussed in this chapter addressed only single-layer grounded lossy dielectrics. It can be extended to include multi-layer substrates and superstrates, with conductors of finite conductivity, so complex microstrip antenna arrays can be accurately modelled; details may be found in [1, 14]. (The half-space problem can also of course be addressed – this was the subject of Sommerfeld's original investigations.) Microstrip antennas can be fed via feed pins, side feeds, or aperture coupling; the first two are readily implemented within the electric field MPIE MoM as in this chapter. It is possible to extend the formulation to include magnetic currents as well, which permits aperture coupling to be modelled efficiently.

For other, more general, treatments of stratified media, Chew's work is particularly lucid [8]. Chew takes a slightly different approach, developing the Sommerfeld integral as a sum (spectrum) of cylindrical waves, and using plane-wave theory to handle stacked layers. His treatment is oriented more at buried antennas (or scatterers) than microstrip structures, reflecting a geophysical background. Ishimaru and Kong both provide coverage of these stratified media in their textbooks [16, 17]. (The transmission matrix formulation widely used for multi-layered media was formulated by Kong in an earlier book.) The latter is especially concise, perhaps too much so for introductory reading. Again, the emphasis is on half-space problems rather than microstrip structures. None of these references considers the numerical evaluation of the integrals in any detail.

Work continues to be published on quite fundamental issues on this topic. Work on wires penetrating interfaces between different media was published by Burke and Miller [18] and was implemented in NEC-3 and NEC-44. An important generalization of this was Michalski and Zheng's work [19, 20], which permitted arbitrary conducting objects to penetrate the interfaces between dielectrics, using the RWG basis functions for the surface discretization. A very comprehensive invited review paper by Michalski discussed handling the "tails" of Sommerfeld integrals [21]. Improved methods for efficient evaluation of the functions also continue to appear [22, 23]. Some aspects of the extension of the MPIE discussed in this chapter to problems involving both electric and magnetic surface currents are discussed in [24]; an attractive feature of this treatment is that it permits very efficient modelling of slots in ground planes.

References

[1] J. R. Mosig, "Integral equation technique," in *Numerical Techniques for Microwave and Millimetre-wave Passive Structures* (T. Itoh, ed.), Chapter 3. New York: Wiley, 1989.

[2] J. Mosig and F. Gardiol, "General integral equation formulation for microstrip antennas and scatterers," *Proc. IEE (H)*, **132**, 424–432, December 1985.

[3] D. M. Pozar, *Microwave Engineering*. New York: Wiley, 2nd edn., 1998.

[4] E. Yamashita and R. Mittra, "Variational method for the analysis of microstrip lines," *IEEE Trans. Microwave Theory Tech.*, **16**, 251–256, April 1968.

[5] R. C. Booton, *Computational Methods for Electromagnetics and Microwaves*. New York: Wiley, 1992.

[6] J. Schwinger, L. L. DeRaad, K. A. Milton and W.-Y. Tsai, *Classical Electrodynamics*. Reading, MA: Perseus Books, 1998.

[7] D. B. Davidson and J. T. Aberle, "An introduction to spectral domain method of moments formulations," *IEEE Antennas Propagat. Mag.*, **46**, 11–19, June 2004.

[8] W. C. Chew, *Waves and Fields in Inhomogeneous Media*. New York: van Nostrand Reinhold, 1990.

[9] W. H. Press, S. A. Teukolsky, W. Vettering and B. R. Flannery, *Numerical Recipes: the Art of Scientific Computing*. Cambridge: Cambridge University Press, 3rd edn., 2007.

[10] M. R. Spiegel, *Mathematical Handbook of Formulas and Tables*. New York: McGraw-Hill, 1968.

[11] J.-M. Jin, *The Finite Element Method in Electromagnetics*. New York: Wiley, 2nd edn., 2002.

[12] D. M. Pozar and D. H. Schaubert, "Scan blindness in infinite phased arrays of printed dipoles," *IEEE Trans. Antennas Propagat.*, **32**, 602–610, June 1984.

[13] J. J. van Tonder and U. Jakobus, "Full-wave analysis of arbitrarily shaped geometries in multilayered media," in *Proceedings of the 14th International Zurich Symposium on Electromagnetic Compatibility*, pp. 459–464, February 2001.

[14] J. R. Mosig, R. C. Hall and F. E. Gardiol, "Numerical analysis of microstrip patch antennas," in *Handbook of Microstrip Antennas* (J. R. James and P. S. Hall, eds.). London: Peter Peregrinus (on behalf of IEE), 1989.

[15] V. W. Hansen, *Numerical Solution of Antennas in Layered Media*. Taunton: Research Studies Press, 1989.

[16] A. Ishimaru, *Electromagnetic Wave Propagation, Radiation and Scattering*. Engelwood Cliffs, NJ: Prentice-Hall, 1991.

[17] J. A. Kong, *Electromagnetic Wave Theory*. New York: Wiley, 1986.

[18] G. J. Burke and E. K. Miller, "Modeling antennas near to and penetrating a lossy interface," *IEEE Trans. Antennas Propagat.*, **32**, 1040–1049, October 1984.

[19] K. A. Michalski and G. Zheng, "Electromagnetic scattering and radiation by surfaces of arbitrary shape in layered media, part I: Theory," *IEEE Trans. Antennas Propagat.*, **38**, 335–344, March 1990.

[20] K. A. Michalski and G. Zheng, "Electromagnetic scattering and radiation by surfaces of arbitrary shape in layered media, part II: Implementation and results for continguous half-spaces," *IEEE Trans. Antennas Propagat.*, **38**, 345–352, March 1990.

[21] K. A. Michalski, "Extrapolation methods for Sommerfeld integral tails," *IEEE Trans. Antennas Propagat.*, **46**, 1405–1418, October 1998.

[22] J. R. Mosig and A. A. Melcón, "Green's functions in lossy layered media: integration along the imaginary axis and asymptotic behaviour," *IEEE Trans. Antennas Propagat.*, **51**, 3200–3208, December 2003.

[23] V. Kourkoulos and A. Cangellaris, "Accurate approximation of Green's functions in planar stratified media in terms of a finite sum of spherical and cylindrical waves," *IEEE Trans. Antennas Propagat.*, **54**, 1568–1576, May 2006.

[24] M. Schoeman and P. Meyer, "On the structure and packing of the moment matrix in problems supporting simultaneous electric and magnetic surface currents," *Microwave Optical Technol. Lett.*, **41**, June 500–505, 2004.

Assignments

A7.1 Develop a code to replicate the results in Figs. 7.4 and 7.9, and then Figs. 7.10, 7.11 and 7.12.

A7.2 Using this, develop an MoM code for a thin printed dipole and repeat the results of Fig. 7.13.

A7.3 **As an advanced task** Instead of using the quasi-static approximation of Eq. (7.82), evaluate this rigorously as well.

8 The method of moments and stratified media: practical applications of a commercial code

8.1 Printed antenna and microstrip technology: a brief review

Microstrip patch antennas are an example of a large class of modern antennas known as "printed antennas." Microstrip was originally developed in the early 1950s as a transmission line, and the first publication on using this structure as a radiator appears to have been by Deschamp in 1953 [1, Section 1.1]. Almost 20 years then passed until the first patent of the modern microstrip antenna was registered in 1973 by Munson, although the structure was independently discovered in at least one other location.[1]

Microstrip antennas are generally constructed using the same photo lithographic process used to create printed circuit boards. In their simplest form, radiation is due primarily to energy leaking out of the cavity formed by the patch located close to a ground plane; physically, the patch is simply a very wide microstrip line. For the basic rectangular patch, the radiation from two opposite sides reinforces, whereas that from the other two sides cancels. The patch is usually supported on a dielectric substrate of some form, primarily for structural reasons. Typical materials are Teflon and glass-reinforced plastics, as used in printed circuit board technology. Typical material properties for these are ϵ_r in the range from 2–2.5, and $\tan\delta$ from 0.0004–0.002. High-ϵ_r substrates such as alumina ceramics produce physically small patches, but with very limited bandwidth. Typical material properties in this case are: ϵ_r 9.7–10.3, $\tan\delta \approx 0.0004$. For some applications, plastic foam substrates have been used. These materials (sometimes using cheap materials such as expanded polystyrene tiles) have properties close to free space: $\epsilon_r \approx 1.05$, and $\tan\delta \approx 0.0008$.

Popular shapes are the original rectangular shape, which is still the most common, as well as square and circular patches. Patches are usually fed either from the side, typically using a microstrip line, or from below, using either a feed pin (usually the center pin of a coaxial cable) or aperture coupling. It is particularly easy to manufacture arrays using this technology (compared with wire antennas, for instance), since the corporate feed network can share the same substrate as the antenna. High-performance antennas usually split the feed network and the antenna onto two separate layers, to improve bandwidth

[1] In 1972, at the National Institute for Defence Research, Council for Scientific and Industrial Research, Pretoria, South Africa. Unfortunately, the only references are internal classified memoranda and reports by C. A. van der Neut and A. Dubbelman.

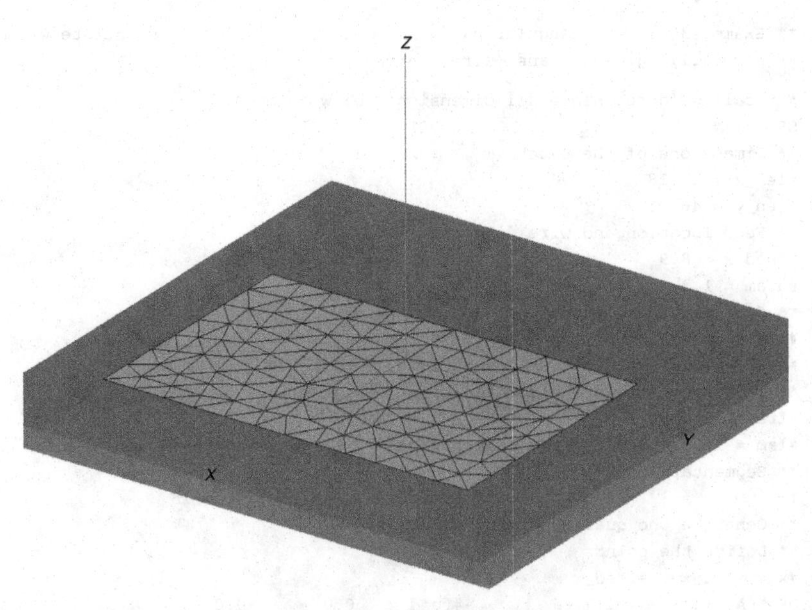

Figure 8.1 FEKO model of a rectangular patch antenna on a grounded substrate at $\lambda_d/15$ discretization.

and minimize unwanted radiation from the feed network. Even these are far easier to manufacture than a waveguide or wire array.

The main advantages of the technology are the following: it can be readily integrated with microwave circuitry; the antennas are flat, and can be conformed to surfaces, since the substrates can be moderately flexible; and it is at least potentially cheap, although high-quality substrates are not. The main drawbacks are limited bandwidth and power-handling capability. The former is the more serious problem in most applications and extensive research has focussed on the use of more complex geometries (doubled-stacked patches, for instance) in an attempt to increase this.

To read more about this class of antennas, the very comprehensive introductory discussion in [2, Chapter 14] can be recommended. Coverage is also available in [3, Section 5.8]. For serious designers, [1] is essential reading.

8.2 A simple patch antenna

In this example, a simple patch antenna is analyzed. The antenna is fed from below using an offset "feed pin" – this a quite typical arrangement. The offset is used to obtain matching; the patch has its highest impedance at the edges, and lowest impedance in the middle. A rectangular patch will exhibit two orthogonal resonances, at frequencies where the length or width corresponds to $\approx 0.48\lambda_d$ ($\lambda_d = \sqrt{\epsilon_r}\,\lambda_0$ is the wavelength in the substrate dielectric). In this case, the feed is offset in the x-direction, so the relevant resonance should be expected at about $\lambda_0 \approx \sqrt{\epsilon_r}\,2.08 \cdot 31.18$ mm, i.e. around 3.1 GHz. The geometry is illustrated in Fig. 8.1. It was generated using FEKO and is based on one of the examples shipped with the code.

```
** Example30a: A rectangular patch antenna on a dielectric substrate with
** a metallic ground plane (wire pin feed)

** Scaling factor since all dimensions below in mm
SF   1                        0.001
** Dimensions of the patch
#len_x = 31.18
#len_y = 46.75
** Feed location and wire diameter
#feed_x = 8.9
#diam = 1.3
** Substrate parameters
#h = 2.87      ** Height
#epsr = 2.2    ** Relative permittivity
** Frequency (for the discretisation)
#freq = 3.0e9
#lam = 1000 * #c0 / #freq / sqrt(#epsr)     ** Wavelength in mm
** Segmentation parameters
IP                            #diam/2   #lam/15   #lam/15
** Generate one quarter of the structure
** Define the points
#x = #len_x - #feed_x
DP   A                       -#feed_x   0.0       0.0
DP   B                        #x        0.0       0.0
DP   C                        #x        #len_y/2  0.0
DP   D                        0.0       0.0       0.0
DP   E                       -#feed_x   #len_y/2  0.0
DP   N                        0.0       0.0       -#h
** Patch
BT   D    B    C
BQ   D    C    E    A
** Symmetry to create the full structure
SY   1    0    3    0
** Feed wire with label 1
LA   1
BL   N    D
** End of geometry
EG   1    0    0    0    0

** Substrate (with ground plane)
GF   10   1              0              1.0       1.0
                              #h        #epsr     1.0

** Voltage source at feed point
A2   1    -1                  1.0       0.0       0.0       0.0       -#h
** Frequency loop in order to compute the impedance
FR   17   0                   2.8e9               3.2e9
** Change the line above as shown below to run with FEKO LITE
** FR   10   0                   2.8e9               3.2e9
** Just compute the impedance, no output of surface currents
OS   0
** Far-field pattern at center frequency
FR   1    0                   3.0e9
FF   1    73   1    1         0         0         5
FF   1    73   1    1         0         90        5
** End
EN
```

Figure 8.2 PREFEKO file for the rectangular microstrip patch.

> **Modelling hints – microstrip antennas**
>
> The PREFEKO model is shown in Fig. 8.2. A few points in this file require comment. Firstly, the feed pin must contact a node on the triangular mesh of the patch. This problem has been encountered before; the solution is explicitly to introduce a node on the patch at this point. (Once again, we comment that this is quite a general issue with MoM codes.) Half the patch is then generated using a triangle and a quadrilateral both of which include this feed pin node; the entire patch is then obtained by imaging in the $y = 0$ plane as usual (the feed pin lies on this plane of symmetry).
>
> The properties of the substrate are defined using the `GF` card. Here, we use the planar multi-layer option 10, for a grounded single dielectric substrate. The other parameters are comprehensively described in the FEKO manual and do not require further comment.

Results for the reflection coefficient of the patch are given in Fig. 8.3, showing computations for both $\lambda_d/15$ and $\lambda_d/25$. (In Chapter 3, the antenna was also analyzed using the FDTD, and it was noted that the resonance was just under 3 GHz for a converged solution.) The antenna is well matched at 2.97 GHz. Compared to our simple estimate above, this is an error of around 4%, but it should be emphasized that that was a very crude approximation. The −10 dB impedance bandwidth is about 100 MHz, or 3%. A simple formula for the bandwidth of microstrip antennas predicts a bandwidth of around h/λ_0, which corresponds well with this result for this $h = 2.87$ mm thick substrate at $\lambda_0 \approx 100$ mm.

8.3 Mutual coupling between microstrip antennas

In a number of practical applications, radiating antennas are located sufficiently close to one another that significant amounts of energy couple between antennas. This is known as *mutual coupling*.[2] In a typical antenna array, this is an important parameter to establish, since it determines the *active impedance* – also known as the driving point impedance. This is the impedance at each port of the antenna, taking into account mutual coupling from all the other antennas. In a simple two-element array, the formula is

$$Z_a \equiv \frac{V_1}{I_1} = Z_{11} + Z_{12}\frac{I_2}{I_1} \tag{8.1}$$

If both elements are fed with equal amplitude and phase excitations (i.e. $I_2 = I_1$), the mutual coupling term Z_{12} adds to the self-impedance term Z_{12}. Alternatively, if the antennas are not part of an array, but connected to different RF systems, mutual coupling can result in undesired energy leaking between the systems. This leads into the field of radiated EMC.

[2] Mutual coupling is used for two related, but not identical, physical parameters. In the one case, it refers to the mutual impedance or admittance. In the other, it refers to the energy coupled from one port to another. The specific usage is usually clear from the context.

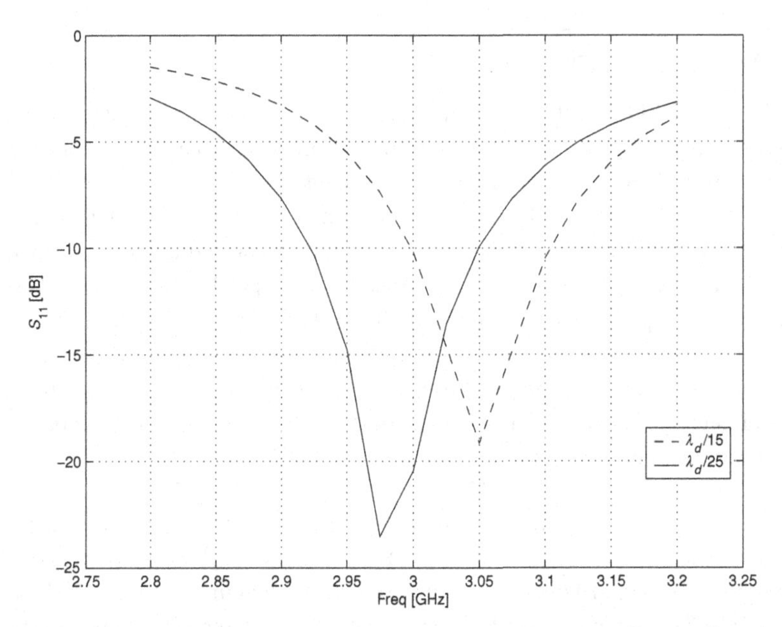

Figure 8.3 Reflection coefficient of the rectangular patch antenna for two discretizations.

Mutual coupling, in terms of voltage (or power) transfer, is complicated by possible mismatches at both transmitter and receiver. The general formula is quite complex, but if both antennas are well matched (in the same Z_0) then S_{12} (or S_{21}, which is identical in reciprocal systems) is the voltage transfer ratio. This can be seen from

$$V_1^- = S_{11}V_1^+ + S_{12}V_2^+ = S_{12}V_2^+|_{S_{11}=0} \tag{8.2}$$

Since microstrip patch antennas are frequently used in an array, it is an interesting exercise to compute the mutual coupling. We are fortunate in that good measured data are available [4]. Jedlicka *et al.* measured the mutual coupling between two patch antennas in both the E-plane (radiating edges adjacent) and the H-plane (non-radiating edges adjacent). The former results in far stronger coupling than the latter, so we will compute E-plane coupling. The elements were $L = 10.57$ cm (radiating edge) $\times$ $W = 6.55$ cm rectangular patch antennas. The substrate thickness was 0.1575 cm, with $\epsilon_R = 2.5$. The loss parameter $\tan\delta$ was not specified, and we will assume it was negligible. The measured resonance frequency was 1.410 GHz. The patches were pin fed. The feed point impedance at the edge of a patch is quite high, and can be reduced by moving the pin a distance x_0 from the edge. This feed pin offset was not specified in the original article, but can be computed as follows. The maximum resistance is approximated by Munson's value:

$$R_m \approx 60\lambda_0/W \tag{8.3}$$

and the input resistance at feed point position x_0 in from the patch edge is

$$R_\in \approx R_m \cos^2\left(\frac{\pi x_0}{L}\right) \tag{8.4}$$

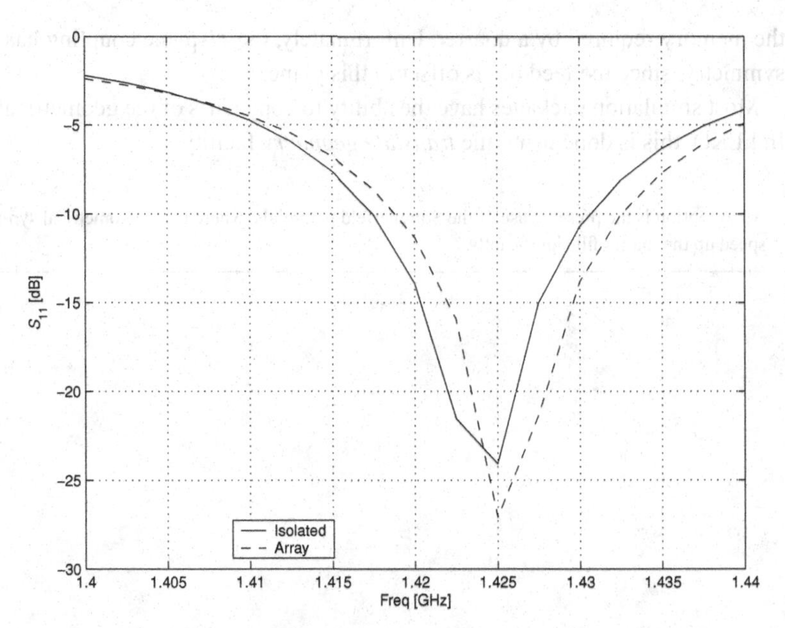

Figure 8.4 Reflection coefficient of the rectangular patch antenna in [4].

For this patch, $R_m \approx 195\,\Omega$ and $x_0/L \approx 0.33$ for a 50 Ω match. Since this is an approximate value, some fine-tuning is necessary with the simulation package to establish the optimal x_0/L as about 0.31. This produces a resonant frequency of $f_r = 1.425$ GHz, around 1% higher than the measured center frequency. Such differences are very common for narrowband structures; the most probable source of error is uncertainty of the exact value of ϵ_R, which is usually easily of this order unless very high quality (and hence expensive) substrates are used. Figure 8.4 shows the computed reflection coefficient. Results are given for both the isolated element case here, as well as the array case, with another patch one wavelength away (terminated in a matched load). As before, the predicted bandwidth of around 0.7% agrees quite well with the computed −10 dB bandwidth of just under 1%.

With the design of the basic patch finalized, the patch is replicated to generate another patch (Fig. 8.5).

Modelling hints

As can be appreciated from the preceding material on the Sommerfeld potentials, the computational cost of this formulation is quite high. Symmetry should be exploited as far as possible to reduce this. For this E-plane coupling problem, there is a plane of magnetic symmetry in the plane containing the feed pins. Note that many codes support both geometrical modelling, which is largely a modelling aid, and does not usually significantly reduce computational cost,[a] as well as field (electric or magnetic) symmetry, which does. (See also Section 6.7.) For this example, using symmetry correctly reduced the number of unknowns and the memory required by a quarter. Unfortunately, the H-plane coupling has *no* field symmetry, since the feed pin is offset in this plane.

Most simulation packages have the ability to copy parts of the geometrical model. In FEKO, this is done using the *translate geometry* facility.

[a] An exception is the present case of the Sommerfeld potentials, where using geometrical symmetry can speed up the matrix fill significantly.

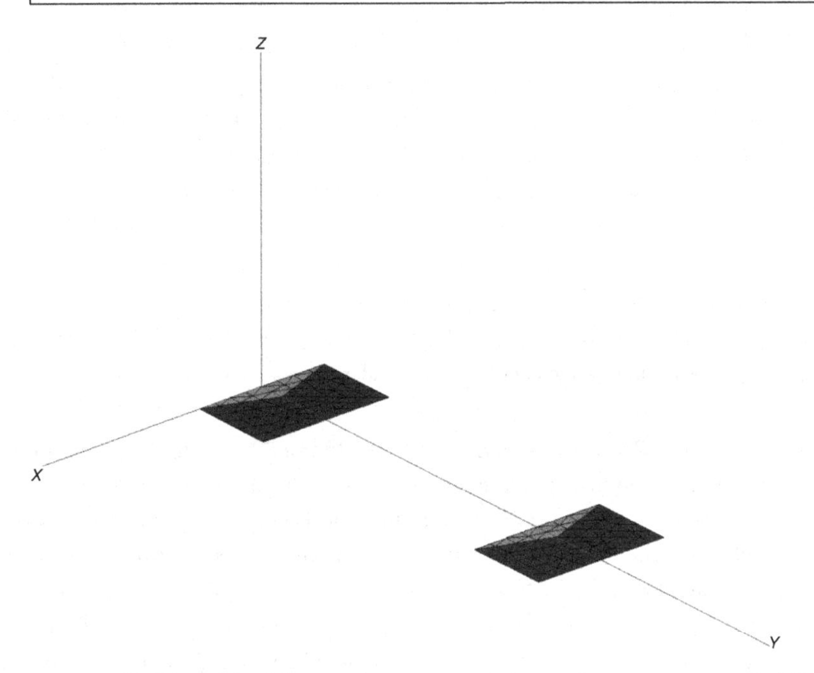

Figure 8.5 FEKO model of the two-element rectangular patch antenna array as in [4], for 1λ spacing. Only the patches are shown.

Computing the mutual coupling is a little tedious; one specifies the inter-element spacing, runs the code at f_r, extracts S_{12} and then repeats the process for the next spacing. Results computed using FEKO with a $\lambda_0/15$ discretization are given in Fig. 8.6. (Note that the distance referred to here (and throughout this section) is the distance between adjacent edges, as in [4], rather than the inter-element spacing of array theory.) A convergence check was performed on the $D = 0.2\lambda_0$ case using a $\lambda_0/25$ mesh which confirmed that $\lambda_0/15$ is quite adequate. There are differences between the measured and computed data, at most around 2 dB, but this is to be expected. One reason for this discrepancy is the sensitive nature of this parameter. Figure 8.7 shows S_{21} as a function of frequency; clearly, very small changes in frequency can easily result in the type of discrepancy noted in Fig. 8.6, in either measurement or computation. Another possibility is the experimental setup, whereby dielectric spacers were inserted as the inter-element spacing increased; this is clearly only an approximation of a continuous substrate. Finally, data for the same problem computed by Mosig *et al.* [5, Fig. 8.27] also show differences of a similar type between measured and computed data, although

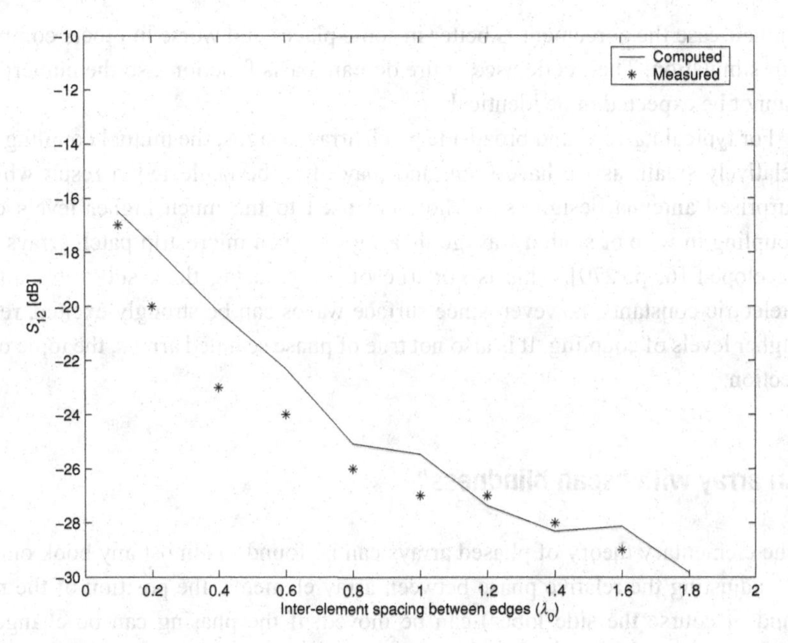

Figure 8.6 S_{12} for the rectangular patch antennas in the text. Measured data from [4].

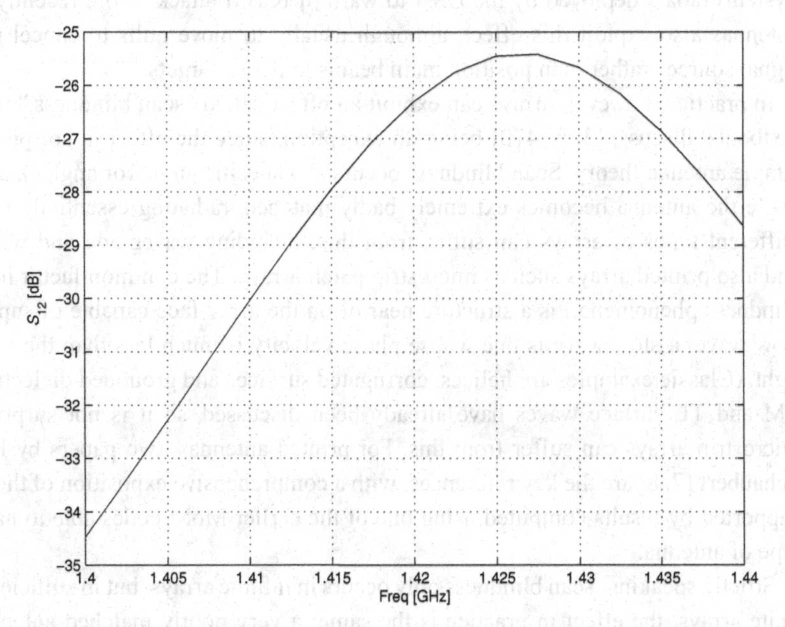

Figure 8.7 E-plane mutual coupling between two patches, one wavelength apart, showing strong frequency dependence.

in their case the agreement is better in some places and worse in others compared with our simulation. Their code used entire domain basis functions, so the numerical results cannot be expected to be identical.

For typical narrowband broadside patch array designs, the mutual coupling levels are relatively small, as we have seen, and may often be neglected, a result which rather surprised antenna designers – who were used to the much higher levels of mutual coupling in wire or slotted waveguide arrays – when microstrip patch arrays were first developed [6, p. 270]. This is *not* true of arrays using thick substrates and/or high dielectric constants, however, since surface waves can be strongly excited, resulting in higher levels of coupling. It is also not true of phase scanned arrays, the topic of the next section.

8.4 An array with "scan blindness"

The elementary theory of phased arrays can be found in almost any book on antennas. By adjusting the relative phase between array elements, the position of the main lobe (and of course the side lobes) can be moved; if the phasing can be changed (either manually or electronically) the beam can be "steered." Phased arrays, as such antennas are called, were a crucial defense technology throughout the Cold War, with one of the most dramatic examples of the technology being the DEWS (Distant Early Warning System) radars deployed by the USA to warn of ICBM attack. More recently, "smart" antennas also exploit this effect, although usually to move nulls to cancel undesired signal sources rather than position main beams to detect targets.

In practice, however, arrays can exhibit an effect called "scan blindness," which few textbooks discuss, [3, p. 470] being an exception, since the effect is not predicted by simple antenna theory. Scan blindness occurs at a specific angle (or angles), and at this angle the antenna becomes extremely badly matched, radiating essentially no energy. Different types of arrays can suffer from this, including waveguide and wire arrays, and also printed arrays such as microstrip patch arrays. The common factor in the scan blindness phenomenon is a structure near or on the array face capable of supporting a slow wave; a slow wave is one whose phase velocity is much less than the velocity of light. (Classic examples are helices, corrugated surfaces and grounded dielectric slabs.) TM and TE surface waves have already been discussed, so it is not surprising that microstrip arrays can suffer from this. For printed antennas, two papers by Pozar and Schaubert [7, 8] are the key references, with a comprehensive exposition of the problem supported by results computed using one of the earlier MoM codes able to handle this type of antenna.

Strictly speaking, scan blindness only occurs in infinite arrays, but in sufficiently large finite arrays, the effect in practice is the same: a very poorly matched antenna which hardly radiates. It is possible to formulate the problem in the spectral domain to produce an infinite array [7, 8], but most commercial codes cannot do this. To demonstrate the effect, we will study a large array of thin printed strip dipoles. We use this structure, rather than patches, to permit a larger array to be simulated. Symmetry should also be

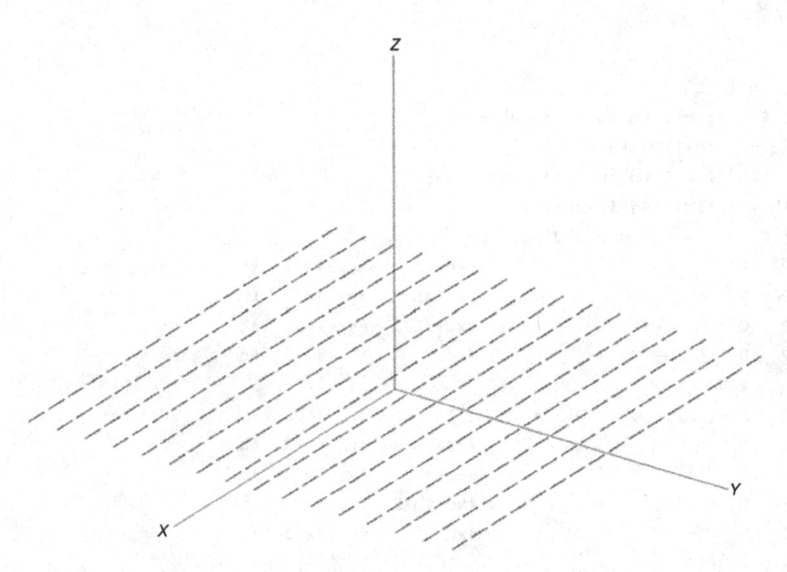

Figure 8.8 256-element printed dipole array.

used as far as possible to increase the effective array size; unfortunately, the phasing of the feeds required to scan the array limits the use of symmetry.

An example of an array produced in FEKO is shown in Fig. 8.8. Each element is a strip dipole, length $L = 0.39\lambda_0$ and width $W = 0.002\lambda_0$, with substrate thickness $h = 0.19\lambda_0$ and relative permittivity $\epsilon_R = 2.55$, as in [7].

> ### Modelling hints – generating a large array
>
> Generating the array can be an exercise in programming; in FEKO, perhaps the simplest approach is to use two nested `FOR` loops, the inner loop generating the dipoles in the E-plane, the outer loop generating "lines" of these dipoles. The key loops are shown in Fig. 8.9; the variable *#a* is the inter-element spacing, and *#N* is the square root of the number of elements – the array is square. Similar ideas could also be used in other simulation packages supporting some form of scripting.

It is interesting firstly to study the effect of the array environment on the element. The concept of active impedance has already been introduced in Eq. (8.1) for two-element arrays. For an N element array, the active impedance of element i is

$$Z_{a_i} \equiv \frac{V_i}{I_i} = Z_{ii} + \sum_{j=1}^{N, j \neq i} Z_{ij} \frac{I_j}{I_i} \tag{8.5}$$

When array feeds are used, this is the impedance automatically computed by codes such as FEKO – although it is not *explicitly* called the active impedance. In Fig. 8.10, the reflection coefficient of an isolated element is compared with that of an element in the center of a 16-element array. (The elements were discretized at around $\lambda_0/50$ for this figure.) Note that the resonance frequency moves upwards by around 10% due to the

```
.
.
.
#yc = #a/2
!! FOR #j = 1 to #N/2 ** Outer loop
#xc = (-#N+1)/2*#a
!! FOR #i = 1 to #N   ** Inner loop
#lb = (2*#N)*(#j-1)+2*#i-1
** Generate the strip dipole antenna
DP   A                         #xc       #yc-#w/2  #h
DP   B                         #xc+#L/2  #yc-#w/2  #h
DP   C                         #xc+#L/2  #yc+#w/2  #h
DP   D                         #xc       #yc+#w/2  #h
LA   #lb
BP   A    B    C    D
DP   E                         #xc       #yc-#w/2  #h
DP   F                         #xc-#L/2  #yc-#w/2  #h
DP   G                         #xc-#L/2  #yc+#w/2  #h
DP   H                         #xc       #yc+#w/2  #h
LA   #lb+1
BP   E    F    G    H
#xc = #xc+#a
!!NEXT
#yc = #yc+#a
!!NEXT
SY   1    0    3    0    #N^2
.
.
.
** Set up array feeds.
!! FOR #k = 0 to 10 ** Start of phase angle loop
#thet = RAD(0+#k*5) ** scan angle theta in radians
#delfz = #k_0 * #a * sin(#thet)
#lb1 = 1
#lb2 = 2
** Impose progressive phase shift in voltage in E-plane (phi=zero)
!! FOR #j = 1 to #N ** Outer loop
#phs = 0 ** re-set phase to zero for each constant-y iteration
!! FOR #i = 1 to #N ** Inner loop
!!IF (#j = 1) and (#i=1) THEN
**   This is the first feed point, new feed (to zero all others).
AE   0    #lb1 #lb2 0         1.0        DEG(#phs) 75
!!ELSE
**   Additional feedpoints - add to sources.
AE   1    #lb1 #lb2 0         1.0        DEG(#phs) 75
!!ENDIF
#phs  = #phs + #delfz
#lb1 = #lb1+2
#lb2 = #lb2+2
!! NEXT
!! NEXT
```

Figure 8.9 Key components of the PREFEKO file used to generate the printed dipole array.

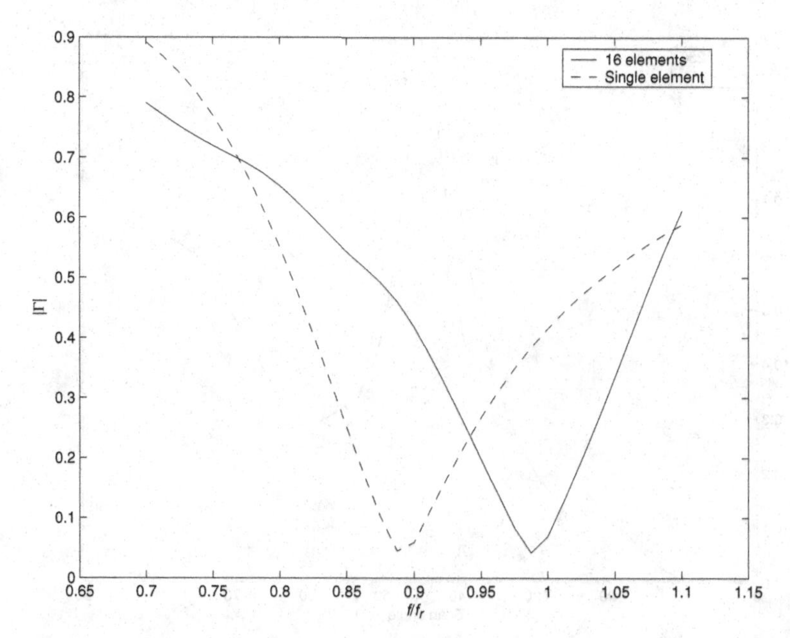

Figure 8.10 Reflection coefficient (in a $Z_0 = 75\ \Omega$ system) versus frequency for both an isolated element and a central element in a 16-element array.

array enviroment. To compute this result, voltages of the same magnitude and phase were applied to each element. Note that this does *not* guarantee a uniformly illuminated array! The reason is that it is the *currents* which determine the radiation pattern, and since the active impedance differs from element to element, so does the resulting current.

Now, the effect of scan angle can be determined. For an $m \times n$ array of sources, to scan a beam an angle θ_s, ϕ_s off broadside requires that the m, nth source should be phased as

$$e^{jk_0(ma \sin\theta_s \cos\phi_s + nb \sin\theta_s \sin\phi_s)} \tag{8.6}$$

This assumes that the array axes are aligned with the x- and y-axes, as in Fig. 8.8, and that the spacing along these axes is a and b respectively. For reasons discussed in detail in [7], only the E-plane scan (the xz-plane in Fig. 8.8, i.e. $\phi = 0$) exhibits scan blindness, and our simulation will only investigate this plane of scan. We also assume that the inter-element spacings are equal in both planes, that is, $a = b$. Hence the *progressive* phase advance (or delay) to add to each element in this plane is

$$\Delta = k_0 a \sin\theta_s \tag{8.7}$$

No progressive phase shift is required in the other plane.

The results of a simulation of a 256-element array are shown in Fig. 8.11. The inter-element spacing is $a = 0.5\lambda_0$ in both planes. Except where mentioned, the substrate was assumed lossless. These results were computed using a $\lambda_0/25$ discretization, which gave acceptable accuracy. For comparison, results computed using an infinite array code [7] are also plotted. The agreement is surprisingly good, and demonstrates how scan

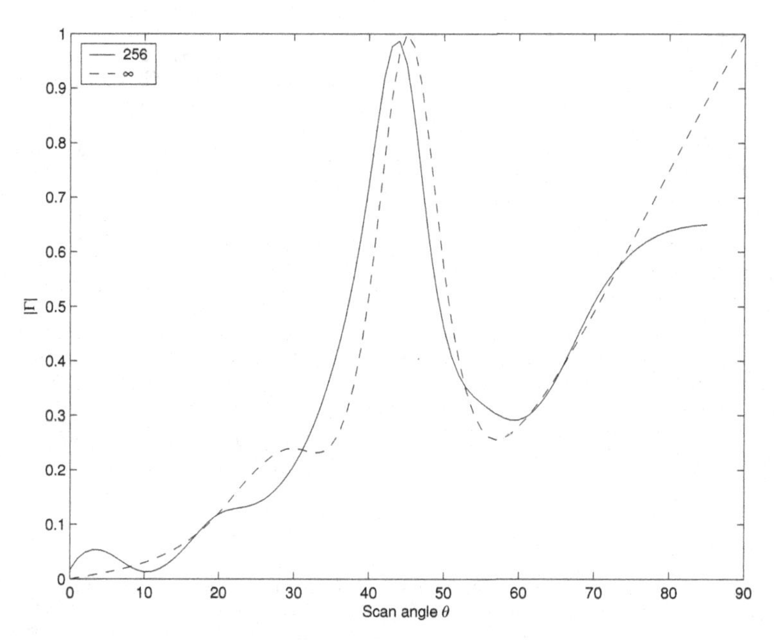

Figure 8.11 Reflection coefficient versus scan angle for a 256-element array and an infinite array; the latter data are from [7].

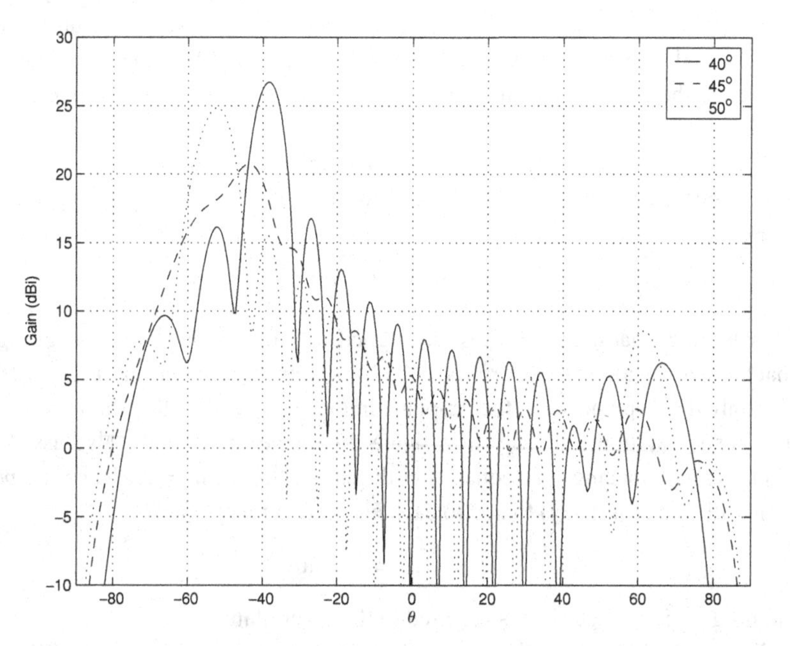

Figure 8.12 E-plane radiation patterns for scan angles of 40°, 45° and 50°, showing scan blindness at 45°.

blindness can impact on a finite antenna which is of a quite practical size. (Although not shown, an 8×8 array gives a similar result, although the reflection coefficient peak is somewhat lower.) Note that, as in [7], all reflection coefficients for this array are referred to a $Z_0 = 75\,\Omega$ system.

Radiation patterns for scan angles of 40°, 45° and 50° for this 256-element array are shown in Fig. 8.12. (The phasing actually produces a scan angle of −40°, −45° and −50°; we note this and do not mention it again.) For this computation, a small amount of loss was added to the substrate; $\tan\delta = 0.002$ was used, which is representative of a good low-loss Teflon-fiberglass substrate. Figure 8.12 plots gain, so substrate loss is taken into account; it is clear that the array works very poorly at $\theta_s = 45°$. In reality, the situation is even worse, since the gain G does not take into account the mismatch loss $(1 - |\Gamma|^2)$ which the antenna presents to the source. The literature on antennas does not seem to adopt a consistent approach to incorporating this effect; some authors [2] incorporate it into antenna efficiency. The product of $G(1 - |\Gamma|^2)$ is also sometimes referred to as realized gain. At the blind scan angle, $\Gamma \approx 1$ so the product $G(1 - |\Gamma|^2)$ is almost zero. At $\theta_s = 50°$, the pattern has improved again, although the peak gain is not quite as large as at 40°. The reason for this is no doubt that the magnitudes and phases of the element currents on the outside of the array differ significantly from those of the central elements due to the different active impedances, and this effect becomes more pronounced as the scan angle increases off broadside.

Modelling hints – array feeds

When modelling an array, the feeds need to be imposed. Most MoM codes permit a number of sources to be used. Most sources are essentially *impressed voltage* feeds: a voltage is specified at each feed point – or port in network theory. The code then computes the resulting current, and from this, the impedance at the port. Multiple feeds simply augment the right-hand side (or forcing) vector $\{V\}$ of the generalized MoM impedance matrix $[Z]\{I\} = \{V\}$. Note that active impedance calculations, unlike mutual power coupling ones, are not affected by the terminating impedance(s) at the other ports.

If some type of loop structure is used to apply feeds – for instance, for different scan angles – *it is very important to ensure that all the previous feeds are zeroed.* How this is done varies from code to code; in FEKO, one tags the first source as a new source and the code zeros all previous ones.

Modelling hints – a useful equivalence for strip dipoles

A strip dipole is often very thin in comparison to its length, and currents are thus essentially constrained to flow along its length in much the same way as with a wire dipole. In this case, it is possible to model the strip as a thin dipole. For a strip width W, the equivalent wire radius a to use is $a = W/4$ [3]. This simplifies the model and reduces the number of unknowns required.

> **Modelling hints – periodic boundary conditions**
>
> Commercial codes are starting to feature periodic boundary conditions, and these can permit the very efficient simulation of infinite arrays by meshing only the prototypical "unit cell" – which would be a single dipole (and possibly the dielectric and ground, depending on the Green function supported) in this case. The periodicity of a structure can be defined in one or two dimensions with appropriate specification of lattice vectors; these do not have to be orthogonal, and can form skew unit cells. The unit cell does not even have to bound the geometry, as features that cross the unit cell boundary can be treated with appropriate half-basis functions. FEKO, for instance, supports periodic boundary conditions, but only for the free-space Green function, so one would need to use surface equivalence to mesh the dielectric surface in the unit cell, as well as meshing the ground plane. (Since the unit cells are usually not large in terms of wavelengths, this would not appear to be a major computational cost, but the matrix fill time can be significant.)

8.5 A concluding discussion of stratified media formulations

The printed dipole array example concludes this chapter on the practical application of Sommerfeld potentials. Before leaving this topic, it is worth briefly mentioning the issue of memory requirements and runtimes. For the former, there is little overhead when using the Sommerfeld formulation, since the memory requirement is still dominated by the matrix storage, which remains N^2, with N the number of unknowns. For the 256-element array, the FEKO simulation with $h/\lambda_0 = 1/25$ had 4864 basis functions, but the use of symmetry resulted in only 2432 unknowns. This required around 183 MB of RAM to store. The statistics for runtime are interesting: this is a moderately large problem in MoM terms, and in a free-space environment one would expect the matrix solution time to start dominating the runtime. In this case, however, using the Sommerfeld potentials, the time required to compute the impedance matrix elements exceeded the time required to solve the linear system by a factor of around 15.[3] By comparison, for a free-space problem with the same number of unknowns (and the same memory requirement), the ratio was around two and a half. Note that we are comparing run-times for problems with the *same number of unknowns*, to get an idea of the cost of the Sommerfeld potentials compared to the free-space Green function. This is *not* equivalent to running the simulation with a grounded substrate with $\epsilon_R = 1$, i.e. vacuum. In this case, one needs to image the patches and the feeds in the ground plane, using symmetry, so the equivalent problem will have more unknowns. Hence *for the same physical problem* of a grounded "dielectric" (actually vacuum) slab, the Sommerfeld formulation is actually

[3] Actual "wall clock" runtimes are so dependent on computer technology that, in common with much of this book, we prefer to use ratios where possible.

little more costly.[4] Of course, this is not relevant in practice, since most substrates have dielectric constants significantly larger than unity, and there is no alternative but the Sommerfeld approach.

Summarizing the development in this and the preceding chapter, stratified media MoM formulations are theoretically complex, challenging to implement but potentially very efficient. This is largely due to *only* the metallic regions of the antenna (wire, patch, feed network, etc.) being discretized – hence quite large microstrip antennas can be modelled. In the context of RF and microwave engineering, the most important contemporary application of this theory is to printed antenna technology, of which microstrip is the most commercially important type, and our examples have concentrated on this technology. Historically, terrestrial broadcasting, especially LF, MF and HF was another important application – indeed, this prompted Sommerfeld's original work – but with the exception of some specialized military systems, this is hardly a dominant technology at present. Subsurface imaging is another significant contemporary application; however, real grounds are not always well stratified, and even if so, the stratifications may not be parallel with the ground–air interface.

In concluding this chapter, some final points should be noted. Firstly, the Sommerfeld–MoM assumes an *infinitely* large substrate on a similarly infinite ground plane. Hence, such MoM programs do *not* provide any information about the effects of *finite* substrates/grounds. Also, many programs based on the Sommerfeld potentials are not truly general purpose. There are theoretical reasons for this: the near-fields are typically obtained via interpolation tables, the far-fields via asymptotic integrals, which may neglect some terms. Using such a program, especially for fields very close to interfaces, may result in anomalies; see for example [9]. However, for the purpose most commercial codes are designed for, usually microstrip and printed structures, the codes are generally robust and accurate.

References

[1] J. R. James and P. S. Hall, eds., *Handbook of Microstrip Antennas*. London: Peter Peregrinus (on behalf of IEE), 1989.

[2] C. A. Balanis, *Antenna Theory: Analysis and Design*. New York: Wiley, 2nd edn., 1997.

[3] W. L. Stutzman and G. A. Thiele, *Antenna Theory and Design*. New York: Wiley, 2nd edn., 1998.

[4] R. P. Jedlicka, M. T. Poe and K. R. Carver, "Measured mutual coupling between microstrip antennas," *IEEE Trans. Antennas Propagat.*, **29**, 147–149, January 1981.

[5] J. R. Mosig, R. C. Hall and F. E. Gardiol, "Numerical analysis of microstrip patch antennas," in *Handbook of Microstrip Antennas* (J. R. James and P. S. Hall, eds.). London: Peter Peregrinus (on behalf of IEE), 1989.

[6] D. M. Pozar and D. H. Schaubert, *Microstrip Antennas: The Analysis and Design of Microstrip Antennas and Arrays*. New York: IEEE Press, 1995.

[4] Note that many codes – FEKO, for instance – will treat this case as an error if using the Sommerfeld formulation, and one is forced to use free-space imaging as above.

[7] D. M. Pozar and D. H. Schaubert, "Scan blindness in infinite phased arrays of printed dipoles," *IEEE Trans. Antennas Propagat.*, **32**, 602–610, June 1984.
[8] D. M. Pozar and D. H. Schaubert, "Analysis of an infinite array of rectangular microstrip patches with idealized probe feeds," *IEEE Trans. Antennas Propagat.*, **32**, 1101–1107, October 1984.
[9] D. B. Davidson and H. d. T. Mouton, "Validation of, and limitations on, the use of NEC-4 for radiation from antennas buried with a homogeneous half-space," *ACES J.*, **13** (2), 302–309, 1998.

9 A one-dimensional introduction to the finite element method

9.1 Introduction

The finite element method (FEM) is one of the best-known methods for the solution of partial differential equations in applied mathematics and computational mechanics. It is a method for solving a differential equation subject to certain boundary values, and in its modern form originated in the field of structural mechanics during the late 1950s; the first specific usage of the term "element" is due to no lesser a person than Courant. In common with the MoM, its historical antecedents are far older than this, in this case dating back to the nineteenth century and the variational methods first described by Lord Rayleigh. It is very widely and routinely used in structural mechanics today, as well as in computational fluid dynamics, computational thermodynamics, the numerical solution of Schrödinger's equation, field problems in general, and of course, in electromagnetics.

An historical aside – Courant and the finite element method

The finite element method as presently accepted can be credited to Courant – whom we have already encountered in the context of the Courant limit for the FDTD method. The published version of his 1942 address to the American Mathematical Society contained an appendix added after the talk, to show by example how variational methods could be put to wider use in potential theory. He used piecewise linear approximations, on a set of triangles which he called "elements" – and thus the method was born [1, p. 5].

With the background we have now acquired with the FDTD and MoM, readers will recognize many features in common with both of these methods in the treatment to follow; indeed, they will probably not be surprised to learn all three can be formulated within a weighted residual setting (although how to do this for the FDTD is not immediately obvious). In common with the MoM, the core idea is to replace some unknown function on a domain by an ensemble of elements, with *known* shape but *unknown* amplitude. Unlike the basic FDTD, where the approximation of the $\boldsymbol{E}$ and $\boldsymbol{H}$ fields is always done on a rectangular, staggered grid, the FEM permits very general geometrical elements to be used and (usually) only uses one grid. The most widely used elements are known as *simplicial* – this simply means line elements in 1D, triangular in 2D and tetrahedral in 3D. Nonetheless, rectangular, prismatic and even curvilinear elements also find widespread application. Since the improved geometrical modelling made possible especially by triangular or tetrahedral meshes is one of the major features distinguishing the FEM from the FDTD, our study of the FEM will be largely restricted to these elements. Interested readers may find treatments of other element shapes in the references.

Similar to the FDTD, but unlike the MoM, the FEM is based on a local description of the field quantities, derived from the differential equation description of the Maxwell equations, and does not automatically incorporate the Sommerfeld radiation condition.[1] In practice, this means some form of mesh termination scheme is required. The easiest is usually an absorbing boundary condition of some type. (However, it is also possible to use an "exact" termination scheme using the MoM on the boundary. This is covered in Chapter 12.) In common with the FDTD, and due to the differential equation basis of the two methods, the FEM permits very straightforward treatment of material discontinuities. Unlike the FDTD, the FEM approximates the function, whereas the FDTD approximates the differential operators (at least on the surface – as will be seen later, the methods are not as dissimilar as they may initially appear). An FEM solution of a wave problem is usually formulated as a "weak" form; the reason for this name will become clear later, but in essence the differentiability requirements of the basis functions are relaxed.

The FEM was first applied in electromagnetics during the late 1960s, at much the same time as the initial work using the MoM and FDTD. The two earliest applications were independently published by Silvester, and by Arlett, Bahrani and Zienkiewicz. Some of the history of the FEM in electromagnetics may be found in [1, p. 5]. However, this promising start was arrested during the 1970s and early 1980s because of a problem called "spurious modes," which, combined with substantial computational cost and complex coding, held back widespread adoption of the FEM in electromagnetics. Fortunately, there was a major theoretical breakthrough with *edge elements* in 1980s, which led to a far greater understanding of the spurious mode problem, and the introduction of largely effective solutions. This improved theoretical understanding, combined with the widespread availability of very powerful computers, and increasing interest in wave interaction with non-metallic structures, has made the FEM a major analysis tool of contemporary CEM.

9.2 The variational boundary value problem: the transmission line problem revisited

The FEM can be derived via two different, but equivalent, procedures. On the one hand, there is the traditional variational approach – more fully, the variational boundary value problem (VBVP) – which we will follow in this introductory chapter. This is also known as the Ritz, or Rayleigh–Ritz method. On the other hand, there is the Galerkin weighted residual formulation, already encountered in Chapter 4, and increasingly used

[1] This should not be confused with the Sommerfeld *potentials* for stratified media.

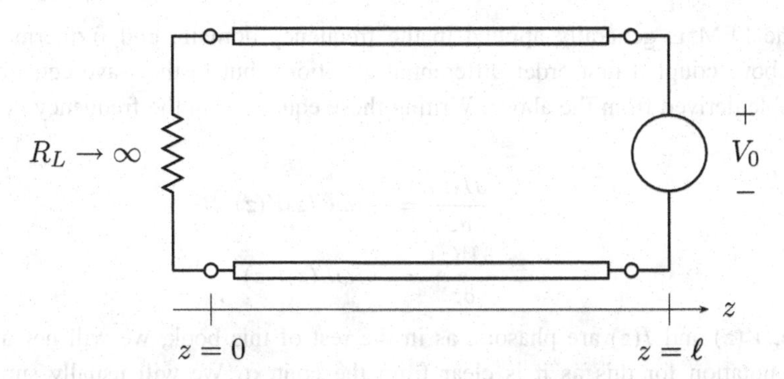

Figure 9.1 The one-dimensional transmission line problem, with zero source impedance.

in the published literature. Whilst the latter is more direct at the formulation level, the former is used by most textbooks, mainly as it allows one to focus on the problem on an element-by-element basis, combining the elements into the overall system via assembly-by-elements. Also, incorporating the boundary conditions is somewhat less obvious in the Galerkin approach. Nonetheless, this alternate approach is so important that we will revisit the FEM from this perspective later, in particular in Section 10.3.

9.2.1 The model problem

To illustrate the application of the method, we will revisit the one-dimensional transmission line problem of Section 2.4. To simplify matters initially, we will firstly set $R_S = 0$ (so that the source voltage V_0 appears directly across the input terminals) and secondly have $R_L \to \infty$, i.e. an open-circuited load. This imposes especially simple boundary conditions, which we will formalize shortly. We will also interchange the source and load, so that the load now appears at $z = 0$ and the source[2] at $z = \ell$. Although a little unconventional in terms of network theory, this will make our initial FEM coding somewhat easier. The problem is illustrated in Fig. 9.1.

In Section 2.4, we applied the FDTD to this, working directly with the first-order differential equations in the time domain, viz. Eqs. (2.13) and (2.14) (the telegraphist's equations), repeated here:

$$\frac{\partial I(z,t)}{\partial z} = -C(z)\frac{\partial V(z,t)}{\partial t} \tag{9.1}$$

$$\frac{\partial V(z,t)}{\partial z} = -L(z)\frac{\partial I(z,t)}{\partial t} \tag{9.2}$$

As before, $L(z)$ and $C(z)$ are the inductance and capacitance per unit length (in H/m and F/m) respectively. Although usually constant, they may possibly vary with length (for instance in a tapered transmission line), and we indicate this explicitly here.

[2] The notation has also been changed slightly compared to that of Chapter 2, using ℓ now for the length of the transmission line, as h is traditionally the mesh size in finite element analysis.

The FEM is generally applied in the frequency domain, and furthermore not to the above coupled first-order differential equations, but to the wave equation in one variable derived from the above. Writing these equations in the frequency domain, we have:

$$\frac{\partial I(z)}{\partial z} = -j\omega C(z)V(z) \tag{9.3}$$

$$\frac{\partial V(z)}{\partial z} = -j\omega L(z)I(z) \tag{9.4}$$

Here, $V(z)$ and $I(z)$ are phasors; as in the rest of this book, we will not use a specific notation for this as it is clear from the context. We will usually suppress the z-dependence, and simply write V and I, as there is only one position variable in the analysis (and similarly, we will write only L and C). $\omega = 2\pi f$ is as usual the angular frequency.

Firstly dividing Eq. (9.4) by $1/L(z)$, then differentiating with respect to z, and finally using Eq. (9.3) to eliminate I, the one-dimensional wave equation is obtained. Note that as only z remains in the analysis, the differentials are now written as total, not partial, derivatives:

$$\frac{d}{dz}\left(\frac{1}{L}\frac{dV}{dz}\right) + \omega^2 CV = 0 \tag{9.5}$$

Note that the first term has been written thus to permit L to vary with position. (When the wave equation is usually derived, it is assumed that L and C are constant, so that the one-dimensional wave equation is then written as $\frac{d^2V}{dz^2} + \frac{\omega^2}{v^2}V = 0$, with $v = (LC)^{-1/2}$.)

The relevant boundary conditions are:

$$\begin{aligned} V(z = \ell) &= V_0 \\ \left.\frac{dV}{dz}\right|_{z=0} &= 0 \end{aligned} \tag{9.6}$$

The former is a homogeneous Dirichlet boundary condition, the latter a homogeneous Neumann one. The latter condition is derived from Eq. (9.4), noting that $I(z = 0) = 0$.

9.2.2 The equivalent variational functional

At this stage, we will state without proof that the stationary point of the following functional corresponds to the solution of Eq. (9.5), subject to the boundary conditions of Eq. (9.6):

$$\frac{1}{2}\int_\ell \left[\frac{1}{L}\left(\frac{dV}{dz}\right)^2 - \omega^2 CV^2\right] dz \tag{9.7}$$

Before proceeding with our finite element solution, a few points are worth noting. Firstly, using Eq. (9.4), this functional can be rewritten as

$$-\frac{1}{2}\omega^2 \int_\ell \left[LI^2 + CV^2\right] dz \tag{9.8}$$

The two parts of the integrand are respectively the stored magnetic and electrical energy per unit length, so this is clearly an energy-related functional. In accordance with many such physical phenomena, the distribution of voltage and current is such that this quantity is rendered stationary. Note that it may be minimized, maximized or a point of inflexion.

The second point to make is that the solution V of Eq. (9.5) is required to be twice-differentiable, whereas V in Eq. (9.7) need only be differentiable once. Such a relaxation leads to the name "weak form" for the equivalent variational functional approach.

The alert reader may well question how boundary conditions other than those of Eq. (9.6) can be handled; additional terms in the functional are required, and this will be addressed subsequently.

Finally, one might ask: why the name "variational functional"? This originates in the calculus of variations, a rich mathematical field little studied nowadays; some key results will be discussed subsequently.

9.2.3 The finite element approximation of the functional

The finite element solution proceeds by approximating $V(z)$ in the functional. The first requirement is to approximate the geometry with a collection of elements – in this case, by simply subdividing the length of the line into N elements, in the same fashion as in Chapter 4. (In two and three dimensions, there are a variety of choices for the geometry of the element, such as triangles and quadrilaterals for the former, but in one dimension the choice is obvious – a line segment.) Given the geometrical approximation, we now approximate the function. As already noted, the function must be at least once differentiable, so a function of at least first order will be required. These requirements relate to each element. There are also some requirements relating to the ensemble (set) of elements. Firstly, to satisfy continuity requirements, the function should also be continuous at element boundaries. Secondly, the ensemble of elements must also satisfy the overall boundary conditions (in this case, those in Eq. (9.6)). Both of these issues will be dealt with subsequently.

Before proceeding further, the following nomenclature should be noted, which is widely used in finite element theory and practice. "Local" nodes, numbering schemes, coordinates, etc., apply to a prototype (or master) element – here, one of the N lines which together approximate the transmission line – considered in isolation, frequently on a local coordinate system of unit length. On the other hand, "global" quantities refer to the *connected* system, taking into account the actual geometry. Note that not only are nodes numbered, but also elements, and some scheme is required to keep track of the relationship between global elements, the associated global node numbers and coordinates, and the local nodes in each element. This is essentially trivial in the example analysed in this chapter, but is in general a non-trivial issue.

We will introduce the following local basis function of first order to approximate the voltage on the element lying between z_l and z_r:

$$\tilde{V}^e = \alpha_l(z)V_l + \alpha_r(z)V_r \tag{9.9}$$

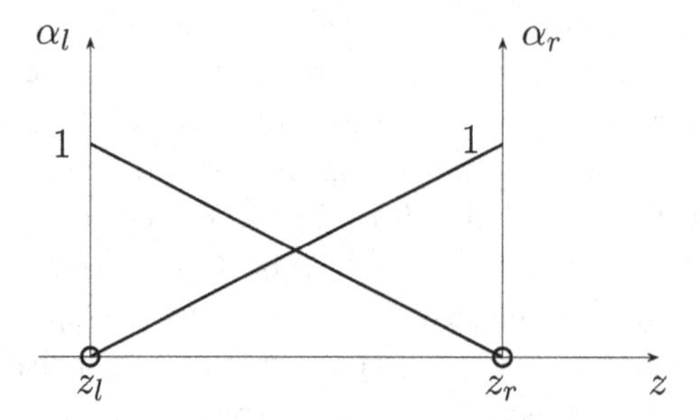

Figure 9.2 The two first-order interpolation functions α_l and α_r.

with

$$\alpha_l(z) = \frac{z_r - z}{z_r - z_l} \tag{9.10}$$

$$\alpha_r(z) = \frac{z - z_l}{z_r - z_l} \tag{9.11}$$

The interpolation functions $\alpha_l(z)$ and $\alpha_l(z)$ are illustrated in Fig. 9.2. Here, V_l and V_r are the voltages on the left node and right node respectively – *and they are the unknowns to be determined in our solution process*. They are also often known in FEA as the *degrees of freedom*. The voltage is now an approximation of the actual solution, and we use $\tilde{V}$ to indicate this. The superscript e reminds us that this is defined on element e. The denominators in the expressions for $\alpha_l(z)$ and $\alpha_r(z)$, viz. $z_r - z_l$, is the length of each element; to keep things simple at this stage, we will assume that line of length ℓ has been discretized into N elements, each of length h_e.[3] If the lengths are all the same, then clearly $h_e = \ell/N$. Substituting our linear approximation for V into the functional, we obtain for element e:

$$\begin{aligned} F^e(\tilde{V}^e) &= \frac{1}{2}\int_{z_l}^{z_r}\left[\frac{1}{L}\left(\frac{d\tilde{V}^e}{dz}\right)^2 - \omega^2 C(\tilde{V}^e)^2\right] dz \\ &= \frac{1}{2}\int_{z_l}^{z_r}\left[\frac{1}{L}\left(\frac{d}{dz}\left[\alpha_l(z)V_l + \alpha_r(z)V_r\right]\right)^2 - \omega^2 C\left[\alpha_l(z)V_l + \alpha_r(z)V_r\right]^2\right] dz \end{aligned} \tag{9.12}$$

This can be written in matrix form as follows:

$$F^e(\tilde{V}^e) = \{V_l V_r\}\left[\tfrac{1}{L_e}\mathcal{S} - \omega^2 C_e \mathcal{T}\right]\left\{\begin{matrix} V_l \\ V_r \end{matrix}\right\} \tag{9.13}$$

[3] L is widely used to represent this in the literature, but it could be confused with the inductance per unit length in this case.

9.2.4 Evaluating the elemental matrices

The elemental matrices $\mathcal{S}$ and $\mathcal{T}$ appear so frequently in finite element analysis that they have commonly accepted names (but not symbols – there is no uniformity in the literature on this. The present chapter follows the notation of [2]). In elasticity problems in structural mechanics, where the FEM originated, the former is called the *stiffness* matrix, and the latter the *mass* matrix. Although obviously without physical meaning in electromagnetics, these terms are increasingly penetrating the electromagnetics literature. Better terms, not linked to the specific field of application, would be *Dirichlet matrix* and *metric* respectively [2, p. 137], but these terms are not widely encountered. The $\mathcal{S}$ matrix has another interesting interpretation as the discretized version of the curl operator, and in the computational electromagnetics literature it is sometimes called the curl matrix.

The elements of these matrices are given as follows:

$$\mathcal{S}_{ij} = \int_{z_l}^{z_r} \frac{d\alpha_i}{dz}\frac{d\alpha_j}{dz}\, dz \tag{9.14}$$

$$\mathcal{T}_{ij} = \int_{z_l}^{z_r} \alpha_i \alpha_j \, dz \tag{9.15}$$

with indices i and j taking both values l and r.

At this stage, we will introduce normalized local coordinates ξ_r and ξ_l, permitting us to write the above in a general form, independent of the specific global geometry (this is another concept very widely used in finite element analysis, although we should caution that it is not appropriate in some advanced applications, such as curvilinear elements, where each element has to be handled individually):

$$\xi_r \equiv \frac{z - z_l}{z_r - z_l} = \frac{z - z_l}{h_e} = \alpha_r(z) \tag{9.16}$$

$$\xi_l = 1 - \xi_r = \alpha_l(z) \tag{9.17}$$

These coordinates have the property $\xi_r + \xi_r = 1$, i.e. they are *dependent* coordinates. In one dimension, it is clearly sufficient to retain only one, and we chose $\xi = \xi_r$, which is zero at the left-hand node, and attains a value of unity at the right-hand node, varying linearly inbetween. Once again, there is no uniform notation – or even nomenclature – in the literature for such normalized local coordinates. The symbol λ_i is also widely used in the computational electromagnetics literature. Such coordinates are often called simplex coordinates, but one will also find the name "barycentric" in use, and, in two dimensions, they are often called "area coordinates" for reasons that will become clear later. These coordinates and their derivatives (especially the gradient in two and three dimensions) have interesting and important properties, which will be studied in some detail in the next chapter.

Using these local coordinates, we could have written

$$\begin{aligned}\tilde{V}^e &= \xi_l V_l + \xi_r V_r \\ &= (1-\xi_r)V_l + \xi_r V_r \\ &= (1-\xi)V_l + \xi V_r \end{aligned} \tag{9.18}$$

where in the last line, we have made the notational simplification $\xi = \xi_r$ as above, and then continued to rewrite the functional in local coordinates. However, we already have Eqs. (9.14) and (9.15) in global coordinates, so we will instead rewrite these in normalized form. From $d\xi/dz = 1/h_e$, the integration variable becomes $dz = h_e d\xi$, and Eqs. (9.14) and (9.15) become:[4]

$$\begin{aligned}\mathcal{S}_{ij} &= h_e \int_0^1 \left(\frac{d\alpha_i}{d\xi}\frac{d\xi}{dz}\right)\left(\frac{d\alpha_j}{d\xi}\frac{d\xi}{dz}\right) d\xi \\ &= \frac{1}{h_e}\int_0^h \frac{d\alpha_i}{d\xi}\frac{d\alpha_j}{d\xi} d\xi \end{aligned} \tag{9.19}$$

$$\mathcal{T}_{ij} = h_e \int_0^1 \alpha_i \alpha_j dz d\xi \tag{9.20}$$

Introducing normalized matrices $[S]$ and $[T]$, we can write

$$[\mathcal{S}] = \frac{1}{h_e}[S] \tag{9.21}$$

$$[\mathcal{T}] = h_e [T] \tag{9.22}$$

with entries

$$S_{ij} = \int_0^1 \frac{d\alpha_i}{d\xi}\frac{d\alpha_j}{d\xi}\, d\xi \tag{9.23}$$

$$T_{ij} = \int_0^1 \alpha_i \alpha_j\, d\xi \tag{9.24}$$

For this linear element, the entries are very simple to evaluate. We will consider the evaluation of S_{lr}, S_{rr}, T_{lr} and T_{rr} in detail:

$$\begin{aligned} S_{lr} &= \int_0^1 \frac{d\alpha_l}{d\xi}\frac{d\alpha_r}{d\xi}\, d\xi \\ &= \int_0^1 (-1)(1)\, d\xi \\ &= -1 \end{aligned} \tag{9.25}$$

[4] The term h_e – or $1/h_e$, depending on the direction of transformation – is actually the Jacobian of the coordinate transformation between local and global coordinates. For curvilinear geometries, the expression becomes considerably more complex, usually being position-dependent, which prevents the use of normalized matrices, as noted earlier.

$$S_{rr} = \int_0^1 \frac{d\alpha_r}{d\xi}\frac{d\alpha_r}{d\xi}\,d\xi$$
$$= \int_0^1 (1)^2\,d\xi \tag{9.26}$$
$$= +1$$

$$T_{lr} = \int_0^1 \alpha_l\alpha_r\,d\xi$$
$$= \int_0^1 (1-\xi)(\xi)\,d\xi$$
$$= \left[\frac{1}{2}\xi^2 - \frac{1}{3}\xi^3\right]\Big|_0^1 \tag{9.27}$$
$$= \frac{1}{6}$$

$$T_{rr} = \int_0^1 \alpha_r\alpha_r\,d\xi$$
$$= \int_0^1 (\xi)^2\,d\xi \tag{9.28}$$
$$= \frac{1}{3}$$

It is easily shown that the matrix is symmetric, and also that the diagonal entries are the same. In short, the normalized matrices have the following form:

$$[S] = \begin{bmatrix} 1 & -1 \\ -1 & 1 \end{bmatrix} \tag{9.29}$$

$$[T] = \frac{1}{6}\begin{bmatrix} 2 & 1 \\ 1 & 2 \end{bmatrix} \tag{9.30}$$

9.2.5 Assembling the system

Equation (9.13) represents the discretized energy-related functional in each element. Now, the elements must be brought together – a process generally referred to as assembly-by-elements in the literature. There are various ways of doing this, and the approach which is followed here is presented for educational purposes. As will be discussed later, real FEM codes use the concepts, but, for reasons of efficiency, the actual implementation is usually somewhat different.

For the local and global numbering scheme as given in Fig 9.3, we can connect the disconnected nodes to the global (connected) nodes using a *connection matrix* $[C]$:

$$\{V\}_{\text{dis}} = [C]\{V\}_{\text{con}} \tag{9.31}$$

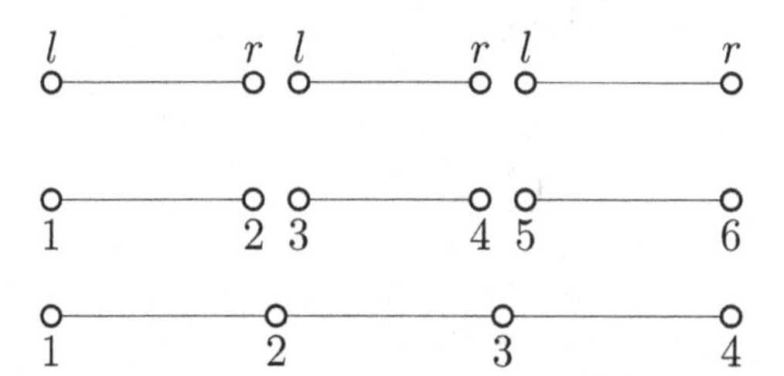

Figure 9.3 Local (top), disconnected (center) and global connected (bottom) numbering schemes for the one-dimensional finite element example.

As an example, for the three-element system in Fig 9.3, this would be:

$$\begin{Bmatrix} V_1 \\ V_2 \\ V_3 \\ V_4 \\ V_5 \\ V_6 \end{Bmatrix}_{\text{dis}} = \begin{bmatrix} 1 & & & \\ & 1 & & \\ & 1 & & \\ & & 1 & \\ & & 1 & \\ & & & 1 \end{bmatrix}_{\text{con}} \begin{Bmatrix} V_1 \\ V_2 \\ V_3 \\ V_4 \end{Bmatrix}_{\text{con}} \tag{9.32}$$

Importantly, note that by doing this we are enforcing the inter-element boundary condition here, namely that of continuous voltage.

The functional for the whole ensemble of elements may now be written in quadratic form:

$$F(\tilde{V}) = \{V\}^T_{\text{dis}} \begin{bmatrix} \frac{S}{h_1 L_1} - \omega^2 h_1 C_1 T & & & \\ & \frac{S}{h_2 L_2} - \omega^2 h_2 C_2 T & & \\ & & \ddots & \\ & & & \frac{S}{h_N L_N} - \omega^2 h_2 C_2 C_N T \end{bmatrix} \{V\}_{\text{dis}} \tag{9.33}$$

where T is the transpose operation. The coefficient matrix is block-diagonal; however, this is not in its final form. It can be written more compactly as

$$F(\tilde{V}) = \{V\}^T_{\text{dis}}[\mathcal{M}]_{\text{dis}}\{V\}_{\text{dis}} \tag{9.34}$$

Writing this now in terms of the connected voltage matrix (which contains the unknowns, the nodal values of the voltage), we obtain

$$F(\tilde{V}) = \{V\}^T_{\text{con}}[\mathcal{C}]^T[\mathcal{M}]_{\text{dis}}[\mathcal{C}]\{V\}_{\text{con}} \tag{9.35}$$

which suggests that the matrix $[\mathcal{M}]$,

$$[\mathcal{M}] = [\mathcal{C}]^t[\mathcal{M}]_{\text{dis}}[\mathcal{C}] \tag{9.36}$$

can be viewed as the coefficient matrix associated with the connected problem, which is

$$F(\tilde{V}) = \{V\}^T_{\text{con}}[\mathcal{M}]\{V\}_{\text{con}}. \tag{9.37}$$

Let us now assume that the transmission line properties $L_e = L$ and $C_e = C$ are constant, and that the finite elements are of uniform length $h_e = h$. For the global numbering scheme adopted, carrying out the mutiplication in Eq. (9.37) yields the following system, which in this specific case is tridiagonal:

$$[\mathcal{M}] = \frac{1}{L}[\mathcal{C}]^t[\mathcal{S}][\mathcal{C}] - \omega^2 C[\mathcal{C}]^t[\mathcal{T}][\mathcal{C}] \tag{9.38}$$

The constitutive parts of this matrix are given by

$$\frac{1}{L}[\mathcal{C}]^t[\mathcal{S}][\mathcal{C}] = \frac{1}{hL}\begin{bmatrix} 1 & -1 & & & & \\ -1 & 2 & -1 & & & \\ & -1 & 2 & -1 & & \\ & & & \ddots & & \\ & & & -1 & 2 & -1 \\ & & & & -1 & 1 \end{bmatrix} \tag{9.39}$$

and

$$C[\mathcal{C}]^t[\mathcal{T}][\mathcal{C}] = \frac{hC}{6}\begin{bmatrix} 2 & 1 & & & & \\ 1 & 4 & 1 & & & \\ & 1 & 4 & 1 & & \\ & & & \ddots & & \\ & & & 1 & 4 & 1 \\ & & & & 1 & 2 \end{bmatrix} \tag{9.40}$$

It is important to note that this tridiagonal structure, with largely repetitive entries, is emphatically *not* a general feature of finite element analysis. Usually, each element's contribution will be different, weighted with its length and material properties. The assumption of uniform mesh length h, and uniform L and C, are the reason that the entries on and around the diagonal are identical, except for those on the first and last rows. The reason that the matrix is tridiagonal is due to the first-order nature of the elements combined with the global numbering scheme adopted. In two- and three-dimensional analysis, such a matrix is usually impossible to arrange.

It is also worth noting that the presence of the $-\omega^2$ term in the second part of Eq. (9.38) results in some problematic issues regarding the final matrix, as each entry may be positive or negative, depending on the size of this term. Typical applications of the FEM to the Laplace or Poisson equation for electrostatics result in a matrix which is positive-definite, but this is not the case here, where the coefficient matrix is indefinite. This has implications for the solvers.

9.2.6 Rendering the functional stationary and solving the problem

The last step necessary is to make the functional stationary. This entails differentiating with respect to each degree of freedom, which in this case is the set of the nodal voltages *which are free to vary*. Now, finally, we have to take into account the boundary conditions on the original domain – in this transmission line problem, the source and load ends. Here, there is one nodal voltage which is *prescribed*, corresponding to the source voltage V_0.

With the source at $z = \ell$, on the right-hand side of the line, this corresponds to global node V_N. This is an *essential* boundary condition, and must be explicitly set. The reader may wonder about the load end, $z = 0$, with nodal value V_1; the boundary condition there was a homogeneous Neumann one, and it turns out that it is a *natural* boundary condition of this finite element approach, which means that it will be satisfied automatically. This is an interesting and important property of the FEM, and we expand on this considerably later.

So, we must differentiate Eq. (9.37) with respect to $V_1, V_2, \ldots, V_{N-1}$. There are efficient methods of doing this with vectors, and we address these in the following chapter, but, for now, we will expand Eq. (9.37) and carry out the differentiation explicitly. Expanding the quadratic form in Eq. (9.37) results in

$$\begin{aligned} F = {} & M_{11}V_1^2 + M_{12}V_1V_2 \\ & + M_{21}V_2V_1 + M_{22}V_2^2 + M_{23}V_2V_3 \\ & + M_{32}V_3V_2 + M_{33}V_3^2 + M_{34}V_3V_4 \\ & \vdots \\ & \cdots + M_{N,N-1}V_{N-1}V_N \\ & + M_{N-1,N}V_NV_{N-1} + M_{N,N}V_N^2 \end{aligned} \tag{9.41}$$

This is now differentiated with respect to $V_1, V_2, \ldots, V_{N-1}$ in turn – but as noted, *not* with respect to V_N, which is prescribed – and each set equal to zero. Using symmetry, $M_{ij} = M_{ji}$, collecting the result into a matrix equation, and simplifying, one obtains the following $N-1 \times N-1$ system:

$$\begin{aligned} \frac{\partial F}{\partial\{V_1, V_2, \ldots, V_{N-1}\}}^T &= \begin{bmatrix} M_{11} & M_{12} & & \\ M_{21} & M_{22} & M_{23} & \\ & \ddots & & \\ & & & M_{N-1,N-1} \end{bmatrix} \begin{Bmatrix} V_1 \\ V_2 \\ \vdots \\ V_{N-1} \end{Bmatrix} \\ &= \begin{Bmatrix} 0 \\ 0 \\ \vdots \\ -M_{N-1,N}V_N \end{Bmatrix} \end{aligned} \tag{9.42}$$

This is a linear system which can be solved using standard tools. For one-dimensional systems with small to moderate numbers of degrees of freedom, no thought need be given to computational efficiency, but it is important to note that, in general, the solution process is the most computationally expensive part of the overall FEM approach. Developing efficient solvers has attracted much attention throughout the history of the FEM.

9.2.7 Coding the FEM

Developing a MATLAB code for this problem is very simple, and a function which provides the core computational elements of such a code is given in Fig. 9.4. The input parameters are the number of elements `N_elem`, the mesh size `h`, the angular frequency `omega`, and the inductance `L` and capicitance `C` per unit length, all in standard

```
function [V_tot] = FEM_1D_solver(N_elem,h,omega,L,C,V_in)
N = N_elem+1; % number of nodes.
% Construct  M:
S = zeros(N); % Pre-allocate matrices.
T = zeros(N);
for ii=1:N;
    S(ii,ii) = 2;
    T(ii,ii) = 4;
    if ii > 1
        S(ii,ii-1) = -1;
        T(ii,ii-1) = 1;
    end
    if ii < N
        S(ii,ii+1) = -1;
        T(ii,ii+1) = 1;
    end
end
S(1,1) = 1; % Correct first and last diagonal elements
S(N,N) = 1;
S = 1/(h*L)*S; % Scale entries
T(1,1) = 2;
T(N,N) = 2;
T = -omega^2*h*C*T/6;
M = S+T;
F = zeros(N-1,1); % Forcing vector
F(N-1) = -M(N-1,N)*V_in; % Voltage at source end.
V = M(1:N-1,1:N-1)\F;
% Post-process
V_tot = zeros(N,1);
V_tot(1:N-1) = V;
V_tot(N) = V_in;
```

Figure 9.4 MATLAB code stub for the one-dimensional FEM analysis of a transmission line.

SI units. The output is a vector with the $N-1$ degrees of freedom augmented by the Nth prescribed value. Note that MATLAB has special commands for dealing with diagonal matrices, which have not been used here, to keep the code as general as possible.

9.2.8 Results and rate of convergence

As always, when a new code has been developed, it must be verified and validated. This problem has a very simple analytical solution; it is the standing wave $V(z) = V_0 \cos \beta(z - \ell)$, where $z - \ell$ accounts for the source at the right-hand side of the line.

Results computed with this code are compared to the analytical results in Fig. 9.5. As with the results in Chapter 2, $L = C = 1$, so the characteristic impedance is $1\,\Omega$ and the velocity of propagation is 1 m/s. $f = 1$ Hz, so the wavelength is 1 m. The figure shows results computed with two and eight elements. Clearly, the former gives a poor approximation of the problem, and the natural boundary condition (homogeneous Neumann, viz. $dV/dz = 0$) on the left-hand side is very poorly satisfied. However, with only eight elements, a relatively good solution has already been obtained.

As is frequently the case with CEM results, differences between the CEM formulation and analytical results become difficult to discern with the eye as the mesh is refined, and a quantitative assessment of the error is required. In this case, the root-mean-square of the norm of the error (computed from the linear interpolation of the FEM solution, at a number of points within each element) is shown in Fig. 9.6. From theoretical considerations, the expected rate of convergence is quadratic in h. One can understand this intuitively in that the linear approximation is complete to first order, so error terms

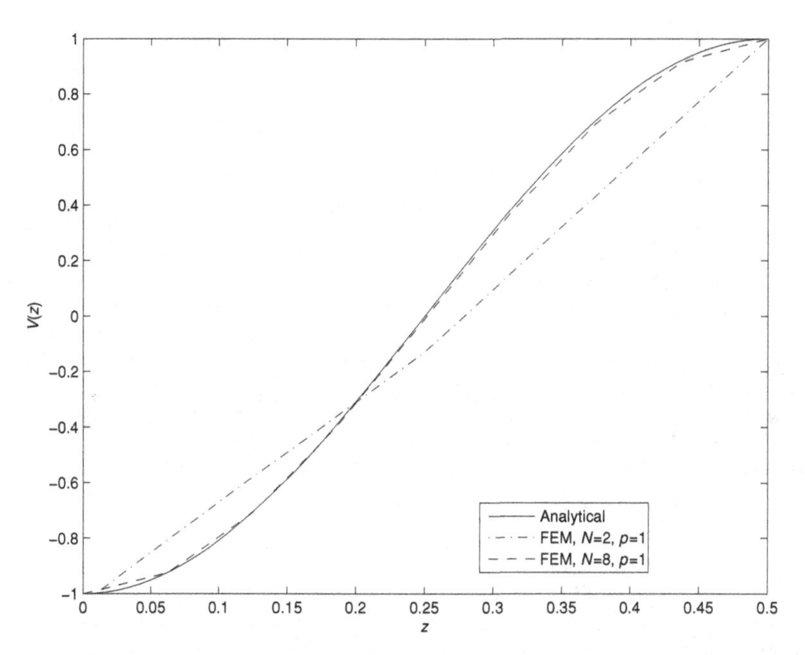

Figure 9.5 Results computed using FEM code compared with analytical solution. N is the number of elements, p the polynomial order of the elements.

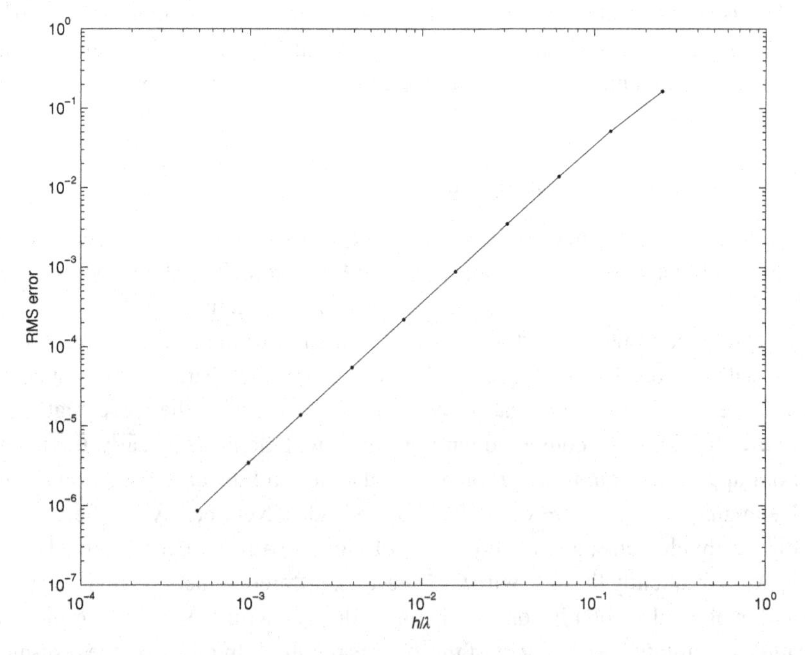

Figure 9.6 Rate of convergence with mesh size of the FEM code using first-order (linear) elements. The slope of the curve is 1.97.

could be expected to be of $\mathcal{O}(h^2)$. (Although the result is correct, a more rigorous analysis is required to establish this result formally.) A first-order polynomial fit to the log–log graph gives a slope of 1.97, very close indeed to the quadratic convergence expected. (See also Section 1.9.) Importantly, this rate of convergence *only* holds when the solution is smooth. The presence of singularities (or sharp discontinuities) slows convergence. In such cases, making the mesh smaller in the vicinity of the non-smooth region is required to restore convergence.

9.3 Improving and generalizing the FEM solution

The results as presented are quite satisfactory, but there are a number of limitations which can be addressed. Two obvious ones are: the use of higher-order finite elements; and the incorporation of more general boundary conditions. We will briefly discuss both.

9.3.1 Higher-order elements

One of the strongest features of the FEM is that is provides a rigorous framework for incorporating higher-order approximations, especially of the unknown functions, but also of the geometry. Almost from the first applications of the FEM, there was interest in higher-order elements and it has occupied the attention of many researchers. Whilst it has frequently proven premature in computational electromagnetics to pronounce a topic as fully solved, for higher-order nodal elements this would indeed appear to be the case. An especially comprehensive treatment of higher-order nodal elements for electromagnetics may be found in [2], and it is the basic reference on this topic. We will present a rather intuitive approach to a second-order scheme, to illustrate the ideas, leaving the interested reader to pursue the topic in more detail from sources such as [2].

The first-order elements used to date in this chapter have an especially simple form when written in simplex (normalized) coordinates: $\alpha_l = \xi_1; \alpha_r = \xi_2$. Reviewing their properties briefly, they are interpolatory polynomials, interpolating to unity at one node, and zero at the other. This must now be extended to second order. The usual method of developing such an element is to insert a mid-point node into each element, halfway between the end nodes. It can be shown that for this case, a suitable set of second-order interpolatory polynomials is

$$\alpha_l = 2\xi_1(\xi_1 - 1/2) \tag{9.43}$$

$$\alpha_c = 4\xi_1\xi_2 \tag{9.44}$$

$$\alpha_r = 2\xi_2(\xi_2 - 1/2) \tag{9.45}$$

with the FEM approximation within element e now given by

$$\tilde{V}^e = \alpha_l V_l + \alpha_c V_c + \alpha_r V_r \tag{9.46}$$

The subscripts l and r refer as before to the left- and right-hand nodes, and c to the mid-point node.

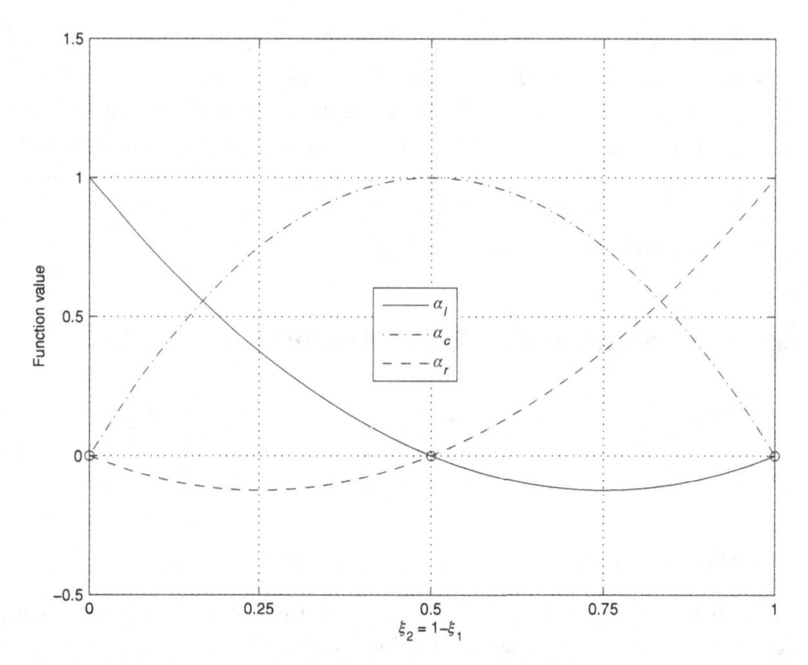

Figure 9.7 The three second-order interpolatory polynomials.

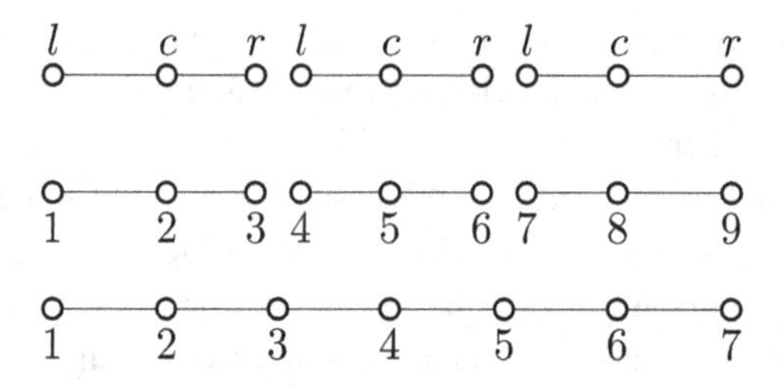

Figure 9.8 Local (top), disconnected (center) and global connected (bottom) numbering schemes for the one-dimensional second-order finite element scheme with mid-point nodes.

By inspection of Eq. (9.43), α_l is unity at $\xi_1 = 1$ (the left-hand node), and zero at both $\xi_1 = 1/2$ (the mid-point node) and $\xi_1 = 0$ (the right-hand node). Similarly, α_c interpolates to the mid-point node, and α_r interpolates to the right-hand node. Since both simplex coordinates are linear functions of z, it is apparent that the overall approximation in Eq. (9.46) is of second order. The three second-order interpolatory polynomials are illustrated in Fig. 9.7, clearly illustrating these properties. It may not be immediately apparent from Eqs. (9.43) to (9.45), but it is easily shown[5] that this is a complete second-order function, with constant, linear and quadratic terms in z.

Repeating the previous analysis, we find that the main change is that we are now dealing with $[S]$ and $[T]$ matrices of dimension 3×3, and the overall number of degrees of freedom has approximately doubled. The local, disconnected and global (connected) numbering schemes is an extension of the previous one, as in Fig. 9.8.

[5] See the end-of-chapter problems for this chapter.

Once again, the elements of the $[S]$ and $[T]$ elemental matrices must be evaluated. As an example, we consider the evaluation of T_{rr}:

$$T_{rr} = \int_0^1 (\alpha_r)^2 \, d\xi = \int_0^1 \xi^2(1-\xi)^2 \, d\xi = \frac{4}{30} \tag{9.47}$$

It is easily shown that $T_{rr} = T_{lr}$, but T_{cc} will have a different value; similar arguments applied to the off-diagonal terms show that there are in total four distinct terms to evaluate once symmetry is taken into account. With basis functions of second order, the integration, whilst still easily performed, starts becoming more tedious. For higher-order elements, the use of symbolic manipulation packages starts becoming essential. Fortunately, this has been done in closed form to reasonably high order in the literature; results for line elements up and including fourth order may be found in [2, p. 466]. For second order, the results are:

$$[S] = \frac{1}{3}\begin{bmatrix} 7 & -8 & 1 \\ -8 & 16 & -8 \\ 1 & -8 & 7 \end{bmatrix} \tag{9.48}$$

$$[T] = \frac{1}{30}\begin{bmatrix} 4 & 2 & -1 \\ 2 & 16 & 2 \\ -1 & 2 & 4 \end{bmatrix} \tag{9.49}$$

For assemblying the system matrix $[M]$, it is possible to repeat the analysis using the connection matrix. However, it is worthwhile outlining a simpler approach (and indeed, the one usually used in actual FEM codes). Revisiting Eq. (9.40), note that $[T]$, for instance, can be written as follows:

$$C[C]^t[T][C] = \frac{hC}{6}\begin{bmatrix} 2 & 1 & & & & \\ 1 & 4 & 1 & & & \\ & 1 & 4 & 1 & & \\ & & & \ddots & & \\ & & & 1 & 4 & 1 \\ & & & & 1 & 2 \end{bmatrix}$$

$$= \frac{hC}{6}\begin{bmatrix} T_{ll}^{(1)} & T_{lr}^{(1)} & & & & \\ T_{rl}^{(1)} & T_{rr}^{(1)} + T_{ll}^{(2)} & T_{lr}^{(2)} & & & \\ & T_{rl}^{(2)} & T_{rr}^{(2)} + T_{ll}^{(3)} & T_{lr}^{(3)} & & \\ & & & \ddots & & \\ & & & T_{rl}^{(N-1)} & T_{rr}^{(N-1)} + T_{ll}^{(N)} & T_{lr}^{(N)} \\ & & & & T_{rl}^{(N)} & T_{rr}^{(N)} \end{bmatrix} \tag{9.50}$$

```
.
.
.
N = 2*N_elem+1; % number of nodes.
S = zeros(N); % Pre-allocate system matrices.
T = zeros(N);
% Set-up elemental matrices
Se = 1/3*[7 -8 1; -8 16 -8; 1 -8 7];
Te = 1/30*[4 2 -1; 2 16 2;-1 2 4];

% Assemble system matrices by element
for ii=1:N_elem; % loop over elements
    glob_l = 2*ii-1; % global node on left of element
    S(glob_l:glob_l+2,glob_l:glob_l+2) = S(glob_l:glob_l+2,glob_l:glob_l+2) + Se;
    % Add elemental contribution to global matrix
    T(glob_l:glob_l+2,glob_l:glob_l+2) = T(glob_l:glob_l+2,glob_l:glob_l+2) + Te;
end
%keyboard
S = 1/(h*L)*S; % Uniform elements, scaling can be done on global system
T = -omega^2*h*C*T;
M = S+T;
F = zeros(N-1,1); % Forcing vector
% Contribution of [Mfp][Vp]
F(N-2) = -M(N-2,N)*V_in;
F(N-1) = -M(N-1,N)*V_in;
V = M(1:N-1,1:N-1)\F;
.
.
.
```

Figure 9.9 MATLAB code stub for the one-dimensional FEM analysis of a transmission line.

The superscript (n) indicates the global element number. Wherever nodes are shared between elements, there are contributions to the overall matrix from each element. In one dimension, the only nodes that are shared are the left and right nodes – and this is true of higher-order elements as well. Thus, only the diagonal entries are affected when the matrix is assembled. Hence, even with the addition of a mid-point node, provided that the numbering scheme of Fig. 9.8 is used, the above provides the basis for the assembly-by-elements code stub given in Fig. 9.9, which simply adds each element's matrix into the global matrix, relying on the fact that, in Fig. 9.8, the global node to the left of the element n is $2n - 1$.

When the elements are not of uniform length, and/or the material properties are no longer constant, the matrix above is given instead by

$$\frac{1}{6}\begin{bmatrix} m^{(1)}T_{ll}^{(1)} & m^{(1)}T_{lr}^{(1)} & & & & \\ m^{(1)}T_{rl}^{(1)} & m^{(1)}T_{rr}^{(1)} + m^{(2)}T_{ll}^{e2} & m^{(2)}T_{lr}^{(2)} & & & \\ & m^{(2)}T_{rl}^{(2)} & m^{(2)}T_{rr}^{(2)} + m^{(3)}T_{ll}^{e3} & m^{(3)}T_{lr}^{e3} & & \\ & & & \ddots & & \\ & & & & m^{(N)}T_{rl}^{(N)} & m^{(N)}T_{rr}^{(N)} \end{bmatrix} \tag{9.51}$$

where the mesh and material properties in the element are represented by $m^{(n)} = h^{(n)}C^{(n)}$. In this case, as each element's $[S]$ and $[T]$ matrices are computed, they are added into the global matrix, at the appropriate global row and column entry, with the appropriate scaling by element length and material property.

It should be noted that a large number of the entries in the final global matrix $\mathcal{M}$ are zero. These are termed "structural zeros." They are due to the local nature of the FEM; degrees of freedom which are not connected do not contribute to the system matrix. Clearly, as the number of degrees of freedom increase, so the matrix becomes increasingly *sparse*. Although we have not exploited this here, the sparse system is one of the most attractive features of the FEM (compared to the method of moments, for instance) and a discussion on exploiting this will be found in a later chapter.

For more general numbering schemes, the global node corresponding to each local node must be found, usually via a connectivity array derived from the mesh. In two – and especially three – dimensions, establishing the element connectivity, and hence the connection between local and global numbering schemes, is a non-trivial part of a typical FEM code. The entries will also not cluster together around the diagonal as in this one-dimensional example, but will rather have entries some distance from the diagonal – this distance defines the *bandwidth* of the matrix. The larger the bandwidth, the worse the "fill-in" will be if the linear system is solved by direct factorization, and methods to reduce this have been extensively studied in general finite element theory, under the generic name of "re-ordering." Several are available in MATLAB, including the `symrcm` script, which implements the well-known reverse Cuthill–McKee reordering.

One last issue which needs attention is the forcing vector. We will state without proof (which we defer to the next chapter) that the forcing vector is given by $[\mathcal{M}_{fp}]\{V_p\}$. Here, f refers to free nodes, and p to prescribed ones – in the present case, only the last node on the right. The submatrix $[\mathcal{M}_{fp}]$ is a rectangular matrix of dimension $N - 1 \times 1$, where N is the number of nodes, and vector $\{V_p\}$ is of length 1 in this case. Hence the forcing vector becomes

$$\begin{Bmatrix} 0 \\ 0 \\ \vdots \\ -M_{N-2,N} V_N \\ -M_{N-1,N} V_N \end{Bmatrix} \tag{9.52}$$

for the second-order elements.

Results computed with a MATLAB code implementing this second-order element are shown in Fig. 9.10. The much greater accuracy obtained with the second-order solution is impressive. A comparison of rates of convergence is given in Fig. 9.11. The second-order element solution converges at the theoretically expected rate of $\mathcal{O}(h^3)$. The second-order solution obviously has more degrees of freedom, N, and it is this which largely determines the computational cost of the method. However, even when plotted against N as in Fig. 9.12 rather than mesh size, the advantage of the higher-order elements is clear; for any given N, the second-order solution has lower error, and the rate of convergence is faster.[6]

[6] The slopes in Figs. 9.11 and 9.12 actually have the same magnitude, although are of opposite sign. This is because N and h are inversely related.

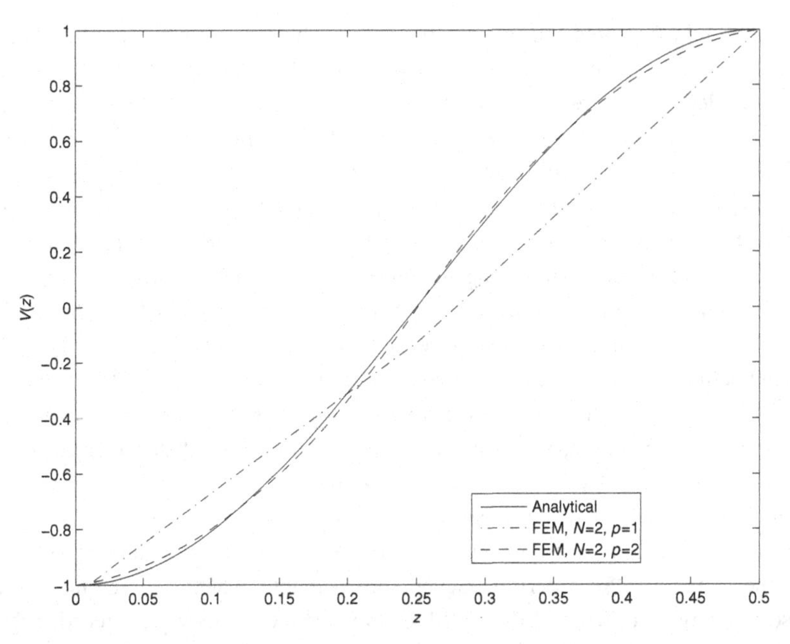

Figure 9.10 A comparison of the first- and second-order FEM solutions of the transmission line problem, also compared with the analytical solution. N is the number of elements, p the polynomial order of the elements.

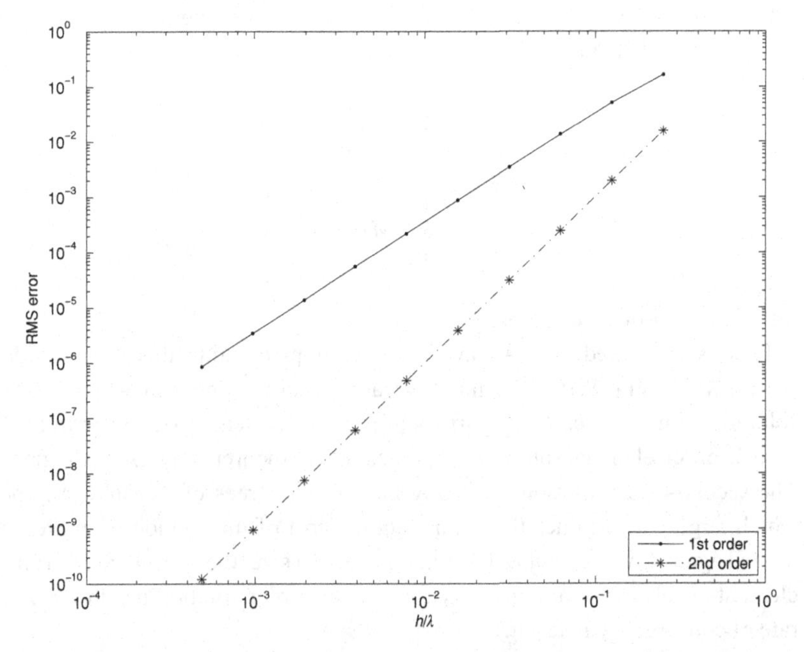

Figure 9.11 Comparison of the rate of convergence of the first-order and second-order elements versus the normalized mesh size h/λ. The slope of the curves are 1.97 and 3.00 respectively.

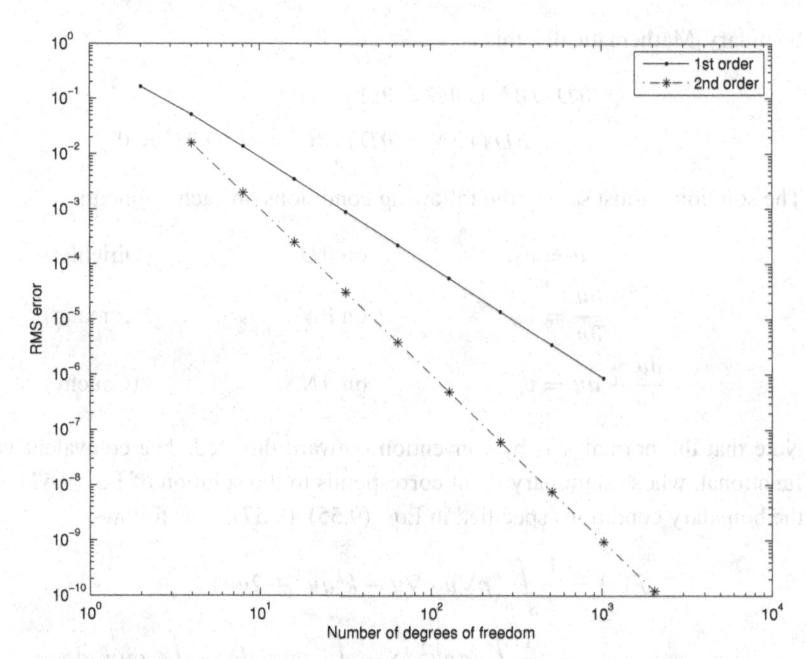

Figure 9.12 Comparison of the rate of convergence of the first-order and second-order elements versus the number of degrees of freedom N.

9.3.2 More general boundary conditions

Before completing this chapter, one issue still needs to be addressed, namely procedures for handling more complex boundary conditions. To handle this in a general fashion, and serve as a bridge to the next chapter, it is useful to examine the following differential equation, which has been termed the *inhomogeneous scalar Helmholtz equation* [2]:

$$\nabla \cdot (p\nabla u) + k^2 qu = g \tag{9.53}$$

Here, u is the scalar variable representing the desired solution. The problem is defined on domain Ω, and material properties are represented by $p(x, y, z)$ and $q(x, y, z)$. The quantity k^2 is a constant, which may or may not be known. $g(x, y, z)$ is a specified driving function. Boundary conditions are specified on the surface $\partial\Omega$ which encloses the domain – also known as the closure. (When the domain Ω is two dimensional, $\partial\Omega$ will be a line, and for a one-dimensional domain, $\partial\Omega$ will be the end-points.)

It is clear that in the context of our one-dimensional transmission line problem, $u = V$, the voltage; $p = 1/L$, the inverse of inductance per unit length; $q = C$, the capacitance per unit length; $k^2 = w^2$, the known angular frequency of excitation; and $g = 0$. The excitation for the transmission line entered via an inhomogeneous Dirichlet boundary condition representing the applied voltage at one end.

The boundary conditions will also be specified in general form. Let the boundary $\partial\Omega$ be composed of three distinct, non-overlapping (but connected) subsets ∂D, ∂N and ∂C which together make up $\partial\Omega$. Each part may consist of multiple, non-overlapping parts. Importantly, a boundary condition of some type must be specified on each part of the boundary. Mathematically, this is

$$\partial D \cup \partial N \cup \partial C = \partial \Omega$$
$$\partial D \cap \partial N = \partial D \cap \partial C = \partial C \cap \partial N = 0 \tag{9.54}$$

The solution u must satisfy the following conditions on each segment:

$$u = u_D \quad \text{on } \partial D \quad \text{(Dirichlet)} \tag{9.55}$$

$$\frac{\partial u}{\partial n} = v_N \quad \text{on } \partial N \quad \text{(Neumann)} \tag{9.56}$$

$$\frac{\partial u}{\partial n} + au = v_C \quad \text{on } \partial N \quad \text{(Cauchy)} \tag{9.57}$$

Note that the normal $\hat{n}$ is by convention outward directed. The equivalent variational functional, whose stationary point corresponds to the solution of Eq. (9.53), subject to the boundary conditions specified in Eqs. (9.55)–(9.57), is as follows:

$$\mathcal{F}(u) = \frac{1}{2}\int_{\Omega}\left(p\nabla u \cdot \nabla u - k^2 q u^2 + 2gu\right)$$
$$+\frac{1}{2}\int_{\partial C} apu^2 dS - \int_{\partial C} puv_C dS - \int_{\partial \Omega} puv_N dS \tag{9.58}$$

Note that with a homogeneous Neumann boundary condition (i.e. $v_N = 0$), the relevant integral vanishes, as already stated earlier in this chapter. Note also that the Dirichlet boundary conditions are imposed explicitly on the solution, as already seen, and hence do not appear in the variational functional.

With this, it is now easy to see how to handle both a general load (and not just the open circuit considered), as well as a source impedance. Consider first a general load. Ohm's law at the load end, $V(z = 0) = -Z_L I(z = 0)$ (note the negative sign associated with current in the load, to be consistent with current flow in the $+z$ direction as in Eqs. (9.3) and (9.4)), combined with the appropriate telegraphist's equation Eq. (9.4), gives

$$\left.\left(\frac{\partial V}{\partial z} - \frac{j\omega L}{Z_L}V\right)\right|_{z=0} = 0 \tag{9.59}$$

which is immediately recognized as a homogeneous Cauchy boundary condition, with $a = +j\omega L/Z_L$ (note that the normal is in the $-z$-direction at the left-hand side of the line, hence $\partial u/\partial n = -\partial V/\partial z$) and $v_C = 0$. Similarly, from $v(z = \ell) = V_0 + Z_s I(z = \ell)$, we obtain

$$\left.\left(\frac{\partial V}{\partial z} + \frac{j\omega L}{Z_S}V\right)\right|_{z=0} = \frac{j\omega L}{Z_S}V_0 \tag{9.60}$$

which is recognized as a general Cauchy boundary condition, with $a = j\omega L/R_L$ and $v_C = (j\omega L/Z_S)V_0$. (Once again, care is needed with the sign of the current.) Note that unlike the FDTD, it is as easy to deal with reactive loads as with resistive ones, as we are working in the frequency domain, and this has been indicated here by using Z_L and Z_S rather than restricting ourselves to R_L and R_S as in the FDTD example.

In this one-dimensional case, the boundary integrals terms reduce simply to the integrand evaluated at the boundary point. When the functional is rendered stationary, those which are linear in u, such as $\int_{\partial\Omega} puv_N dS$ in Eq. (9.58), augment the right-hand side of the set of equations. (Obviously, the sign of the term changes as it is moved across to the RHS.) Note that the degree(s) of freedom associated with ∂C – or ∂N – are still treated as free to vary. However, the term $\frac{1}{2}\int_{\partial C} apu^2 dS$, which is quadratic in u, contributes ap to the LHS matrix entry corresponding to the relevant degree of freedom. Once again, the relevant degree(s) of freedom associated with ∂C are treated as free to vary.

In practical terms, with a load impedance of Z_L at $z = 0$, a term of $ap = j\omega/Z_L$ is added to the first entry in the system matrix in Eq. (9.42). Note that with an open circuit termination, this term vanishes, as expected.[7] Since the relevant v_C is zero in this case, the RHS vector of Eq. (9.42) is not affected.

With a source impedance of Z_S and source voltage V_0 at $z = \ell$, a term of $ap = j\omega/Z_S$ is added to the last entry in the system matrix in Eq. (9.42). The last nodal voltage V_N is no longer prescribed; hence, the system matrix is becomes of dimension $N \times N$, and the RHS vector is similarly enlarged, with the $(N-1)$th term now also zero. A term $pv_C = (j\omega/Z_S)V_0$ is entered at the Nth position of the RHS vector. Having a source enter the FEM formulation via a Cauchy boundary condition, rather than imposed as a Dirichlet one, is quite common in electromagnetic applications.

For the second-order element solution, the boundary conditions are treated in the same fashion. With the interpolatory nodal functions of the preceding section, there is still only one degree of freedom at each end of the line, so the above prescription can be applied without change.

Using a matched source and load, viz. $Z_S = Z_L = Z_0 = \sqrt{L/C}$, provides a very simple method of testing the code, as the magnitude of the voltage should be $V_0/2$. Note that the voltage along the line, $V(z)$, is now of course complex-valued, as the wave is now a travelling one, with a linearly varying phase increasing with z for this matched example – recall that with the source on the right and the load on the left, the wave is travelling in the $-z$-direction, with phase $e^{+j\beta z}$ – and one must bear this in mind when plotting the voltage.

9.4 Further reading

The standard reference for finite elements in electromagnetics since its first publication in 1983 (revised in 1990, and again in 1996) has long been the text by Silvester and Ferrari [2]; although there are newer sources now available, it can still be particularly strongly recommended for introductory reading. Although the transmission line example

[7] However, with a short circuit termination, there is a problem. The solution of this is explored as an excercise.

presented here represents a different physical problem (being a dynamic analysis of a lossless line, compared to a DC analysis of a lossy line), the development in this chapter owes much to the introductory chapter in that source, both in the development of the material, and in terms of notation. It has a particularly strong coverage of nodal elements – indeed, Silvester's original work on higher-order nodal elements, dating back to the 1970s, is widely cited in the general finite element literature – and also a very good chapter on the electromagnetics of finite elements. As already mentioned, in addition to a very comprehensive theoretical development, it contains tabulated elemental entries for line, triangle and tetrahedral nodal elements, to order four, three and two respectively. An extensive list and discussion of other FEM references in electromagnetics may be found at the end of the next chapter.

9.5 Conclusions

In this chapter, we have developed a finite element solution to a simple transmission line problem, using both first-order and then second-order elements, and generalizing to arbitrary source and load terminations. As the problem was posed in the frequency domain, it turned out to be simpler to compute than the FDTD solution of Chapter 2. However, for each frequency point at which the solution is desired, the FEM code must be re-run, whereas a single run of the FDTD code can provide wideband data. One of the strongest points of the FEM is the relative ease with which higher-order basis functions can be incorporated, both theoretically and in coding, and we demonstrated this by outlining the development of a second-order solution too. By contrast, higher-order FDTD has proven a difficult problem, and higher-order MoM is usually complicated by the requirement to evaluate especially the self-terms with much greater accuracy than is usually required for a low-order solution. Another very strong point of the FEM is its ability to use different element shapes; this is not relevant in one dimension, but this will be clearly demonstrated in the following chapters.

In the next chapter, we extend finite element analysis to two dimensions, focussing on the vector, rather than scalar, wave equation. It will be seen that a different type of element, namely the *edge* (or vector) element is the element of choice for vector wave applications. We will also revisit some issues which we deferred in this chapter for later study, in particular the properties of variational functionals, and the development of the FEM from the weighted residual (Galerkin) perspective.

References

[1] P. P. Silvester and G. Pelosi, *Finite Elements for Wave Electromagnetics*. New York: IEEE Press, 1994.

[2] P. P. Silvester and R. L. Ferrari, *Finite Elements for Electrical Engineers*. Cambridge: Cambridge University Press, 3rd edn., 1996.

Problems and assignments

Problems

P9.1 Prove the stated symmetry properties of the $[S]$ and $[T]$ matrices for the first-order element. (Hint: consider $S_{ll} = \int_0^1 \frac{d\alpha_l}{d\xi}\frac{d\alpha_l}{d\xi}\, d\xi$. Noting that $\xi = \xi_r$, change the variable of integration to $\xi_l = 1 - \xi_r$, and thus show that $S_{ll} = S_{rr}$.)

P9.2 Show that Eq. (9.42) follows from Eq. (9.41).

P9.3 Evaluate several elements of the $[S]$ and $[T]$ matrices (Eqs. (9.48) and (9.49)) for the second-order element. Also demonstrate their symmetry properties.

P9.4 This problem shows formally the completeness of the second-order finite element used. These elements, given without proof in Eq. (9.46), and using the interpolatory polynomials of Eqs. (9.43) to (9.45), can be derived as follows. Firstly, start by writing the desired complete second-order function for an element of unit length:

$$\tilde{V}^e = a^e + b^e\xi + c^e\xi^2 \tag{9.61}$$

For the three nodes (l, c and r), $\xi = 0$, $1/2$ and 1 respectively. Now write the above equation in terms of the three nodal voltages:

$$V_l = a^e$$

$$V_c = a^e + \frac{b^e}{2} + \frac{c^e}{4}$$

$$V_r = a^e + b^e + c^e$$

Then solve for a^e, b^e and c^e in terms of V_l, V_c and V_r. From this, and using $\xi = \xi_2$ and $\xi_1 = 1 - \xi_2$, show that Eq. (9.46) follows.

P9.5 As commented in the text, when either the load or source impedance is zero, the formulation in Section 9.3.2 experiences problems. Suggest ways to work around this, such as the use of alternate methods to impose the boundary conditions.

Assignments

A9.1 Implement the second-order code outlined in Section 9.3.1 and obtain results similar those in Fig. 9.10.

A9.2 Implement the full solution with source and load impedances as outlined in Section 9.3.2, and plot the magnitude and phase of the voltage along the line for a matched system.

10 The finite element method in two dimensions: scalar and vector elements

10.1 Introduction

In the preceding chapter, an introduction to the finite element method was provided by way of a one-dimensional problem. In the course of that development, a number of core features of a typical finite element analysis and FEM code were presented, including the concepts of the variational boundary value problem (VBVP) – which is solved instead of the original differential equation, the importance of boundary conditions, assembly-by-elements, rates of convergence and higher-order elements. Whilst very useful indeed for didactic purposes, the one-dimensional introduction does not permit one to address a number of important issues, which can indeed be addressed in two dimensions. The most important of these is the necessity of a new type of element, originally known as an *edge* element, but now generally called a *vector* element, where the degrees of freedom no longer reside at element nodes, but rather along element edges (in their lowest-order form, as edge elements), on faces, and (in three dimensions) over the volume of the element.

However, before vector elements are addressed, there are still some very useful topics to discuss with scalar (nodal) elements in two dimensions, and the first part of this chapter will revisit some topics which were deferred from the previous one, as well as demonstrate an application of a two-dimensional solver to a quasi-static problem (the quasi-TEM analysis of a microstrip transmission line), where the electric fields can be adequately represented as the gradient of the scalar electric potential ϕ. The opportunity will also be taken to demonstrate the power of extrapolation for improving the accuracy of results with very little computational overhead. Although extrapolation cannot always be applied, when rates of convergence are known, or can be estimated from computed data, this can be very useful, in particular when convergence rates are slow, as is the case when the solution is not smooth.

We will also revisit the one-dimensional formulation in this chapter, and discuss the alternate Galerkin formulation, which was briefly alluded to in the previous chapter. Although the variational formulation remains the method of choice to explain the FEM, the Galerkin method is increasingly widely used in the literature, in particular at formulation level, and it is important to understand that approach too.

10.2 Finite element solution of the Laplace equation in two dimensions using scalar elements

10.2.1 The variational boundary value problem approach

Consider the following partial differential equation (PDE) in two dimensions:

$$\nabla \cdot \epsilon \nabla \phi = 0 \tag{10.1}$$

For linear, isotropic media, we have $\epsilon = \epsilon_r \epsilon_0$, and this is equivalently

$$\nabla \cdot \epsilon_r \nabla \phi = 0 \tag{10.2}$$

In a materially homogeneous region, this reduces to the Laplace equation:

$$\nabla^2 \phi = 0 \tag{10.3}$$

With a PDE, boundary conditions must of course be specified. For a second-order PDE such as this, the following on the *closure* (boundary) are necessary and sufficient for a unique solution:[1]

- A value of function ϕ is specified – this is a *Dirichlet* boundary condition. If $\phi = 0$, this is called a *homogeneous* boundary condition.
- A value of the normal derivation, $\partial\phi/\partial n$, is specified – this is a *Neumann* boundary condition. Again, if $\partial\phi/\partial n = 0$, this is called a *homogeneous* Neumann boundary condition.
- A linear combination of the above is specified – $\epsilon(\partial\phi/\partial n) + \gamma\phi = q$. This is known as a *mixed* boundary condition (also sometimes as a Cauchy boundary condition); the Neumann boundary condition is a special case of this with $\gamma = 0$.

Note that these may be mixed in any ratio along the boundary: the boundary may be entirely Dirichlet, or entirely Neumann,[2] or entirely mixed, or some combination of these along different sections of the boundary. However, they must be *disjoint* – that is, more than one may not be simultaneously specified along the same part of the boundary.

Note also that in the context of the inhomogeneous Helmholtz equation, Eq. (9.53) introduced in the previous chapter, $p = \epsilon$, $u = \phi$, $k^2 = 0$, $g = 0$, and there is no term corresponding to q.

Before proceeding further, we will briefly revise some finite element terminology. With finite elements, we usually employ basis functions which span only a small part of the domain – subsectional as opposed to entire domain, in MoM parlance. This region is generally known as the *element*, and the basis function is also frequently called the *shape function*. The term *elemental function* is also sometimes encountered. (It is generally accepted that the term "finite element" comes from this geometrical decomposition into finite regions – as opposed to infinite elements, which are also sometimes used – but it has also been attributed to the finite energy in an element.) As with the MoM, a

[1] Note that for higher-order PDEs, additional boundary conditions are required.

[2] In which case, the PDE can be solved only to within an unknown constant.

variety of shape functions have been used. For problems that can be worked in terms of a scalar unknown, the most useful are generally families of complete polynomial interpolation functions. Later in the chapter, we will see another type of incomplete polynomial function, which is not interpolatory, but is very widely used, namely the edge-based element, which comes into its own when the working variable is a vector.

The equivalent variational functional

As in Chapter 9, we will approach the finite element method from the variational functional perspective. Instead of directly solving Eq. (10.1), we are going to work with an equivalent problem, namely an energy *functional*, whose stationary point (a minimum in this case) corresponds to the solution of the PDE. For Eq. (10.1), a suitable functional is

$$W(\phi) = \frac{1}{2}\iint \epsilon(\nabla\phi)^2 \, dS \tag{10.4}$$

This will be stated without proof for the present – subsequently we will return to this, since the proof yields important information about the boundary conditions. Note that this functional represents the energy $\frac{1}{2}\iint \boldsymbol{D}\cdot\boldsymbol{E}\, dS$. Note also that the function ϕ in the original equation had to be at least twice differentiable; in the above, it need only be once differentiable. Due to this "weakening" of the requirements on the function, this is sometimes called the *weak formulation*. For a linear, isotropic medium, we have $\epsilon = \epsilon_r\epsilon_0$ and since we are eventually going to set the derivative of W to zero, we can just as well divide out by ϵ_0 at this stage, leaving only the ϵ_r term:

$$W(\phi) = \frac{1}{2}\iint \epsilon_r(\nabla\phi)^2 \, dS \tag{10.5}$$

(When computing terms such as capacitance from this, ϵ_0 must of course be restored.)

The shape functions

In one dimension, the only choice to make is the shape of the *basis* function, but in two and three dimensions, both the shape function and the geometrical shape of the element can be chosen. The most popular choices in two dimensions are *triangular* and *quadrilateral* elements; for reasons already discussed, our focus will be on triangular elements, although rectangular elements will be used to introduce some ideas regarding vector elements. Assuming that the geometrical region (the domain) has been decomposed into elements – later, ways of doing this will be discussed – we note that Eq. (10.4) is valid within each element, and in the following we will initially focus on this energy functional on an element-by-element basis.

Zero-order elements (the equivalent of the pulse basis functions we used for the first MoM example back in Chapter 4) cannot be differentiated even once, so are not admissable in this problem. Hence, we will start with first-order elements. In this case, the approximating function can be written as

$$\phi \approx a + bx + cy \tag{10.6}$$

The constants a, b and c are, of course, what we require the FEM eventually to compute for us. However, it is more convenient to write this in a form where the unknowns are the potentials at the three triangle nodes, or in other words:

$$\phi \approx \alpha_1(x, y)\phi_1 + \alpha_2(x, y)\phi_3 + \alpha_3(x, y)\phi_3 \tag{10.7}$$

This assumes the existence of suitable functions $\alpha_1(x, y)$, $\alpha_2(x, y)$ and $\alpha_3(x, y)$; their properties will emerge shortly.

Noting $\phi_1 = a + bx_1 + cy_1$ and similarly for the other two nodes, we have

$$\begin{bmatrix} \phi_1 \\ \phi_2 \\ \phi_3 \end{bmatrix} = \begin{bmatrix} 1 & x_1 & y_1 \\ 1 & x_2 & y_2 \\ 1 & x_3 & y_3 \end{bmatrix} \begin{bmatrix} a \\ b \\ c \end{bmatrix} \tag{10.8}$$

Inverting the nodal coordinate matrix, we find:

$$\begin{bmatrix} a \\ b \\ c \end{bmatrix} = \begin{bmatrix} 1 & x_1 & y_1 \\ 1 & x_2 & y_2 \\ 1 & x_3 & y_3 \end{bmatrix}^{-1} \begin{bmatrix} \phi_1 \\ \phi_2 \\ \phi_3 \end{bmatrix} \tag{10.9}$$

Now we have:

$$\phi = [1 \quad x \quad y] \begin{bmatrix} 1 & x_1 & y_1 \\ 1 & x_2 & y_2 \\ 1 & x_3 & y_3 \end{bmatrix}^{-1} \begin{bmatrix} \phi_1 \\ \phi_2 \\ \phi_3 \end{bmatrix} \tag{10.10}$$

which may be rewritten as

$$\phi = \sum_{i=1}^{3} \phi_i \alpha_i(x, y) \tag{10.11}$$

with

$$\alpha_1 = \frac{1}{2A}[(x_2y_3 - x_3y_2) + (y_2 - y_3)x + (x_3 - x_2)y] \tag{10.12}$$

and A the triangle area (which is conveniently half the determinant of the nodal coordinate matrix).[3] The other functions α_2 and α_3 are obtained by cyclic interchange of the indices, modulus three.

Note that the functions α_i are interpolatory on the three vertexes (nodes): i.e. unity at node i, and zero at the other nodes. (Once again, we caution that not all the finite elements we will study have this property.)

[3] Note here that A is a *signed* quantity, whose sign depends on whether the nodes are numbered clockwise or anticlockwise. See Section 10.8.2 for further discussion.

Manipulating the energy term

Substituting Eq. (10.11) into Eq. (10.4), the following is obtained for the energy in an element e:

$$W^e = \frac{1}{2}\iint_{S^e} \epsilon_r^e \nabla\phi \cdot \nabla\phi \, dS$$
$$= \frac{1}{2}\sum_{i=1}^{3}\sum_{j=1}^{3} \epsilon_r^e \phi_i \left[\iint_{S^e} \nabla\alpha_i \cdot \nabla\alpha_j \, dS\right] \phi_j \tag{10.13}$$

where we have now assumed that the permittivity is constant within element e. This is very compactly written in matrix notation as:

$$W^e = \frac{1}{2}\{\phi\}^T \epsilon_r^e [S^e]\{\phi\} \tag{10.14}$$

with $\{\phi\}$ the vector of nodal potentials and

$$S_{ij}^e = \iint_{S^e} \nabla\alpha_i \cdot \nabla\alpha_j \, dS \tag{10.15}$$

The issue of notation and nomenclature of these matrices has already been addressed in Chapter 9. As in that chapter, we will use $[S]$ for this matrix, and $[T]$ for another frequently encountered matrix. (This notation is due to Silvester and Ferrari [1]. Unfortunately, there is no standard notation in this regard in the literature. Savage and Peterson, for instance, use $[E]$ and $[F]$ respectively [2], as does Jin [3].) Note also that we will not distinguish in this chapter between $[S]$ and $[\mathcal{S}]$; in Chapter 9, the former was used to indicate a normalized matrix, the latter for the matrix incorporating the appropriate geometrical scaling. As normalized matrices will not be used in this chapter, it is not necessary to distinguish between them in the present work.[4]

The expressions are simple to evaluate, for example,

$$S_{12}^e = \frac{1}{4A}[(y_2 - y_3)(y_3 - y_1) + (x_3 - x_2)(x_1 - x_3)] \tag{10.16}$$

and the other matrix entries can be obtained by cyclic permutation of the indices, modulo 3.

Connecting the elements

As in the preceding chapter, we have initially worked in isolation, considering the element on its own; now, the elements need to be connected, in the same way that the one-dimensional elements had to be. Each element has nodes numbered *locally* from one to three. In practice of course, there will be a (perhaps very large) number of elements, with nodes numbered according to some *global* numbering scheme as discussed in Chapter 9. (It is worth commenting that mapping local to global information, and vice versa, usually requires a significant amount of book-keeping in the average FEM code.)

[4] This should not be interpreted as meaning that normalized, or universal, matrices are not useful; indeed, for higher-order elements, they are of great utility. It is simply that, for first-order nodal elements, the expressions are so simple that they can be easily evaluated directly.

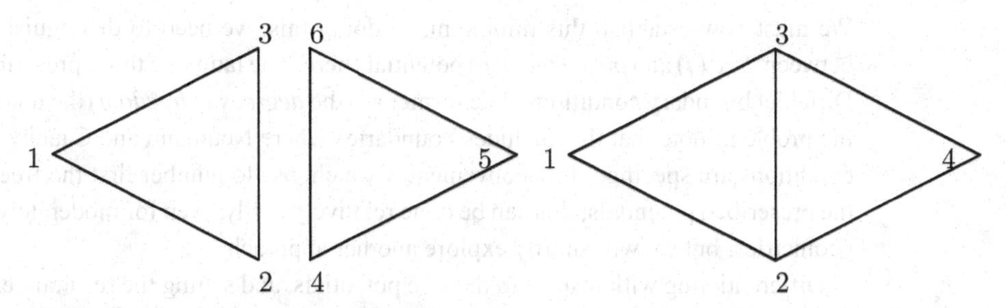

Figure 10.1 Two triangular elements, disconnected (left) and connected (right).

We need some method to *connect* the elements; various approaches are available. At present, as in Chapter 9, we will assume the existence of a *connection matrix* which tells us how to map the unconnected nodes to the connected mesh. As a simple example, see Fig. 10.1, which shows two such triangles. The connection matrix for this system is

$$[C] = \begin{bmatrix} 1 & & & \\ & 1 & & \\ & & 1 & \\ & 1 & & \\ & & & 1 \\ & & 1 & \end{bmatrix} \tag{10.17}$$

and thus

$$\{\phi_{\text{dis}}\} = [C]\{\phi_{\text{con}}\} \tag{10.18}$$

with

$$\{\phi_{\text{dis}}\} = \{\phi_1\,\phi_2\,\phi_3\,\phi_4\,\phi_5\,\phi_6\}_{\text{dis}}^T$$
$$\{\phi_{\text{con}}\} = \{\phi_1\,\phi_2\,\phi_3\,\phi_4\}_{\text{con}}^T \tag{10.19}$$

Although this may be belabouring the obvious, this connection matrix ensures that the potential at each node is the same on all elements sharing that node. (This seems simple and obvious, but we will see that in the context of vector fields, this may not always be desirable.)

Using this, the resulting equation for the energy in the whole system is

$$W = \frac{1}{2}\{\phi_{\text{con}}\}^T[S]\{\phi_{\text{con}}\} \tag{10.20}$$

$$[S] = [C]^T \begin{bmatrix} \epsilon_r^{(1)}\,[S^{(1)}] & \\ & \epsilon_r^{(2)}\,[S^{(2)}] \end{bmatrix} [C] \tag{10.21}$$

As shown in Eq. (10.21), note that the stiffness matrix of each element must be multiplied by the relative permittivity associated with the element before assembly.

However, the formulation is not completed yet. It will be recalled that it is the solution which *minimizes* the variational functional which corresponds to the solution of the PDE, and all we have here is an expression for the energy in the connected elements.

We must now establish this minimum. In doing this, we need to distinguish formally between *free* (f) and *prescribed* (p) potentials here. The latter are those prescribed by the Dirichlet boundary conditions. The former are the *degrees of freedom* (the unknowns) in the problem; note that this includes boundaries where Neumann and Cauchy boundary conditions are specified. It is convenient if we choose to number first the free and then the prescribed potentials; this can be done relatively easily, even for moderately complex geometries, but we will shortly explore another approach.

Differentiating with respect to the free potentials, and setting the resultant expression to zero, one obtains

$$\frac{\partial W}{\partial\{\phi_f\}} = \frac{\partial}{\partial\{\phi_f\}}\left(\{\phi_f^T \phi_p^T\}\begin{bmatrix} S_{ff} & S_{fp} \\ S_{pf} & S_{pp}\end{bmatrix}\begin{Bmatrix}\phi_f \\ \phi_p\end{Bmatrix}\right) = 0 \tag{10.22}$$

Expanding the quadratic, differentiating, and then using $[S_{fp}] = [S_{pf}]^T$, yields:

$$\begin{bmatrix} S_{ff} & S_{fp}\end{bmatrix}\begin{Bmatrix}\phi_f \\ \phi_p\end{Bmatrix} = 0 \tag{10.23}$$

or, more conveniently,

$$[S_{ff}]\{\phi_f\} = -[S_{fp}]\{\phi_p\} \tag{10.24}$$

Compare this with Eq. (9.42); the RHS vector in that equation is $-[S_{fp}]\{\phi_p\}$, noting that $-M_{N-1,N}V_N$ is the only non-zero entry.

Once again, this is a system of linear equations which can be solved using standard techniques. Here we should note that the matrices $[S_{ff}]$ and $[S_{fp}]$ are sparse, containing only entries where nodes are shared by elements; for initial implementation work we need not exploit this, but FEM codes for practical applications must, or much of the benefit of the FEM is lost. Note also that these terms include ϵ_r^e in the S matrix elements as in Eq. (10.21).

A mathematical aside – partial differentiation of matrices and vectors

Since the free potentials are most conveniently written as a vector, it is useful to note that vectors can be differentiated much as scalars; in particular, the following are useful:

$$\frac{\partial c\{x\}^T}{\partial\{x\}} = c[I] \tag{10.25}$$

$$\frac{\partial\{z\}^T}{\partial\{z\}} = \frac{\partial\{z\}}{\partial\{z\}^T} = \frac{\partial\{y\}}{\partial\{y\}^T} = \frac{\partial\{y\}^T}{\partial\{y\}} = [I] \tag{10.26}$$

For the quadratic form:

$$\frac{\partial\{x\}^T[A]\{x\}}{\partial\{x\}} = 2[A]\{x\} \tag{10.27}$$

For the linear form:

$$\frac{\partial\{x\}^T[B]}{\partial\{x\}} = [B] \tag{10.28}$$

For the scalar products:

$$\frac{\partial\,[\{y\}\{z\}]}{\partial\{x\}} = \frac{\partial\,[\{z\}\{y\}]}{\partial\{x\}} = \frac{\partial\{z\}^T}{\partial\{x\}}\{y\}^T + \frac{\partial\{y\}}{\partial\{x\}}\{z\} \tag{10.29}$$

In the above, c is a scalar constant, $\{x\}$ and $\{z\}$ are column vectors, $\{y\}$ is a row vector (and the elements of $\{y\}$ and $\{z\}$ may be functions of elements of $\{x\}$), and $[A]$ and $[B]$ are conformable matrices, with $[A], [B] \neq f(x_i)$ and $[I]$ is the identity matrix. This result greatly simplifies the analytical work required in minimizing the functional. Such identities can be proven by expanding the vector expression into its components, and then differentiating with respect to each of them in turn. A good reference to read more on this topic is [4, Appendix B]; further identities may also be found there.

Coding hints – FEM data structures

Note that in practice one rarely numbers all the nodes in an unconnected fashion first; instead, node 4 on the right-hand triangle would probably be referenced using some data structure of the form `element(m)%nodeone`, with `m` the element number, in a language such as FORTRAN 90/95 which supports *derived data types* – i.e. objects of a type defined by the user. The % in FORTRAN 90/95 is a *component selector*, and returns the component called `nodeone` from the `m`th entry in derived data type `element`. In MATLAB (which does not support this type of derived data structure), one might have a variable named `element_nodeone(m)`, or use a two-dimensional array of the form `element_nodes(m,local_node_num)`; there are a variety of possibilities.

Furthermore, even if used, the connection matrix is also not stored as explained here; the reason is that it is highly sparse and could be stored far more efficiently in some type of *compressed storage* scheme.

10.2.2 Some practical issues: assembling the system

In FEM parlance, the process of filling the finite element system matrix is frequently known as *matrix assembly*. For practical codes, it is generally convenient to loop over the elements rather than the nodes (recall that the degrees of freedom are the nodal potentials for this first-order scheme). This is known as *assembly-by-elements*. For a particular global degree of freedom i, any element that contains this node will contribute to the matrix. For triangles, this number depends on the mesh. We will now discuss two practical methods that simplify this matrix assembly process.

Connecting the system

The connection matrix is useful for explaining the method, but inconvenient in practice. Practical programs do this essentially by inspection, as already outlined in Section 9.3.1. A global numbering system is adopted from the start. As each element's $[S]$ matrix is computed, it is entered into the global matrix. A formal method for doing this has been described in [1, pp. 51–53], but essentially one simply adds the contributions of each elemental matrix at the appropriate global row and column entry. Once again, note that *sparsity* has not yet been exploited.

Handling the boundary conditions

Repeating the matrix equation to solve

$$[S_{ff}]\{\phi_f\} = -[S_{fp}]\{\phi_p\}$$

we see that the prescribed boundaries form the right-hand side of the matrix equation. This assumes that free unknowns have been numbered first, then prescribed unknowns, as already briefly mentioned. Entries of the form S_{ff} (i.e. both nodes free) are entered into the system matrix; entries of the form S_{fp} (i.e. one node free) are multiplied by the prescribed potential and entered into the right-hand side vector. Entries of the form S_{pf} and S_{pp} play no role. (Actually, $[S_{pf}]^T = [S_{fp}]$ and this is implicitly included during the minimization process, when this is exploited.)

Another method has been described in the literature [1, pp. 49–50] which is useful when it is *not* possible, or very inconvenient, to number first free then prescribed elements; it uses dummy entries, and increases the matrix size slightly. We will briefly outline the method, with reference to Fig. 10.1, and assume that nodes 3 and 4 are prescribed, with nodes 1 and 2 free to vary. From Eq. (10.24), the system to solve is of the form

$$\begin{bmatrix} S_{11} & S_{12} \\ S_{21} & S_{22} \end{bmatrix} \begin{Bmatrix} \phi_1 \\ \phi_2 \end{Bmatrix} = - \begin{bmatrix} S_{13} & S_{14} \\ S_{23} & S_{24} \end{bmatrix} \begin{Bmatrix} \phi_3 \\ \phi_4 \end{Bmatrix} \tag{10.30}$$

(Note that for a typical nodal finite element, $S_{13} = 0$, but we retain the entry to illustrate the method.)

The prescribed potentials can be expressed in the following identity:

$$\begin{bmatrix} D_{33} & \\ & D_{44} \end{bmatrix} \begin{Bmatrix} \phi_3 \\ \phi_4 \end{Bmatrix} = \begin{bmatrix} D_{33} & \\ & D_{44} \end{bmatrix} \begin{Bmatrix} \phi_3 \\ \phi_4 \end{Bmatrix} \tag{10.31}$$

$[D]$ is any definite diagonal matrix – usually, but not always, the unit matrix $[I]$. This equation simply states that the potentials must have their prescribed value. By combining Eqs. (10.30) and (10.31), one obtains

$$\begin{bmatrix} S_{11} & S_{12} & & \\ S_{21} & S_{22} & & \\ & & D_{33} & \\ & & & D_{44} \end{bmatrix} \begin{Bmatrix} \phi_1 \\ \phi_2 \\ \phi_3 \\ \phi_4 \end{Bmatrix} = \begin{bmatrix} -S_{13} & -S_{14} \\ -S_{23} & -S_{24} \\ D_{33} & \\ & D_{44} \end{bmatrix} \begin{Bmatrix} \phi_3 \\ \phi_4 \end{Bmatrix} \tag{10.32}$$

```
[node_free_flag,num_free_nodes] = free_nodes(a,b,h,w);
node_pre_flag = not(node_free_flag); % Prescribed nodes
[node_prenz_flag,num_prenz_nodes] = prescr_nodes(a,b,h,w);
.
.
for ielem=1:NUM_ELEMENTS % Assemble by elements, using alternate
                          % approach to handling prescribed boundaries.
   %for ielem=1:1
   trinodes = ELEMENTS(ielem,:);
   [S_elem] = s_nodal( NODE_COORD(trinodes(1),1),NODE_COORD(trinodes(1),2),...
       NODE_COORD(trinodes(2),1),NODE_COORD(trinodes(2),2),...
       NODE_COORD(trinodes(3),1),NODE_COORD(trinodes(3),2) );
   S_elem = S_elem*eps_r(ielem); % Scale by material properties
   % Assemble into global matrix.
      for jnode = 1:3
      jj = ELEMENTS(ielem,jnode);
      for knode = 1:3
         kk = ELEMENTS(ielem,knode);
         if node_free_flag(jj) && node_free_flag(kk)
         % i.e. both free
            S_mat(jj,kk) = S_mat(jj,kk)+S_elem(jnode,knode);
         elseif node_free_flag(jj) && node_pre_flag(kk)
         % i.e. one free, one prescribed
            b_vec(jj) = b_vec(jj) - S_elem(jnode,knode)*phi_pre(kk);
            S_mat(kk,kk) =1;
            b_vec(kk) = phi_pre(kk);
         end
      end
   end
end
```

Figure 10.2 MATLAB code stub of a two-dimensional FEM code, showing assembly-by-elements with free and prescribed potentials mixed together.

This still does not appear to be very useful, until one considers renumbering the potentials. For instance, let vertices 1-2-3-4 be renumbered $2'$-$4'$-$1'$-$3'$ where the prime indicates the renumbered node (i.e. the free potentials are now $2'$ and $4'$, and the prescribed potentials $1'$ and $3'$). The renumbered system is now

$$\begin{bmatrix} D_{1'1'} & & & \\ & S_{2'2'} & & S_{2'4'} \\ & & D_{3'3'} & \\ & S_{4'2'} & & S_{4'4'} \end{bmatrix} \begin{Bmatrix} \phi_{1'} \\ \phi_{2'} \\ \phi_{3'} \\ \phi_{4'} \end{Bmatrix} = \begin{Bmatrix} D_{1'1'}\phi_{1'} \\ -S_{2'1'}\phi_{1'} - S_{2'3'}\phi_{3'} \\ D_{3'3'}\phi_{3'} \\ -S_{4'1'}\phi_{1'} - S_{4'3'}\phi_{3'} \end{Bmatrix} \tag{10.33}$$

Provided that the other matrix diagonal entries are not of an entirely different order of magnitude, having $[D] = [I]$ works well (if so, then using the average of the diagonal entries has been proposed). Dropping the primes, one obtains

$$\begin{bmatrix} 1 & & & \\ & S_{22} & & S_{24} \\ & & 1 & \\ & S_{42} & & S_{44} \end{bmatrix} \begin{Bmatrix} \phi_1 \\ \phi_2 \\ \phi_3 \\ \phi_4 \end{Bmatrix} = \begin{Bmatrix} \phi_1 \\ -S_{21}\phi_1 - S_{23}\phi_3 \\ \phi_3 \\ -S_{41}\phi_1 - S_{43}\phi_3 \end{Bmatrix} \tag{10.34}$$

From this, the algorithm is clear: matrix entries involving two free nodes are entered into the left-hand side system matrix. Matrix entries involving a prescribed node, however, result in an entry on the diagonal on the LHS matrix, and augment the RHS vector with the negated product of the relevant $[S]$-free-prescribed matrix entry and the prescribed potential. A code stub which uses this approach is shown in Fig. 10.2. The functions `free_nodes` and `prescr_nodes` return flags indicating whether nodes are free or inhomogeneous (i.e. non-zero, `nz` in the code) prescribed, for certain geometrical parameters `a`, `b`, `h` and `w` – to be described shortly; the `not(node_free_flag)` operation[5] finds *all* the prescribed nodes. Although the code only deals with the case of node `jj` free and node `kk` prescribed, since the loop cycles through all values of `jnode` and `knode`, the complementary case will either be taken care of subsequently, or has already been dealt with. The routine `s_nodal` returns the $[S]$ matrix for the element, and the vector `eps_r` contains ϵ_r for each element.

This approach has the benefit that the nodes do not need to be numbered in any particular fashion; however, the overall matrix is increased in size from only the free nodes to all nodes. In practice, it is often not possible to number free nodes first, and a code working with the alternate option (that of only free nodes) must renumber the nodes internally first, which adds some book-keeping overhead. Usually, the prescribed nodes are on the boundary, so in two dimensions the extra overhead is not significant, but in three dimensions it can be, and the overhead of the book-keeping is often worthwhile. Both approaches are used in practice.

10.2.3 An application to microstrip

Microstrip technology was discussed in some detail in Chapter 8, and the Green function for a grounded dielectric slab was derived in the spectral domain in Section 7.3. Had we pursued this, we could have developed an MoM solution to the problem (and, as mentioned there, details of such an implementation may indeed be found in the literature [5]). As mentioned in the opening paragraph in that section, for the dominant quasi-TEM mode of propagation, a quasi-static analysis provides useful information. In particular, the equivalent dielectric constant and the characteristic impedance of the line can be determined.

For a FEM solution, the unbounded domain shown in Fig. 7.1 must be converted to a bounded domain. The easiest way to do this is to place the microstrip within a PEC box, as shown in Fig. 10.3. The PEC box provides homogenous Dirichlet boundary conditions on the exterior boundary. The quasi-TEM mode has electric symmetry vertically about the center of the structure, and only the right half thereof is included in the FEM model. Electric symmetry for a quasi-static solver in the electric potential implies a homogeneous Neumann boundary condition, which in FE terms implies a natural boundary condition. The mesher used generates first a rectangular mesh, which is then split into triangles – more details are given in Section 10.8. A rather coarse mesh in shown in Fig. 10.4. Results computed using a somewhat finer mesh are shown in Fig. 10.5. (Note that as capacitance per unit length is invariant with geometrical scale, the dimensions are given in dimensionless form.)

The parameters of interest from a circuit viewpoint are the characteristic impedance and the effective dielectric constant. These can be computed by running the code twice, to find two values for the capacitance per unit length, once with the dielectric constant of the substrate set to its actual value, which yields C, and then repeating the analysis

[5] `not` is the Boolean NOT operation in MATLAB.

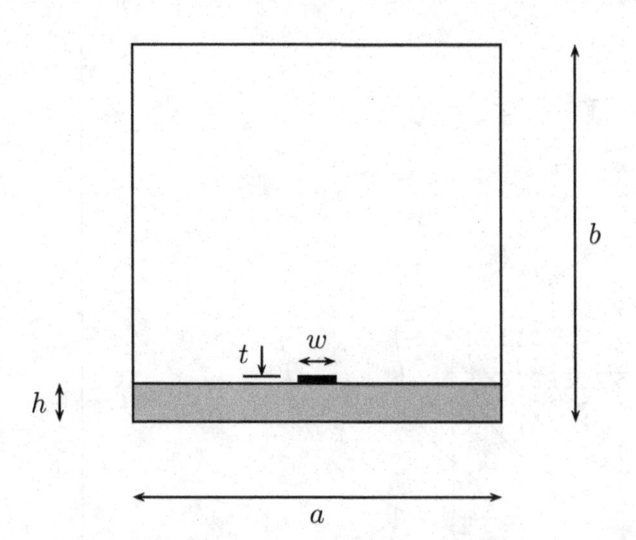

Figure 10.3 The boxed microstrip model used in the text. h is the thickness of the substrate, w and t the width and thickness of the center conductor, and a and b the width and height of the PEC box.

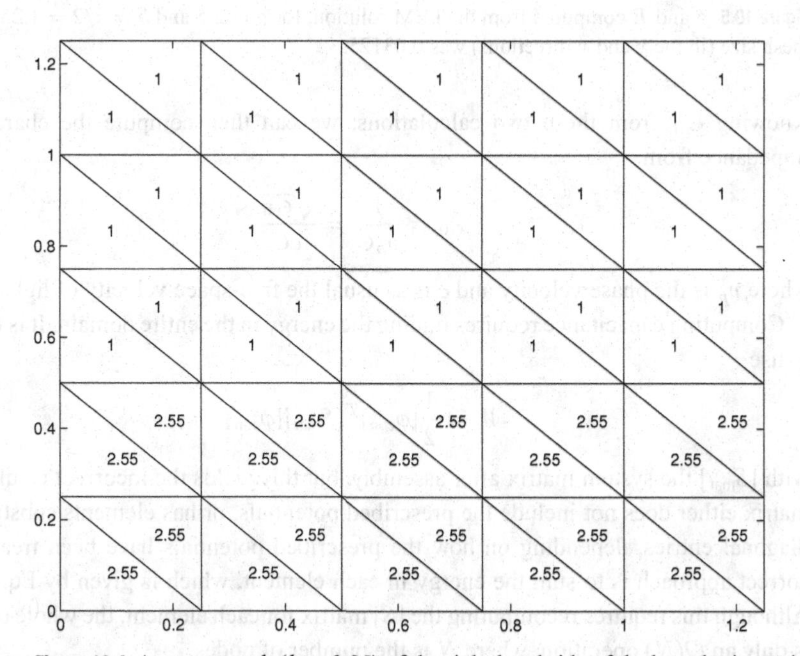

Figure 10.4 A coarse mesh ($h = 0.25$) of the right-hand side of the boxed microstrip structure, showing the elements and ϵ_r in each element.

with a free space dielectric, which yields C_0. Since capacitance is proportional to the effective dielectric constant of an equivalent material homogeneously filling the region, we have that

$$\epsilon_{\text{eff}} = \frac{C}{C_0} \tag{10.35}$$

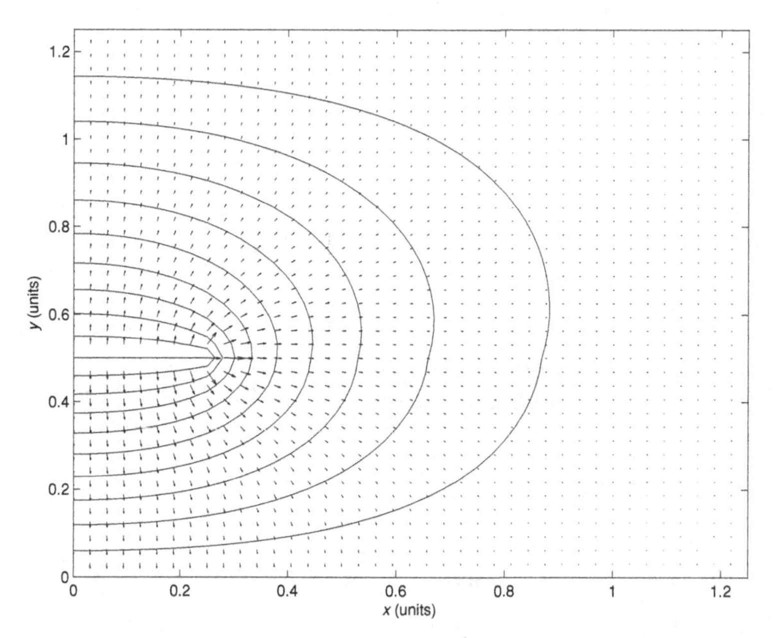

Figure 10.5 ϕ and $\boldsymbol{E}$ computed from the FEM solution, for $a = 2.5$ and $b = a/2 = 1.25$. The mesh size (in the x and y directions) was 0.03125.

Knowing ϵ_{eff} from these two calculations, we can then compute the characteristic impedance from

$$Z_0 = \frac{1}{v_p C} = \frac{\sqrt{\epsilon_{\text{eff}}}}{cC} \tag{10.36}$$

where v_p is the phase velocity and c is as usual the free-space velocity of light.

Computing capacitance requires finding the energy in the entire domain. It is tempting to use

$$W = \frac{1}{2}\{\phi_{\text{con}}\}^T[S_{\text{mat}}]\{\phi_{\text{con}}\} \tag{10.37}$$

with $[S_{\text{mat}}]$ the system matrix after assembly, but this yields the incorrect result, as this matrix either does not include the prescribed potentials, or has elements substituted by diagonal entries, depending on how the prescribed potentials have been treated. The correct approach is to sum the energy in each element, which is given by Eq. (10.14). Although this requires recomputing the $[S]$ matrix for each element, the whole operation is only an $\mathcal{O}(N)$ operation, where N is the number of nodes.

The results for this problem converge rather slowly as the mesh is refined. Using the extrapolation procedure presented in Section 1.9, the rate of convergence can be established as only of order h. Extrapolation to zero mesh size can be applied to both C and C_0, giving the results in Tables 10.1 and 10.2 for two box sizes.[6] Formulas for

[6] One should be careful to extrapolate the quantity obtained directly from the FEM solution whose rate of convergence has been established – capacitance in this case – and *not* a derived quantity such as ϵ_{eff} or Z_0.

Table 10.1 Extrapolated results for boxed microstrip, $a = 2.5$, $b = 1.25$

h (mm)	0.1250	0.0625	0.03125	$\rightarrow 0$
ϵ_{eff}	1.8537	1.8509	1.8496	1.8481
$Z_0(\Omega)$	73.6285	77.0304	78.8037	80.7112

Table 10.2 Extrapolated results for boxed microstrip, $a = 5$, $b = 2.5$

h (mm)	0.1250	0.0625	0.03125	$\rightarrow 0$
ϵ_{eff}	1.9705	1.9632	1.9597	1.9556
$Z_0(\Omega)$	80.4007	83.8117	85.5939	87.4894

effective dielectric constant and characteristic impedance in the literature give values of 1.990 and 89.8 Ω respectively, so the extrapolated values for the larger box give an error of around 1.7% and 2.6% respectively. The results could no doubt be further improved by increasing the size of box again.

Usually, one would expect more rapid convergence, especially for capacitance. With a complete first-order element, a convergence rate of $\mathcal{O}(h^2)$ would be expected for the potential, and since the capacitance is computed from the energy, which is a variational functional of the potential, one might naively expect a rate of convergence of $\mathcal{O}(h^4)$ for capacitance, whereas in reality it is only $\mathcal{O}(h)$. The reason for this disappointingly slow rate of convergence is the singular nature of the fields at the edge of the center conductor. To improve the result more rapidly, one would need to selectively refine the mesh in this region, a process known as *mesh adaptation*. This topic is discussed later in this book. This is an area where the FEM is highly advantageous, as selectively refining FDTD meshes has proven very difficult.

10.2.4 More on variational functionals

Earlier, we mentioned that the equivalence between the PDE and the variational functional lies at the heart of the variational FEM approach. Having now seen a basic FEM formulation developed, we need to return to the theoretical underpinnings of the method. We will work with the more general Poisson PDE, which includes a source term, for a homogeneous region:

$$\nabla^2 \phi = -\frac{\rho}{\epsilon} \tag{10.38}$$

where ρ is the source, and the boundary conditions on $S \equiv S_1 + S_2$, as before, are Dirichlet on S_1 and Neumann on S_2. There does not appear to be a systematic process to construct variational functionals from PDEs (the reverse process is called Euler's method), and usually one instead shows that the proposed variational functional has the required properties. Thus we propose that the following variational functional has an extremal point, which corresponds to the solution of the Poisson equation above, with the required boundary conditions:

$$W(\phi) = \frac{1}{2}\iint \nabla\phi \cdot \nabla\phi \, dS - \iint \phi\frac{\rho}{\epsilon}\, dS \tag{10.39}$$

and we will then show that it indeed has these properties.

Proving the equivalence of the functional and PDE

We will now apply what is known as a variational analysis. We postulate the following:

$$\phi' = \phi + \theta h \tag{10.40}$$

where ϕ' is the trial solution, ϕ is the solution of the PDE, h is some (differentiable) function (which, rather importantly, must be zero at prescribed boundaries since by definition ϕ is known there) and θ is a (real valued) perturbation parameter. This is then substituted into the variational functional, Eq. (10.39):

$$W(\phi + \theta h) = W(\phi) + \theta \iint \nabla\phi \cdot \nabla h \, dS - \theta \iint \frac{\rho}{\epsilon} h \, dS + \frac{1}{2}\theta^2 \iint \nabla h \cdot \nabla h \, dS$$

The first and last terms are always greater than or equal to zero (*positive semi-definite* in mathematical terms). The term in θ is the *first variation*; what we must now show is that this is zero. To do this, we will use *Green's theorem*, which is essentially multi-dimensional integration by parts:

$$\iint_S u\nabla^2 v \, dS = \oint_C u(\nabla v) \cdot d\mathbf{C} - \iint_S \nabla u \nabla v \, dS \tag{10.41}$$

Using this, one finds:

$$\iint \nabla\phi \cdot \nabla h \, dS = \oint h\frac{\partial\phi}{\partial n}\, dC - \iint h\nabla^2\phi \, dS \tag{10.42}$$

Now, a subtle argument is introduced. The contour integral must be zero to eliminate the first variation. Clearly, on S_1, $h = 0$ by definition, since the value of ϕ is known. If $\partial\phi/\partial n = 0$ on S_2, then we have achieved our aim. This, of course, is just the homogeneous Neumann boundary condition.

The other surface integral term yields

$$-\iint h\nabla^2\phi \, dS = \iint \frac{\rho}{\epsilon} h \, dS \tag{10.43}$$

since ϕ is the solution of the PDE. This cancels with the other term in θ. Hence, the first variation is zero, subject to either Dirichlet boundary conditions on S_1 or homogeneous Neumann boundary conditions on S_2, and we have shown what we set out to achieve.

In the finite element procedure, we actually perform the operation in the inverse order. Minimizing[7] the energy functional, by differentiating with respect to the free potentials, is equivalent to forcing the first variation to zero; given prescribed boundary conditions on S_1, we then *naturally* enforce homogeneous Neumann boundary conditions on S_2.

[7] In general, one should rather speak of rendering the functional stationary, or finding the extremal point, but for this problem the functional is indeed minimized.

(It is worth noting that the latter boundary condition is enforced in an *average* sense on S_2; that is, it is not exactly enforced at each point on S_2.)

Summary of boundary conditions

The issue of boundary conditions is so important with the FEM that it deserves to be highlighted. There are two types of boundary conditions most frequently encountered in elementary FEM analysis:

- Dirichlet boundary conditions: these are *essential* and must be explicitly set.
- Homogeneous Neumann boundary conditions: these are *natural* and are implicitly enforced. An homogeneous Neumann boundary condition corresponds to a symmetry plane; it is often used to reduce the computational domain.

The reason that it is so important to be aware of this is that even if one is only using an FEM code and has no intention of ever writing one, code developers assume that users know this. In particular, it is very important to realize that an unset boundary condition is not an error in the FEM process: it is a natural homogeneous Neumann boundary condition.

As mentioned earlier, other boundary conditions may also be encountered, including inhomogeneous Neumann boundary conditions and mixed boundary conditions. The extension of the functional to these boundary conditions has been discussed in Section 9.3.2.

Boundary conditions at material interfaces

One of the great strengths of the FEM is that handling inhomogeneous regions is very simple. There are, however, one or two subtleties worth highlighting. The boundary conditions on the electrostatic potential at the interface between regions 1 and 2, with appropriate dielectric constants, are

$$\phi_1 = \phi_2 \tag{10.44}$$

$$\epsilon_{r1}\frac{\partial \phi_1}{\partial n} = \epsilon_{r2}\frac{\partial \phi_2}{\partial n} \tag{10.45}$$

The former comes from the requirement of tangential electric field continuity, the latter from normal electric flux continuity.

When assemblying the system, we force potentials to be continuous at a material interface, hence satisfying the former. It turns out that the latter is a natural boundary condition of the variational approach. This is an important point. To show this, one starts with

$$\nabla \cdot \epsilon \nabla \phi = -\rho \tag{10.46}$$

and the variational functional

$$W(\phi) = \frac{1}{2}\iint \epsilon \nabla \phi \cdot \nabla \phi \, dS - \iint \phi \rho \, dS \tag{10.47}$$

and proceeds with an analysis along exactly the same lines as before, but with the domain split in two.[8] Two additional terms then appear in the first variation, representing the flux continuity condition at the interface. From the stationarity requirement, flux continuity follows (for details, refer to [3, Section 3.2]).

Within a code, the above is usually entirely invisible to the user.

10.2.5 The Poisson equation: incorporating a source term

Including the term $-\int \phi\rho \, dS$ representing the source in the functional Eq. (10.47) results in a new matrix, $[T]$. (As mentioned in Chapter 9, this is sometimes called the *mass matrix*.) The (known) source term ρ is discretized using the same interpolation scheme as ϕ, i.e. first-order triangular finite elements in this case, *but with known coefficients*. The entries in $[T]$ are computed from

$$T_{ij}^e = \iint_{S_e} \alpha_i \alpha_j \, dS \tag{10.48}$$

with α the nodal interpolation functions as before.

The result is a matrix equation of the following form:

$$[S_{ff}]\{\phi_f\} = \frac{1}{\epsilon_0}[T]\{\rho\} - [S_{fp}]\{\phi_p\} \tag{10.49}$$

where ϵ_r^e is included within the $[S]$ system sub-matrices as in Eq. (10.21). If the domain is filled with homogeneous material, with ϵ_r constant throughout, this could be written as

$$[S'_{ff}]\{\phi_f\} = \frac{1}{\epsilon_r\epsilon_0}[T]\{\rho\} - [S'_{fp}]\{\phi_p\} \tag{10.50}$$

where the prime indicates that ϵ_r has not been included in the matrices.

It is interesting to note that the inhomogeneous part of the PDE ($\{\rho\}$) plays the same role in the finite element system matrices as the inhomogeneous part of the boundary conditions ($\{\phi_p\}$).

10.2.6 Discussion

This completes our introductory discussion of the method. An obvious extension for the Laplace (and Poisson) equations is to introduce higher-order elements, using quadratic, cubic, quartic or even higher. This has been very comprehensively addressed in [1], and for static problems works very well. However, for dynamic problems, our main interest, we need to introduce a different type of element, called variously the edge element, vector element or Whitney element, so we will not pursue scalar elements any further. In order to address vector elements, we need to introduce a concept briefly mentioned in Chapter 9 and widely used in FEM analysis, namely simplex coordinates. Before doing

[8] The extension to an arbitrary number of different materials is obvious.

this, however, we will take a small theoretical diversion and discuss an alternate method of formulating the FEM in the next section, namely the Galerkin approach.

10.3 The Galerkin (weighted residual) formulation

As already mentioned Section 9.2, the FEM can be derived via two different, but equivalent, procedures. On the one hand, there is the variational boundary value problem approach used extensively in this and the preceding chapter. On the other hand, there is the Galerkin weighted residual formulation, already encountered in Chapter 4.[9]

In terms of our previous work on the method of weighted residuals in Chapter 4, the linear operator for the Laplace equation, Eq. (10.1), is $\mathcal{L} = \nabla \cdot \epsilon \nabla$; the unknown function $f = \phi$ and the forcing function $g = 0$. (Again, in this case, the mathematical term homogeneous is sometimes used, this time in the context of the PDE, rather than the boundary conditions, or the material composition of the domain.) We proceed as with the MoM, by introducing basis functions, weighting functions W and an inner product. The unknown function (the potential ϕ in this case) is expanded as

$$\phi \approx \sum_{i=1}^{N} a_n h_n \tag{10.51}$$

A Galerkin approach is followed, with weighting functions w_m of the same form as the basis functions:

$$W = \sum_{m=1}^{M} w_m \tag{10.52}$$

and an inner product is defined for this two-dimensional problem as

$$\langle a, b \rangle = \iint_S ab \, dS \tag{10.53}$$

Hence, as before, a linear system is obtained, with entries of the following form for the m, nth system matrix element:

$$\langle w_m, \mathcal{L} a_n h_n \rangle \tag{10.54}$$

Boundary conditions are treated in the same fashion as with the VBVP approach, viz. using Green's theorem to move one differential operator to the weighting function; the resulting boundary term permits incorporation of the boundary conditions.

At this stage, this looks so similar to the MoM that one might wonder why the FEM is regarded as a different method. (Indeed, a number of workers in the 1980s unified the methods thus.) Although in general terms there are indeed similarities at this very fundamental functional analysis level, there are very substantial practical differences which justify treating the methods separately. The most important is that the operator $\mathcal{L}$ is now a *differential* as opposed to an integral (or integro-differential) operator; this

[9] Readers who are not working through this book sequentially might wish to read Chapter 4 at this stage.

means that only elements in close geometrical proximity have non-zero system matrix entries, and hence a very large number of the matrix entries are zero. Mathematically, this is a *sparse* matrix; the MoM with integral equations generates full matrices; the efficient solution of these linear systems requires different solvers. Furthermore, FEM matrix entries are generally easily evaluated, being integrals of polynomials, whereas the MoM entries contain singular or near-singular terms resulting from the Green function. Finally, another important difference is that with the MoM integral equation formulation, the boundary conditions are built into the formulation; with the FEM, these must be explicitly imposed.

There are some subleties about the Galerkin formulation which only fully emerge when one implements it. Firstly, at Dirichlet boundaries, the function value is known, and no weighting function is required. Secondly, at other boundaries, a "half"-basis function is needed. Also, the global numbering scheme is adopted from the start, and the "assembly" is now by basis functions, rather than elements. For these reasons, although the CEM literature increasingly uses the Galerkin formulation to pose the FEM problem, in practice, ideas from the VBVP approach, in particular "assembly-by-elements," continue to be very useful.

These ideas will now be put in concrete form by (re-)deriving the finite element formulation for the one-dimensional wave equation of Chapter 9; there, the derivation follows the traditional variational boundary value problem approach. (Although the theme of this chapter is two-dimensional FEM, the ideas are especially easily illustrated in one dimension.) For simplicity, the model problem of Section 9.2 will be used again. We start as before with the PDE (actually an ODE in this case) and boundary conditions, Eq. (9.5), repeated here for convenient reference:

$$\frac{d}{dz}\left(\frac{1}{L}\frac{dV}{dz}\right) + \omega^2 CV = 0 \tag{10.55}$$

with boundary conditions

$$\begin{aligned} V(z = \ell) &= V_0 \\ \left.\frac{dV}{dz}\right|_{z=0} &= 0 \end{aligned} \tag{10.56}$$

Firstly, the residual of Eq. (4.52) is formed; since the present ODE is homogeneous (i.e. source g in Eq. (4.52) is zero), the LHS is directly the residual $\mathcal{R}$. This residual is now weighted with the weighting (or test) function W and integrated over the domain. Mathematically, we are forming the inner[10] product of W and $\mathcal{R}$. As noted, W must be zero at prescribed (Dirichlet) boundaries, as the residual is zero there by definition. (This is easy to arrange when W is discretized.) The problem is now the following:

$$\int_0^\ell \left[W\frac{d}{dz}\left(\frac{1}{L}\frac{dV}{dz}\right) + \omega^2 WCV\right] = 0 \tag{10.57}$$

[10] Sometimes in electromagnetics, symmetric.

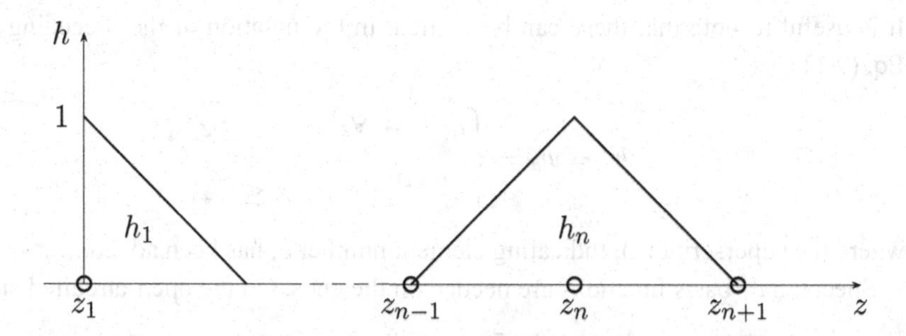

Figure 10.6 Triangular basis functions 1 and n of the Galerkin formulation. As shown, the first basis function is only a half-function; see text for discussion.

Now, integration by parts, viz.

$$\int_{x_1}^{x_2} u \frac{dv}{dx} dx = uv|_{x_1}^{x_2} - \int_{x_1}^{x_2} \frac{du}{dx} v dx \tag{10.58}$$

is used to transfer one derivative from the function V to the weighting function W:

$$\int_0^\ell \left[-\left(\frac{dW}{dz}\right)\left(\frac{1}{L}\frac{dV}{dz}\right) + \omega^2 WCV \right] dz + W\left(\frac{1}{L}\frac{dV}{dz}\right)\bigg|_0^\ell = 0 \tag{10.59}$$

From Eq. (10.56), $dV/dz = 0$ at $z = 0$, and $W(z = \ell) = 0$ since V is prescribed there, it is clear that the boundary terms make no contribution in this case and the equation can be rewritten as

$$\int_0^\ell \left[\left(\frac{dW}{dz}\right)\left(\frac{1}{L}\frac{dV}{dz}\right) - \omega^2 WCV \right] dz = 0 \tag{10.60}$$

Now, W and V must be discretized:

$$W \approx \sum_{m=1}^{N} w_m \tag{10.61}$$

$$V \approx \sum_{n=1}^{N} V_n h_n \tag{10.62}$$

First-order (triangular) basis functions, as in Fig. 10.6, will be used. As a Galerkin formulation is used, the expansion and weighting functions will be identical (other than for one special case noted below); the basis functions for $n = 2, \ldots, N-1$ are

$$h_n = w_n = \begin{cases} \dfrac{z - z_{n-1}}{z_n - z_{n-1}} & \forall z_{n-1} \leq z \leq z_n \\[2ex] \dfrac{z_{n+1} - z}{z_{n+1} - z_n} & \forall z_n \leq z \leq z_{n+1} \end{cases} \tag{10.63}$$

It is useful to note that these can be written, in the notation of the preceding chapter, Eq. (9.11), as

$$h_n = w_n = \begin{cases} \alpha_r^{(n)} & \forall z_{n-1} \leq z \leq z_n \\ \alpha_l^{(n+1)} & \forall z_n \leq z \leq z_{n+1} \end{cases} \tag{10.64}$$

where the superscript (e), indicating element number e, has been added.

Special half-basis functions are needed on the ends. On the open-circuited side:

$$h_1 = w_1 = \frac{z_2 - z}{z_2 - z_1} = \alpha_l^{(1)} \quad \forall z_1 \leq z \leq z_2 \tag{10.65}$$

and at the load end:

$$h_N = \frac{z - z_{N-1}}{z_N - z_{N-1}} = \alpha_r^{(N)} \quad \forall z_{N-1} \leq z \leq z_N \tag{10.66}$$

and as a special case:

$$w_N = 0 \quad \forall z_{N-1} \leq z \leq z_N \tag{10.67}$$

w_N is zero to satisfy the requirement that the weighting function must vanish at this prescribed boundary.

Substitution of Eqs. (10.61) and (10.62) into Eqs. (10.59) produces a matrix equation of the form

$$[\mathcal{M}]\{V\} = 0 \tag{10.68}$$

The entries of the system matrix $[\mathcal{M}]$ are given by

$$\mathcal{M}_{m,n} = \int_0^\ell \left[\left(\frac{dw_m}{dz} \right) \left(\frac{1}{L} \frac{dh_n}{dz} \right) - \omega^2 w_m C h_n \right] dz \tag{10.69}$$

The restricted support (i.e. limited region where they are non-zero) of the subdomain basis functions implies that a large number of entries will be zero. Indeed, there will only be non-zero entries for $|m - n| \leq 1$; i.e. the matrix is tridiagonal. (Note that although sparseness is a general property the FEM system matrix, the tridiagonal structure is specific to this one-dimensional problem, with first-order basis functions, with nodes numbered sequentially as in Fig. 10.6.) Due to the symmetry of the matrix entries (with one exception, to be handled shortly), only $\mathcal{M}_{m,m}$, $\mathcal{M}_{m,m-1}$ and $\mathcal{M}_{m,m+1}$ need be computed explicitly. The matrix entry $\mathcal{M}_{mm}$, for instance, is given by

$$\mathcal{M}_{mm} = \frac{1}{L} \int_{z_{m-1}}^{z_m} \left(\frac{d\alpha_r^{(m)}}{dz} \right) \left(\frac{d\alpha_r^{(m)}}{dz} \right) dz - \omega^2 C \int_{z_{m-1}}^{z_m} \alpha_r^{(m)} \alpha_r^{(m)} dz \tag{10.70}$$

$$+ \frac{1}{L} \int_{z_m}^{z_{m+1}} \left(\frac{d\alpha_l^{(m+1)}}{dz} \right) \left(\frac{d\alpha_l^{(m+1)}}{dz} \right) dz - \omega^2 C \int_{z_m}^{z_{m+1}} \alpha_l^{(m+1)} \alpha_l^{(m+1)} dz \tag{10.71}$$

Written in this form, it is clear that one is simply directly summing the contributions from *elements* m and $m + 1$ in this step; by comparison with the variational development of the previous chapter, this was done in the assembly-by-elements process when the disjoint elements were connected. Previously computed results for the integrals, Eq. (9.29) and (9.30) are directly applicable; assuming again that all elements are of the same length h, the result for the diagonal entries of $[\mathcal{M}]$ is

$$\mathcal{M}_{m,m} = \frac{1}{hL}(1+1) - \omega^2 hC\left(\frac{2}{6}+\frac{2}{6}\right) \tag{10.72}$$

$$= \frac{2}{hL} - \omega^2 hC\frac{4}{6} \tag{10.73}$$

To make the similarity clearer, $4/6$ has not been reduced to $2/3$ in the above.

The integral for the off-diagonal element $\mathcal{M}_{m,m+1}$ contains contributions from either elements m or $m+1$, but not both, and is

$$\mathcal{M}_{m,m+1} = -\frac{1}{hL} - \omega^2 hC\frac{1}{6} \tag{10.74}$$

The symmetry properties give the other off-diagonal elements:

$$\mathcal{M}_{m,m+1} = \mathcal{M}_{m+1,m} = \mathcal{M}_{m-1,m} = \mathcal{M}_{m,m-1} \tag{10.75}$$

For the first element, the half-basis function of Eq. (10.65) gives

$$\mathcal{M}_{1,1} = \frac{1}{hL} - \omega^2 hC\frac{2}{6} \tag{10.76}$$

and for the last element, the half-basis function of Eq. (10.66) and the zero function of Eq. (10.67) give

$$\mathcal{M}_{N-1,N} = -\frac{1}{hL} - \omega^2 hC\frac{1}{6} \tag{10.77}$$

$$\mathcal{M}_{N,N} = \mathcal{M}_{N,N-1} = 0 \tag{10.78}$$

Substituting this into Eq. (10.68), and moving the prescribed nodal value V_N to the RHS, exactly the same system as Eq. (9.42) results. The notation $\mathcal{M}$ has been used for the system matrix in this Galerkin development, but it is now seen that the matrices are identical, so we could have used the preceding chapter's notation, viz. M, in the discussion.

This presentation some interesting points requiring consideration. Firstly, how general is the equivalence noted above, between the Galerkin formulation and the variational functional approach of Chapter 9? The above analysis was limited to the one-dimensional wave equation of Eq. (10.55), subject to the boundary conditions of Eq. (10.56).

The answer to this requires some knowledge of concepts from functional analysis; it is well known that for a self-adjoint positive definite linear operator that discretization via either the variational functional approach or the Galerkin method leads to the same matrix equation. Unfortunately, many operators in high-frequency electromagnetics, such as the vector wave equation with lossy materials, do not have these properties. Using a symmetric inner product (see Section 4.5), it has been shown that this equivalence can be extended to a number of these problems. Readers wanting to pursue this in more depth are referred to the discussion in Jin [3, Chapters 2 and 6].

Secondly, given the equivalance between the methods (at least for many problems of interest), since the Galerkin method does not require knowledge of an equivalent variational functional, whose stationary point is then sought, the Galerkin method would appear to be the more straightforward procedure, so one might well wonder why most books, including the present one, focus so much on the variational method. Indeed, at formulation level, the Galerkin method is very attractive for its simplicity, and furthermore, it is clearly advantageous when no variational functional is available.

The answer here is as follows. When using traditional subdomain basis functions as in the development here, the core difference is that whereas the variational method focusses on an element, with basis functions defined locally on that element, the Galerkin method focusses on the entire domain, with basis functions with limited support (in the present case, over two connected line segments). Hence, with this method, the global numbering scheme must be known and taken into account from the start when computing the matrix entries, whereas with the traditional variational approach, this is done at the assembly-by-elements stage. Whilst in the very simple one-dimensional problem shown here, there is little difference in programming complexity between these approaches (and indeed, the Galerkin might even seem simpler, given the absence of a connection matrix), in two and especially three dimensions it is easier to focus on elements rather than basis functions from the viewpoint of developing FEM programs. Again, readers wanting a more in-depth discussion are referred to Jin [3, Section 2.4]; he concludes the discussion with the following insightful comment: "In conclusion, the previous [variational] formulation is preferred when one describes the procedure of the finite element method in practice, whereas the present one [Galerkin] is more illustrative when the basic principle of the method is concerned."

A mathematical aside – properties of linear operators

A self-adjoint linear operator is one for which $\langle \mathcal{L}\phi, \psi\rangle = \langle \mathcal{L}\phi, \psi\rangle$. ϕ and ψ are elements of the underlying vector space. Note that boundary conditions play a role in determining whether an operator is self-adjoint or not; see, for instance, [3, Section 6.1] for a discussion of the self-adjointedness of the vector wave equation. For a positive definite operator, $\langle \mathcal{L}\phi, \phi\rangle > 0 \; \forall \phi \neq 0$. As ϕ is an arbitrary non-zero element of the vector space, physical insight can be useful in determining whether an operator has this property or not. For a extended discussion of functional analysis, see [6] in the context of electromagnetics, and [7] in the context of finite element theory in general.

10.4 Simplex coordinates

Simplex coordinates – also known as homogeneous or barycentric (or in 2D, area) coordinates – provide an entirely local geometrical description within a triangle (in 2D) or tetrahedron (in 3D). This is very convenient, since it allows much of the work required to be done once (on what is often called the "parent" triangle) and then with some simple geometrical scaling, it can be applied to any triangle or tetrahedron. They are intimately linked to *simplicial elements* – the simplest possible geometrical shape in the space, that is line elements in one dimension, triangles in two dimensions and tetrahedra in three dimensions. (The concept can be extended to higher dimensions, but loses any geometrical interpretation.)

In general, simplex coordinates are defined as the ratios of lengths (1D), areas (2D) or volumes (3D) that a point in the interior (or on the boundary) splits the line/triangle/tetrahedron into. The size $\sigma(S)$ of a simplex S is defined as

$$\sigma(S) = \frac{1}{N!}\begin{vmatrix} 1 & x_1^{(1)} & x_1^{(2)} & \dots & x_1^{(N)} \\ 1 & x_2^{(1)} & x_2^{(2)} & \dots & x_2^{(N)} \\ \vdots & & & & \\ 1 & x_{N+1}^{(1)} & x_{N+1}^{(2)} & \dots & x_{N+1}^{(N)} \end{vmatrix} \tag{10.79}$$

where superscripts denote space directions and subscripts denote vertices.

10.4.1 Simplex coordinates in one, two and three dimensions

In one dimension, we have

$$\lambda_1 = \frac{\sigma(S_1)}{\sigma(S)} = \frac{\begin{vmatrix} 1 & x \\ 1 & x_2 \end{vmatrix}}{L} = \frac{x_2 - x}{L}$$

$$\lambda_2 = \frac{\sigma(S_2)}{\sigma(S)} = \frac{\begin{vmatrix} 1 & x_1 \\ 1 & x \end{vmatrix}}{L} = \frac{x - x_1}{L}$$

These express the ratios of length from the right and left nodes respectively to point x, to the total length of the element. These are frequently encountered in MoM analysis as local coordinates, and were already encountered in Chapter 9 in the context of one-dimensional FEM.

In two dimensions, we have

$$\lambda_1 = \frac{\sigma(S_1)}{\sigma(S)} = \frac{\begin{vmatrix} 1 & x & y \\ 1 & x_2 & y_2 \\ 1 & x_3 & y_3 \end{vmatrix}}{2A} = \frac{(x_2y_3 - x_3y_2) + (y_2 - y_3)x + (x_3 - x_2)y}{2A} \tag{10.80}$$

This represents the ratio of the area of the triangle P23 to 123 – see Fig. 10.8, in Section 10.7.3. It will be noted that $\lambda_1 = \alpha_1$, the first-order interpolatory function used in our earlier analysis, Eq. (10.12), indicated how convenient the simplex coordinates are for functions defined over a triangle. There are three simplex coordinates in 2D: λ_1, λ_2 and λ_3, describing the three area ratios.

In three dimensions, we have

$$\lambda_1 = \frac{\sigma(S_1)}{\sigma(S)} = \frac{\begin{vmatrix} 1 & x & y & z \\ 1 & x_2 & y_2 & z_2 \\ 1 & x_3 & y_3 & z_3 \\ 1 & x_4 & y_4 & z_4 \end{vmatrix}}{6V} \tag{10.81}$$

This represents the ratio of the volume of the tetrahedron P234 to the volume of the element.

There are four simplex coordinates in 3D: λ_1, λ_2, λ_3 and λ_4, describing the four volumetric ratios.

Equation (10.80) assumes that the three vertices defining the triangle are two-dimensional, and hence the triangle lies in the two-dimensional plane. Sometimes, the vertices are defined in three dimensions; in this case, the triangle "floats" in three-dimensional space. One obvious case is when the triangle is one face of a tetrahedron, in which case one of the simplex coordinates computed from Eq. (10.81) will be zero, and the other three are the required simplex coordinates. However, in applications involving surface modelling (see, for instance, Chapter 6), this is not the case. Formulas for the computation of simplex coordinates of a triangle floating in 3D space may be found in [8, Section 21.3].

Readers are also reminded of Eq. (6.37), which provides a connection between simplex coordinates and Cartesian coordinates. (The generalization of this formula to 3D, using four simplex coordinates and the four vertices of a tetrahedron, is obvious.)

10.4.2 Some properties of simplex coordinates

Aside from the interpretation as the ratio of sizes, simplex coordinates have other important properties. Some of these are as follows:

- The coordinates are normalized, thus $\sum_{i=1}^{N+1} \lambda_i = 1$.
- In two and three dimensions, the *gradient* of each simplex coordinate is a constant, and normal to the relevant edge (2D) or face (3D). In 2D, for example:

$$\nabla \lambda_i = \frac{l_i}{2A} \hat{n}_i \tag{10.82}$$

  with A the area of the triangle, l_i the length of edge i, and $\hat{n}_i$ the normal to edge i. This property is extensively exploited in vector elements, of which more later.
- Because of the normalization, $0 \leq \lambda_i \leq 1 \, \forall i$. This can be a useful and quick test to see whether a point lies inside or outside an element.

10.5 The high-frequency variational functional

For electrodynamic problems, subject to the deterministic vector wave equation,

$$\nabla \times \frac{1}{\mu_r} \nabla \times \boldsymbol{E} - k_0^2 \epsilon_r \boldsymbol{E} = -jk_0 Z_0 \boldsymbol{J} \tag{10.83}$$

with $\boldsymbol{J}$ a source internal to domain Ω and k_0 the free-space wavenumber, the equivalent variational functional which must be rendered stationary is

$$F(\boldsymbol{E}) = \int_\Omega \left[\frac{1}{\mu_r} (\nabla \times \boldsymbol{E}) \cdot (\nabla \times \boldsymbol{E}) - k_0^2 \epsilon_r \boldsymbol{E} \cdot \boldsymbol{E} \right] d\Omega + jk_0 Z_0 \int_\Omega \boldsymbol{E} \cdot \boldsymbol{J} \, d\Omega \tag{10.84}$$

This is the general functional for lossy isotropic materials; see [3, Chapter 6] for further discussion of this and extensions to anistropic media. It assumes either homogeneous Dirichlet or Neumann boundary conditions or a mixture of the two on the boundary of domain Ω.

A closely related functional for the source-free vector wave equation

$$\nabla \times \frac{1}{\mu_r} \nabla \times \boldsymbol{E} - k_i^2 \epsilon_r \boldsymbol{E} = 0 \tag{10.85}$$

is the following:

$$F(\boldsymbol{E}) = \int_\Omega \left[\frac{1}{\mu_r} (\nabla \times \boldsymbol{E}) \cdot \nabla \times \boldsymbol{E}) - k_i^2 \epsilon_r \boldsymbol{E} \cdot \boldsymbol{E} \right] d\Omega \tag{10.86}$$

subject to the same boundary conditions. In this case, the solution is the set of eigenvalues k_i and associated eigenvectors $\boldsymbol{E}_i$; if the domain contains lossy materials, the eigenvalues are complex.

In order to show the above properties, one proceeds in a fashion similar to the Poisson equation, using a vector Green's theorem for the double-curl operator. The details are available in [1, 3] and although more complex than the Poisson case, the method is the same, so we will not repeat them here.

This form (often called the curl-curl form) has been used for high-frequency FEM analysis for many years. However, although it appears fairly straightforward to discretize, it turned out to have a number of problems which occupied analysts for some years. One of the most important advances was the introduction of vector (edge) elements in the late 1980s, and this is the topic of the next section.

10.6 The null space of the curl operator and spurious modes

One of the supposed strengths of the FEM was its accuracy, in particular when compared to a method such as the FDTD, until serious problems with "spurious modes" were found using standard (node-based) FEM for electromagnetic eigenvalue problems (we will define these later). The traditional, nodal FEM approach, typical of structural mechanics, deals with a vector field by approximating each component separately:

$$E_x \approx \sum E_{x_i} f(x, y, z) \tag{10.87}$$

with $f(x, y, z)$ a standard basis function such as those we have already seen (although extended to three dimensions). This was then repeated for E_y and E_z and substituted into Eq. (10.84) or (10.86). As Silvester and Pelosi comment [9, p. 8]:

> The first approach (nodal elements) may be called the structural mechanics approach ... at least some theory and much practical experience should be transferable to electromagnetics. Further, it has the appeal of simplicity and familiarity. The same approximating functions can be called upon to serve for both scalar and vector cases, and the vectorial coefficients have clear meaning as component representations of $\boldsymbol{E}$ or $\boldsymbol{H}$ ... the structural mechanics approach has one major flaw for electromagnetic field analysis: *it doesn't work very well.* The reason is simply that the fields that occur in structural mechanics and those encountered in EM are fundamentally different. The electromagnetic field vectors not only obey the Maxwell curl equations, but they are also constrained by the divergence equations.

Before discussing some of the more intricate details of spurious modes, we note an immediate and practical problem with the nodal approach: since the field is approximated by its values at the nodes, if we use the method we used for the static problems for connecting the elements (that is, all values at a node are set equal on all the elements which share the node), then the result is that we force *all* components of the field to be continuous. At an interface between two different types of material, only the *tangential* components of $\boldsymbol{E}$ or $\boldsymbol{H}$ should be continuous. If the material boundary happens to coincide with a plane parallel to one of the coordinate axes, then it would not be too difficult to arrange that we do this with only the tangential field components, leaving additional degrees of freedom to permit the normal field to be discontinuous. But in general, we are unlikely to be so fortunate, and the material interface will create a very tricky problem indeed.

By comparison, the vector FEM approach approximates the full vector field:

$$\boldsymbol{E} \approx \sum e_{ij} \boldsymbol{w}_{ij} \tag{10.88}$$

with an edge-based *vector* function such as:

$$\boldsymbol{w}_{ij} = \lambda_i \nabla \lambda_j - \lambda_j \nabla \lambda_i \tag{10.89}$$

As before, λ_i is the simplex coordinate with respect to node i. This is then used to discretize Eq. (10.84) or (10.86).

It is far from immediately apparent why what appears to be a minor change in approach should yield significantly better solutions – after all, the vector basis functions are simply another way of representing the vector nature of the problem. In order to understand this, we need first to look a little more carefully at the high-frequency functionals. We will start with the eigenvalue problem, where the problems originate.

Following the standard discretization and substitution of the basis functions, the stationary points of the functional, Eq. (10.86), correspond to solutions of the following generalized eigenmatrix equation:

$$[S]\{e_i\} = k_i^2 [T]\{e_i\} \tag{10.90}$$

where $[S]$ and $[T]$ represent the discretized versions of the first and second terms in Eq. (10.86). The eigenvalues k_i represent the resonance frequencies of the cavity, and the vectors $\{e_i\}$ the eigenvalues, i.e. the various resonant modes (or eigenmodes).

Various approaches are now possible. A particularly revealing one is to note that the divergence constraint,

$$\nabla \cdot \epsilon \boldsymbol{E} = 0 \tag{10.91}$$

is implied within the functional, but in *frequency dependent* form. We can see this by taking the divergence of both sides of Eq. (10.85); noting the vector identity $\nabla \cdot \nabla \times \boldsymbol{a} \equiv 0 \ \forall \ \boldsymbol{a}$, it is clear that

$$k_i^2 \nabla \cdot \epsilon_r \boldsymbol{E} = 0 \tag{10.92}$$

For the dynamic case (that is, $k_i \neq 0$) the divergence equation is indeed satisfied.

The problem, however, enters via the *other* possibility for satisfying this equation, namely $k_i = 0$. In this case, the divergence equation *is no longer necessarily satisfied*. This corresponds of course to the static case, where $\boldsymbol{E} = -\nabla V$, and we note (since $\nabla \times \nabla V = 0 \ \forall V$) the theoretically infinite number of solutions of the form of the field as the gradient of a potential and zero eigenvalue $\{(\nabla V, 0)\}$ also satisfies the vector wave equation, constituting its *null-space* (also known as kernel, abbreviated *ker*, in some of the literature).

A particularly elegant example of such a null-space eigenmode for the well-known rectangular waveguide problem was given by [10], and it is so illuminating that it is worth repeating here. Peterson and Wilton considered the classic eigenvalue problem of a rectangular waveguide with PEC walls, dimensions a by b, with the solutions

$$\boldsymbol{E}_{mn} = -\hat{x}\frac{n\pi}{b}\cos\frac{m\pi x}{a}\sin\frac{n\pi y}{b} + \hat{y}\frac{m\pi}{a}\sin\frac{m\pi x}{a}\cos\frac{n\pi y}{b} \tag{10.93}$$

with eigenvalues:

$$k_{mn}^2 = \left(\frac{m\pi}{a}\right)^2 + \left(\frac{n\pi}{b}\right)^2 \tag{10.94}$$

This is very well known and features prominently in almost any undergraduate electromagnetic text. However, these texts never mention that there is *another* valid solution of the vector wave equation, viz. the static solution:

$$\boldsymbol{E}_{mn}^{\text{null}} = \hat{x}\frac{m\pi}{a}\cos\frac{m\pi x}{a}\sin\frac{n\pi y}{b} + \hat{y}\frac{n\pi}{b}\sin\frac{m\pi x}{a}\cos\frac{n\pi y}{b} \tag{10.95}$$

with eigenvalues $k_{mn} = 0$.

These null-space solution(s) look almost identical to the waveguide solutions, but are critically different – note that they can be written in the form

$$\boldsymbol{E}_{mn}^{\text{null}} = \nabla\left(\sin\frac{m\pi x}{a}\sin\frac{n\pi y}{b}\right) \tag{10.96}$$

Also very importantly, unlike Eq. (10.93), these static solutions do *not* have zero divergence, as can quickly be established by inspection. Because the eigenvalues of these null space modes are zero, these are simply rejected as unwanted solutions when one does an analytical solution of the problem (using separation of variables, for instance). However – and this is a critical point! – the standard high-frequency variational curl-curl functional admits these solutions, and *the finite element procedure will also compute them*. (Unless, that is, one can modify the functional to exclude these solutions – there has been success with such approaches and we will mention this again later, but the formulation is somewhat more involved.)

So, to summarize, due to the properties of the high-frequency variational functional, the finite element procedure will produce not only the wanted, dynamic eigenvalues and eigenvectors, but also a number of "zero" eigenvalues and associated static eigenvectors. Since the finite element solution is of course approximate, the "zero" eigenvalues will not be exactly zero, but may shift up in frequency. If their values become sufficiently large, they may creep into the range of the dynamic eigenvalues and we will no longer be able to distinguish between the dynamic eigenvalues and these (very poor) approximations of zero. In this case, we have a "spurious mode" (also known as a spurious solution) – an eigenvalue and associated eigenvector in the high-frequency range, but not satisfying the divergence criteria and hence entirely unphysical. Such a mode can be more formally defined as *numerical solutions of the eigenproblem that do not converge to any physical mode of the electromagnetic resonator modelled as the mesh is refined* [11]. (In a sense, all the null-space modes are spurious, as they do not correspond to valid solutions of the full Maxwell equations, but in the literature the term is reserved for modes with non-zero eigenvalues.)

Some of the earlier papers on edge elements, as they were then known, can be confusing in places. One may be left with the impression that edge elements entirely eliminate the null-space modes. This is not correct – edge elements still compute these modes, but with better fidelity, so that they do not corrupt the desired range of eigenmodes. There have been other approaches which aim to eliminate the spurious modes entirely, but edge elements do *not* accomplish this.

Regarding deterministic problems, since $k_i \neq 0$ will have been set in a deterministic problem, the numerical process, now being capable of reproducing an irrotational mode spectrum (i.e. the null-space), instead ensures that such a modal content is absent [1, p. 313].[11]

It is interesting that spurious modes were not encountered in the FDTD community. The reason is that the Yee grid implicitly satisfies Gauss' laws (the divergence criteria).

Although the literature tends to indicate that, from a theoretical point, spurious modes have been a closed issue for many years, this is not entirely true. Careful mathematical study by Caorsi, Fernandes and Raffetto has disproved some earlier theories in this regard, advancing a new theory; the overview paper by Fernandes and Raffetto [11] is particularly illuminating, and given that it is essentially about the mathematics of the issue, quite accessible for an engineering readership. Importantly, that paper reviews a number of tetrahedral and hexahedral elements, showing that those in widespread use

[11] There is another school of thought on this topic. It has been argued that the driven solution can be viewed as a sum of eigenvectors, and hence incorrect eigenvectors may also corrupt a deterministic problem [12, p. 408]. In any case, by either argument, edge elements also lead to better solutions for deterministic problems.

satisfy the requirements for being spurious-free in their theory. Readers are cautioned that this is not necessarily true of other elements, such as prismatic and pyramidal elements; indeed, the present author encountered spurious modes in two published higher-order pyramidal elements.

10.7 Vector (edge) elements

10.7.1 An historical perspective

What are now called vector elements, but were originally known as "edge-based" elements, date back to the 1980s in CEM, although the underlying ideas of the structure of the electromagnetic field date back to 1957 and what are known as Whitney forms. In 1980, the French mathematician J. C. Nedelec published a paper that has since become the canonical reference in this field [13], although, ironically, he did not define the edge-based element itself; instead, the paper investigates the structure of the polynomial spaces which the basis functions should span in a highly mathematical format, which is not readily accessible to electronic engineers. (He was clearly influenced by earlier ideas of Riavart and Thomas [14] and it is useful to read their paper before attempting to read Nedelec's.) Some of the earliest work in electrical engineering is due to Bossavit [15]; Barton and Cendes [16] were among the first to address high-frequency electromagnetics with edge elements and their derivation is the one now generally given. Another type of related element, also a vector element, was the hexahedral element, originally introduced by van Welij in its lowest-order straight-sided form in 1985 [17], and in generalized form by Crowley *et al.* in 1988 [18]. Cendes' subsequent work produced one of the first higher-order tetrahedral elements [19]. Webb and Forghani's work on hierarchal tetrahedral elements was the standard reference for many years [20], until succeeded by Webb's later work [21]. Those elements became the standard basis for higher-order schemes; recently, some other variations have been published, for instance [22].

During the 1990s, many researchers made excellent use of these elements and also advanced the theory underlying them. The following is only a selection of the work: Lee, *et al.* introduced a formulation for dispersion analysis in waveguides which remains in widespread use today [23, 24] (and will be discussed in detail shortly); Lee and Mittra worked on cavity eigenvalue problems [25] (and this paper remains useful today, since it contains analytical expressions for the elemental matrices); Gedney and Navsariwala [26], Dibben and Metaxas [27], and Lee *et al.* [28] used edge elements for time domain analysis; Savage and Peterson introduced alternative higher-order tetrahedral elements in [2]; Jin, Volakis, Kempel and their students made significant contributions to applications, especially cavity backed patches (this work is well summarized in [3, 29]), and also in new hierarchal elements [30]; Dyczij-Edlinger *et al.* made advances in understanding the impact of the low-frequency ill-conditioning of the curl-curl formulation [31]; Graglia *et al.* made progress with interpolatory as opposed to hierarchal elements [32]; more recently, the present author extended work on waveguide analysis using higher-order mixed and complete elements [33], and with Botha, worked on error estimation [34, 35].

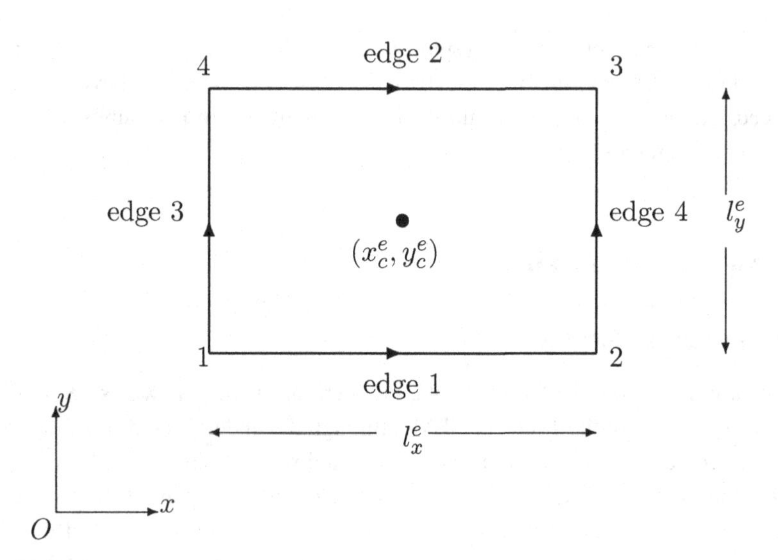

Figure 10.7 The rectangular edge element. Based on [3, Fig. 8.1].

10.7.2 Theory of vector elements

With this historical background, we now return to the elements. Before we study them in detail, we will first look at the impact they had on CEM. Although much of the early literature concentrates on the "spurious mode" problem, there are practical reasons which make these elements very useful in analysis. Firstly, for the lowest-order elements, the degrees of freedom are proportional to the tangential electric field along an edge (and hence the widely used name, edge elements); we will show this shortly. Thus tangential continuity is very simple to enforce. Secondly, flux continuity is a natural boundary condition. Thirdly, it is easier to model corners, or other regions where the field becomes singular, since there is no nodal value at the singularity. Finally, they greatly ameliorated the problems with spurious modes: we will return to this subsequently.

Vector elements are most easily introduced using a two-dimensional vector element for the *rectangular* element, shown in Fig. 10.7. The field is approximated as

$$\boldsymbol{E}_e \approx \sum_{i=1}^{4} \boldsymbol{N}_i^e E_i^e \tag{10.97}$$

Here, $\boldsymbol{N}_i^e$ is the *vector* basis function and E_i^e is a scalar *degree of freedom*, the tangential field along the ith edge in this case. The vector functions $\boldsymbol{N}_i^e$ are given by:

$$\boldsymbol{N}_1^e = \frac{1}{l_y^e}\left(y_c^e - y + \frac{l_y^e}{2}\right)\hat{x} \tag{10.98}$$

$$\boldsymbol{N}_2^e = \frac{1}{l_y^e}\left(y - y_c^e + \frac{l_y^e}{2}\right)\hat{x} \tag{10.99}$$

$$N_3^e = \frac{1}{l_x^e}\left(x_c^e - x + \frac{l_x^e}{2}\right)\hat{y} \tag{10.100}$$

$$N_4^e = \frac{1}{l_x^e}\left(x - x_c^e + \frac{l_x^e}{2}\right)\hat{y} \tag{10.101}$$

with $(x_c^e; y_c^e)$ the coordinates of the center of the element, and l_x^e and l_y^e the element lengths in the x- and y-directions respectively.

Now, note the following: N_1^e is zero on edge 2 (since $y = y_c^e + l_y^e/2$ everywhere on edge 2) and it is unity on edge 1; also, it is purely tangential ($\hat{x}$-directed) along this edge. On edges 3 and 4 it increases linearly from the top to the bottom, and it is purely normal ($\hat{x}$-directed) along these edges. One quickly establishes that N_2^e has the same properties, but with edges 1 and 2 interchanged, and that N_3^e and N_4^e also have similar properties, but obviously with x and y interchanged. In short, these basis functions provide a *mixed-order* approximation of the field – on the edges, the approximation is constant tangentially, and linear normally. (Indeed, these elements are frequently called CT/LN elements, constant tangential/linear normal.) Note also that due to these properties, E_1^e is the tangential field along edge 1, and similarly E_2^e, E_3^e and E_4^e are the tangential fields along edges 2, 3 and 4 respectively. These are the degrees of freedom for this element. *Very importantly*, these properties permit enforcing tangential continuity *without* affecting the normal components, and this is precisely the boundary condition required by $\boldsymbol{E}$ or $\boldsymbol{H}$ fields, or indeed any 1-forms in the language of differential forms.

A mathematical aside – differential forms

Some of the work on vector elements uses the mathematics of differential forms – Bossavit is one of the main proponents of this [36]. Although the ideas can be readily understood without any knowledge of this field, it is useful to know a little of the terminology, as it is increasingly penetrating the CEM literature:

- 0-forms: this is a scalar function with functional but not derivative continuity, an example being the electric static potential ϕ.
- 1-forms: these are vector functions with tangential but not normal continuity, such as $\boldsymbol{E}$. These are also known as polar, or true, vectors, and are time-even under time reversal.
- 2-forms: these are vector functions with normal but not tangential continuity, such as $\boldsymbol{B}$ and $\boldsymbol{J}$. These are also known as axial vectors, or pseudo-vectors, and are time-odd under time reversal.
- 3-forms: discontinuous scalar functions, such as $\nabla \cdot \boldsymbol{D}$.

For an elegant discussion of polar versus axial vectors, and time symmetry, Feynmann's chapter on this is a classic [37, Chapter 52].

Note that this element is *not* by design interpolatory, although for this lowest-order element it can be made thus.[12] The degrees of freedom (E_1^e, E_2^e, E_3^e and E_4^e) represent field quantities along an edge; indeed, in Nedelec's original work, they are defined as integrals of the tangential field component along the edge, i.e. the average tangential field value. This is quite different to the nodal elements discussed earlier.

We should also comment that there are a variety of names for this element, including mixed order; "first" order; "half-th order", H_0(curl); and as already mentioned, constant tangential/linear normal (CT/LN). This last is especially insightful and is the present author's preference.

These elements have other additional significant properties. Interestingly, by taking $\hat{z} \times \boldsymbol{N}_i^e$, another class of elements is derived with the complementary property of providing *normal* continuity; these are useful for problems involving flux or current, i.e. 2-forms. Furthermore, we have already seen that the full-wave functional has a term of the form $\int_{S^e} \nabla \times \boldsymbol{E} \cdot \nabla \times \boldsymbol{E}\, dS$. It is important to note that $\hat{x}$-directed terms linear in x do *not* contribute to this term; i.e. these would be "wasted" degrees of freedom, which have been removed from these elements. This observation, at heart, was the core of Nedelec's contribution. Finally, *within* the element, the approximated $\boldsymbol{E}$ field has *zero divergence*. (Recall that this is not explicitly enforced in the curl-curl functional.) Because the spurious modes are associated with solutions with non-zero divergence, many early papers on vector elements concentrated on this property. Whilst low-order vector elements are indeed divergence free *within* the elements, the divergence is discontinuous at element boundaries, and furthermore, a number of successful vector elements are *not* divergence free. (Indeed, an argument has be made that since one is *not* removing the spurious modes, but computing them more accurately, the element should not be divergence free!) The superior suppression of spurious modes is now understood to be due to a better approximation of the null-space of the vector wave equation, that is, the zero frequency solutions we discussed above. The vector elements do a better job of representing these static $\nabla\phi$ eigenmodes; the reason is that the tangential-continuity-only of the vector elements admits a larger number of functions in the null-space. We noted earlier that ϕ should be continuous, implying that $\nabla\phi$ must be tangentially continuous (which is all that is imposed by edge elements), but the natural boundary condition permits the normal derivative to exhibit the correct jump discontinuity at material interfaces. Webb's 1993 paper remains one of best introductory discussions of edge-based elements [38]; the more recent paper by Fernandes and Raffetto [11] has already been mentioned and is an important starting point for further reading.

10.7.3 Vector elements on triangles – the Whitney element

Our preceding discussion considered rectangular elements. As mentioned on several occasions, one of the main advantages of the FEM over the FDTD is the geometrical modelling flexibility afforded by triangular and tetrahedral elements in two and three

[12] The degrees of freedom have been interpreted as the tangential field value at the center of the relevant edge by some researchers who have worked with interpolatory vector elements.

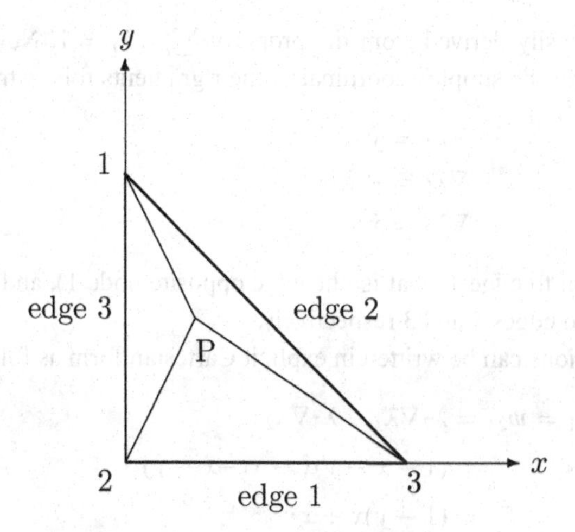

Figure 10.8 The right-angled parent triangle.

dimensions respectively, so it is important to understand how the same properties can be obtained for these types of elements.

Vector elements on simplicial elements are defined in terms of simplex coordinates. Again, these have acquired a variety of names during their development, including Whitney, Nedelec, Bossavit or simply edge-based elements. In its lowest-order form, the element has the following definition:

$$\boldsymbol{w}_{ij} = \lambda_i \nabla \lambda_j - \lambda_j \nabla \lambda_i \tag{10.102}$$

There are three such elements per triangle, or six per tetrahedron, each associated with the edge from node i to node j, as will now be demonstrated.

The Whitney element is the basis for *all* vector simplicial elements, both interpolatory and hierarchal, so its properties are of great importance. Firstly, an obvious question is, why does it have this specific form? To answer this, it is useful to study the right-angled triangle shown in Fig. 10.8, of unit length along the x- and y-axes. (It is also a useful exercise in understanding simplex coordinates.) The simplex coordinates are the ratios as follows:

$$\begin{aligned}\lambda_1 &= \frac{\text{area}_{\triangle P23}}{\text{area}_{\triangle 123}}\\ &= \frac{1/2 \text{ base} \times \text{height}}{1/2}\\ &= y\end{aligned} \tag{10.103}$$

since the area of triangle 123 is 1/2, and the base of triangle P23 is unity and its height is y.

Similarly,

$$\begin{aligned}\lambda_2 &= 1 - (x + y)\\ \lambda_3 &= x\end{aligned} \tag{10.104}$$

The expression for λ_2 is easily derived from the property $\sum_{i=1}^{3} \lambda_i = 1$. Now that we have explicit expressions for the simplex coordinates, their gradients follow trivially:

$$\nabla\lambda_1 = \hat{y} \tag{10.105}$$

$$\nabla\lambda_2 = -\hat{x} - \hat{y} \tag{10.106}$$

$$\nabla\lambda_3 = \hat{x} \tag{10.107}$$

We note that $\nabla\lambda_1$ is normal to edge 1 (that is, the edge opposite node 1), and similarly $\nabla\lambda_2$ and $\nabla\lambda_3$ are normal to edges 2 and 3 respectively.

Now, the Whitney functions can be written in explicit Cartesian form as follows:

$$\begin{aligned} \boldsymbol{N}_1 = \boldsymbol{w}_{23} &= \lambda_2\nabla\lambda_3 - \lambda_3\nabla\lambda_2 \\ &= (1 - x - y)\hat{x} - x(-\hat{x} - \hat{y}) \\ &= (1 - y)\hat{x} + x\hat{y} \\ \boldsymbol{N}_2 = -\boldsymbol{w}_{13} &= -y\hat{x} + x\hat{y} \\ \boldsymbol{N}_3 = \boldsymbol{w}_{12} &= -y\hat{x} + (-1 + x)\hat{y} \end{aligned} \tag{10.108}$$

These are illustrated in Fig. 10.9.

Due to the simple form of these functions on this right-angled parent element, we can immediately establish some of the crucial features of these functions. Let us focus on $\boldsymbol{N}_1 = \boldsymbol{w}_{23}$. Along edges 2 and 3, this function is purely normal, and increases linearly from node 1 to node 2 along edge 3, and similarly from node 1 to node 3 along edge 2. Along edge 1, it has both tangential and normal components. These are easily separated on this right-angled parent element; on edge 1, they are the $\hat{x}$ and $\hat{y}$ components respectively, that is, $(1 - y)|_{y=0} = 1$ and x respectively. Thus, on this edge, the tangential component is constant, and the normal component is linear. In short, $\boldsymbol{N}_1 = \boldsymbol{w}_{23}$ is a basis function with a constant tangential component on edge 1, and linear normal components along all the edges. The same is easily shown for the other two basis functions. Hence, this Whitney element has the same mixed-order CT/LN behavior as the rectangular element studied earlier. Furthermore, suitable degrees of freedom are again the average tangential fields along each edge. It is also immediately obvious from Eq. (10.108) that the divergence of the Whitney functions is zero.

An important note: although we have established these properties on a right-angled parent element, they are *generally* true for Whitney elements on *any* triangle; we will not however show this now. (Some further discussion on the Whitney element may be found in Appendix A.)

Another important point: what of the normal field components? The boundary condition in this case is normal *flux* continuity; it turns out that this is a natural boundary condition of the variational process, and hence is automatically satisfied at material interfaces [3, Section 5.8.3].

It is an interesting question to ask *why* this function might originally have been proposed. Firstly, as already noted, the gradient of a simplex coordinate is *constant*, and is directed perpendicular to the edge opposite the relevant node. Hence, using the gradient of the simplex coordinates promises a method to separate normal and tangential

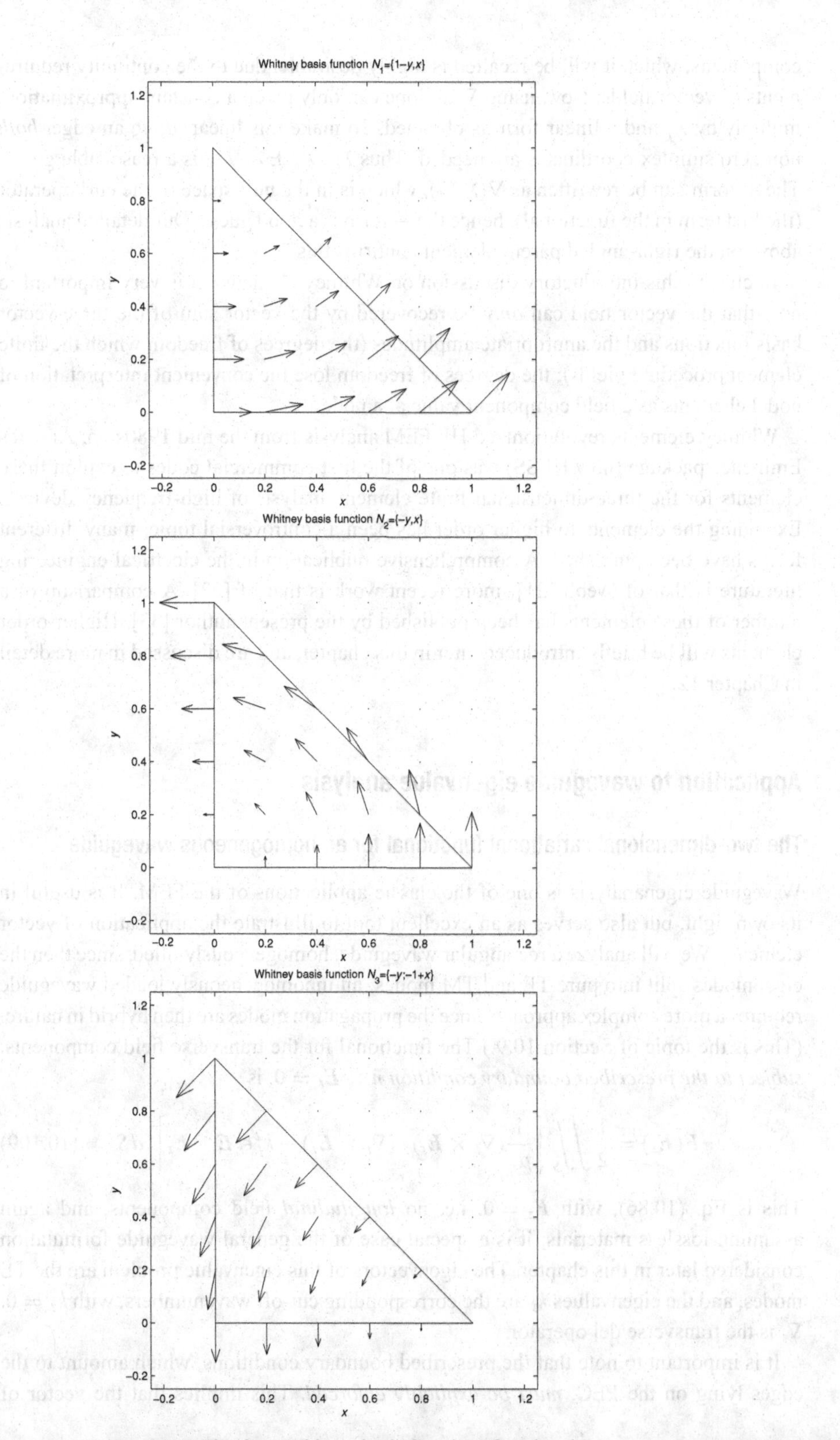

Figure 10.9 The three Whitney basis functions for triangles.

components, which it will be recalled is highly desirable, due to the continuity requirements of vector fields. Now, using $\nabla\lambda_i$ alone can only given a constant approximation; multiply by λ_j and a linear form is obtained. To make this linear *along* an edge, *both* non-zero simplex coordinates are needed. Thus $\lambda_i \nabla\lambda_j \pm \lambda_j \nabla\lambda_i$ is a reasonable guess. The $+$ form can be rewritten as $\nabla(\lambda_i\lambda_j)$, which is in the null-space of the curl operator (the first term in the functional), hence the $-$ form is a good guess. Our detailed analysis above on the right-angled parent element confirms this.

In closing this introductory discussion on Whitney elements, it is very important to note that the vector field can *only* be recovered by the vector sum of the three vector basis functions and the appropriate amplitudes (the degrees of freedom which the finite element procedure yields); the degrees of freedom lose the convenient interpretation of nodal elements as a field component value at a node.

Whitney elements revolutionized HF FEM analysis from the mid 1980s on; Ansoft's Eminence package (now HFSS) was one of the first commercial codes to exploit these elements for the three-dimensional finite element analysis of high-frequency devices. Extending the elements to higher order has been a controversial topic; many different forms have been published. A comprehensive publication in the electrical engineering literature is that of Webb [21]; more recent work is that of [22]. A comparison of a number of these elements has been published by the present author [33]. Higher-order elements will be briefly introduced later in this chapter, and are discussed in more detail in Chapter 12.

10.8 Application to waveguide eigenvalue analysis

10.8.1 The two-dimensional variational functional for an homogeneous waveguide

Waveguide eigenanalysis is one of the classic applications of the FEM. It is useful in its own right, but also serves as an excellent tool to illustrate the application of vector elements. We will analyze a rectangular waveguide, homogeneously filled, since then the eigenmodes split into pure TE and TM modes; an inhomogeneously loaded waveguide requires a more complex approach since the propagation modes are then hybrid in nature. (This is the topic of Section 10.9.) The functional for the transverse field components, *subject to the prescribed boundary condition* $\hat{n} \times \boldsymbol{E}_t = 0$, is

$$F(\boldsymbol{E}_t) = \frac{1}{2}\iint_S \left[\frac{1}{\mu_r}(\nabla_t \times \boldsymbol{E}_t)\cdot(\nabla_t \times \boldsymbol{E}_t) - k_i^2 \epsilon_r \boldsymbol{E}_t \cdot \boldsymbol{E}_t\right] dS \qquad (10.109)$$

This is Eq. (10.86), with $E_z = 0$, i.e. no *longitudinal* field components, and again assuming lossless materials; it is a special case of the general waveguide formulation considered later in this chapter. The eigenvectors of this eigenvalue problem are the TE modes, and the eigenvalues k_i are the corresponding cut-off wavenumbers, with $k_z = 0$. ∇_t is the transverse del operator.

It is important to note that the prescribed boundary conditions, which amount to the edges lying on the PEC, *must be explicitly enforced.* This implies that the vector of unknowns, $\{e\}$, in the generalized eigenvalue problem:

$$[S]\{e\} = k^2[T]\{e\} \tag{10.110}$$

would appear to include *prescribed*, i.e. *zero*, values. This is incorrect. It may be shown that this equation includes *only* contributions from the free edges, i.e.

$$[S_{ff}]\{e_f\} = k^2[T_{ff}]\{e_f\} \tag{10.111}$$

To derive this, write the discretized functional *before it is rendered stationary* as

$$F\{e\} = \{e_f e_p\}^T \begin{bmatrix} S_{ff} S_{fp} \\ S_{pf} S_{pp} \end{bmatrix} \{e_f e_p\} \tag{10.112}$$

Now, differentiating with respect to the free edges, and then applying the prescribed boundary condition $\{e_p\} \equiv 0$, one obtains Eq. (10.111).

This does of course require (globally) numbering the free edges first, and then the prescribed edges. If using a connection matrix approach, another renumbering matrix could be used afterwards to implement this. Alternatively, during matrix assembly, any entries corresponding to prescribed edges can simply be removed from the system.

If the TM modes are sought, then Eq. (10.109) must be solved with $\boldsymbol{H}_t$ as the working variable. In this case, homogeneous Neumann boundary conditions are appropriate – i.e. *no* explicit boundary conditions need be set at all.

This problem is especially easy to solve using rectangular elements, but since we would like to illustrate the application of the Whitney elements, we will use triangular elements. Firstly, we will need a mesh of such elements, but we will defer consideration of this until later, and concentrate initially on the theoretical analysis. For each element, we need the elemental (that is, stiffness $[S]$ and mass $[T]$) matrix elements. Using simplex coordinates, we can evaluate these quite easily.

10.8.2 Explicit formula for the elemental matrix entries

Before deriving expressions for the elemental matrices, it is worth briefly reviewing the two approaches which have been used. The approach we will use is essentially a direct approach, where we evaluate the simplex coordinates in terms of the Cartesian coordinates of the actual element. The other approach uses the right-angled parent element of Fig. 10.8, and computes the matrices for this element; a coordinate transformation is then performed to the actual element, and the inverse of the Jacobian of this transformation is used to scale the matrix elements. The former approach is that of Lee and Mittra, who published some of the first explicit formulas in [25] for tetrahedral CT/LN elements (these formulas were extended by the present author to diagonally anisotropic materials in [39]), Savage and Peterson, who presented a very useful alternative formulation in [2], and Jin [3]. The latter approach is best exemplified by [12]. Savage and Peterson's approach leads to particularly compact expressions, and is the one we will use here. The following is based on their work, but simplified to triangles, using notation consistent with that of this chapter, and using the standard Whitney elements. (Savage and Peterson further scale the elements by the edge lengths.)

Recall that the variational formulation requires the evaluation of two matrices:

$$S_{ij} = \iint_S \nabla_t \times \boldsymbol{N}_i \cdot \nabla_t \times \boldsymbol{N}_j \, dS \tag{10.113}$$

and

$$T_{ij} = \iint_S \boldsymbol{N}_i \cdot \boldsymbol{N}_j \, dS \tag{10.114}$$

With z the direction of propagation, the $\nabla_t \times$ and ∇_t operators reduce to the two-dimensional operators in the xy-plane, which we will imply in the following.

The CT/LN elements are given by $\boldsymbol{N}_i = \boldsymbol{w}_{i1,i2} = \lambda_{i1}\nabla\lambda_{i2} - \lambda_{i2}\nabla\lambda_{i1}$ per edge. Here, $i1$ and $i2$ are the endpoints of edge i. The local triangular numbering scheme is as already discussed.

Now, the three simplex coordinates λ_i are given by

$$\lambda_i = a_i + b_i x + c_i y \tag{10.115}$$

and the gradient thereof by

$$\nabla\lambda_i = b_i \hat{x} + c_i \hat{y} \tag{10.116}$$

The actual coefficients $\{a_i; b_i; c_i\}$ may be computed by inverting the coordinate matrix

$$\begin{bmatrix} b_1 & c_1 & a_1 \\ b_2 & c_2 & a_2 \\ b_3 & c_3 & a_3 \end{bmatrix} = \begin{bmatrix} x_1 & x_2 & x_3 \\ y_1 & y_2 & y_3 \\ 1 & 1 & 1 \end{bmatrix}^{-1} \tag{10.117}$$

This equation may be obtained by writing Eq. (10.115) for each node i and noting that $\lambda_i = 1$ at node i. Now the following two vectors are defined for nodes i and j:

$$\begin{aligned} \boldsymbol{v}_{ij} &= \nabla\lambda_i \times \nabla\lambda_j \\ &= (b_i c_j - b_j c_i)\hat{z} \\ &= -\boldsymbol{v}_{ji} \end{aligned} \tag{10.118}$$

This vector is easily computed once $\{b_i; c_i\}$ are known. Similarly we define

$$\begin{aligned} \phi_{ij} &= \nabla\lambda_i \cdot \nabla\lambda_j \\ &= b_i b_j + c_i c_j \end{aligned} \tag{10.119}$$

Note that both $\boldsymbol{v}_{ij}$ and ϕ_{ij} are constant within a triangle, and hence may be taken outside integrals in which they appear.

Consider the evaluation of the curl-curl term, Eq. (10.113):

$$\begin{aligned} \nabla \times \boldsymbol{N}_i &= \nabla \times (\lambda_{i1}\nabla\lambda_{i2} - \lambda_{i1}\nabla\lambda_{i2}) \\ &= \nabla \times (\lambda_{i1}\nabla\lambda_{i2}) - \nabla \times (\lambda_{i2}\nabla\lambda_{i1}) \\ &= 2\nabla\lambda_{i1} \times \nabla\lambda_{i2} \\ &= 2\boldsymbol{v}_{i1,i2} \end{aligned} \tag{10.120}$$

From the second to third line in the above, the vector identities $\nabla \times (\phi \boldsymbol{A}) = \phi \nabla \times \boldsymbol{A} + \nabla \phi \times \boldsymbol{A}$ and $\nabla \times \nabla \phi \equiv 0$ have been used.

Using this, Eq. (10.113) becomes:

$$\begin{aligned} S_{ij} &= 4 \iint_S \boldsymbol{v}_{i1,12} \cdot \boldsymbol{v}_{j1,j2} \, dS \\ &= 4A \boldsymbol{v}_{i1,12} \cdot \boldsymbol{v}_{j1,j2} \end{aligned} \tag{10.121}$$

Note that the widely used expression for element area in terms of the determinant of the coordinate matrix,

$$2A' = \begin{vmatrix} 1 & x_1 & y_1 \\ 1 & x_2 & y_2 \\ 1 & x_3 & y_3 \end{vmatrix} \tag{10.122}$$

actually yields a potentially signed area A', whose sign depends on the sense (clockwise or anticlockwise) of the coordinate numbering. A in the above is the *unsigned* area of the element, that is $A = |A'|$.

The second term that appears in Eq. (10.114) requires the computation of dot products:

$$\begin{aligned} \boldsymbol{N}_i \cdot \boldsymbol{N}_j &= (\lambda_{i1} \nabla \lambda_{i2} - \lambda_{i2} \nabla \lambda_{i1}) \cdot (\lambda_{j1} \nabla \lambda_{j2} - \lambda_{j2} \nabla \lambda_{j1}) \\ &= [\lambda_{i1} \lambda_{j1} (\nabla \lambda_{i2} \cdot \nabla \lambda_{j2}) - \lambda_{i1} \lambda_{j2} (\nabla \lambda_{i2} \cdot \nabla \lambda_{j1}) \\ &\quad - \lambda_{i2} \lambda_{j1} (\nabla \lambda_{i1} \cdot \nabla \lambda_{j2}) + \lambda_{i2} \lambda_{j2} (\nabla \lambda_{i1} \cdot \nabla \lambda_{j1})] \end{aligned} \tag{10.123}$$

Using the notation of Eq. (10.119), this can be written as

$$\boldsymbol{N}_i \cdot \boldsymbol{N}_j = [\lambda_{i1} \lambda_{j1} \phi_{i2,j2} - \lambda_{i1} \lambda_{j2} \phi_{i2,j1} - \lambda_{i2} \lambda_{j1} \phi_{i1,j2} + \lambda_{i2} \lambda_{j2} \phi_{i1,j1}] \tag{10.124}$$

Thus the associated matrix elements become:

$$\begin{aligned} T_{ij} &= \phi_{i2,j2} \iint_S \lambda_{i1} \lambda_{j1} \, dS - \phi_{i2,j1} \iint_S \lambda_{i1} \lambda_{j2} \, dS \\ &\quad - \phi_{i1,j2} \iint_S \lambda_{i2} \lambda_{j1} \, dS + \phi_{i1,j1} \iint_S \lambda_{i2} \lambda_{j2} \, dS \end{aligned} \tag{10.125}$$

Using the general integration formula for integrals in simplex coordinates (see Appendix D):

$$\iint_S \lambda_1^i \lambda_2^j \lambda_3^k \, dS = \frac{2! \, i! \, j! \, k!}{(2 + i + j + k)!} A \tag{10.126}$$

the expression for T_{ij} may be simplified (note that $0! \equiv 1$). In Eq. (10.125), each integral involves integration over two simplex coordinates, possibly identical. These can be expressed in matrix form as

$$M_{ij} = \iint_S \lambda_i \lambda_j \, dS = \frac{1}{12} \begin{bmatrix} 2 & 1 & 1 \\ 1 & 2 & 1 \\ 1 & 1 & 2 \end{bmatrix} \tag{10.127}$$

Using this, Eq. (10.125) reduces to

$$T_{ij} = A[\phi_{i2,j2} M_{i1,j1} - \phi_{i2,j1} M_{i1,j2} - \phi_{i1,j2} M_{i2,j1} + \phi_{i1,j1} M_{i2,j2}] \tag{10.128}$$

10.8.3 Coding

We now have all the theory we need. However, finite element codes require a lot of "house-keeping" – the unstructured nature of finite element meshes is both their strong point (permitting very accurate local geometrical modelling) and a significant complication (since a lot of lists need to be generated and then maintained). We will now discuss a number of these issues.

Edge and node numbering schemes

With an FEM code, adopting sensible local and global numbering conventions *and then using these consistently* is absolutely essential. The local edge numbering scheme we discussed earlier (whereby the edge number corresponds to the node opposite) is not widely used in practice. The following is the most widely used in the literature:

Edge	Local edge number
e_{12}	1
e_{13}	2
e_{23}	3

In the above, e_{ij} is the edge directed from node i to node j. It is important to note that although the degree of freedom associated with the edge is a scalar, it is nonetheless *signed*.

A convention that can be recommended is *first* to sort the nodes in each element into ascending global order. This ensures that when edges are assigned, they are always directed from lower to higher node numbers, and thus the edges shared by two or more elements always have the same sign. All the local edge numbering schemes in use in the literature are consistent [2, 25] (taking into account that some number from 0 and some from 1). (Note that the *sign* of the edges is not, however, consistent: for example, edge 3 in [3, Fig. 8.2] has the opposite sense to that above.)

Global edge numbers are assigned from 1 upwards; within an element, global edges are incremented in the same pattern as the local edges. To illustrate this by example, element $e1$ will always contain edges 1, 2 and 3 (although not necessarily global nodes 1, 2, 3 and 4, of course, since these are assigned by the mesher); if element $e2$ shares its first edge with element $e1$, then its remaining edges will be globally numbered 4 and 5, local edges e_{13}^{e2} and e_{23}^{e2} respectively.

The above sounds more complex than it is, as is often the case with finite element data structures, and becomes clear when coding.

Data structures

Before programming starts, it is useful to establish the major data structures that will be needed. For a mesh with N_n nodes, N_e elements and E edges, the major data structures required will include at least the following:

vertices Dimensioned as $(N_n, 2)$. This stores the (x, y) coordinates of each vertex (node).

elements (or *nodes*) Dimensioned as $(N_e, 3)$. This stores the three nodes associated with each triangular element.

edge_nodes Dimensioned as $(E, 2)$. This stores the global nodes that each edge connects.

materials Dimensioned as (N_e). This stores the material number. Another (usually very much smaller) data structure will be required to store the actual constitutive parameters for each material.

dof Dimensioned as E for Whitney elements. These are the degrees of freedom.

Two major data structures omitted here (deliberately) are the $[S]$ and $[T]$ matrices for the system. For initial work, these can simply be stored as full matrices, but to exploit fully the power of the FEM, sparse storage schemes must be used. This is discussed in Chapter 12.

The above data structures are accessed so frequently that they should be globally accessible. In MATLAB, this is done using the `global` statement. In FORTRAN 90, one uses modules.

Meshing

For the beginner, this often seems the most challenging task. For two-dimensional problems, however, one can build quite satisfactory meshes by hand, or using publicly available software such as `gmsh`. The easiest way of generating triangular meshes for a rectangular domain is first to divide the domain into smaller rectangles, and then to split each of these further into two triangles. An example of such a mesh is shown in Fig. 10.10. (Also shown on this plot are global node and element numbers; the manner in which these are assigned is essentially arbitrary, and the finite element code should be able to handle this.) It is also easy to automate this type of meshing procedure. In the context of this, Delaunay triangulation [8, pp. 1133ff] deserves mention, although the algorithm will not be further addressed here: such a triangulation has the largest minimum angles amongst all triangulations of a set of points. This is important, as it may be shown from theoretical considerations that the error in a FEM solution is inversely related to the smallest angle.

Book-keeping

The issue of making the edges has already been discussed. The book-keeping required does not end here, however. One also needs connection information (the equivalent of the connection matrix discussed earlier). For a "regular" triangular mesh such as that of Fig. 10.10, it is clear that an edge can be connected to at most two triangles, but, in general, no such assumption can be made.

Building the interconnectivity data is primarily a problem in list-searching. The simplest method of doing this is for *each* edge, to search through all elements and see whether the edge nodes coincide. This is *not* a good idea for large meshes, since this is

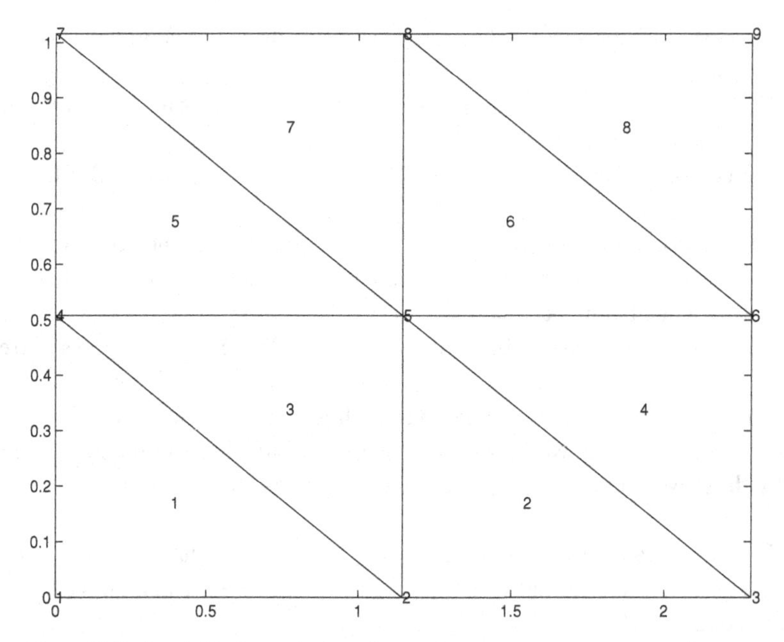

Figure 10.10 An eight-element triangular mesh.

an $\mathcal{O}(EN) \sim \mathcal{O}(N^2)$ operation, but for small meshes it works. Real codes use additional node-element lists to accelerate the search.

One also needs some type of renumbering scheme, so that the free edges may be numbered first. An approach which works is first to flag each edge as free or prescribed. In the present case, simply checking whether the nodal coordinates of the edge coincide with $x = 0$, $x = a$, $y = 0$ or $y = b$ is sufficient, but in general this can also be quite a complex search. Once this has been done, an index list is then built which gives the original global edge number for each degree of freedom. Again, this sounds more complex than it actually is. With these data, and with the convention that shared edges have the same sign, matrix assembly proceeds very quickly.

Solving the eigenvalue problem

From a mathematical viewpoint, the most complex part of the finite element analysis (and certainly the most computationally expensive) is actually the solution of the generalized eigenvalue problem represented by Eq. (10.90), repeated here:

$$[S]\{e_i\} = k_i^2[T]\{e_i\}$$

Fortunately, modern scientific programming environments such as MATLAB make this very simple; for instance, in MATLAB, the function `eig` solves this with one command, and the function `eigs` is available for some sparse matrix eigenproblems. (Similar routines are available in LAPACK, if using FORTRAN 90 or C, although calling them requires a little more work.) What emerges from the analysis is a set of eigenvalues, each with its associated eigenvectors.

As should be anticipated from our earlier discussion, this vector element FEA includes static modes. (This very important point is often not mentioned explicitly, and causes novices no end of problems.) Interestingly, it is possible to *predict* the number of such modes. The idea is the following. For the Whitney element, the curl of the field is represented by a constant. For the null-space of the eigenvalue problem, where the field can be represented by a potential, this potential function must thus be linear. The obvious approximation of a linear potential using nodal elements would require one degree of freedom per unconstrained (free) node. One of the solutions is actually the trivial solution $\boldsymbol{E} = 0$ (corresponding to a constant potential) and must be discounted (since it is also a valid, albeit trivial, solution of the dynamic problem) and thus the dimension of the null-space, K, is the number of unconstrained nodes minus one for Whitney elements. Hence K can be very large. In the two-dimensional case, the ratio of edges to nodes tends to around three, so almost one-third of computed eigenvalues are actually null-space ones.

In practice, the trivial solution is also irrelevant. So, once the eigenvalue problem has been solved, we must first sort the computed eigenvalues into ascending order, then count the number of free *nodes*, i.e. $K + 1$, and then finally, eigenvalue $K + 2$ is the first eigenvalue of interest. (Again, this type of operation is very easily implemented in MATLAB, using the `sort` function.)

Post-processing

Once the finite element analysis is complete, the vector degrees of freedom need to be post processed to yield meaningful field data. As has been commented previously, unlike interpolatory nodal-based elements, where a degree of freedom typically represents a field component at a particular node, hierarchal vector elements only reconstruct a physically meaningful field when summed together. In this case, the eigenvector corresponding to a particular eigenvalue does not in itself directly represent a field. Given the degrees of freedom and the corresponding basis functions, the field $\boldsymbol{E}(x, y)$ can be computed at any point within the element.

For this, one needs to compute directly the sum of the Whitney elements within each element, that is:

$$\boldsymbol{E}^e(x, y) = E_{12}^e \boldsymbol{w}_{12} + E_{13}^e \boldsymbol{w}_{13} + E_{23}^e \boldsymbol{w}_{23} \tag{10.129}$$

with E_{ij}^e the degrees of freedom and w_{ij} the basis functions. (Here, it is worthwhile pointing out that some authors include the appropriate edge lengths in the basis function, e.g. $w_{ij} = \ell_{ij}(\lambda_i \nabla\lambda_j - \lambda_j \nabla\lambda_i$.) The reason this is sometimes done is that the degree of freedom is then the tangential field at each edge. In this case, the $[S]$ and $[T]$ matrix entries are scaled appropriately [2], and the basis functions in Eq. (10.129) must of course *also* include the edge length. This is obvious, but easy to overlook, since the lengths are often implied but not consistently retained in some of the literature.)

All the theory needed for this has already been presented. The simplex coordinates for point (x, y) are computed from its basic definition as in Eq. (10.80), expanded here

Table 10.3 First eight transverse electric modes in a standard X-band waveguide, giving the cut-off wavenumber and frequency

Mode	k_c (rad/m)	f_c (GHz)
TE_{10}	137.43	6.5573
TE_{20}	274.86	13.1146
TE_{01}	309.21	14.7539
TE_{11}	338.38	16.1455
TE_{30}	412.28	19.6719
TE_{21}	413.71	19.7401
TE_{31}	515.35	24.5899
TE_{40}	549.71	26.2292

for all three coordinates:

$$
\lambda_1 = \frac{\begin{vmatrix} 1 & x & y \\ 1 & x_2 & y_2 \\ 1 & x_3 & y_3 \end{vmatrix}}{2A}
$$
$$
\lambda_2 = \frac{\begin{vmatrix} 1 & x_1 & y_1 \\ 1 & x & y \\ 1 & x_3 & y_3 \end{vmatrix}}{2A}
$$
$$
\lambda_3 = \frac{\begin{vmatrix} 1 & x_1 & y_1 \\ 1 & x_2 & y_2 \\ 1 & x & y \end{vmatrix}}{2A} \tag{10.130}
$$

The gradients are computed as in Section 10.8.2, using specifically Eq. (10.116).

10.8.4 Results

The eigenvalues can be put into one-to-one correspondence with the analytically known eigenvalues. For a standard X-band guide, with $a = 22.86$ mm and $b = 10.16$ mm, the first eight TE eigenmodes are listed in Table 10.3.

The relative error of the eigenvalues computed with the FEM compared to the analytical results is shown in Fig. 10.11. Clearly, refining the mesh has the desired result of decreasing the error. Individually, the eigenmodes display different convergence with, for instance, the seventh eigenmode (TE_{31}) being accurately computed by even the very coarse 16-element mesh. This behavior has been observed in many implementations, and what is usually studied is an average error. In Fig. 10.12, the result for the average

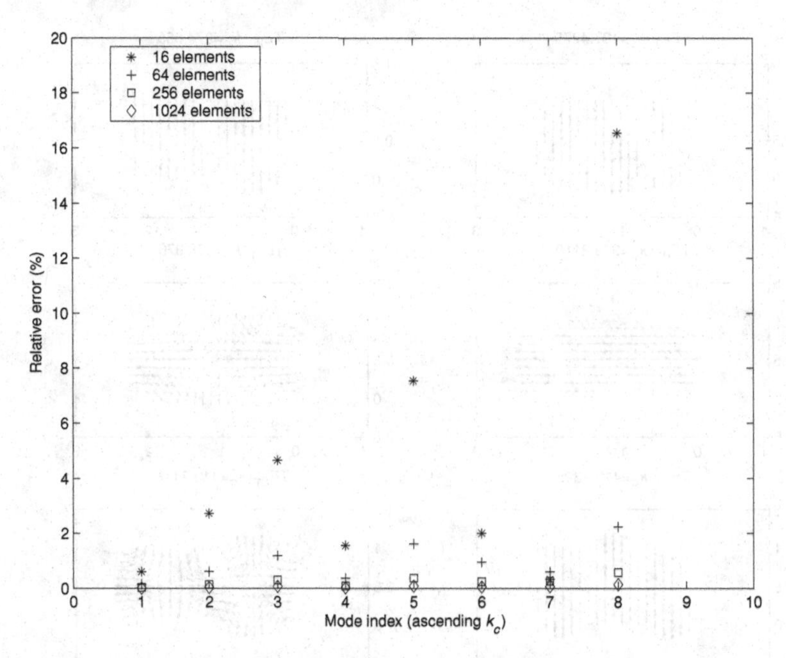

Figure 10.11 The relative error in the first eight eigenmodes.

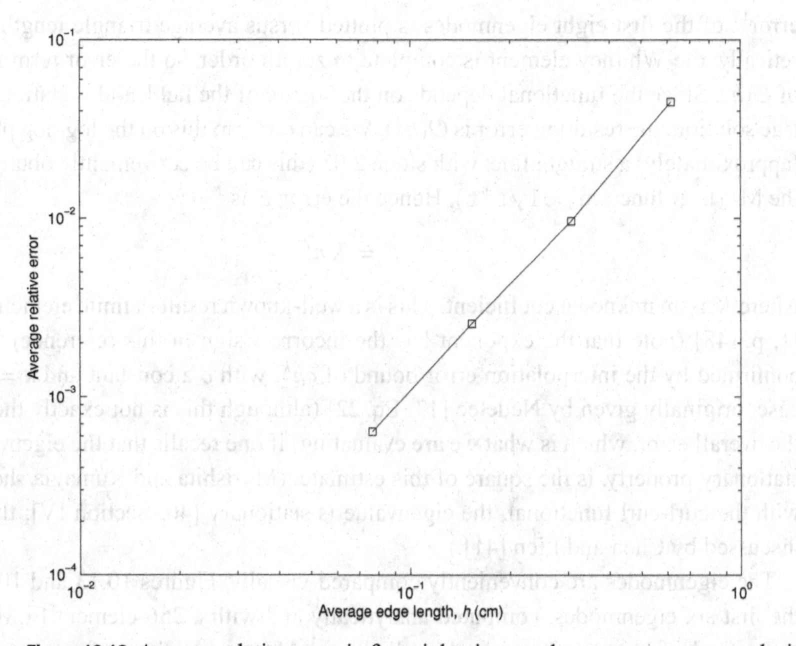

Figure 10.12 Average relative error in first eight eigenmodes versus average mesh size h (cm).

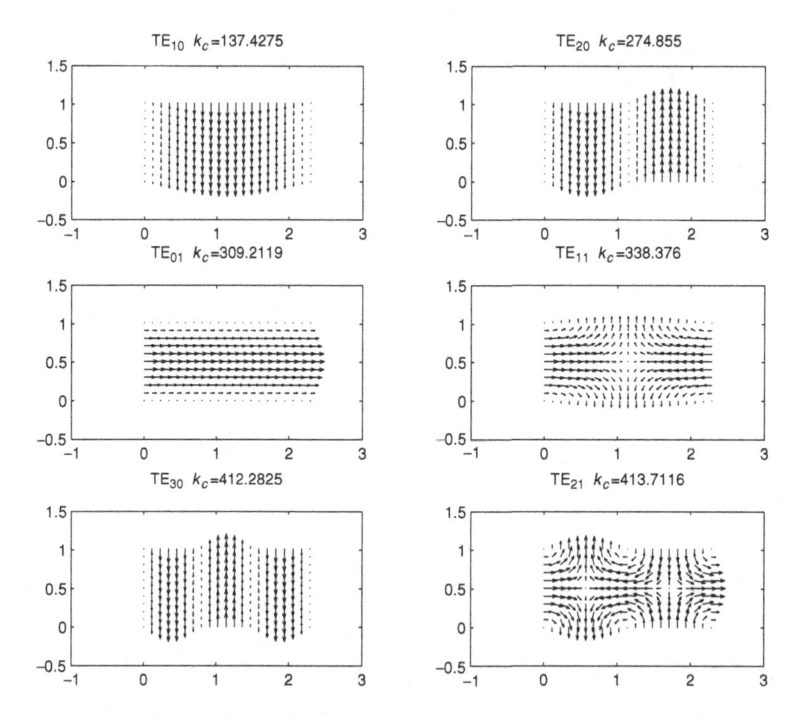

Figure 10.13 Quiver plot of the first six eigenmodes, computed analytically.

error[13] of the first eight eigenmodes is plotted versus average triangle length h. Theoretically, the Whitney element is complete to zeroth order, so the error term should be of $\mathcal{O}(h)$. Since the functional depends on the square of the field, and is stationary at the true solution, the resulting error is $\mathcal{O}(h^2)$. We can confirm this on the log–log plot; this is (approximately) a straight line, with slope 2.03 (this can be conveniently obtained using the MATLAB function `polyfit`). Hence the error E is

$$E = Kh^2 \tag{10.131}$$

where K is an unknown coefficient. This is a well-known result in finite element analysis [1, p. 148] (note that the exponent has the incorrect sign in this reference). It is also confirmed by the interpolation error bound of ch^k, with c a constant and $k = 1$ in this case, originally given by Nedelec [13, Eq. 22] (although this is not exactly the same as the overall error, which is what we are evaluating) if one recalls that the eigenvalue, as a stationary property, is the square of this estimate. (Morishita and Kumagia showed that with the curl-curl functional, the eigenvalue is stationary [40, Section IV]; this is also discussed by Chen and Lien [41].)

The eigenmodes are conveniently compared visually. Figures 10.13 and 10.14 show the first six eigenmodes, computed analytically and with a 256-element FEM solution respectively. These results were plotted with the MATLAB `quiver` function. Note that

[13] The corresponding figure in the first edition, Fig. 9.7, had a small error in the computation of the error which has been corrected here.

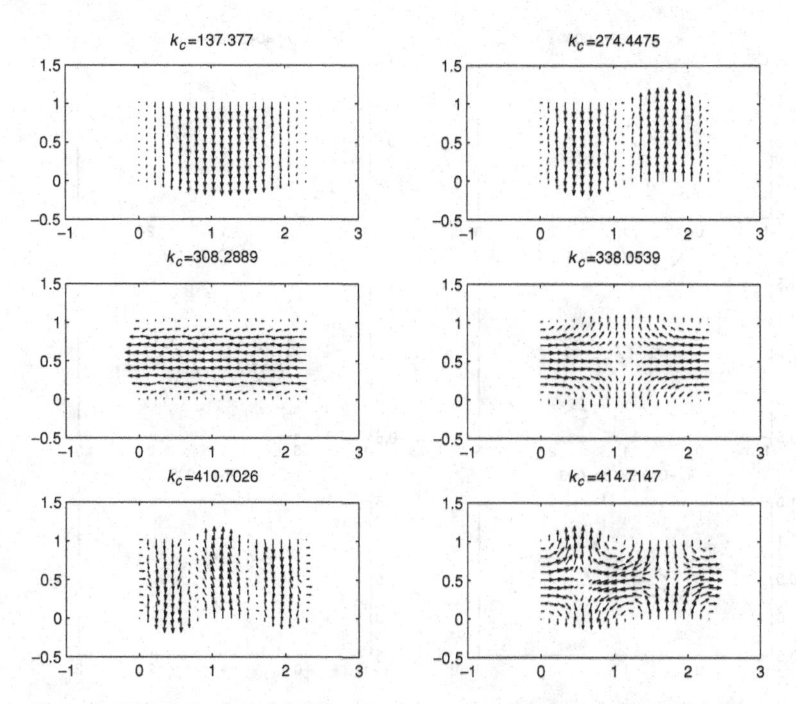

Figure 10.14 Quiver plot of the first six eigenmodes, computed with a 256-element FEM solution.

the sign of the eigenmode is essentially arbitrary; for instance, the TE_{01} eigenmode has been computed with opposite sign by the analytical and finite element methods. Also, for interest, the first six null-space eigenmodes are shown in Fig. 10.15. Some wavenumbers appear to be complex; this is simply due to taking the square root of numbers approximating zero, but slightly negative. There are 105 such eigenvalues and associated eigenmodes, in a problem with 360 degrees of freedom. One notes that, in general, these modes satisfy the boundary condition of zero tangential field, but cannot of course be recognized as traditional TE modes.

10.8.5 Degenerate modes

A *degenerate mode* occurs when there are multiple eigenvalues with the same numerical value, corresponding to different modal distributions across the waveguide cross-section. In square waveguide ($a = b$) all modes are degenerate, as a 90° rotation leaves the geometry unchanged. The finite element analysis of this section correctly resolves such modes, but not always with the modal patterns as expected. For instance, when the first two modes are computed in square waveguide, the modes which the FEM computes are as shown in Fig. 10.16; when summed and differenced, the conventional TE_{10} and TE_{01} mode patterns emerge, as shown in Fig. 10.17. As a linear combination of degenerate eigenvectors is clearly also an eigenvector, this is not unexpected, but can appear anomalous when first encountered.

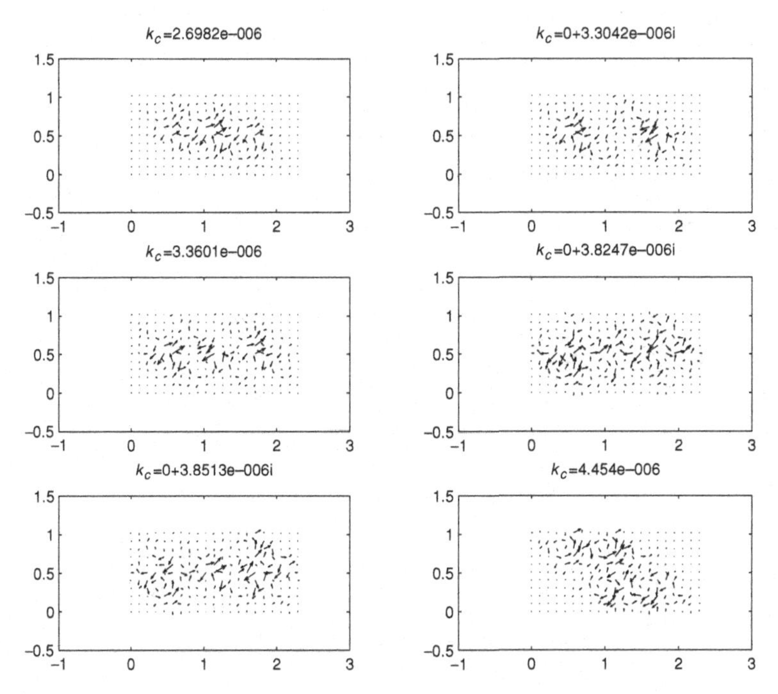

Figure 10.15 Quiver plot of the first six null-space eigenmodes, computed with a 256-element FEM solution.

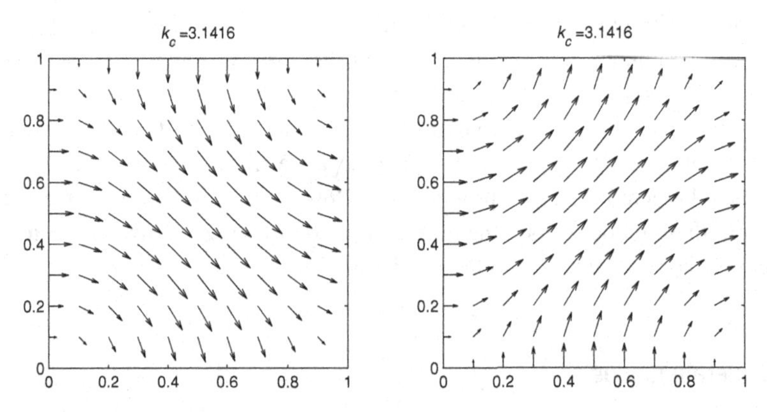

Figure 10.16 Quiver plot of the first two eigenmodes, computed with a 128-element FEM solution using LT/QN elements.

Similarly, in circular waveguide, the first two TE modes (TE_{11} and TE_{31}) are both degenerate, with multiplicity two. The TE_{11} eigenmodes are shifted 90° with respect to each other, corresponding to the solutions depending on $\cos\phi$ and $\sin\phi$ respectively. (They are, however, immediately recognizable as TE_{11} eigenmodes.) The TE_{31} eigenmodes are shifted 45° with respect to each other, corresponding to $\cos 2\phi$ and $\sin 2\phi$. However, the fifth eigenmode, TE_{01}, is ϕ-invariant and, indeed, only one appears in the FEM solution.

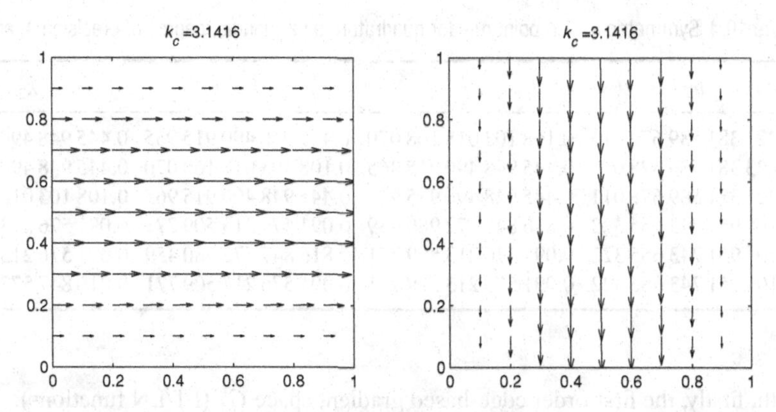

Figure 10.17 Quiver plot of the sum and difference of the first two eigenmodes.

10.8.6 Higher-order vector elements

As with nodal elements, increasing the polynomial order of the basis functions can greatly increase the rate of convergence of a FEM solution. Unlike nodal elements, where the use of the Silvester polynomials (based on Lagrangian interpolatory polynomials) yields unique higher-order elements,[14] with vector elements a large number of different higher-order elements have been proposed. This topic will be discussed in more depth in the next chapter but, for now, we will use the hierarchal basis functions proposed by Webb [21]. The idea is to first add another three basis functions to make a complete first-order element and then add further elements to make an incomplete second-order one; it turns out that a further two functions are needed for this. (This is not an obvious result; it was originally derived by Nedelec [13].) These elements are also widely known as linear tangential/linear normal (LT/LN) and linear tangential/quadratic normal (LT/QN) respectively. As with the Whitney elements, the additional three LT/LN basis functions each have tangential components only along one edge, so the degrees of freedom associated with them are also edge-based, and are shared between elements in the same way. The LT/QN functions have no tangential projections on the edges, so are *not* shared between triangular elements – but are of course on the faces connecting three-dimensional tetrahedra.

Mathematically, to produce a mixed second-order element, we enrich the space of edge-based CT/LN Whitney functions W_1^e:

$$
\begin{aligned}
\boldsymbol{W}_1^{(e1)} &= \lambda_1 \nabla \lambda_2 - \lambda_2 \nabla \lambda_1 \\
\boldsymbol{W}_1^{(e2)} &= \lambda_1 \nabla \lambda_3 - \lambda_3 \nabla \lambda_1 \\
\boldsymbol{W}_1^{(e3)} &= \lambda_2 \nabla \lambda_3 - \lambda_3 \nabla \lambda_2
\end{aligned}
\tag{10.132}
$$

[14] There are other approaches to nodal elements, such as Hermitian polynomials, which can provide continuity of both the function and its derivatives, but the Silvester polynomial approach is the most widely use in electromagnetics.

Table 10.4 Symmetric $n = 6$ point rule for quadrature on a triangle, degree of precision 4, after [42]

w_i	λ_{1i}	λ_{2i}	λ_{3i}
0.223 381 589 678 011	0.108 103 018 168 070	0.445 948 490 915 965	0.445 948 490 915 965
0.223 381 589 678 011	0.445 948 490 915 965	0.108 103 018 168 070	0.445 948 490 915 965
0.223 381 589 678 011	0.445 948 490 915 965	0.445 948 490 915 965	0.108 103 018 168 070
0.109 951 743 655 322	0.816 847 572 980 459	0.091 576 213 509 771	0.091 576 213 509 771
0.109 951 743 655 322	0.091 576 213 509 771	0.816 847 572 980 459	0.091 576 213 509 771
0.109 951 743 655 322	0.091 576 213 509 771	0.091 576 213 509 771	0.816 847 572 980 459

with, firstly, the first-order edge-based gradient space G_1^e (LT/LN functions):

$$\begin{aligned} \boldsymbol{G}_1^{(e1)} &= \nabla(\lambda_1\lambda_2) = \lambda_1\nabla\lambda_2 + \lambda_2\nabla\lambda_1 \\ \boldsymbol{G}_1^{(e2)} &= \nabla(\lambda_1\lambda_3) = \lambda_1\nabla\lambda_3 + \lambda_3\nabla\lambda_1 \\ \boldsymbol{G}_1^{(e3)} &= \nabla(\lambda_2\lambda_3) = \lambda_2\nabla\lambda_3 + \lambda_3\nabla\lambda_2 \end{aligned} \tag{10.133}$$

and, secondly, the second-order face-based rotational space R_2^f (LT/QN functions):

$$\begin{aligned} \boldsymbol{R}_2^{(f1)} &= \lambda_2\lambda_3\nabla\lambda_1 + \lambda_1\lambda_3\nabla\lambda_2 - 2\lambda_1\lambda_2\nabla\lambda_3 \\ \boldsymbol{R}_2^{(f2)} &= \lambda_3\lambda_1\nabla\lambda_2 + \lambda_2\lambda_1\nabla\lambda_3 - 2\lambda_2\lambda_3\nabla\lambda_1 \end{aligned} \tag{10.134}$$

The properties of these basis functions have been outlined above, and will be discussed in more detail in the next chapter (see Section 11.2).

In terms of coding higher-order functions, the traditional school of thought has been to evaluate the elemental matrices in closed form, using formula similar to those in Section 10.8.2. However, the time spent filling the matrix entries (which is of $\mathcal{O}(N)$, i.e. linear in the number of elements) is usually not the dominant part of the overall FEM solution process. Especially when developing experimental codes, where speed is not an over-riding priority, evaluating the matrix entries using quadrature (numerical integration) is an attractive option. Since the integrands are polynomials, appropriate integration rules can perform the integration exactly (obviously given the constraints of finite precision arithmetic, but the same consideration also applies to closed-form expressions). It is perhaps surprising how few quadrature points are required – a fourth-order polynomial can be exactly integrated on a triangle with an only six-point rule. As the integrand in the mass matrix of an LT/QN element is at most of fourth order (as the product of two second-order basis functions), and the stiffness matrix integrand is at most of second order, this rule is sufficient to exactly integrate an LT/QN element. An appropriate rule is given in Table 10.4; the quadrature formula for a triangle of unit area is

$$\int f(\lambda_1, \lambda_2, \lambda_3) d\Omega = \sum_{i=1}^{n} w_i f(\lambda_{1i}, \lambda_{2i}, \lambda_{3i}). \tag{10.135}$$

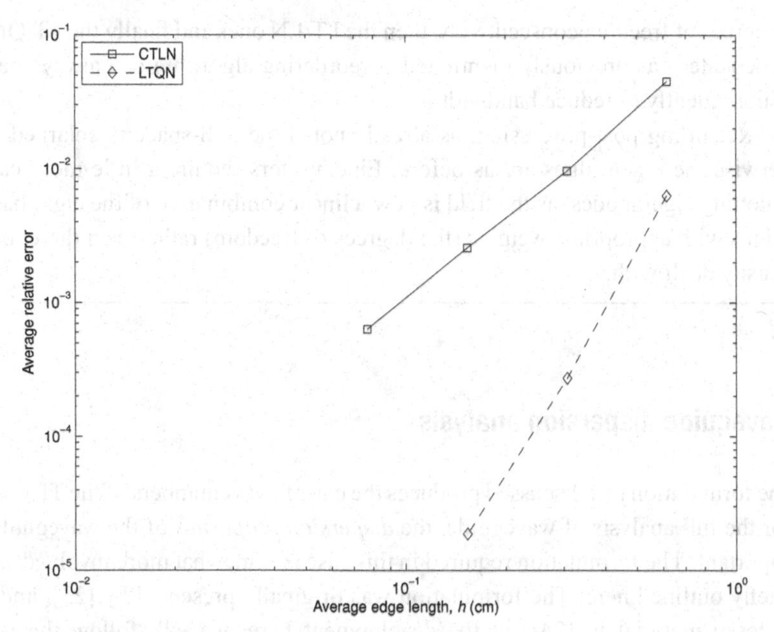

Figure 10.18 Average relative error in first eight eigenmodes versus average mesh size h (cm) for CT/LN and LT/QN elements.

The curl of the vector basis functions should be pre-computed analytically when using quadrature. As the elements are hierarchal, the curl of the CT/LN functions ($\boldsymbol{W}_1^{(ei)}$) has as already computed. The LT/LN functions ($\boldsymbol{G}_1^{(e1)}$) are in the null-space of the curl operator, and have zero curl. The curl of the LT/QN functions are given in Eq. (11.27) in the next chapter.

Results are shown in Fig. 10.18. Convergence rates are 4.20 for the LTQN elements, vs. 2.03 for the CTLN elements respectively, demonstrating the expected doubling of convergence rates. Comments similar to the higher-order nodal elements of the previous chapter, regarding error versus degrees of freedom rather than h, pertain here as well.

Coding hints – data structures and related issues for higher-order elements

In particular with two-dimensional analysis, higher-order elements require little in the way of new data structures. Clearly, if an edge does not lie on a prescribed boundary, then both the CT/LN degree of freedom and the LT/LN degree of freedom are free to vary. The two LT/QN degrees of freedom, which are "face-based," are always free to vary in this type of analysis. So, once the CT/LN degrees of freedom have been flagged as either free or prescribed, the LT/LN ones are treated identically, and the issue does not arise at all with those of LT/QN order, which are always free.

In terms of numbering the degrees of freedom, from the viewpoint of keeping matrix bandwidth to a minimum, the degrees of freedom would best be numbered on an element-by-element basis, but for simple codes numbering all the CT/LN degrees of freedom consecutively, then the LT/LN ones and finally the LT/QN ones is adequate – as previously mentioned, a reordering algorithm can always be applied subsequently to reduce bandwidth.

Regarding post-processing, as already noted the null-space is enlarged, but otherwise the eigenvalues are as before. Eigenvectors require a little more care when plotting eigenmodes, as the field is now a linear combination of the eight basis functions with appropriate weights (the degrees of freedom) rather than three, but this is easily dealt with.

10.9 Waveguide dispersion analysis

The formulation just discussed produces the cut-off wavenumbers of the TE eigenmodes. For the full analysis of waveguide, the *dispersion properties* of the waveguide are very important. The formulation required in this case is somewhat more involved, and will be briefly outlined here. The formulation was originally presented by [23], and extended to lossy material in [24]. In the development here, we will follow the notation of [3, pp. 288–290].

10.9.1 A vector formulation based on the transverse and axial field components

Consider the vector Helmholtz equation for an inhomogeneously filled guide:

$$\nabla \times \frac{1}{\mu_r} \nabla \times \boldsymbol{E} - k_0^2 \varepsilon_r \boldsymbol{E} = 0 \text{ in } \Omega \tag{10.136}$$

with boundary conditions

$$\hat{n} \times \boldsymbol{E} = 0 \quad \text{on } \Gamma_1 \tag{10.137}$$

$$\hat{n} \times \nabla \times \boldsymbol{E} = 0 \quad \text{on } \Gamma_2 \tag{10.138}$$

Ω is the waveguide cross-section; the boundary comprises electric wall Γ_1 (and magnetic wall Γ_2, if needed, for symmetry). Assuming known z-dependence as $\boldsymbol{E}(x, y, z) = \boldsymbol{E}(x, y)e^{-jk_z z}$, the standard variational functional can be written as

$$F(\boldsymbol{E}) = \frac{1}{2} \iint_S \left[\frac{1}{\mu_r} (\nabla_t \times \boldsymbol{E}_t) \cdot (\nabla_t \times \boldsymbol{E}_t)^* - k_0^2 \epsilon_r \boldsymbol{E} \cdot \boldsymbol{E}^* \right. \\ \left. + \frac{1}{\mu_r} (\nabla_t E_z + jk_z \boldsymbol{E}_t) \cdot (\nabla_t E_z + jk_z \boldsymbol{E}_t)^* \right] d\Omega \tag{10.139}$$

Here, ∇_t denotes the transverse del operator, $\boldsymbol{E}_t$ the transverse component of the field, and E_z is the z-component of the field. This functional can be discretized to yield an eigenvalue system that can be solved for k_0^2 for a given k_z. In practical device design it is usual to specify the operating frequency (i.e. k_0) and then solve for propagation constant k_z. As originally discussed in [23], this can be solved by introducing the following change of variables:

$$e_t = k_z \boldsymbol{E}_t \tag{10.140}$$

$$e_z = -jE_z \tag{10.141}$$

Substituting these in Eq. (10.139) and multiplying it by k_z^2, one obtains

$$F(\boldsymbol{e}) = \frac{1}{2}\iint_S \left\{ \frac{1}{\mu_r}(\nabla_t \times \boldsymbol{e}_t) \cdot (\nabla_t \times \boldsymbol{e}_t)^* - k_o^2 \epsilon_r \boldsymbol{e}_t \cdot \boldsymbol{e}_t^* + k_z^2 \left[\frac{1}{\mu_r}(\nabla_t e_z + \boldsymbol{e}_t) \cdot (\nabla_t e_z + \boldsymbol{e}_t)^* - k_0^2 \epsilon_r e_z e_z^* \right] \right\} d\Omega \tag{10.142}$$

Discretization of this for a given k_0 results in a system with k_z^2 as its eigenvalues.

Discretizing this using the FEM, one uses *vector* (edge) elements for the transverse field $\boldsymbol{e}_t$, and *nodal* elements for the axial field e_z. The discretized functional is

$$F = \frac{1}{2}\sum_{e=1}^{M} \left(\{e_t^e\}^T [A_{tt}^e] \{e_t^e\}^* + k_z^2 \begin{Bmatrix} e_t^e \\ e_z^e \end{Bmatrix}^T \begin{bmatrix} B_{tt}^e & B_{tz}^e \\ B_{zt}^e & B_{zz}^e \end{bmatrix} \begin{Bmatrix} e_t^e \\ e_z^e \end{Bmatrix}^* \right) \tag{10.143}$$

The elemental matrices are given by:

$$[A_{tt}^e] = \iint_{\Omega^e} \left[\frac{1}{\mu_r^e} \{\nabla_t \times \boldsymbol{N}^e\} \cdot \{\nabla_t \times \boldsymbol{N}^e\}^T - k_0^2 \epsilon_r^e \{\boldsymbol{N}^e\} \cdot \{\boldsymbol{N}^e\}^T \right] d\Omega \tag{10.144}$$

$$[B_{tt}^e] = \iint_{\Omega^e} \frac{1}{\mu_r^e} \{\boldsymbol{N}^e\} \cdot \{\boldsymbol{N}^e\}^T d\Omega \tag{10.145}$$

$$[B_{tz}^e] = \iint_{\Omega^e} \frac{1}{\mu_r^e} \{\boldsymbol{N}^e\} \cdot \{\nabla_t N^e\}^T d\Omega \tag{10.146}$$

$$[B_{zt}^e] = \iint_{\Omega^e} \frac{1}{\mu_r^e} \{\nabla_t N^e\}^T \cdot \{\boldsymbol{N}^e\} \, d\Omega \tag{10.147}$$

$$[B_{zz}^e] = \iint_{\Omega^e} \left[\frac{1}{\mu_r^e} \{\nabla_t N^e\} \cdot \{\nabla_t N^e\}^T - k_0^2 \epsilon_r^e \{N^e\} \cdot \{N^e\}^T \right] d\Omega \tag{10.148}$$

Most of these matrices are immediately recognized as combinations of either the stiffness or mass matrices for vector or nodal elements,[15] and the preceding theory in this chapter can be directly applied. The only non-standard term required is one involving the transverse divergence of the axial field component and the transverse vector, and it can be handled in much the same fashion as before, either integrated analytically or using quadrature.

Writing the functional for the entire connected system as before, Eq. (10.143) can be rewritten as

$$F = \frac{1}{2}\{e_t\}^T [A_{tt}] \{e_t\}^* + \frac{1}{2} k_z^2 \begin{Bmatrix} e_t \\ e_z \end{Bmatrix}^T \begin{bmatrix} B_{tt} & B_{tz} \\ B_{zt} & B_{zz} \end{bmatrix} \begin{Bmatrix} e_t \\ e_z \end{Bmatrix}^* \tag{10.149}$$

[15] For this reason, the matrices have been named $[A]$ and $[B]$, respectively, in the notation of [3].

and after rendering the functional stationary, the following generalized eigenvalue problem is obtained:

$$\begin{bmatrix} A_{tt} & 0 \\ 0 & 0 \end{bmatrix} \begin{Bmatrix} e_t \\ e_z \end{Bmatrix} = -k_z^2 \begin{bmatrix} B_{tt} & B_{tz} \\ B_{zt} & B_{zz} \end{bmatrix} \begin{Bmatrix} e_t \\ e_z \end{Bmatrix} \tag{10.150}$$

Subsequent work by Lee generalized this formulation to include lossy materials [24]. The formulation remains essentially unchanged, except that the complex conjugation is removed in Eq. (10.139) (and in the following equations) so that the functional becomes the following:

$$\begin{aligned} F(\boldsymbol{e}) = \frac{1}{2} \iint_S \Bigg\{ & \frac{1}{\mu_r} (\nabla_t \times \boldsymbol{e}_t) \cdot (\nabla_t \times \boldsymbol{e}_t) - k_o^2 \epsilon_r \boldsymbol{e}_t \cdot \boldsymbol{e}_t \\ & + k_z^2 \left[\frac{1}{\mu_r} (\nabla_t e_z + \boldsymbol{e}_t) \cdot (\nabla_t e_z + \boldsymbol{e}_t) - k_0^2 \epsilon_r e_z e_z \right] \Bigg\} d\Omega \end{aligned} \tag{10.151}$$

The loss is incorporated via complex permittivity ϵ_r as follows:

$$\epsilon_r = \epsilon_r'(1 - j \tan \delta) = \epsilon_r' \left(1 - j \frac{\sigma}{\omega \epsilon_0}\right) \tag{10.152}$$

Clearly, the entries of the system matrices become complex, and the eigensystem to be solved is now complex-valued. The eigenvalues are also complex-valued; as formulated here, the real part is the phase constant (k_z or β) and the imaginary part is the attenuation constant α.

In closing this section on the formulation, note that Peterson *et al.* present essentially the same analysis using the *magnetic* field as the working variable in [43, Section 9.11].

10.9.2 The cut-off eigenanalysis formulation

At cut-off, the axial wavenumber k_z reduces to zero. Hybrid modes decouple into either TE or TM modes [3, p. 242] and, as will be seen below, the TE and TM modes can be computed separately, as done in Section 10.8. Revisiting Eq. (10.139), it is apparent that with $k_z = 0$, the functional simplifies to the following (note that the later formulation of [24] is used here, i.e. without the complex conjugation):

$$\begin{aligned} F(\boldsymbol{E}) = \frac{1}{2} \iint_S \Bigg[& \frac{1}{\mu_r} (\nabla_t \times \boldsymbol{E}_t) \cdot (\nabla_t \times \boldsymbol{E}_t) - k_c^2 \epsilon_r \boldsymbol{E}_t \cdot \boldsymbol{E}_t \\ & + \frac{1}{\mu_r} (\nabla_t E_z) \cdot (\nabla_t E_z) - k_c^2 \epsilon_r E_z \cdot E_z \Bigg] d\Omega \end{aligned} \tag{10.153}$$

Note that k_z must be set to zero in the *unscaled* system of Eq. (10.139) rather than the scaled system of Eq. (10.142); the scaling of Eq. (10.141) zeros the transverse fields at cut-off, which is not appropriate here.

Discretizing this using vector elements for the transverse field components $\boldsymbol{E}_t$ and scalar elements for the axial field component E_z, one obtains

$$F = \frac{1}{2} \sum_{e=1}^{M} \left(\begin{Bmatrix} E_t^e \\ E_z^e \end{Bmatrix}^T \begin{bmatrix} S_{tt}^e & 0 \\ 0 & S_{zz}^e \end{bmatrix} \begin{Bmatrix} E_t^e \\ E_z^e \end{Bmatrix} + k_c^2 \begin{Bmatrix} E_t^e \\ E_z^e \end{Bmatrix}^T \begin{bmatrix} S_{zz}^e & 0 \\ 0 & T_{zz}^e \end{bmatrix} \begin{Bmatrix} E_t^e \\ E_z^e \end{Bmatrix} \right) \tag{10.154}$$

Note that E^e is used for the elemental degrees of freedom, rather than e^e, to remind one that there is no scaling in this eigensystem (and similarly E^z instead of e^z). The elemental matrices are given by the standard stiffness and mass matrices for vector and scalar elements respectively:

$$\left[S_{tt}^e\right] = \iint_{\Omega^e} \frac{1}{\mu_r^e}\{\nabla_t \times N^e\} \cdot \{\nabla_t \times N^e\}^T d\Omega \tag{10.155}$$

$$\left[T_{tt}^e\right] = \iint_{\Omega^e} \epsilon_r^e\{N^e\} \cdot \{N^e\}^T \tag{10.156}$$

$$\left[S_{zz}^e\right] = \iint_{\Omega^e} \frac{1}{\mu_r^e}\{\nabla_t N^e\} \cdot \{\nabla_t N^e\}^T d\Omega \tag{10.157}$$

$$\left[T_{zz}^e\right] = \iint_{\Omega^e} \epsilon_r^e\{N^e\} \cdot \{N^e\}^T \tag{10.158}$$

Applying the usual variational procedure, the following generalized eigenvalue problem is obtained:

$$\begin{bmatrix} S_{tt} & 0 \\ 0 & S_{zz} \end{bmatrix} \begin{Bmatrix} E_t \\ E_z \end{Bmatrix} = k_c^2 \begin{bmatrix} T_{tt} & 0 \\ 0 & T_{tt} \end{bmatrix} \begin{Bmatrix} E_t \\ E_z \end{Bmatrix} \tag{10.159}$$

Due to the decoupling between the TE and TM modes, this can equivalently be written as two smaller systems:

$$[S_{tt}]\{E_t\} = k_{c,TE}^2[T_{tt}]\{E_t\} \tag{10.160}$$

$$[S_{zz}]\{E_z\} = k_{c,TM}^2[T_{tt}]\{E_z\} \tag{10.161}$$

These systems may be solved separately. The first system produces as eigenvalues the square of the cut-off frequencies for the TE modes, and as eigenmodes the associated TE modal distributions for the *transverse* field components; the second, respectively, the square of the cut-off frequencies for the TM modes, and the associated TM modal distributions for the *axial* field components. (Some post-processing would be required to extract the transverse field distributions.) Interestingly, if a comprehensive error analysis is undertaken, the TM modes converge at the same rate as the TE modes, but display higher absolute error.

10.9.3 Homogeneously filled guides: TE modes only

For homogeneously filled guides, for the TE modes,[16] the axial field component is zero. This immediately affords considerable simplification in the formulation above, and Eq. (10.150) reduces to

$$[A_{tt}]\{e_t\} = -k_z^2[B_{tt}]\{e_t\} \tag{10.162}$$

[16] Such guides of course support TM modes as well; this can be computed using the transverse-field only formulation, but using the *magnetic* fields as working variable. Alternately, to compute both TE and TM modes, the full formulation is required.

(Note that by making the further simplification $k_z = 0$, one obtains the cutoff wavenumber eigenvalue problem already discussed in detail in Section 10.8, viz. $[S_{tt}][e_t] = k^2[T_{tt}][e_t]$.)

For the TE-only analysis, from Eqs. (10.144) and (10.145), it is clear that the system matrices needed can be assembled directly from the $[S]$ and $[T]$ matrices as

$$\left[A_{tt}^e\right] = \frac{1}{\mu_r}[S_{tt}^e] - k_0^2\epsilon_r[T_{tt}] \tag{10.163}$$

$$\left[B_{tt}^e\right] = \frac{1}{\mu_r}[T_{tt}^e] \tag{10.164}$$

10.9.4 Eigensolution

Solving either the lossless or lossy system with a significant number of elements is not trivial. A major issue is that $[A_{tt}^e]$ matrix is indefinite, as it is formed from the difference of the stiffness matrix of the vector elements and the mass matrix of the nodal elements, scaled by wavenumber squared. A rather elegant scaling originally proposed in [23] assists in this regard. The idea is to rewrite Eq. (10.162) in such a way that the dominant modes become dominant eigenvalues [23]. For lossless waveguides,[17] there is an upper bound $\theta^2 = k_0^2\mu_{r,\max}\epsilon_{r,\max}$ for the eigenvalue k_z^2, where $\mu_{r,\max}$ and $\epsilon_{r,\max}$ are the maximum relative permeability and permittivity of the materials within the waveguide (this corresponds to the propagation constant of a TEM wave in homogeneous medium with these material parameters). Equation (10.162) is rewritten as follows:

$$[B_{tt}]\{e_t\} = \frac{\theta^2}{\theta^2 - k_z^2}\left[B_{tt} + \frac{A_{tt}}{\theta^2}\right]\{e_t\} \tag{10.165}$$

Direct expansion verifies that it is identical to Eq. (10.162). The motivation for this is that the most dominant propagating modes, with largest positive k_z, will result in the $\lambda = \frac{\theta^2}{\theta^2 - k_z^2}$ term being larger; Lanczos eigenvalue solvers (such as those in ARPACK) converge more rapidly for the largest eigenvalues. A similar procedure can be applied to the full problem for an inhomogenously filled waveguide, in which case the full set of eigen-equations is now

$$\begin{bmatrix} B_{tt} & B_{tz} \\ B_{zt} & B_{zz} \end{bmatrix}\begin{Bmatrix} e_t \\ e_z \end{Bmatrix} = \frac{\theta^2}{\theta^2 - k_z^2}\begin{bmatrix} B_{tt} + \frac{A_{tt}}{\theta^2} & B_{tz} \\ B_{zt} & B_{zz} \end{bmatrix}\begin{Bmatrix} e_t \\ e_z \end{Bmatrix} \tag{10.166}$$

10.9.5 Results: a half-filled dielectric loaded rectangular waveguide

With the verification in the preceding section completed, one can now address an inhomogeneously loaded guide. An example is the half-filled guide of [3, Fig. 7.5, p. 243], shown in Fig. 10.19. The waveguide has a width:height ratio of 2:1 (slightly less than the X-band guide), and the $\epsilon_r = 4$, $\mu_r = 1$ loading occupies the lower half of the guide.

[17] In [24], it is suggested that the same scaling, using the lossless values, be used in the lossy case.

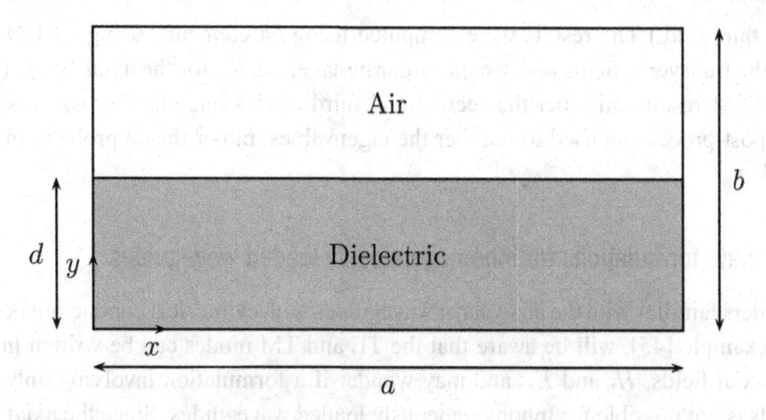

Figure 10.19 Half-filled rectangular waveguide; $b = a/2$, $d = b/2$, $\epsilon_r = 4$, $\mu_r = 1$. After [3, Fig. 7.5].

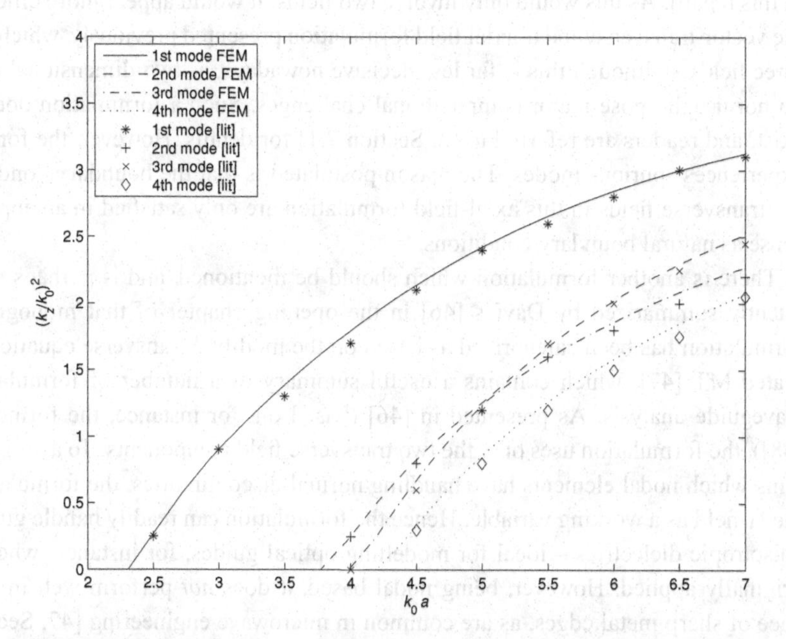

Figure 10.20 Dispersion curves for the first four modes in loaded rectangular waveguide, using the full inhomogeneously loaded waveguide formulation. 465 degrees of freedom.

Results are compared to curves[18] given in the literature [3, Fig. 8.8, p. 291]. (Methods such as transverse resonance [44, Section 3.9] can be used to compute the cutoff frequency of the guide. However, this method cannot yield the full field solution.) The results in the literature are given for the plane of symmetry treated as a magnetic wall [3, Fig. 8.8a, p. 291] and as an electric wall [3, Fig. 8.8b, p. 291]. These correspond to even and odd modes in standard waveguide terminology. These have been combined in Fig. 10.20. (In passing, note that the lowest and highest modes are even, the second

[18] The data points have been read off the graphs by hand, so some small error must be anticipated in this process.

and third odd.) The results were computed using 64 elements, using LT/QN elements for the transverse fields and second-order nodal elements for the axial fields. (Note that the FEM results mislabel the second and third modes once they cross; this is due to the post-processing used to number the eigenvalues, rather than a problem in the FEM itself.)

10.9.6 Alternate formulations for inhomogeneously loaded waveguides

Readers familiar with the analysis of waveguides with cylindrical conducting boundaries, for example [45], will be aware that the TE and TM modes can be written in terms of the axial fields, H_z and E_z, and may wonder if a formulation involving only the axial fields is not possible for inhomogeneously loaded waveguides. Since the axial fields are tangential to material discontinuities, nodal elements should not experience problems in this regard. As this would only involve two fields, it would appear more efficient than the vector-transverse/nodal-axial field formulation presented previously, which involves three fields – although this is far less decisive nowadays, as two-dimensional problems do not usually pose major computational challenges. Such a formulation does indeed exist, and readers are referred to [3, Section 7.1] for details. However, the formulation experiences spurious modes. The reason postulated is that the boundary conditions on the transverse fields in this axial-field formulation are only satisfied in an approximate sense as natural boundary conditions.

There is another formulation which should be mentioned, and is perhaps most eloquently summarized by Davies [46] in the opening chapter of that monograph. His formulation has been categorized as based on the modified transverse equation, abbreviated MT [47], which contains a useful summary of a number of formulations for waveguide analysis. As presented in [46] (based on, for instance, the formulation in [48]), the formulation uses only the two transverse field components. To avoid the problems which nodal elements have handling normal discontinuities, the formulation uses the H-field as a working variable. Hence the formulation can readily handle guides with anisotropic dielectrics – ideal for modelling optical guides, for instance, where it was originally applied. However, being nodal based, it does *not* perform well in the presence of sharp metal edges, as are common in microwave engineering [47, Section VI]. Another point mentioned by Jin [3, p. 261] is the following: in the presence of PEC boundaries, although the natural boundary condition ($\hat{n} \times \nabla \times \boldsymbol{H}$) is indeed enforced in the usual weak sense, this may not be sufficiently accurate. Hence explicitly enforcing this may be required. A major advantage of this approach when proposed was that only two field components were required, but, as noted above, this advantage is less significant nowadays.

10.10 Further reading

This chapter has focussed heavily on vector finite elements; the explanations of the properties of the elements reflect what might be called the current orthodoxy. It should be mentioned that there has been criticism of these elements from some quarters, most stridently from Mur [49]; some of his criticisms have been addressed by [11]. One should note that his criticism is heavily influenced by his work on magnetostatic problems, where the permeability can vary enormously from element to element and vector elements may indeed exhibit serious problems due to this. Recall also our earlier discussion about material interfaces and field continuity, and the problems with node-based elements, in Section 10.6; de Lager and Mur were able to introduce a node-based element which can indeed handle material discontinuities [50]. However, a decade after this work appeared, this element had not been applied to three-dimensional high-frequency analysis, so it seems likely that the current vector elements will continue to dominate finite element analysis for high-frequency electromagnetics. On the topic of spurious modes, work by Vardapetyan and Demkowicz has addressed the problem at a quite fundamental level, introducing Lagrange multipliers in the functional; [51, 52] is representative of their work.

More generally, the reader is fortunate that there are a number of excellent and reasonably current texts on the FEM available. Silvester and Ferrari's book [1] (first published in 1983, approximately doubling in length with the 1990 second edition, and increasing again in length significantly with the 1996 third edition) was for years the only reference in the field, and the current edition contains good coverage of high-frequency topics, in addition to extensive coverage of statics and magnetostatics. Although last revised over a decade back, it remains an excellent reference in the field, and it continues to serve as a standard reference on especially nodal elements. Jin's revised text [3] is probably the book of first choice for high-frequency electromagnetics, which it concentrates on exclusively, and it can be highly recommmended. Volakis *et al.*'s text also focusses on high-frequency applications, and contains much useful information on various elements [29]. Pelosi *et al.*'s book lives up to its name, and is a good starter text [53]; it has a particularly good treatment of waveguide analysis, and readers wanting to learn more about the material on waveguide analysis covered at an introductory level in this chapter will find that reference of considerable value. Peterson *et al.*'s book is somewhat more general in scope than just the FEM, but provides particularly deep coverage of coupled FEM/MoM formulations [43]. The text by Salazar-Palma *et al.* [12] is more of a research monograph; it concentrates primarily on interpolatory elements. The coverage is more theoretical than the other texts discussed here, and it is especially useful as preparatory reading if one intends working through mathematical papers such as Nedelec's. The book by Bondeson *et al.* is relatively recent [54]; despite its relatively short length, it provides a comprehensive discussion of many important topics in both FEA and CEM in general, and is similar to the present text in being one of very few books to address the FDTD, FEM and MoM. The text by Monk [55] is far more mathemetical in nature and would repay careful study for those wanting to delve deeper into the mathematical structure underlying finite elements in electromagnetics.

Two other very useful sources are the 1996 anthology edited by Silvester and Pelosi [9]; the extensive annotations are especially useful for putting the work in perspective, and the anthology contains a number of earlier papers which are otherwise hard to come by. Its age is starting to show somewhat, and a number of important papers have appeared since the anthology was published (and have been referenced in this chapter) but these are generally easily accessible. The collection edited by Itoh *et al.* [56] (also in 1996) contains a number of important contributions; in the context of vector elements, [10] deserves particular mention. For readers who would like to embark on their own three-dimensional implementation (a topic covered at introductory level in the following chapter), two papers will be of considerable interest, since they provide an eminently practical viewpoint on finite element coding. The first is by the present author [57]; the second reflects experience by Kempel's group [58], and was written specifically to complement the former. In [57], a number of practical issues are discussed, but mesh generation and linear algebra are only very briefly considered. In [58], an excellent overview of the many meshing packages available is provided, as well as a discussion of some sparse matrix solution routines. Sparse matrix schemes are on the one hand essentially an entirely practical problem, but on the other their efficient use is essential for commercial codes – we will briefly discuss this in the next chapter. The book by Duff *et al.* [59] is the standard reference on this. It has to be commented that specifically the topic of sparse matrices is not comprehensively treated in the CEM literature on finite elements.

A comprehensive discussion on simplex coordinates may be found in [1]. Coverage of this, as well as many useful elements of computational geometry applied to triangles, including Delaunay triangulation, may be found in the encyclopedic [8, Chapter 21].

There is an enormous literature on finite elements in general, reflecting its industry-standard status in especially computational mechanics. For many years one of the standard references has been that of Zienkiewicz [60], now in its sixth edition and with two co-authors. A text which can be strongly recommended for time-domain FEA (discussed in the next chapter) is [61]. The current author was first taught the method from [4], and it remains a useful textbook despite its age. For more on basic FEM texts, see [1].

10.11 Conclusions

This chapter has considered the finite element method in (primarily) two dimensions, with particular focus on high-frequency electromagnetic field solutions. Using the variational formulation, the basic method was introduced for the scalar Laplace equation, and an application to the quasi-static analysis of microstrip was shown. The alternate Galerkin approach to developing the FEM was discussed, and it was compared in detail to the variational formulation using the one-dimensional problem of the preceding chapter as a vehicle. After a brief introduction to simplex coordinates, a discussion of vector (edge) elements for the vector wave equation followed. An eigenvalue problem was solved, and used to illustrate ideas both about the theory of finite element solutions of the vector wave equation, as well as a plethora of practical issues which one must address when writing an actual finite element code. The application of a mixed second-order vector element was outlined. Finally, the question of dispersion analysis of waveguide was discussed theoretically. Two-dimensional finite element codes require only moderate coding complexity, and it is quite realistic to attempt development of such a code oneself, including the use of higher-order elements.

The extension to three dimensions will be discussed in the following chapter, along with a variety of more advanced topics on the FEM in the final chapter. The next chapter starts with an illustrative but relatively simple 3D application, namely finding the eigenvalues of a highly conducting rectangular cavity using the 3D Whitney element. The extension of vector elements to higher order will be comprehensively revisited in the context of 3D, and the application of these will be illustrated by way of both an eigenvalue and a deterministic problem (the latter being an obstacle in a rectangular waveguide, analyzed using both commercial and research codes). Mesh termination schemes will be discussed, including the first-order absorbing boundary conditions. In the final chapter, the FEM/MoM hybrid formulation is introduced. Then a time domain formulation of the finite element method for the vector wave equation is outlined, and the connection with the FDTD established. The issue of sparse matrix storage schemes and solution methods is considered, before finishing the coverage with an introduction to the field of error estimation and mesh adaptation.

References

[1] P. P. Silvester and R. L. Ferrari, *Finite Elements for Electrical Engineers*. Cambridge: Cambridge University Press, 3rd edn., 1996.

[2] J. S. Savage and A. F. Peterson, "Higher-order vector finite elements for tetrahedral cells," *IEEE Trans. Microwave Theory Tech.*, **44**, 874–879, June 1996.

[3] J.-M. Jin, *The Finite Element Method in Electromagnetics*. New York: Wiley, 2nd edn., 2002.

[4] D. H. Norrie and G. de Vries, *An Introduction to Finite Element Analysis*. New York: Academic Press, 1978.

[5] D. B. Davidson and J. T. Aberle, "An introduction to spectral domain method of moments formulations," *IEEE Antennas Propagat. Mag.*, **46**, 11–19, July 2004.

[6] D. G. Dudley, *Mathematical Foundations for Electromagnetic Theory*. New York: IEEE Press, 1994.

[7] B. D. Reddy, *Introductory Functional Analysis with Applications to Boundary Value Problems and Finite Elements*. New York: Springer-Verlag, 1998.

[8] W. H. Press, S. A. Teukolsky, W. Vettering and B. R. Flannery, *Numerical Recipes: the Art of Scientific Computing*. Cambridge: Cambridge University Press, 3rd edn., 2007.

[9] P. P. Silvester and G. Pelosi, *Finite Elements for Wave Electromagnetics*. New York: IEEE Press, 1994.

[10] A. F. Peterson and D. R. Wilton, "Curl-conforming mixed-order edge elements for discretizing the 2D and 3D vector Helmholtz equation," in *Finite Element Software for Microwave Engineering* (T. Itoh, G. Pelosi and P. P. Silvester, eds.), Chapter 5. New York: Wiley, 1996.

[11] P. Fernandes and M. Raffetto, "Characterization of spurious-free finite element methods in electromagnetics," *COMPEL – The International Journal for Computation and Mathematics in Electrical and Electronic Engineering*, **21**, 147–164, 2002.

[12] M. Salazar-Palma, T. K. Sarkar, L. E. García-Castillo, T. Roy and Djordjević, *Iterative and Self-Adaptive Finite-Elements in Electromagnetic Modelling*. Boston, MA: Artech House, 1998.

[13] J. C. Nedelec, "Mixed finite elements in $\Re^3$," *Numerische Mathematik*, **35**, 315–341, 1980.
[14] P. A. Riavart and J. M. Thomas, "A mixed finite element method for 2nd order elliptic problems," in *Mathematical Aspects of the Finite Element Method* (I. Galligani and E. Mayera, eds.), vol. 606 of *Lecture Notes on Mathematics*, pp. 293–315. New York: Springer-Verlag, 1977.
[15] A. Bossavit, "Finite elements for the electricity equation," in *The Mathematics of Finite Elements and Applications* (J. R. Whiteman, ed.), pp. 85–91. London: Academic Press, 1982.
[16] M. L. Barton and Z. J. Cendes, "New vector finite elements for three-dimensional magnetic field computation," *J. Appl. Phys.*, **61**, 3919–2921, April 1987.
[17] J. S. van Welij, "Calculation of eddy currents in terms of H on hexahedra," *IEEE Trans. Magn.*, **21**, 2239–41, 1985.
[18] C. W. Crowley, P. P. Silvester and H. Hurwitz, "Covariant projection elements for 3D vector field problems," *IEEE Trans. Magn.*, **24**, 397–400, 1988.
[19] Z. J. Cendes, "Vector finite elements for electromagnetic field computation," *IEEE Trans. Magn.*, **27**, 3958–3966, September 1991.
[20] J. P. Webb and B. Forghani, "Hierarchal scalar and vector tetrahedra," *IEEE Trans. Magn.*, **29**, 1495–1498, March 1993.
[21] J. P. Webb, "Hierarchal vector basis functions of arbitrary order for triangular and tetrahedral finite elements," *IEEE Antennas Propagat.*, **47**, 1244–1253, August 1999.
[22] P. Ingelström, "A new set of H(curl)-conforming hierarchical basis functions for tetrahedral meshes," *IEEE Trans. Microwave Theory Tech.*, **54**, 106–114, January 2006.
[23] J.-F. Lee, D.-K. Sun and Z. J. Cendes, "Full-wave analysis of dielectric waveguides using tangential vector finite elements," *IEEE Trans. Microwave Theory Tech.*, **39**, 1262–1271, August 1991.
[24] J.-F. Lee, "Finite element analysis of lossy dielectric waveguides," *IEEE Trans. Microwave Theory Tech.*, **42**, 1025–1031, June 1994.
[25] J.-F. Lee and R. Mittra, "A note on the application of edge-elements for modeling three-dimensional inhomogeneously-filled cavities," *IEEE Trans. Microwave Theory Tech.*, **40**, 1767–1773, September 1992.
[26] S. D. Gedney and U. Navsariwala, "An unconditionally stable finite element time-domain solution of the vector wave equation," *IEEE Microwave Guided Wave Lett.*, **5**, 332–334, October 1995.
[27] D. C. Dibben and R. Metaxas, "Time domain finite element analyis of multimode microwave applicators," *IEEE Trans. Magn.*, **32**, 942–945, May 1996.
[28] J.-F. Lee, R. Lee and A. Cangellaris, "Time-domain finite-element methods," *IEEE Trans. Antennas Propagat.*, **45**, 430–442, March 1997.
[29] J. Volakis, A. Chatterjee and L. Kempel, *Finite Element Method for Electromagnetics: Antennas, Microwave Cicuits and Scattering Applications*. Oxford & New York: Oxford University Press and IEEE Press, 1998.
[30] L. S. Andersen and J. L. Volakis, "Development and application of a novel class of hierarchical tangential vector finite elements for electromagnetics," *IEEE Trans. Antennas Propagat.*, **47**, 112–120, January 1999.
[31] R. Dyczij-Edlinger, G. Peng and J. Lee, "A fast vector-potential method using tangentially continuous vector finite elements," *IEEE Trans. Microwave Theory Tech.*, **46**, 863–868, June 1998.

[32] R. D. Graglia, D. R. Wilton and A. F. Peterson, "Higher order interpolatory vector bases for computational electromagnetics," *IEEE Trans. Antennas Propagat.*, **45**, 329–342, March 1997.

[33] D. B. Davidson, "An evaluation of mixed-order versus full-order vector finite elements," *IEEE Trans. Antennas Propagat.*, **51**, 2430–2441, September 2003.

[34] M. M. Botha and D. B. Davidson, "An explicit a posteriori error indicator for electromagnetic, finite element analysis in 3D," *IEEE Trans. Antennas Propagat.*, **53**, 3717–3725, November 2005.

[35] M. M. Botha and D. B. Davidson, "The implicit, element residual method for a posteriori error estimation in FE-BI analysis," *IEEE Trans. Antennas Propagat.*, **54**, 255–258, January 2006.

[36] A. Bossavit, *Computational Electromagnetism: Variational Formulations, Complementarity, Edge Elements*. San Diego, CA: Academic Press, 1998.

[37] R. P. Feynmann, R. B. Leighton and P. Sands, *The Feynmann Lectures on Physics*, vol. 1. Reading, MA: Addison-Wesley, 1963.

[38] J. P. Webb, "Edge elements and what they can do for you," *IEEE Trans. Magn.*, **29**, 1460–1465, March 1993.

[39] D. B. Davidson, "Comments on and extensions of 'A note on the application of edge-elements for modeling three-dimensional inhomogeneously-filled cavities'," *IEEE Trans. Microwave Theory Tech.*, **46**, 1344–1346, September 1998.

[40] K. Morishita and N. Kumagia, "Unified approach to the derivation of variational expression for electromagnetic fields," *IEEE Trans. Microwave Theory Tech.*, **25**, 34–40, January 1977.

[41] C. H. Chen and C. Lien, "The variational principle for non-self-adjoint electromagnetic problems," *IEEE Trans. Microwave Theory Tech.*, **28**, 878–886, August 1980.

[42] D. A. Dunavant, "High degree efficient symmetrical Gaussian quadrature formulas for the triangle," *Int. J. Numer. Meth. Eng.*, **21**, 1129–1148, 1985.

[43] A. F. Peterson, S. L. Ray and R. Mittra, *Computational Methods for Electromagnetics*. Oxford & New York: Oxford University Press and IEEE Press, 1998.

[44] D. M. Pozar, *Microwave Engineering*. New York: Wiley, 2nd edn., 1998.

[45] S. Ramo, J. R. Whinnery and T. van Duzer, *Fields and Waves in Communication Electronics*. Chichester: Wiley, 3rd edn., 1994.

[46] J. B. Davies, "Complete modes in uniform waveguide," in *Finite Element Software for Microwave Engineering* (T. Itoh, G. Pelosi and P. P. Silvester, eds.), Chapter 1. New York: John Wiley and Sons, 1996.

[47] B. M. Dillon and J. P. Webb, "A comparison of formulations for the vector finite element analysis of waveguides," *IEEE Trans. Microwave Theory Tech.*, **42**, 308–316, February 1994.

[48] F. A. Fernandez, Y. Lu, J. B. Davies and S. Zhu, "Finite element analysis of complex modes in inhomogeneous waveguide," *IEEE Trans. Magn.*, **29**, 1601–1604, March 1993.

[49] G. Mur, "The fallacy of edge elements," *IEEE Trans. Magn.*, **34**, 3244–3247, September 1998.

[50] I. E. de Lager and G. Mur, "Generalized Cartesian finite elements," *IEEE Trans. Magn.*, **34**, 2220–2227, July 1998.

[51] L. Vardapetyan and L. Demkowicz, "hp-adaptive finite elements in electromagnetics," *Computer Meth. Appl. Mechan. Eng.*, **169**, pp. 331–344, 1999.

[52] L. Vardapetyan and L. Demkowicz, "hp-vector finite elements method for the full-wave analysis of waveguides with no spurious modes," *Electromagnetics*, **22**, 419–428, July 2002.

[53] G. Pelosi, R. Coccioli and S. Selleri, eds., *Quick Finite Element Method for Electromagnetic Waves*. Boston, MA: Artech House, 1998.
[54] A. Bondeson, T. Rylander and P. Ingelström, *Computational Electromagnetics*. New York: Springer Science, 2005.
[55] P. Monk, *Finite Element Methods for Maxwell's Equations*. Oxford: Oxford University Press, 2003.
[56] T. Itoh, G. Pelosi and P. P. Silvester, eds., *Finite Element Software for Microwave Engineering*. New York: Wiley, 1996.
[57] D. B. Davidson, "Implementation issues for three-dimensional vector FEM programs," *IEEE Antennas Propag. Soc. Mag.*, **42**, 100–107, December 2000.
[58] A. Awadhiya, P. Barba and L. Kempel, "Finite element method programming made easy???," *IEEE Antennas Propag. Soc. Mag.*, **45**, 73–79, August 2003.
[59] I. S. Duff, A. M. Erisman and J. K. Reid, *Direct Methods for Sparse Matrices*. Oxford: Oxford University Press, 1986.
[60] O. C. Zienkiewicz, R. L. Taylor and J. Z. Zhu, *The Finite Element Method*. Oxford: Elsevier Butterworth-Heinemann, 6th edn., 2005.
[61] T. J. R. Hughes, *The Finite Element Method: Linear Static and Dynamic Finite Element Analysis*. Englewood Cliffs, NJ: Prentice-Hall, 1987. (Dover re-print, 2000.)

Problems and assignments

Problems

P10.1 Derive Eq. (10.16).

P10.2 Verify Eq. (10.24). Hint: from Eq. (10.29), derive an identity for the case where $\{z\} = \{x\}$, and $\{y\}$ is a constant row vector.

P10.3 As outlined in Section 10.2.4, and using Eqs. (10.46) and (10.47) show that flux continuity is a natural boundary condition at a material interface

P10.4 Show that Eq. (10.95) is a solution of Eq. (10.85) (for $k_i^2 = 0$), but not of Gauss's law in a source-free region, viz. $\nabla \cdot (\epsilon \boldsymbol{E}) = 0$.

Assignments

A10.1 Develop a code to compute the potential for boxed microstrip, as in Section 10.2.3, and, from this, the characteristic impedance of microstrip.

A10.2 Develop a code to compute the TE eigenmodes, and investigate the convergence with mesh refinement, as in Section 10.8.

A10.3 Extend the above to include LT/QN elements, as outlined in Section 10.8.6.

A10.4 Develop a code to compute the TM eigenmodes, similar to the development in Section 10.8. (Note that in this case, the problem is formulated in terms of the $\boldsymbol{H}$ field, and one uses natural boundary condition on the waveguide walls.)

11 The finite element method in three dimensions

In this penultimate chapter, the preceding discussion of the finite element method is extended to three dimensions. We start by extending the eigenanalysis of the preceding chapter to three dimensions, using Whitney elements. Although a 3D FEM code is non-trivial to implement, simple problems can nonetheless be addressed, and the eigenanalysis of a PEC cavity is used to illustrate the development of a 3D FEM code. Higher-order vector elements have already been introduced in the preceding chapter for two-dimensional analysis; in this chapter, a more detailed theoretical treatment will be presented for the more general three-dimensional case, and an LT/QN element is applied to the cavity eigenanalysis problem. Following this, the stationary functional formulation for deterministic (driven) problems will be outlined. In this and the preceding chapter, eigenvalue problems have been used to illustrate the FEM; in this chapter, a deterministic three-dimensional problem will also be discussed, namely the analysis of waveguide obstacles. Finite element analysis is ideal for this problem, and good results have been obtained by a number of workers. Results for two waveguide problems computed using FEM codes incorporating higher-order elements will be shown. The chapter concludes with a discussion on the use of absorbing boundary conditions for open-region problems, and a first-order ABC is presented. Treatment of a more sophisticated scheme is deferred to the final chapter.

11.1 The three-dimensional Whitney element

The FEM using vector elements in 3D is in a sense just a straightforward extension of the 2D analysis, and the elemental functions are identical, as seen below; however, the mesh generation and book-keeping problems can become formidable. Nonetheless, by choosing a simple physical problem, developing a special-purpose 3D finite element analysis code can be undertaken. Automatically generating tetrahedral meshes in 3D is a complex task, but using existing tools for this greatly reduces the coding burden. As with all 3D CEM methods, the reader should be cautioned that developing a truly general-purpose 3D FEM code is a challenging task. References that can assist in this regard may be found in Section 10.10.

The three-dimensional Whitney element is exactly the same as in two dimensions:

$$\boldsymbol{w}_{ij} = \lambda_i \nabla \lambda_j - \lambda_j \nabla \lambda_i, \tag{11.1}$$

with the obvious difference that there are now six degrees of freedom per tetrahedron, rather than three per triangle, since a tetrahedron has six edges. This element has exactly the same well-known properties of constant tangential/linear normal field (CT/LN) approximation along edges (hence, of mixed order) as its two-dimensional counterpart and needs no further discussion.

The problem which will be addressed is the eigenanlysis of a resonant cavity, a box of dimensions $a \times b \times d$ with PEC walls. The variational functional for the source-free vector wave equation, Eq. (10.85), repeated here:

$$\nabla \times \frac{1}{\mu_r} \nabla \times \boldsymbol{E} - k_i^2 \epsilon_r \boldsymbol{E} = 0, \tag{11.2}$$

is Eq. (10.86), also repeated here:

$$F(\boldsymbol{E}) = \int_{\Omega} \left[\frac{1}{\mu_r} (\nabla \times \boldsymbol{E}) \cdot \nabla \times \boldsymbol{E}) - k_i^2 \epsilon_r \boldsymbol{E} \cdot \boldsymbol{E} \right] d\Omega \tag{11.3}$$

The solutions sought are the eigenpairs $(k_i^2, \boldsymbol{E}_i)$; k_i is the eigenvalue (cavity resonant frequency) and $\boldsymbol{E}_i$ the associated eigenmode (cavity resonant mode).

11.1.1 Explicit formula for the tetrahedral elemental matrix entries

As in Section 10.8.2, explicit forms for the mass and stiffness matrices may be derived for these elements. That material is repeated here in summary form, generalized to three dimensions, as originally presented in [1] (although for a different higher-order element). The variational formulation requires the evaluation of two matrices:

$$S_{ij} = \iiint_S \nabla \times \boldsymbol{N}_i \cdot \nabla \times \boldsymbol{N}_j \, dS \tag{11.4}$$

and

$$T_{ij} = \iiint_S \boldsymbol{N}_i \cdot \boldsymbol{N}_j \, dS \tag{11.5}$$

The CT/LN elements are given by $\boldsymbol{N}_i = \boldsymbol{w}_{i1,i2} = \lambda_{i1} \nabla \lambda_{i2} - \lambda_{i2} \nabla \lambda_{i1}$ per edge. Here, $i1$ and $i2$ are the endpoints of edge i. The local tetrahedral numbering scheme is an extension of the two-dimensional one, and is shown in Table 11.1. The four simplex coordinates λ_i are given by

$$\lambda_i = a_i + b_i x + c_i y + d_i z \tag{11.6}$$

and the gradient thereof by

$$\nabla \lambda_i = b_i \hat{x} + c_i \hat{y} + d_i \hat{z} \tag{11.7}$$

Table 11.1 Local edge and face numbering convention for 3D tetrahedrons

Local edge numbering			Local face numbering			
Edge	Local nodes		Face	Local nodes		
1	1	2	1	1	2	3
2	1	3	2	1	2	4
3	1	4	3	1	3	4
4	2	3	4	2	3	4
5	2	4				
6	3	4				

The actual coefficients $\{a_i; b_i; c_i; d_i\}$ may be computed by inverting the coordinate matrix:

$$\begin{bmatrix} b_1 & c_1 & d_1 & a_1 \\ b_2 & c_2 & d_2 & a_2 \\ b_3 & c_3 & d_3 & a_3 \\ b_4 & c_4 & d_4 & a_4 \end{bmatrix} = \begin{bmatrix} x_1 & x_2 & x_3 & x_4 \\ y_1 & y_2 & y_3 & y_4 \\ z_1 & z_2 & z_3 & z_4 \\ 1 & 1 & 1 & 1 \end{bmatrix}^{-1} \tag{11.8}$$

This equation may be obtained by writing Eq. (11.6) for each node i and noting that $\lambda_i = 1$ at node i. Now the following two vectors are defined for nodes i and j:

$$\begin{aligned} \boldsymbol{v}_{ij} &= \nabla\lambda_i \times \nabla\lambda_j \\ &= (c_i d_j - c_j d_i)\hat{x} + (b_j d_i - b_i d_j)\hat{y} + (b_i c_j - b_j c_i)\hat{z} \\ &= -\boldsymbol{v}_{ji} \end{aligned} \tag{11.9}$$

This vector is easily computed once $\{b_i; c_i; d_i\}$ are known; in MATLAB, it can be computed directly using the `cross` function. Similarly we define

$$\begin{aligned} \phi_{ij} &= \nabla\lambda_i \cdot \nabla\lambda_j \\ &= b_i b_j + c_i c_j + d_i d_j \end{aligned} \tag{11.10}$$

which again can be computed using the `dot` function in MATLAB. As in the 2D case, both $\boldsymbol{v}_{ij}$ and ϕ_{ij} are constant within a triangle, and may be taken outside integrals in which they appear.

The curl-curl term, Eq. (11.4), yields Eq. (10.120), repeated here:

$$\nabla \times \boldsymbol{N}_i = 2\boldsymbol{v}_{i1,i2} \tag{11.11}$$

Using this, Eq. (11.4) becomes:

$$\begin{aligned} S_{ij} &= 4 \iiint_V \boldsymbol{v}_{i1,12} \cdot \boldsymbol{v}_{j1,j2}\, dV \\ &= 4V\, \boldsymbol{v}_{i1,12} \cdot \boldsymbol{v}_{j1,j2} \end{aligned} \tag{11.12}$$

As noted in the context of the area of a triangle, Eq. (10.122), the expression for tetrahedral volume in terms of the determinant of the coordinate matrix,

$$6V' = \begin{vmatrix} 1 & x_1 & y_1 & z_1 \\ 1 & x_2 & y_2 & z_2 \\ 1 & x_3 & y_3 & z_3 \\ 1 & x_4 & y_4 & z_4 \end{vmatrix} \tag{11.13}$$

also yields a potentially signed volume V', whose sign depends on the sense (clockwise or anticlockwise) of the coordinate numbering. V in Eq. (11.12) is the *unsigned* volume of the element, that is $V = |V'|$.

The integrand of Eq. (11.5) requires the computation of dot products, and yields Eq. (10.123), repeated here:

$$\begin{aligned} \boldsymbol{N}_i \cdot \boldsymbol{N}_j = & [\lambda_{i1}\lambda_{j1}(\nabla\lambda_{i2} \cdot \nabla\lambda_{j2}) - \lambda_{i1}\lambda_{j2}(\nabla\lambda_{i2} \cdot \nabla\lambda_{j1}) \\ & - \lambda_{i2}\lambda_{j1}(\nabla\lambda_{i1} \cdot \nabla\lambda_{j2}) + \lambda_{i2}\lambda_{j2}(\nabla\lambda_{i1} \cdot \nabla\lambda_{j1})] \end{aligned} \tag{11.14}$$

Using the notation of Eq. (11.10), this can be written as

$$\boldsymbol{N}_i \cdot \boldsymbol{N}_j = [\lambda_{i1}\lambda_{j1}\phi_{i2,j2} - \lambda_{i1}\lambda_{j2}\phi_{i2,j1} - \lambda_{i2}\lambda_{j1}\phi_{i1,j2} + \lambda_{i2}\lambda_{j2}\phi_{i1,j1}] \tag{11.15}$$

The associated matrix elements are thus:

$$\begin{aligned} T_{ij} = & \phi_{i2,j2} \iiint_V \lambda_{i1}\lambda_{j1}\, dV - \phi_{i2,j1} \iiint_V \lambda_{i1}\lambda_{j2}\, dV \\ & - \phi_{i1,j2} \iiint_V \lambda_{i2}\lambda_{j1}\, dV + \phi_{i1,j1} \iiint_V \lambda_{i2}\lambda_{j2}\, dV \end{aligned} \tag{11.16}$$

Using the general integration formula for integrals in simplex coordinates (see Appendix D):

$$\iiint_V \lambda_1^i \lambda_2^j \lambda_3^k \lambda_4^l\, dV = \frac{3!\, i!\, j!\, k!\, l!}{(3+i+j+k+l)!} V \tag{11.17}$$

the expression for T_{ij} may be simplified (note that $0! \equiv 1$). In Eq. (11.16), each integral involves integration over two simplex coordinates, possibly identical. These can be expressed in matrix form as

$$M_{ij} = \iiint_V \lambda_i \lambda_j\, dV = \frac{1}{20} \begin{bmatrix} 2 & 1 & 1 & 1 \\ 1 & 2 & 1 & 1 \\ 1 & 1 & 2 & 1 \\ 1 & 1 & 1 & 2 \end{bmatrix} \tag{11.18}$$

Using this, Eq. (11.16) reduces to

$$T_{ij} = V[\phi_{i2,j2}M_{i1,j1} - \phi_{i2,j1}M_{i1,j2} - \phi_{i1,j2}M_{i2,j1} + \phi_{i1,j1}M_{i2,j2}] \tag{11.19}$$

11.1.2 Coding

It will be appreciated that a working 2D code provides an excellent basis for developing a 3D code, as the above is a straightforward extension of the 2D approach. Hence, before attempting a 3D code, developing a working 2D one is strongly advised. All the comments made in the context of two-dimensional elements are equally germane here; however, one should be aware that for general problems, the coding effort is substantially more, by perhaps an order of magnitude. This is due to the complexity of three-dimensional tetrahedral meshes and the much larger problem size required by realistic 3D problems. A specialized application, such as the rectangular cavity eigenanalysis problem at hand, is tractable in a reasonable time, however, and the discussion here focusses on that.

Once again, conventions should be adopted right from the start; since higher-order elements are going to be addressed later, which have degrees of freedom linked to faces as well as edges, the faces must also be numbered. See Table 11.1 for one such convention. As before, e_{ij} is the edge directed from node i to node j, and as in the 2D case, although the degree of freedom associated with the edge is a scalar, it is nonetheless signed. The convention discussed in Section 10.8.3, namely first sorting the nodes in each element into ascending global order, can once again be recommended. Simultaneously, this also ensures that degrees of freedom associated with faces (which are also signed) are compatible between elements. Global edge numbers can be assigned in the same fashion. Similarly, comments made in Section 10.8.3 regarding consistency of the local edge numbering schemes and signs in use in the literature apply here too. Note, however, that the face numbering conventions in the literature are generally *not* consistent. This one follows [1, Table II], but differs from [2], for example.

Data structures

The data structures are very similar to the 2D data structures, but reflect the 3D nature of the problem. For a mesh with N_n nodes, N_e elements, and E edges, some of the major data structures required are:

vertices Dimensioned as $(N_n, 3)$. This stores the (x, y, z) coordinates of each vertex (node).

elements (or *nodes*) Dimensioned as $(N_e, 4)$. This stores the four nodes associated with each tetrahedral element.

edge_nodes Dimensioned as $(E, 2)$. This stores the global nodes that each edge connects.

face_nodes Dimensioned as $(E, 3)$. This stores the global nodes that each face comprises (not needed for a CT/LN implementation, but required for higher-order schemes).

materials Dimensioned as (N_e). This stores the material number. Another (usually very much smaller) data structure will be required to store the actual constitutive parameters for each material.

dof Dimensioned as E for Whitney elements. These are the degrees of freedom.

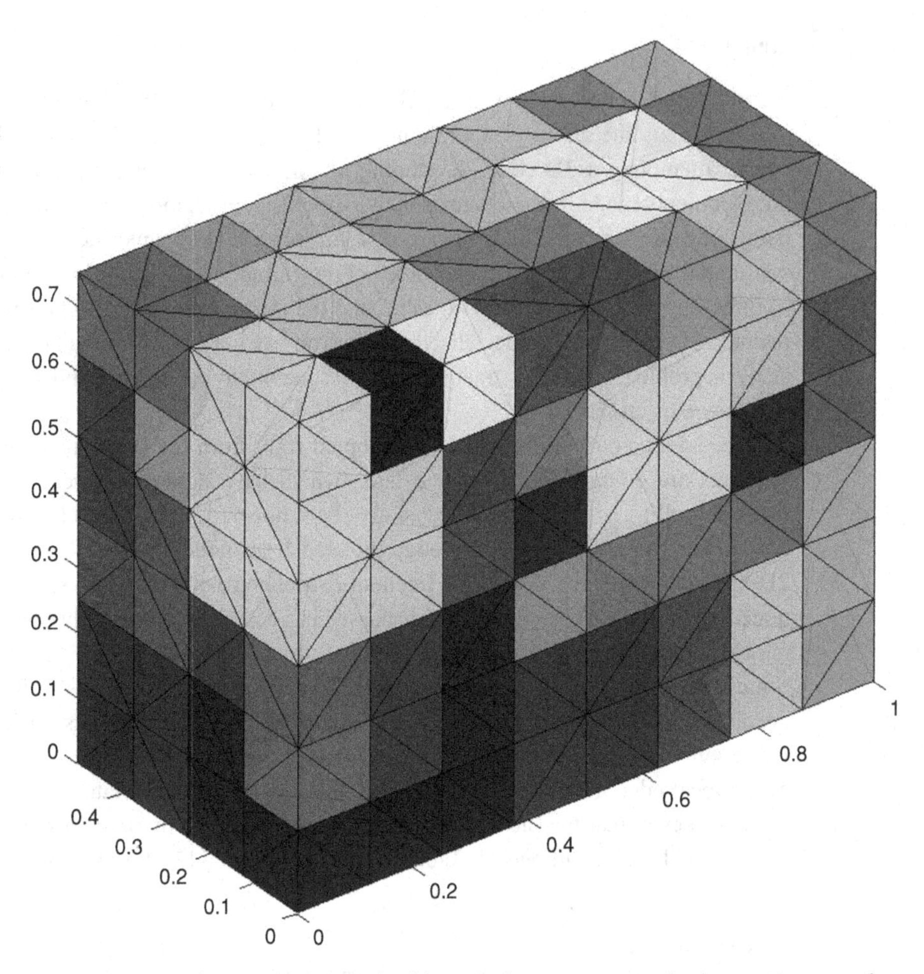

Figure 11.1 A "regular" tetrahedral mesh with 1152 elements, generated using `delaunay3` from a "brick" mesh.

Meshing and book-keeping

In Section 10.8.3, a triangular mesh was created by splitting each element of a rectangular mesh into two triangles. There is a 3D analogy, which splits a rectangular prism of volume abc into five tetradedra, four "corner" elements of equal volume $abc/6$ and a "central" element of volume $abc/3$. (This was shown in [3, Fig. 9.5, p. 276], unfortunately not repeated in the third edition.) However, this generates a "handed" element; it cannot be "glued" onto an identical element to create a volumetric mesh, as the relevant prism faces are split in opposite directions. (It is possible that one could use mirror images of the element to assemble a compatible mesh, but this is not trivial to implement.) An easier way to do this is to use MATLAB's `delaunay3` function, providing as input a list of nodes corresponding to a regular prismatic ("brick") mesh. This then generates compatible tetrahedral elements, typically subdividing each "brick" into six elements. An example is shown in Fig. 11.1. As will be seen, such "regular" meshes are not always desirable; for more general work, public domain software such as `gmsh` may be used to generate a mesh. Saved to a file in a standard format, this mesh can then be imported

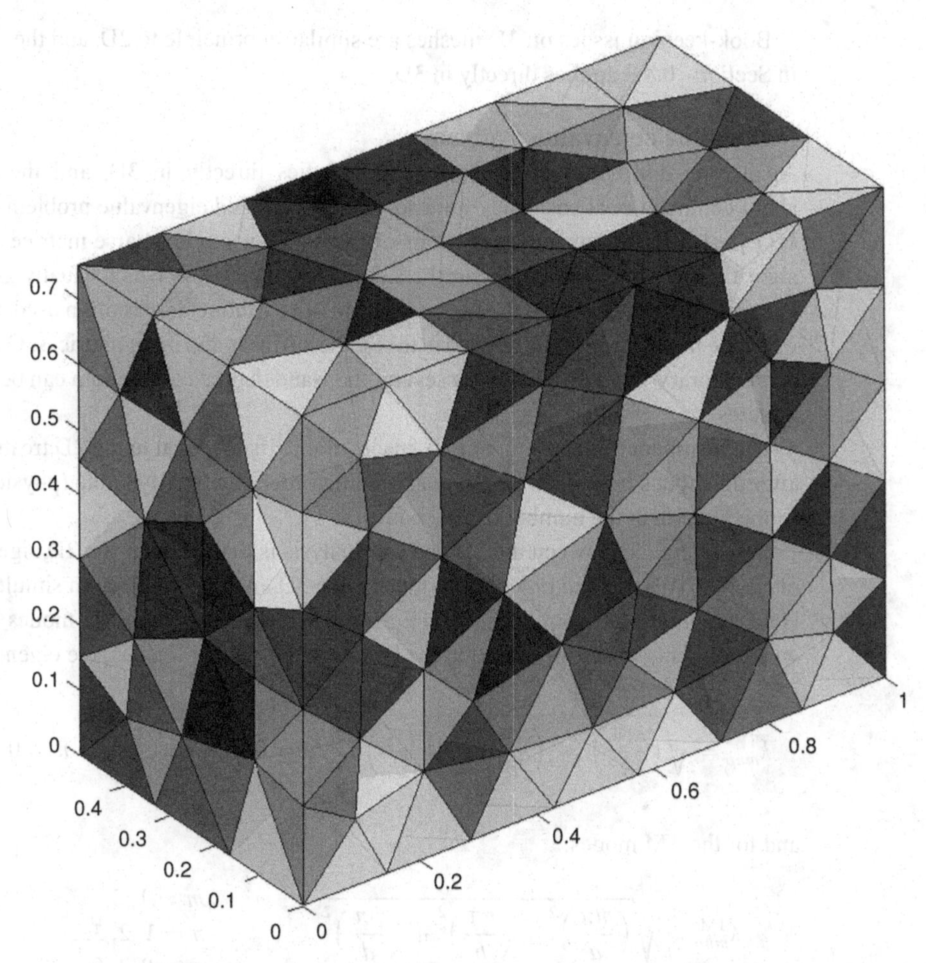

Figure 11.2 A proper tetrahedral mesh with 1001 elements, generated using gmsh, imported into MATLAB via a custom parsing function.

via a simple parser into the analysis software. An example of such a mesh, with a very similar element count to that of Fig. 11.1, is shown in Fig. 11.2. As will be seen shortly, the latter (more irregular) mesh gives better results.[1] Both meshes were visualized using MATLAB's `tetramesh` function.

Before moving off the topic of 3D meshing, it is worth mentioning that, for some applications, triangular prismatic meshes have proven useful. (Such a mesh extrudes a 2D triangular surface mesh in the direction perpendicular to the surface.) Volakis, Kempel and colleagues have been very successful with such meshes for a variety of antenna problems [4, 5]. One might term this $2\frac{1}{2}$D-modelling, although of course the full 3D field solution is obtained.

[1] This is perhaps counter-intuitive initially. One reason is that the quality of an FE mesh is inversely related to minimum angles (the larger, the better), and this is better achieved on an irregular mesh, incorporating some mesh equilibration during the meshing process; another reason is that dispersion effects can sometimes partially cancel on an irregular mesh.

Book-keeping issues on 3D meshes are similar in principle to 2D, and the discussion in Section 10.8.3 applies directly in 3D.

Solving the eigenvalue problem

Again, the discussion in Section 10.8.3 applies directly in 3D, and the comment about computational cost even more so. The generalized eigenvalue problem $[S]\{e_i\} = k_i^2[T]\{e_i\}$ is quite demanding in terms of memory, even with sparse matrices, as many algorithms for this problem factor the RHS matrix, a process which usually generates a lot of fill-in. For simplicity, the full-matrix MATLAB routine `eig` is often used, especially when the main purpose of the code is to test the stiffness and mass matrices. On a typical contemporary PC, a problem with several thousand degrees of freedom can be solved in minutes.

The treatment of the static, or null-space, modes, is identical to the 2D treatment, and after the modes have been sorted in ascending order, the first dynamic (physical) mode is one more than the number of free nodes.

One *difference* between the 2D and 3D analysis is that whereas the 2D eigenanalysis of Section 10.8 yielded *only* the TE modes, the 3D analysis yields both simultaneously. Here, we imply the convention of TE_z and TM_z when labelling 3D modes [6, 7], as widely (but not uniformly) used in the literature. For the TE modes, the eigenvalues are

$$k_{mnp}^{\mathrm{TE}} = \sqrt{\left(\frac{m\pi}{a}\right)^2 + \left(\frac{n\pi}{b}\right)^2 + \left(\frac{p\pi}{d}\right)^2} \quad \left.\begin{array}{l} m = 0, 1, 2, \ldots \\ n = 0, 1, 2, \ldots \\ p = 1, 2, 3, \ldots \end{array}\right\} m = n \neq 0 \tag{11.20}$$

and for the TM modes:

$$k_{mnp}^{\mathrm{TM}} = \sqrt{\left(\frac{m\pi}{a}\right)^2 + \left(\frac{n\pi}{b}\right)^2 + \left(\frac{p\pi}{d}\right)^2} \quad \begin{array}{l} m = 1, 2, 3, \ldots \\ n = 1, 2, 3, \ldots \\ p = 0, 1, 2, \ldots \end{array} \tag{11.21}$$

Note that numerically, the values of the eigenvalues are identical, so degenerate modes will be present; however, the starting indices are not the same. The dominant (lowest order) mode will depend on the relative sizes of a, b and d. For a PEC box $1 \times 0.5 \times 0.75$ [1], the first eight eigenmodes and their resonant wavenumbers are given in Table 11.2.[2] As expected, modes three and four, and five and six, are degenerate, which is always a good test of both the FEM formulation and the ability of the eigensolver to resolve multiple eigenvalues.

With these reference results to validate the implementation, results can be computed on a variety of meshes. A typical result is shown in Fig. 11.3. Using a first-order polynomial fit, the slope of the second curve (the average over eight modes, on the `gmsh` meshes is 2.22, which compares well with the theoretically expected result of 2. The "split-hex" mesh performs less well.

[2] In [1], the second mode is labelled TE_{110}, which is not consistent with the present TE_z notation; this is presumably either an error, or a different convention was used.

Table 11.2 First eight resonant modes of the rectangular cavity resonator in the text, giving mode type and wavenumber

Mode	k_c (rad/m)
TE_{101}	5.235 987 8
TM_{110}	7.024 814 7
TE_{011}	7.551 448 9
TE_{201}	7.551 448 9
TM_{111}	8.178 874 3
TE_{111}	8.178 874 3
TM_{210}	8.885 765 9
TE_{102}	8.947 259 8

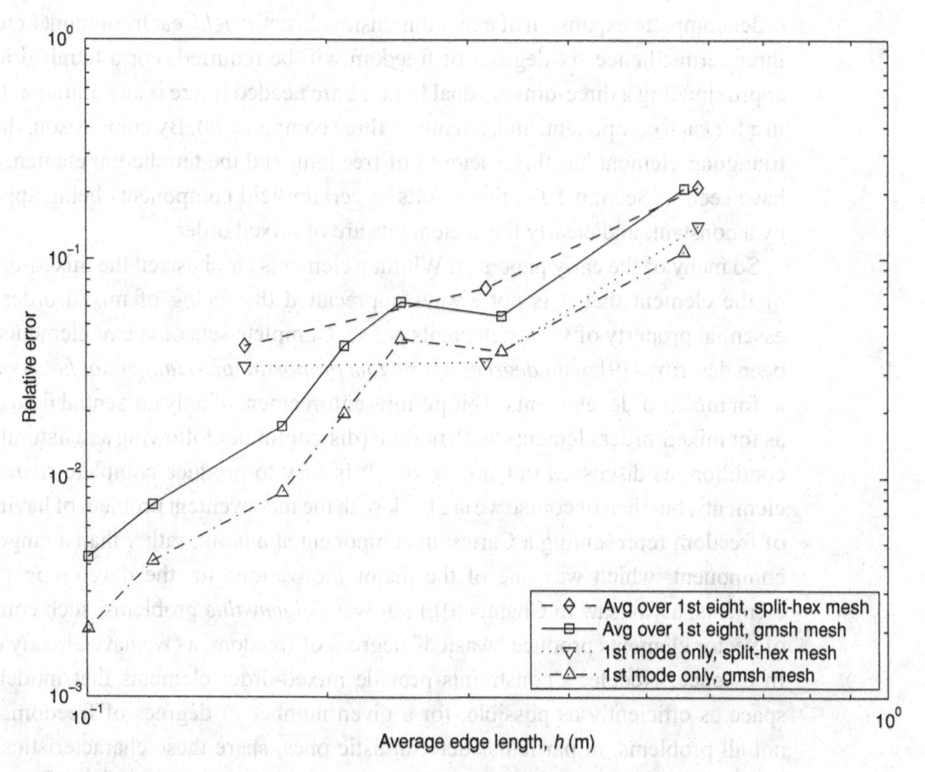

Figure 11.3 Convergence of the CT/LN solution, using two different meshes.

11.2 Higher-order elements

Although extending "edge" elements to higher orders became a topic of interest as soon as the CT/LN elements achieved widespread acceptance, it remained a topic of active research for many years following. Development of such elements raised a number of issues, including: hierarchal versus interpolatory behavior; methods for the construction of the element shape functions; the interpretation of the degrees of freedom; the construction of prototype elemental matrices (analytical versus quadrature); and the efficient iterative solution of the poorly conditioned linear algebra systems which unfortunately often result. Various names have been used: the two-part field description as used in the preceding chapter (e.g. linear tangential/quadratic normal, LT/QN) is particularly insightful and is used here. Before introducing higher-order elements, however, it is worthwhile briefly discussing the question of completeness and vector elements.

11.2.1 Complete versus mixed-order elements

A family of polynomials is *complete* to order N if a linear combination of its members can exactly express any polynomial of degree not exceeding N, but no higher [8, p. 272]. For a complete first-order approximation of a function in x and y, three terms are needed: one constant and two terms linear in x and y respectively. Clearly, for a first-order complete expansion of a two-dimensional vector *field*, each component will require three terms, hence six degrees of freedom will be required. For a tetrahedral element, approximating a three-dimensional field, 12 are needed (there is an additional linear term in z for each component, and, of course, three components). By comparison, the Whitney triangular element has three degrees of freedom, and the tetrahedral element six; as we have seen in Section 10.7, this results in certain field components being approximated by a constant, and clearly these elements are of mixed order.

So many of the early papers on Whitney elements emphasized the mixed-order nature of the element that it is not always appreciated that being of mixed order is not an essential property of vector elements per se. Complete sets of vector elements have also been described [9], *with degrees of freedom proportional to tangential field components*, as for mixed-order elements. This permits enforcement of only tangential field continuity, as for mixed-order elements, with normal (dis)continuity following as a natural boundary condition, as discussed in Chapter 10. (It is easy to produce complete *scalar* or nodal elements, but then of course we are back with the inconvenient problem of having degrees of freedom representing a Cartesian component at a point, rather than a tangential field component, which was one of the major motivations for the development of vector elements, as we saw in Chapter 10.) For wave *eigenvalue* problems, such complete sets of vector elements produce "wasted" degrees of freedom, as we have already discussed. In essence, Nedelec's constraints provide mixed-order elements that model the curl-space as efficiently as possible, for a given number of degrees of freedom. However, not all problems, in particular deterministic ones, share these characteristics. Work by Webb [10] and the present author [11] has indicated that some vector electromagnetic problems are more efficiently analyzed using complete-order vector elements, typically when the solution is dominated by electric fields strongly "gradient" in nature.

11.2.2 Hierarchal vector basis functions

There are presently two competing approaches to higher-order vector elements. One approach is interpolatory; in this case, a degree of freedom is typically associated with

Table 11.3 Webb's hierarchal elements (to second order complete) [10]

CT/LN ($6 \times 1 = 6$ edge-based degrees of freedom)		
Edge-based, $\boldsymbol{W}_1^{(e)}$	1 per edge	$\lambda_i \nabla \lambda_j - \lambda_j \nabla \lambda_i$
Additional LT/LN functions ($6 \times 1 = 6$ extra edge-based degrees of freedom)		
Edge-based, $\boldsymbol{G}_1^{(e)}$	1 per edge	$\nabla \left(\lambda_i \lambda_j\right)$
Additional LT/QN functions ($4 \times 2 = 8$ extra face-based degrees of freedom)		
Face-based, $\boldsymbol{R}_{20}^{(f)}$	2 per face	$\lambda_j \lambda_k \nabla \lambda_i + \lambda_i \lambda_k \nabla \lambda_j - 2\lambda_i \lambda_j \nabla \lambda_k$ for $\{i; j; k\} = \{1; 2; 3\}$ and $\{2; 3; 1\}$
Additional QT/QN functions (6×1 edge-based $+ 4 \times 1$ face-based $= 10$ extra degrees of freedom)		
Edge-based, $\boldsymbol{G}_2^{(e)}$	1 per edge	$\nabla \left(\lambda_i \lambda_j [\lambda_i - \lambda_j]\right)$
Face-based, $\boldsymbol{G}_{20}^{(f)}$	1 per face	$\nabla \left(\lambda_i \lambda_j \lambda_k\right)$

Adapted from [11],

a tangential field at a specific point. The other approach is hierarchal, in which case a specific higher-order set contains all the lower-order basis functions.[3] For mesh refinement/enrichment purposes, hierarchal elements are very useful, and here we consider only the use of such elements, in particular those presented in [10]. (For a comprehensive discussion of interpolatory elements, see [12]. These elements can be used for h-adaptation, but are inconvenient at least for p-adaptation. We will discuss these topics in Section 12.5.) These elemental basis functions are summarized in Table 11.3, along with the number of degrees of freedom per tetrahedron and their respective associations with edges or faces. (The nomenclature "edge-based" or "face-based" is used to indicate that the relevant degree of freedom is associated with the integral of the tangential field over the edge or face respectively; this is discussed in more detail in the next section.) Webb presented the information slightly differently in his paper [10, Tables III, IV and V]; here, the additional gradient-space functions required for the LT/LN and QT/QN elements have been explicitly written as gradients of products of simplex coordinates to highlight this functional dependence. Also shown in the table is the notation used by Webb, and used in Eqs. (10.132)–(10.134). $\boldsymbol{W}_1^{(e)}$ are edge-based Whitney functions, in the space $W_1^{(e)}$; $\boldsymbol{G}_1^{(e)}$ are first-order edge-based gradient functions, in the space $G_1^{(e)}$; $\boldsymbol{R}_{20}^{(f)}$ are second-order face-based rotational functions, in the space $R_{20}^{(f)}$; and $\boldsymbol{G}_{20}^{(f)}$ are second-order face-based rotational functions, in the space $R_{20}^{(f)}$. Note that only the *additional* basis functions required are tabulated, to avoid repetition; i.e. the full second-order QT/QN set of basis functions will include *all* thirty listed. The double-subscript 20 in the face-based second-order functions may appear unnecessary, but it is required for elements of order higher than two. Note also that which two $\boldsymbol{R}_{20}^{(f)}$ functions are chosen of three possible candidates is actually arbitrary, and the restriction on $\{i; j; k\}$ shown

[3] Nodal elements can be both interpolatory and hierarchal; there does not appear to be a proof prohibiting a set of vector elements from having both properties. However, no such vector elements have yet been proposed.

Table 11.4 Comparison of various hierarchal element schemes (to LT/QN)

	CT/LN, all	
Edge-based	1 per edge	$\lambda_i \nabla \lambda_j - \lambda_j \nabla \lambda_i$
	LT/QN, Savage [13]	
Edge-based	1 per edge	$\nabla(\lambda_i \lambda_j)$
Face-based	2 per face	$\lambda_i(\lambda_j \nabla \lambda_k - \lambda_k \nabla \lambda_j)$
	(and $\{j; i; k\}$ but not $\{k; i; j\}$)	
	LT/QN, Webb and Forghani [14]	
Edge-based	1 per edge	$\nabla(\lambda_i \lambda_j)$
Face-based	2 per face	$\lambda_i \lambda_k \nabla \lambda_j$
	(and $\{j; k; i\}$ but not $\{i; j; k\}$)	
	LT/QN, Andersen and Volakis [15]	
Edge-based	1 per edge	$(\lambda_i - \lambda_j)\times$ $(\lambda_i \nabla \lambda_j - \lambda_j \nabla \lambda_i)$
Face-based	2 per face (as for Savage's elements)	$\lambda_i(\lambda_j \nabla \lambda_k - \lambda_k \nabla \lambda_j)$
	LT/QN, Webb [10]	
Edge-based	1 per edge	$\nabla(\lambda_i \lambda_j)$
Face-based	2 per face	$\lambda_j \lambda_k \nabla \lambda_i + \lambda_i \lambda_k \nabla \lambda_j - 2\lambda_i \lambda_j \nabla \lambda_k$ for $\{i; j; k;\} = \{1; 2; 3\}$ and $\{2; 3; 1\}$

After [11], ©2003 IEEE, reprinted with permission.

may be chosen differently. The core issue is that the *same* choice must be made on the faces of connected tetrahedrons.

Webb's approach is elegant in that one progressively enriches the curl space, and then the gradient space. (Earlier proposals did not follow this approach.) For example, moving from CT/LN to LT/LN, one adds elements of the form $\nabla\left(\lambda_i \lambda_j\right)$, one per edge, which is clearly in the gradient space. (The curl of this is identically zero.) This then gives a complete first-order approximation function. Moving from LT/LN to LT/QN, an additional eight face-based degrees of freedom are added, giving 20 vector-based functions and degrees of freedom per tetrahedron.

Note that the lowest-order space, $W_1^{(e)}$, which comprises the Whitney elements, is actually a special case. For the triangular element, this space includes a rotational function, but also two constant functions, which are gradients, and similar comments apply to the tetrahedral element. (It is possible to split these using a tree-cotree decomposition.)

Many other hierarchal elements have been published, in particular of LT/QN order. Some of these are summarized in Table 11.4. This table should serve as a useful summary of some of the various elements in use. Another contribution on hierarchal elements is the work of Sun *et al.* [16]; they use a slightly different approach to construct the elements, but the resulting basis functions are very similar, although not identical, to [10]. Ingelström presented alternate elements in [17]. Most of these (including those of Savage described above) can be seen as variants of the elements originally proposed by Webb and Forghani [14]. (Indeed, not only are these variants on a theme, they are also linear transforms.) As already noted in the context of Webb's elements, all the face elements exclude (arbitrarily) one possible combination of $\{i; j; k\}$; this asymmetry has long been noted, and is required to avoid linearly dependent basis functions.

These elements are generally constructed by "inspection," using the properties of simplex coordinates, and the gradients thereof. Webb's work remains one of the more comprehensive and theoretically motivated developments along these lines to appear in the electrical engineering literature, and has formed the basis for many other subsequent investigations, e.g. [18], which addressed the closely related class of divergence-conforming elements. However, there are some issues with elements of QT/CuN (mixed third order) and higher in his treatment; for a contemporary summary of this, and higher-order elements in general, see [19], which also discusses methods for improving the associated matrix conditioning via scaling of the basis functions.

It is worth investigating the properties of these elements a little further, since some of these are far from trivially obvious. For instance, it is not immediately apparent *why* the higher-order hierarchal elements have degrees of freedom associated with edges, faces or in some cases, with neither of these (the "volume-centered" degrees of freedom).

11.2.3 Properties of hierarchal basis functions

For this, it is useful to return to some basic properties of these elements, as originally laid down by Nedelec [20]. (It should be commented that not all vector elements which have been proposed satisfy his criteria, but those presently under discussion do.) Nedelec focussed on degrees of freedom, rather than basis functions; indeed, his original work simply states the necessary properties, rather than proposing actual basis functions. The degrees of freedom as he defined them are not unique,[4] even for the lowest-order (Whitney) element, although in practice the non-uniqueness is only a matter of a constant for the lowest-order case and does not impact on the code at all. However, as seen in the previous section, a variety of different basis functions have been proposed for higher-order elements.

This is rather cryptically implied in Definition 4 of Nedelec's original work [20]. For "kth" mixed-order elements, the $6k$ edge-based degrees of freedom for 3D elements ($3k$ in 2D) should be given by

$$\int_a \boldsymbol{u} \cdot \hat{t} q \, dC, \qquad \forall q \in P_{k-1} \tag{11.22}$$

where $\boldsymbol{u}$ is a basis function and $\hat{t}$ is the unit vector along edge a. P_k is the linear space of polynomials of degree $\leq k$. For the Whitney element, with $k = 1$, we see that q may only be a constant. In the case of this element, with $(\lambda_i \nabla\lambda_j - \lambda_j \nabla\lambda_i)$ form, this constant is often implicitly unity, and the associated Nedelec degree of freedom (which may be viewed as located at the middle of the relevant edge, although this is not essential) is the tangential field on this edge. We commented earlier in Section 10.7 that it may be shown

[4] The polynomial space described is, however.

that the integral of the tangential component of the Whitney element along an edge is constant; we will now do this.

This proof is rather simple. Integrating the Whitney element along an edge yields two integrals. The first is of the form

$$\int_a \lambda_i(\nabla\lambda_j \cdot \hat{t})\, dC \tag{11.23}$$

and the other has i and j interchanged, and is of opposite sign. Throughout the element, $\nabla\lambda_j$ is constant, and it is perpendicular to the edge opposite node j (this was discussed in Section 10.7). Clearly, $\nabla\lambda_j \cdot \hat{t}$ is thus also a constant along any particular edge. Along the edge directed from nodes i to j, what remains is an integral of a simplex coordinate, varying linearly from 0 to 1, along the edge. The result is $\pm 1/2\ell$, with ℓ the edge length, and the sign depending on the direction of integration. Clearly, this is a constant, as is the other integral. Obviously, incorporating additional constants, such as Nedelec's q, changes only the final constant, which is irrelevant in practice. The result is as in Appendix A,

$$\boldsymbol{E}_{\tan}|_{\text{edge}_i} = \frac{E_i}{\ell_i} \tag{11.24}$$

When $\boldsymbol{E}_{\tan}|_{\text{edge}_i}$ is integrated along edge i, the result is the well-known identity that the appropriate degree of freedom is the tangential field along edge i.

Importantly, on the other two sides, one or the other simplex coordinate will be zero, and the other *entirely normal to the edge*. Thus Eq. (11.22) will yield zero for this term on the other two edges (due to the $\boldsymbol{u} \cdot \hat{t}$ term). The argument is precisely the same for tetrahedra.

Additional edge-based degrees of freedom, as required for Webb's scheme for LT/LN-order elements and higher, of the form

$$\nabla(\lambda_i\lambda_j) = \lambda_i\nabla\lambda_j + \lambda_j\nabla\lambda_i$$

yield exactly the same result – they contribute additional degrees of freedom on edge $\{i; j\}$ and nothing to the other edges. (Note that a different choice of q may be required in this case, otherwise the degree of freedom is zero. A linear function is an obvious possibility, such as a suitable Legendre polynomial. Note also that these functions reside in the space $k = 2$.)

Now, the face-based elements. Nedelec's original definition of the $4k(k-1)$ face-based degrees of freedom for higher-order elements of maximum (but not complete) order k was

$$\iint_f \boldsymbol{u} \times \hat{n} \cdot \boldsymbol{q}\, dS, \qquad \forall \boldsymbol{q} \in (P_{k-2})^2 \tag{11.25}$$

Here, $\hat{n}$ is the unit vector normal to edge f, and the polynomial q is now two dimensional (for $k = 2$, this must be a constant). Let us now see why these additional degrees of freedom, which enrich the curl space for the LT/QN element, are associated *only* with faces. We will consider vector elements of the form $\lambda_i(\lambda_j\nabla\lambda_k - \lambda_k\nabla\lambda_j)$; the Webb LT/QN enrichment in Table 11.3 is a linear combination of two such forms, so the argument includes these. On face i, j, k, one of the simplex coordinates will always be zero on each edge; for example, λ_i is zero on edge $\{j, k\}$, so these do not contribute to the edge-based degrees of freedom. (This extends to faces, e.g. λ_i is zero everywhere on face $\{j, k, l\}$. Hence this basis function will have no tangential projection on any other face.)

Over face $\{i, j, k\}$, the degrees of freedom are thus:

$$\iint_f \lambda_i\lambda_j(\nabla\lambda_k \times \hat{n}) \cdot \boldsymbol{q}\, dS - \iint_f \lambda_i\lambda_k(\nabla\lambda_j \times \hat{n}) \cdot \boldsymbol{q}\, dS \tag{11.26}$$

The $(\nabla\lambda_k \times \hat{n})$ and $(\nabla\lambda_j \times \hat{n})$ terms are constant over this face, as is q, and what remains are two standard integrals in simplex form, proportional to the triangle area and thus constant.

The higher-order elements (quadratic tangential/cubic normal, QT/CuN, etc.) involve additional "volume-centered" degrees of freedom. These each involve products of *all* the simplex coordinates, so are clearly zero on *all* faces and edges. For these basis functions, the associated degree of freedom as defined by Nedelec is a weighted integral over the volume.

11.2.4 Practical impact of higher-order basis functions in an FEM code

The discussion in the preceding section may appear highly theoretical, so it is worthwhile summarizing the practical impact hereof. Finite element codes do not usually actually compute the degrees of freedom as defined by Eqs. (11.22) and (11.25), since this usually serves no particular purpose. The "degrees of freedom" for which an FEM code solves are usually simply the unknowns associated with each basis function; as we have seen, for the Webb elements (and most other properly defined vector elements) these degrees of freedom can be correctly associated with edges, faces or the volume, and for the first two, the degrees of freedom are tangential field projections onto the edge or face, as required by Nedelec's original work. To enforce field continuity correctly, a degree of freedom associated with an edge or face must simply be shared between all connected elements; we discussed this in the context of edges in Section 11.1.2, and also mentioned there that same holds for faces. The node numbering scheme proposed there ensured that the directions were consistent between elements for both edges and faces. Volume-centered degrees of freedom have no projection on the edges or faces and hence are *not* shared by adjoining elements.

Higher-order 3D elements require a little more in the way of new data structures than do their 2D counterparts. There are four different types of degrees of freedom, and keeping track of which is which is important for post-processing purposes (it is not necessary if one only wants to compute eigenvalues). It is also wise to separate flags for edges and faces from the degrees of freedom. As with 2D analysis, if an edge does not lie on a prescribed boundary, then both the CT/LN degree of freedom and the LT/LN degree of freedom are free to vary. Unlike the 2D analysis, faces must also be flagged as free or prescribed, which impacts on the two face-based LT/QN degrees of freedom per face. When assembling the system, some simplification can be made by noting that both edge-based types of degrees of freedom have the same free or prescribed properties on an edge; similarly, both face-based degrees of freedom share the same properties in this regard.

As already noted in Section 10.8.6, from the viewpoint of keeping matrix bandwidth to a minimum, the degrees of freedom would best be numbered on an element-by-element basis, but to start with, numbering all the CT/LN degrees of freedom consecutively, then the LT/LN ones and finally the LT/QN ones is adequate.

Regarding matrix assembly, one can repeat the analysis of Section 11.1.1 and derive analytical expressions for the mixed second-order elements, but even more so than in the case of the 2D analysis, quadrature is very attractive. The equivalent fourth-order rule is given in Table 11.5 for a tetrahedron of unit volume.

Coding hints – quadrature tables

Tabulated quadrature rules need to be used with caution. (Note also that quadrature in higher dimensions is sometimes called cubature.) Firstly, the rules are not always consistent in terms of the normalization. As an example, the rule in Table 11.5 in the original reference [21] differs by a factor of 6 from that tabulated here. The arises from the volume of a right-angled tetrahedron with unit sides, 1/6. Similarly, for a triangle, a factor of 1/2 is often required (the area of right-angled triangle with unit sides), and many rules for quadrature on a line are for the interval $[-1, 1]$. Secondly, some published rules contain errors.

A very easy check on the validity of a table is to sum the weights, which should give unity (or indicate what the normalizing constant is); similarly, for a given quadrature point i, $\sum_{j=1}^{m+1} \lambda_{ji} = 1$, where m is the dimensionality of the problem . More elaborate tests should verify that the rule accurately integrates polynomials of the specified accuracy. Here, the formula for integration over a triangle or tetrahedron, Eq. (D.3) and (D.4) in Appendix D, provide very useful reference results. As an example, the product of four first-order polynomials in all four of the simplex coordinates should yield $\frac{1}{840}V$, on a tetrahedron of volume V.

One point to watch is that some rules include quadrature points on the edges. The author has observed reduced order accuracy when applying these rules to vector elements. This probably due to the special properties along edges of vector elements. (Indeed, with special quadrature rules, this can be exploited to "lump" the mass matrix, a topic which will be discussed in Section 12.3.) Rules with interior points appear to be more reliable for general quadrature.

Finally, the curl of the elements should be pre-computed. The result for first of the two hierarchal Webb functions per face [10] of Table 11.3, $\boldsymbol{R}_{20}^{(f1)i1,i2,i3}$ – with the face triple-node-index added to Webb's notation – is

$$\nabla \times \boldsymbol{R}_{20}^{(f)i1,i2,i3} = -3\lambda_{i2}\boldsymbol{v}_{i1,13} - 3\lambda_{i1}\boldsymbol{v}_{i2,13}. \tag{11.27}$$

Table 11.5 Symmetric $n = 11$ point rule for quadrature on a tetrahedron, degree of precision 4, after [21], including a factor of 6 omitted in that reference

w_i	λ_{1i}	λ_{2i}	λ_{3i}	λ_{4i}
−0.078 933 333 333 333	0.250 000 000 000 000	0.250 000 000 000 000	0.250 000 000 000 000	0.250 000 000 000 000
0.045 733 333 333 333	0.071 428 571 428 571	0.071 428 571 428 571	0.071 428 571 428 571	0.785 714 285 714 286
0.045 733 333 333 333	0.071 428 571 428 571	0.071 428 571 428 571	0.785 714 285 714 286	0.071 428 571 428 571
0.045 733 333 333 333	0.071 428 571 428 571	0.785 714 285 714 286	0.071 428 571 428 571	0.071 428 571 428 571
0.045 733 333 333 333	0.785 714 285 714 286	0.071 428 571 428 571	0.071 428 571 428 571	0.071 428 571 428 571
0.149 333 333 333 333	0.399 403 576 166 799	0.399 403 576 166 799	0.100 596 423 833 201	0.100 596 423 833 201
0.149 333 333 333 333	0.399 403 576 166 799	0.100 596 423 833 201	0.399 403 576 166 799	0.100 596 423 833 201
0.149 333 333 333 333	0.399 403 576 166 799	0.100 596 423 833 201	0.100 596 423 833 201	0.399 403 576 166 799
0.149 333 333 333 333	0.100 596 423 833 201	0.100 596 423 833 201	0.399 403 576 166 799	0.399 403 576 166 799
0.149 333 333 333 333	0.100 596 423 833 201	0.399 403 576 166 799	0.100 596 423 833 201	0.399 403 576 166 799
0.149 333 333 333 333	0.100 596 423 833 201	0.399 403 576 166 799	0.399 403 576 166 799	0.100 596 423 833 201

The curl of the other function, viz. $\boldsymbol{R}_{20}^{(f)i2,i3,i1}$, can be obtained by cyclic permutation of the indices. Analytical expressions for other higher-order elements may be found in [11].[5]

In Section 10.8.6, some coding hints were given for higher-order triangular elements. With higher-order tetrahedral elements, coding is a little more complex, since face-based degrees of freedom may now also be either prescribed or free to vary. The code stub shown in Fig. 11.4 shows one way of doing the assembly. It is not especially elegant, as one needs to handle edge-based and face-based degrees of freedom separately – at least as coded here (and note that the stub does not show the code for the face-face assembly). The data structures will be familiar from Section 11.1.2. The function `sandt3D_LTQN` returns the elemental stiffness and mass matrices and is not discussed in detail here; it uses the quadrature scheme discussed above. The data are packed as follows:

$$[S] = \begin{bmatrix} S_{e1,e1} & S_{e1,e2} & S_{e1,f1} & S_{e1,f2} \\ S_{e2,e1} & S_{e2,e2} & S_{e2,f1} & S_{e2,f2} \\ S_{f1,e1} & S_{f1,e2} & S_{f1,f1} & S_{f1,f2} \\ S_{f2,e1} & S_{f2,e2} & S_{f2,f1} & S_{f2,f2} \end{bmatrix} \tag{11.28}$$

and similarly for $[S]$. Here, $e1$ refers to the six CT/LN (Whitney) edge-based functions, $e2$ to the six additional edge-based LT/LN functions in the first-order gradient space, and $f1$ and $f2$ to the four LT/QN face-based functions in the second-order rotational space of Table 11.3. The data structures `dof_e1`, `dof_e2`, `dof_f1` and `dof_f2` are the edge- and face-based degrees of freedom, generalizing the `dof` data structure of Section 11.1.2. As coded, an entry for a given edge (face) in these structures either returns the global degree of freedom number (for one which is free to vary) or a zero (for a prescribed edge (face)). Note that the MATLAB logical AND function (&) that is used treats any non-zero value as a logical one.[6] Note also that the code stub shown in Fig. 11.4 does not include the material parameters; a general-purpose code must handle this on an element-by-element basis.

Results for a mixed second-order solution (using the hierarchal Webb elements of Table 11.3) for the PEC cavity discussed in Section 11.1.2 are given in Fig. 11.5. The slopes for the average over the first eight curves are 2.22 and 3.71 for the CT/LN and LT/QN elements respectively, comparing well with the theoretically expected values of 2 and 4. (The slopes for the dominant mode are very similar, 2.32 and 3.56 respectively.) Note that refining 3D meshes is not as straightforward as refining 2D meshes; reducing h does not always refine the mesh reasonably uniformly in all directions, hence the error behavior is usually a little more erratic in 3D than in 2D, as shown here. It is the overall trends which are important.

As mentioned in the context of the 2D higher-order implementation in Section 10.8.6, a higher-order solution has more degrees of freedom, N, for the same mesh than the corresponding lower-order one. Nonetheless, as in 2D, the higher-order elements are

[5] The appendix of that paper should read "Curls of other elements."

[6] For coding efficiency, the "short-circuit" && function is used. A short-circuit operator evaluates the second operand only when the result is not fully determined by the first operand.

```
for ielem=1:N
    tetnodes = ELEMENTS(ielem,:);
            [S_elem,T_elem] = sandt3D_LTQN( vertices(tetnodes(1),:),...
                                 vertices(tetnodes(2),:),...
                                 vertices(tetnodes(3),:),...
                                 vertices(tetnodes(4),:) );
            for jedge = 1:6 % edge-edge interactions
                jj_e1 = dof_e1(ELEMENT_EDGES(ielem,jedge));
                jj_e2 = dof_e2(ELEMENT_EDGES(ielem,jedge));
                for kedge = 1:6
                    kk_e1 = dof_e1(ELEMENT_EDGES(ielem,kedge));
                    kk_e2 = dof_e2(ELEMENT_EDGES(ielem,kedge));
                    if jj_e1 && kk_e1  % i.e. both edges free
                        S(jj_e1,kk_e1) = S(jj_e1,kk_e1)+S_elem(jedge,kedge);
                        T(jj_e1,kk_e1) = T(jj_e1,kk_e1)+T_elem(jedge,kedge);
                        S(jj_e2,kk_e1) = S(jj_e2,kk_e1)+S_elem(jedge+6,kedge);
                        T(jj_e2,kk_e1) = T(jj_e2,kk_e1)+T_elem(jedge+6,kedge);
                        S(jj_e2,kk_e2) = S(jj_e2,kk_e2)+S_elem(jedge+6,kedge+6);
                        T(jj_e2,kk_e2) = T(jj_e2,kk_e2)+T_elem(jedge+6,kedge+6);
                        % Use symmetry for e1e2 interaction
                        S(kk_e1,jj_e2) = S(jj_e2,kk_e1);
                        T(kk_e1,jj_e2) = T(jj_e2,kk_e1);
                    end
                end
            end

            % Face-edge and vice-versa interactions.
            for jface = 1:4
                jj_f1 = dof_f1(ELEMENT_FACES(ielem,jface));
                jj_f2 = dof_f2(ELEMENT_FACES(ielem,jface));
                for kedge = 1:6
                    kk_e1 = dof_e1(ELEMENT_EDGES(ielem,kedge));
                    kk_e2 = dof_e2(ELEMENT_EDGES(ielem,kedge));
                    if jj_f1 && kk_e1  % i.e. face and edge free
                        S(jj_f1,kk_e1) = S(jj_f1,kk_e1)+S_elem(jface+12,kedge);
                        T(jj_f1,kk_e1) = T(jj_f1,kk_e1)+T_elem(jface+12,kedge);
                        S(jj_f1,kk_e2) = S(jj_f1,kk_e2)+S_elem(jface+12,kedge+6);
                        T(jj_f1,kk_e2) = T(jj_f1,kk_e2)+T_elem(jface+12,kedge+6);
                        S(jj_f2,kk_e1) = S(jj_f2,kk_e1)+S_elem(jface+16,kedge);
                        T(jj_f2,kk_e1) = T(jj_f2,kk_e1)+T_elem(jface+16,kedge);
                        S(jj_f2,kk_e2) = S(jj_f2,kk_e2)+S_elem(jface+16,kedge+6);
                        T(jj_f2,kk_e2) = T(jj_f2,kk_e2)+T_elem(jface+16,kedge+6);
                        % Use symmetry for face-edge entries:
                        S(kk_e1,jj_f1) = S(jj_f1,kk_e1);
                        T(kk_e1,jj_f1) = T(jj_f1,kk_e1);
                        S(kk_e2,jj_f1) = S(jj_f1,kk_e2);
                        T(kk_e2,jj_f1) = T(jj_f1,kk_e2);
                        S(kk_e1,jj_f2) = S(jj_f2,kk_e1);
                        T(kk_e1,jj_f2) = T(jj_f2,kk_e1);
                        S(kk_e2,jj_f2) = S(jj_f2,kk_e2);
                        T(kk_e2,jj_f2) = T(jj_f2,kk_e2);
                    end
                end
            end
            % Face -face interactions
            .
            .
            .
```

Figure 11.4 MATLAB code stub for the assembly of the 3D FEM system matrix for LT/QN elements.

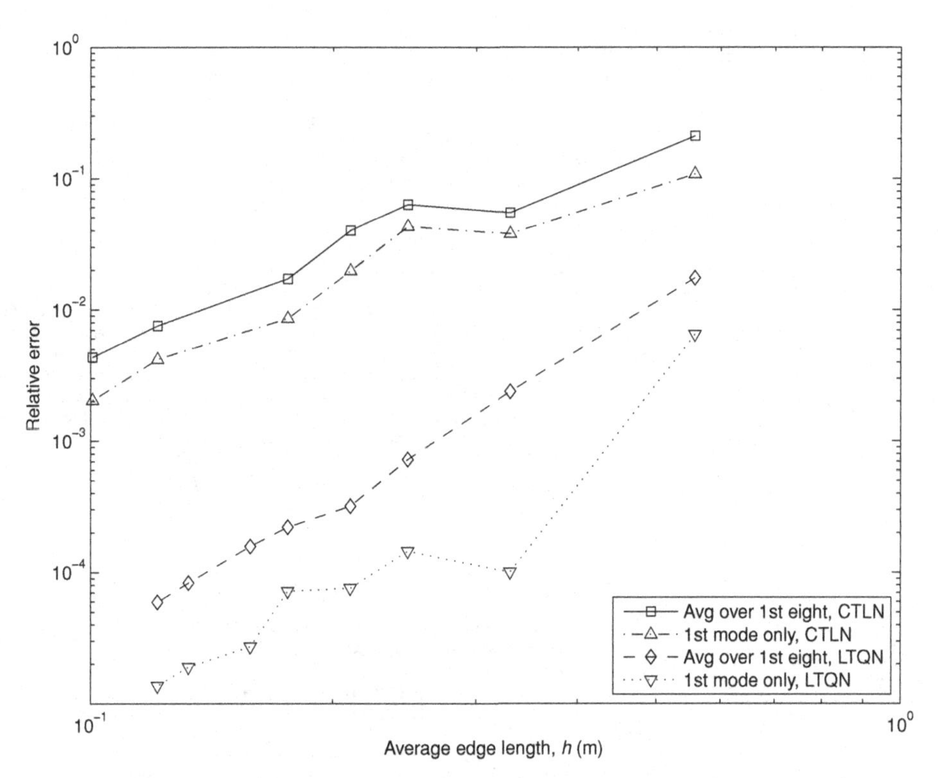

Figure 11.5 Comparison of the convergence of the CT/LN and LT/QN solutions versus edge length, using meshes created using `gmsh`.

clearly superior even when plotted against N, as is done in Fig. 11.6. In particular for fine meshes (large N), it will be noted that the error in an LT/QN solution is usually at least an order of magnitude lower than the corresponding CT/LN one for the same N, providing strong motivation for the use of these elements.

For most practical applications, the extra work involved in implementing a mixed second-order FEM solution is well worth it, with the greatly improved accuracy quickly repaying the additional coding investment. However, for mixed third- and higher-order elements, the trade-off is less clear, and the significant extra coding work involved must be balanced again the required accuracy for the problem at hand. Furthermore, higher-order elements can become extremely ill-conditioned, in particular hierarchal ones, as the element order increases; several published approaches attempt to improve this by partial orthogonalization of the higher-order functions, which adds further coding complexity. From a practical perspective, there are often, if not usually, so many other uncertainties in the CEM modelling process that the extremely high accuracy such elements offer is frequently compromised by other factors. One in particular is the limitation to rectilinear geometries. This can be addressed using curvilinear elements, which are more likely to be useful for general modelling when combined with mixed second-order elements than very high order rectilinear elements. Curvilinear elements are briefly addressed at the end of the next chapter.

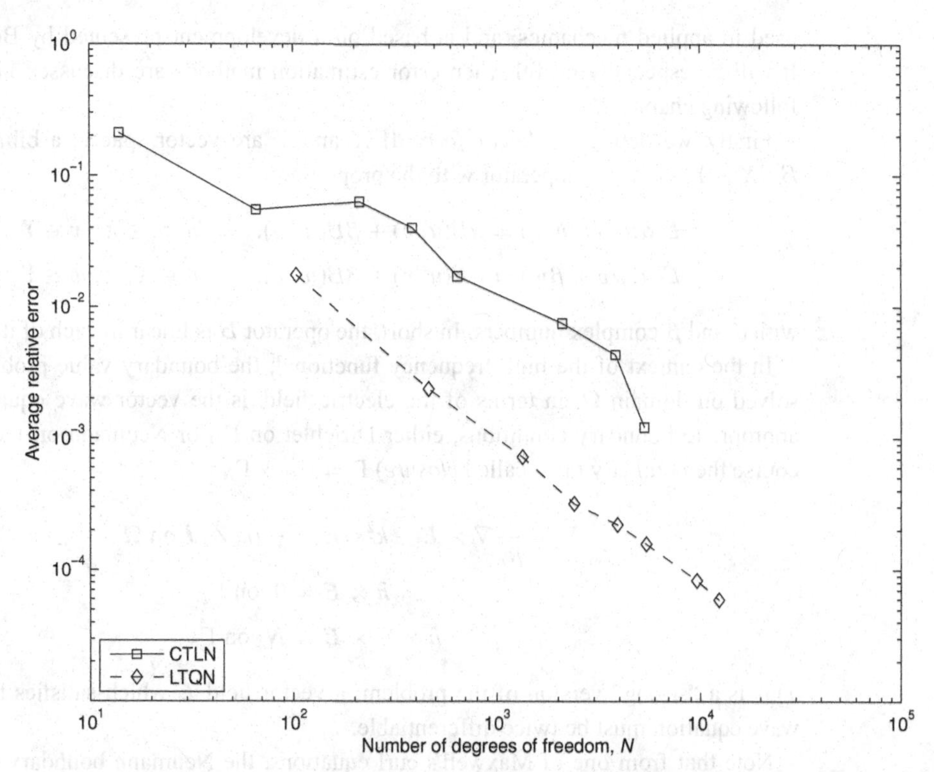

Figure 11.6 Comparison of the convergence of the CT/LN and LT/QN solutions versus number of degrees of freedom, using meshes created using gmsh.

A tricky problem which was surprisingly neglected in the literature until recently was the question of how to match fields when using hierarchal elements to an actual specified field, as required by a Dirichlet boundary condition, for instance. With interpolatory elements, this would be trivial, since each degree of freedom would, by design, correspond to a tangential field component at a point on each element. With hierarchal elements however, this is only uniquely defined for Whitney elements. Webb [22] has very elegantly addressed this issue for higher-order hierarchal elements using the elements in [10]. As we have seen, starting with the conventional Whitney elements, Webb's elements enrich alternately the gradient and curl spaces. Webb exploits this in [22] to match alternately the tangential components of the electric field, and then the normal component of the electric field (the curl space). Since any such matching using hierarchal elements is approximate, he uses a projective approach to improve the accuracy of the matching. This was extended by Petersson and Jin [23].

11.3 The FEM from the variational boundary value problem viewpoint

It is useful at this stage to introduce some further ideas from functional analysis, extending the introduction in Section 4.5. This approach is strongly influenced by the methods used in applied mechanics, and is based on a development presented by Botha [24]. It will be especially useful when error estimation methods are discussed later in the following chapter.

Firstly, we define a *bilinear form*. If X and Y are vector spaces, a bilinear form $\mathcal{B} : X \times Y \to \mathbb{C}$ is an operator with the properties

$$\begin{aligned} \mathcal{B}(\alpha u + \beta w, v) &= \alpha\mathcal{B}(u, v) + \beta\mathcal{B}(w, v), \qquad u, w \in X,\ v \in Y \\ \mathcal{B}(u, \alpha v + \beta w) &= \alpha\mathcal{B}(u, v) + \beta\mathcal{B}(w, v), \qquad u \in X,\ v, w \in Y \end{aligned} \tag{11.29}$$

with α and β complex numbers. In short, the operator $\mathcal{B}$ is linear in each of its "slots."

In the context of the high-frequency functional, the boundary value problem to be solved on domain Ω, in terms of the electric field, is the vector wave equation with appropriate boundary conditions, either Dirichlet on Γ_D or Neumann on Γ_N, with of course the boundary (also called *closure*) $\Gamma = \Gamma_D + \Gamma_N$:

$$\begin{aligned} \nabla \times \frac{1}{\mu_r} \nabla \times \boldsymbol{E} - k_0^2 \varepsilon_r \boldsymbol{E} = -jk_0 Z_o \boldsymbol{J} \text{ on } \Omega \\ \hat{n} \times \boldsymbol{E} = 0 \text{ on } \Gamma_D \\ \hat{n} \times \nabla \times \boldsymbol{E} = \boldsymbol{N} \text{ on } \Gamma_N \end{aligned} \tag{11.30}$$

This is a "strong" version of the problem; a vector field $\boldsymbol{E}$ which satisfies the vector wave equation must be twice differentiable.

Note that from one of Maxwell's curl equations, the Neumann boundary condition can also be written as

$$\hat{n} \times \boldsymbol{H} = \frac{j}{\mu\omega} \boldsymbol{N} \text{ on } \Gamma_N \tag{11.31}$$

Thus, the Neumann boundary condition can be seen equivalently as a constraint on tangential $\boldsymbol{H}$.

Using a method of weighted residuals approach, with an arbitrary testing function $\boldsymbol{W}$, and otherwise proceeding in a very similar manner to that of Section 10.2.4, it may be shown that the following is the "weak" representation of the boundary value problem represented by Eq. (11.30):

$$\begin{aligned} &\iiint_V \left[\frac{1}{\mu_r} (\nabla \times \boldsymbol{E}) \cdot (\nabla \times \boldsymbol{W}) - k_0^2 \epsilon_r \boldsymbol{E} \cdot \boldsymbol{W} \right] dV \\ &\quad - \iint_{\Gamma_D} \frac{1}{\mu_r} (\nabla \times \boldsymbol{E}) \cdot (\hat{n} \times \boldsymbol{W}) \, dS \\ &= - \iint_{\Gamma_N} \frac{1}{\mu_r} \boldsymbol{N} \cdot \boldsymbol{W} dS - jk_0 Z_0 \iiint_V \boldsymbol{J} \cdot \boldsymbol{W} dS \end{aligned} \tag{11.32}$$

with $\hat{n} \times \boldsymbol{E} = 0$ on Γ_D.

A symmetry argument can be used to establish that $\boldsymbol{W}$ must also satisfy the homogeneous boundary condition on Γ_D, so that the surface integral over Γ_D on the left-hand side falls away. Alternately, as in Section 10.3, it can simply be noted that on specified boundaries, no weighting function is required. The final form of the variational boundary value problem is

$$\mathcal{B}(\boldsymbol{E}, \boldsymbol{W}) = \mathcal{L}(\boldsymbol{W}) \qquad \forall\, \boldsymbol{W} \in W,\ \boldsymbol{E} \in W \tag{11.33}$$

The bilinear and linear forms are defined as

$$\mathcal{B}(\boldsymbol{E}, \boldsymbol{W}) = \iiint_V \left[\frac{1}{\mu_r} (\nabla \times \boldsymbol{E}) \cdot (\nabla \times \boldsymbol{W}) - k_0^2 \epsilon_r \boldsymbol{E} \cdot \boldsymbol{W} \right] dV \tag{11.34}$$

$$\mathcal{L}(\boldsymbol{W}) = -\iint_{\Gamma_N} \frac{1}{\mu_r} \boldsymbol{N} \cdot \boldsymbol{W}\, dS - jk_0 Z_0 \iiint_V \boldsymbol{J} \cdot \boldsymbol{W}\, dS \tag{11.35}$$

The space in which the solution and testing vector functions lie is defined as

$$W = \{\boldsymbol{a} \in H(\text{curl}, \Omega) | \hat{n} \times \boldsymbol{a} = 0 \text{ on } \Gamma_D\} \tag{11.36}$$

This is the space of curl-conforming vector basis functions which we have already discussed, with the additional constraint of the homogeneous Dirichlet boundary condition.

In this development, the Neumann boundary condition has been "absorbed" into the variational boundary value problem – via the first term on the right-hand side of Eq. (11.35). (It will be recalled that in Section 10.2.4, a similar result was obtained in the context of a homogeneous Neumann boundary condition.) The Dirichlet boundary condition must however be explicitly enforced via a restriction on the space W. (This sounds more complex than it is; as was seen in Section 10.8, this is implemented in practice by zeroing the prescribed edges.)

With the variational boundary value problem established, one can then proceed to demonstrate that the stationary functional representation of the problem is the following:

$$F(\boldsymbol{E}) = \frac{1}{2}\mathcal{B}(\boldsymbol{E}, \boldsymbol{E}) - \mathcal{L}(\boldsymbol{E}), \qquad \boldsymbol{E} \in W \tag{11.37}$$

This is the familiar curl-curl functional, which we used in the preceding chapter (although the linear term was zero for the eigenvalue problem). Note that this (and indeed the variational boundary value problem from which the stationary functional form is obtained) is known as a "weak" form; the differentiability requirements on the solution space have been reduced (the function $\boldsymbol{E}$ need only be once differentiable now).

11.4 A deterministic 3D application: waveguide obstacle analysis

11.4.1 Introduction

The analysis of waveguide discontinuities has been a canonical problem for analytical, approximate and now numerical approaches since the pioneering work of Marcuvitz and colleagues during World War II. Using variational formulations, and quasi-static approximations of the fields, Marcuvitz *et al.* were able to analyze an extraordinary variety of problems, documented in the classic text originally published in 1951 and now fortunately available again [25]. Subsequently, mode-matching methods were introduced for the analysis of "stepped" discontinuities – i.e. structures where the waveguide modes could be computed in a stepwise fashion, and matched at two-dimensional planes. However, for general, arbitrary discontinuities, and of course those involving non-metallic discontinuities such as dielectrics, differential equation based methods such as the finite element method (FEM) and finite difference time domain (FDTD) method are now the methods of choice. In this section, we will first present the formulation for this, which also affords the opportunity to deal with the more general version of the curl-curl functional as discussed in Section 11.3, and then analyze a waveguide device using both a code developed by the present author, as well as a commercial FEM package.

11.4.2 The waveguide formulation

The formulation to be discussed is a straightforward extension of Jin's approach [26], published by the present author in [27]. His formulation addressed two-port, single-mode analysis, with the waveguide oriented in the $\hat{z}$-direction. Here, general waveguide orientation(s) will be considered. The formulation assumes hollow, rectangular guide at the ports (although the extension to homogeneously filled guide is straightforward). The TE_{10} mode is assumed in the following. In between the ports, in the region to be discretized using finite elements, the waveguide may contain linear, inhomogeneous, lossy, dielectric and/or magnetic material(s); and/or conductors (for instance, posts or irises); and may change orientation (e.g. E-plane bends) or dimension (e.g. E- and/or H-plane steps). The formulation to be used does, however, assume isotropic media. The generalization of the analysis to multiple ports, the inclusion of higher-order modes, and the extension to more general waveguide, will be outlined subsequently.

Formulation overview

The key part of the formulation is to write the electric field at port 1 (S_1) as the sum of the known incident and unknown reflected fields in terms of the (ξ, η, ζ) coordinate system local to the port, with ζ in the local direction of propagation, and set to zero at each port, as follows:

$$\begin{aligned} \boldsymbol{E}(\xi,\eta,\zeta) &= \boldsymbol{E}^{\mathrm{inc}}(\xi,\eta,\zeta) + \boldsymbol{E}^{\mathrm{ref}}(\xi,\eta,\zeta) \\ &= (E_0\boldsymbol{e}_{10}(\xi,\eta)e^{-jk_{\zeta_{10}}\zeta} + RE_0\boldsymbol{e}_{10}(\xi,\eta)e^{+jk_{\zeta_{10}}\zeta})|_{\zeta=0} \end{aligned} \tag{11.38}$$

$\boldsymbol{e}_{10}(\xi, \eta)$ is the relevant waveguide eigenmode (the TE_{10} eigenmode here) and $k_{\zeta_{10}}$ is the modal propagation constant. Note that it is necessary to retain the $e^{-jk_{\zeta_{10}}\zeta}$ term, even though the field is evaluated at $\zeta = 0$, since the boundary condition to be discussed involves the derivative of the field, which must be evaluated *before* setting $\zeta = 0$.

The next key element of the formulation is to convert Eq. (11.38) to a boundary condition of the third type involving both the field and its normal derivative. Such boundary conditions can be incorporated in the bilinear functional, as will be seen shortly. The detail is given in [26, Section 8.5], briefly, the result is

$$\hat{n} \times (\nabla \times \boldsymbol{E}) + \gamma\hat{n} \times (\hat{n} \times \boldsymbol{E}) = \boldsymbol{U}^{\mathrm{inc}} \tag{11.39}$$

with

$$\gamma = jk_{\zeta_{10}}, \qquad \boldsymbol{U}^{\text{inc}} = -2jk_{\zeta_{10}}\boldsymbol{E}^{\text{inc}} \tag{11.40}$$

It should be noted that, in obtaining Eq. (11.39), the transverse only nature of the TE field is exploited. TM modes contain axial $\boldsymbol{E}$ field components, and the boundary condition cannot thus be written for an $\boldsymbol{E}$ field solver. TM mode analysis could be undertaken by using an $\boldsymbol{H}$ field solver.

The same is repeated at port 2, but at that port, there is only an unknown transmitted field:

$$\begin{aligned} \boldsymbol{E}(\xi,\eta,\zeta) &= \boldsymbol{E}^{\text{trans}}(\xi,\eta,\zeta) \\ &= TE_0\boldsymbol{e}_{10}(\xi,\eta)e^{-jk_{\zeta_{10}}\zeta}|_{\zeta=0} \end{aligned} \tag{11.41}$$

Similar comments apply as regards the $e^{-jk_{\zeta_{10}}\zeta}$ term. The boundary condition at port 2 is

$$\hat{n}\times(\nabla\times\boldsymbol{E}) + \gamma\hat{n}\times(\hat{n}\times\boldsymbol{E}) = 0 \tag{11.42}$$

In Jin's original formulation, the phase was referenced to each port. In the present formulation, the transmission coefficient T incorporates the "insertion" phase, i.e. for a section of empty guide length ℓ, T will have phase angle $-k_{z_{10}}\ell$. This produces the same phase that would be measured using a vector network analyzer, with reference planes calibrated at the ports.

The equivalent variational functional (assuming isotropic but possibly lossy materials), subject to these boundary conditions on the ports and $\boldsymbol{E}_{\tan} = 0$ on the perfectly conducting walls, is

$$\begin{aligned} F(\boldsymbol{E}) = \frac{1}{2}\iiint_V &\left[\frac{1}{\mu_r}(\nabla\times\boldsymbol{E})\cdot(\nabla\times\boldsymbol{E}) - k_0^2\epsilon_r\boldsymbol{E}\cdot\boldsymbol{E}\right] dV \\ &+ \iint_{S_1}\left[\frac{\gamma}{2}(\hat{n}\times\boldsymbol{E})\cdot(\hat{n}\times\boldsymbol{E}) + \boldsymbol{E}\cdot\boldsymbol{U}^{\text{inc}}\right] dS \\ &+ \iint_{S_2}\left[\frac{\gamma}{2}(\hat{n}\times\boldsymbol{E})\cdot(\hat{n}\times\boldsymbol{E})\right] dS \end{aligned} \tag{11.43}$$

This can be obtained from the development in Section 11.3. In this case, in the Neumann boundary condition of Eq. (11.30), repeated here,

$$\hat{n}\times\nabla\times\boldsymbol{E} = \boldsymbol{N} \text{ on } \Gamma_N$$

the vector function $\boldsymbol{N}$ is $\boldsymbol{U}^{\text{inc}} - \gamma\hat{n}\times(\hat{n}\times\boldsymbol{E})$; this is substituted into the linear operator $\mathcal{L}$ of Eq. (11.35), and a vector identity is used to shift one of the $\hat{n}\times\hat{n}\times$ operators to the weighting function.

For readers interested in the details of the finite element discretization of this functional, [26] and [27] are recommended.

Computation of the S-parameters

The above formulation produces R and T for port 1 (S_{11} and S_{21}). It must be repeated with an incident field at port 2 to obtain S_{12} and S_{22}. Only the excitation vector changes, so this is simply a question of repeating the matrix solve. For multiple ports, the extension is obvious: T is computed at *each* port, producing one column of the S matrix. The excitation is then repeated at each port to produce other columns.

The S-parameters may be computed directly from the fields on the ports. A more accurate approach uses the orthogonality of the modes to integrate the fields computed over each port [26, Section 8.5]; as an example, for two ports the transmission coefficient is given by

$$T = \frac{2}{abE_0} \iint_{S_2} \boldsymbol{E}(\xi, \eta, \zeta) \cdot \boldsymbol{e}_{10}(\xi, \eta)\, dS \tag{11.44}$$

As before, $\boldsymbol{e}_{10}(\xi, \eta)$ is the relevant waveguide eigenmode; a and b are the waveguide dimensions.

The waveguide formulation: another perspective

The formulation can be viewed as a finite element/boundary integral (FE/BI) formulation, using the waveguide Green function for "exact" mesh termination. (For radiation or scattering problems, FE/BI formulations use the free-space, or sometimes the half-space, Green function, e.g. [26, Section 10.4]; this is discussed in the final chapter.) The current dominant-mode-only analysis uses only the first in the infinite series of modes comprising the waveguide Green function. It is accurate provided that the ports are sufficiently far removed from the discontinuities (assuming, of course, that only the dominant mode is above cut-off). Higher-order modes are easily included in the formulation; this does require re-computing both the left-hand side matrix and right-hand side vector, since the former has one term dependent on the propagation constant, and the latter is obviously dependent on the incident mode shape. The formulation presently assumes hollow waveguide at the ports, i.e. only TE (and TM modes, if an $\boldsymbol{H}$ field solver is also implemented) are included. More exotic modes, or numerically determined ones, could also be incorporated into the formulation (see, for example, [26, Chapter 11]).

11.5 Application to two waveguide discontinuity problems

With the formulation in hand, we will now proceed to analyze two waveguide discontinuity problems. The first demonstrates multi-port analysis; the second demonstrates the use of complete vector basis functions. The latter is based on [11].

11.5.1 Application to a Magic-T

Introduction

The "Magic-T" (see Fig. 11.7) is a 180° hybrid. Such devices are four-port structures, with the following interesting properties. A signal applied to port 1 is evenly split into

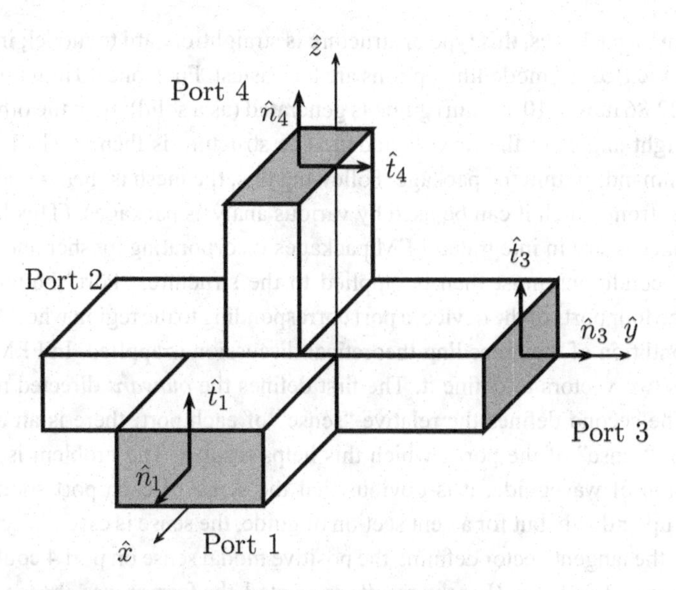

Figure 11.7 The Magic-T hybrid waveguide junction.

two in-phase components at ports 2 and 3, and port 4 is isolated. Conversely, a signal applied to port 4 is evenly split, but with 180° phase difference, between ports 2 and 3, and port 1 is isolated. It can also be operated as a combiner, in which case when input signals are applied to ports 2 and 3, the sum appears at port 1 and the difference at port 4. Ideally, the *S*-parameters of the device are [7, p. 402]:

$$[S] = \frac{-j}{\sqrt{2}} \begin{bmatrix} 0 & 1 & 1 & 0 \\ 1 & 0 & 0 & -1 \\ 1 & 0 & 0 & 1 \\ 0 & -1 & 1 & 0 \end{bmatrix} \tag{11.45}$$

The 180° hybrids can be made in various ways, e.g. microstrip or stripline, the "rat race" being a very popular implementation in planar technologies. The Magic-T is a waveguide implementation of a 180° hybrid [7, p. 411].

This is an example of a structure where the approximate analytical techniques of Marcuvitz *et al.* are unable to provide useful data and a full-wave solution becomes imperative. (A complex equivalent circuit is presented for the Magic-T [25, p. 386], but only measured data at one frequency point are provided.) The behavior of the waveguide Magic-T departs very significantly from the ideal of Eq. (11.45), as will be seen.

Setting up the problem

The setup procedure for the junction as shown in Fig. 11.7 illustrates a number of features one would expect in any RF FEM code. The specific code described in detail is FEMFEKO, which was an experimental FEM code using FEKO-like input and output files; it was not made available for general use, although some features were incorporated within later releases of FEKO. The meshing was done using a commercial FEM mesher,

FEMAP. In most packages, this type of structure is straightforward to model; in FEMAP, for instance, the `Solid` modelling options are the easiest. First, one 40 mm long section of X-band (22.86 mm $\times$ 10.16 mm) guide is generated (as a solid); then the other 40 mm section (at right angles to the first) is added. The structure is then meshed, using the meshing commands within the package. Following this, the mesh is then `export`-ed as a neutral file, from which it can be used by various analysis packages. (This last step is of course unnecessary in integrated FEM packages incorporating mesher and solver.)

Boundary conditions must then be applied to the structure. "Port" boundaries are required at the four ports of the device, a port corresponding to the region where the modal boundary condition of the preceding theoretical discussion is applied. In FEMFEKO, a port requires two vectors to define it. The first defines the *outward* directed normal on each port. The second defines the relative "sense" of each port; there is an ambiguity regarding the "sense" of the ports, which this helps resolve. The problem is that for a straight section of waveguide, it is obvious that the sense of each port should be the same, either up or down, but for a bent section of guide, the sense is essentially arbitrary. For instance, the tangent vector defining the positive modal sense on port 4 could equally well be chosen as $+\hat{y}$ or $-\hat{y}$. (For the results presented, the former was chosen. On ports 1, 2 and 3, $+\hat{z}$ was chosen as the tangent vector.) Various packages deal with this issue in different ways.

This problem was also solved using a commercial package, ANSOFT's HFSS code. Constructing the finite element model is very similar to the procedure described above, but since the mesher is integrated within the package, it is appreciably more user-friendly. Nonetheless, the requirement of correctly specifying boundaries in particular rests with the user. HFSS meshes the structure automatically, and then refines the mesh until a user-specified level of accuracy is reached (usually, a negligible change in S-parameters from one iterative pass to the next).

For the results to be presented, the geometrical primitive cubes which defined port 1 had a length of 40 mm (i.e. approximately 20 mm of guide from the junction), for ports 2 and 3 they were 30 mm (also approximately 20 mm of guide) and for port 4, 30 mm (again, approximately 20 mm of guide). These lengths were based on the results for other waveguide structures; the requirement is that there be sufficient length to allow evanescent modes to die out before the ports. As in our 2D eigenvalue problem, the waveguide was an X-band guide with dimensions 22.86 mm $\times$ 10.16 mm.

Results

This geometry has no simple analytical solution, as already discussed. To obtain data to compare with these results, ANSOFT's HFSS code was used to generate another FEM solution of the problem. The HFSS model was identical in size to the FEMAP (and FEMFEKO) models, so that phase results could also be compared.

The S-parameter data for port 1 are presented in Figs. 11.8 and 11.9. Port 4 is indeed isolated; since S_{41} is very small, some discrepancy between the FEMFEKO and HFSS results is to be expected. Ports 2 and 3 show equal, in-phase, power splitting. Note, however, how far S_{11} departs from the theoretical ideal of no reflection at port 1. A brief consideration of the problem shows that this is not unexpected, since the waveguide fed

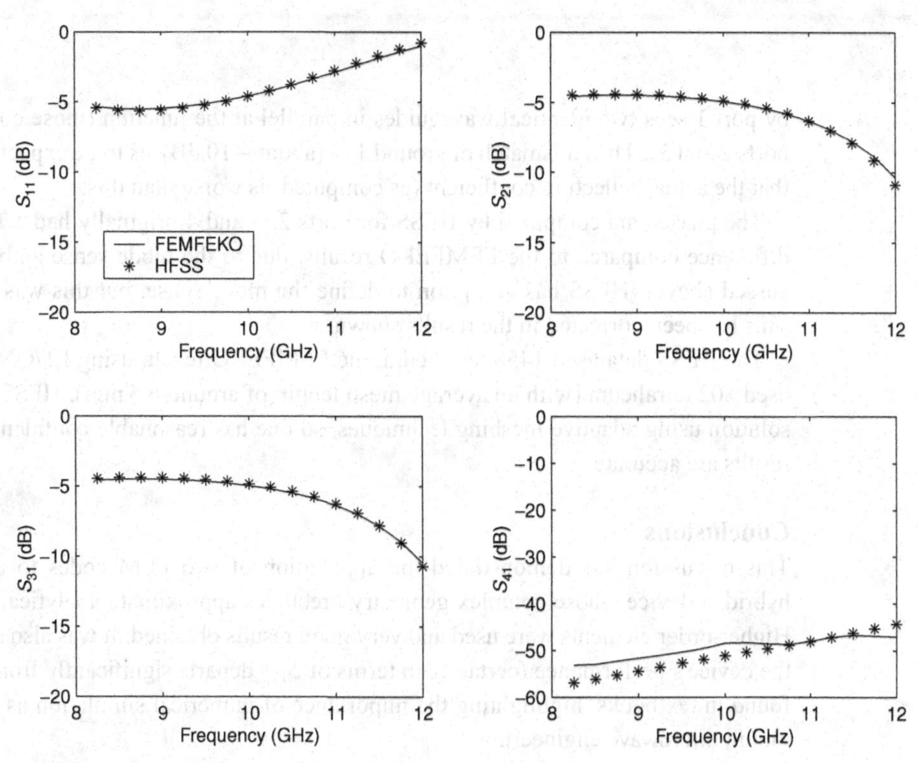

Figure 11.8 *S*-parameters of the Magic-T for port 1; magnitude.

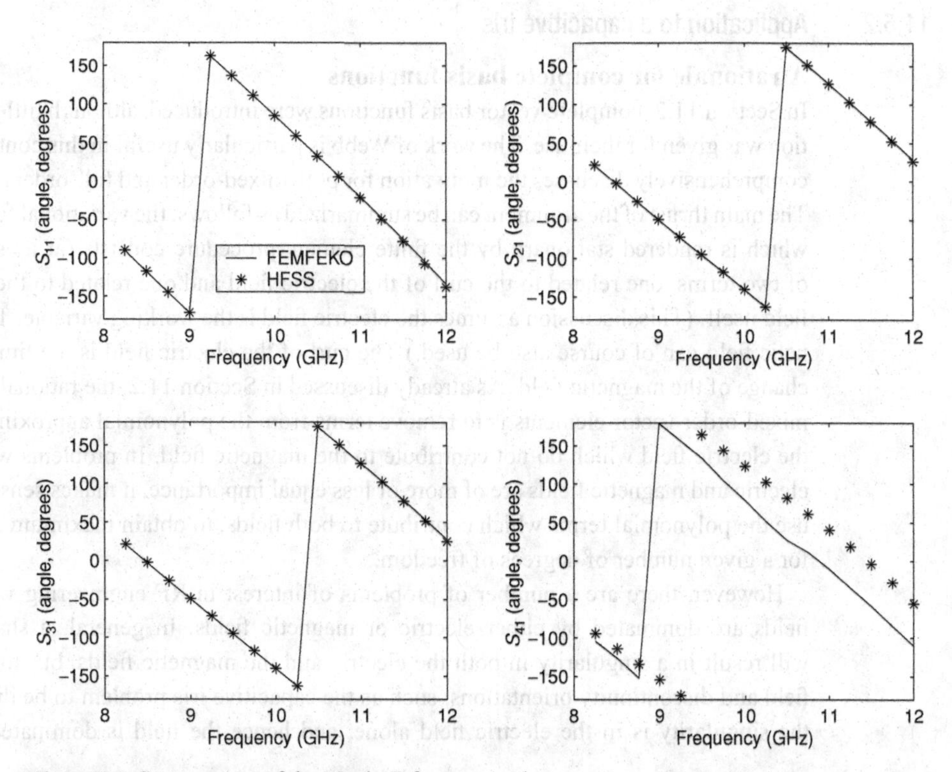

Figure 11.9 *S*-parameters of the Magic-T for port 1; phase.

by port 1 sees two identical waveguides in parallel at the junction (those connected to ports 2 and 3). Thus mismatch of around 1/3 (about -10 dB)[7] is to be expected. We see that the actual reflection coefficient (as computed) is worse than this.

The phase data computed by HFSS for ports 2, 3 and 4 originally had a 180° phase difference compared to the FEMFEKO results, due to the mode sense ambiguity discussed above. (HFSS has an option to define the mode sense, but this was not used.) This has been corrected in the results shown.

The HFSS data used 1458 tetrahedra; the FEMFEKO result, using LT/QN elements, used 802 tetrahedra (with an average mesh length of around 6.5 mm). HFSS refines its solution using adaptive meshing techniques, so one has reasonable confidence that the results are accurate.

Conclusions

This discussion has demonstrated the application of two FEM codes to a Magic-T hybrid, a device whose complex geometry precludes approximate analytical solutions. Higher-order elements were used and very good results obtained. It was also shown that the device's performance (certainly in terms of S_{11}) departs significantly from the ideal found in textbooks, highlighting the importance of numerical simulation as a valuable tool in microwave engineering.

11.5.2 Application to a capacitive iris

A rationale for complete basis functions

In Section 11.2, complete vector basis functions were introduced, although little motivation was given for their use. The work of Webb is particularly useful in this context; [10] comprehensively discusses the motivation for both mixed-order and full-order elements. The main thrust of the argument can be summarized as follows: the variational functional which is rendered stationary by the finite element procedure consists (at its simplest) of two terms, one related to the curl of the electric field and one related to the electric field itself. (This discussion assumes the electric field is the working variable. The magnetic field can of course also be used.) The curl of the electric field is the time rate of change of the magnetic field. As already discussed in Section 11.2, the rationale behind mixed-order vector elements is to remove terms from the polynomial approximation of the electric field which do not contribute to the magnetic field. In problems where the electric and magnetic fields are of more or less equal importance, it makes sense only to use the polynomial terms which contribute to both fields, to obtain maximum accuracy for a given number of degrees of freedom.

However, there are a number of problems of interest in RF engineering where the fields are dominated by either electric or magnetic fields. In general, a sharp edge will result in a singularity in both the electric and the magnetic fields, but for certain field and discontinuity orientations, such as the capacitive iris problem to be discussed, the singularity is in the electric field alone, and hence the field is dominated by the

[7] The reflection coefficient of a system with a load equal to half the characteristic impedance.

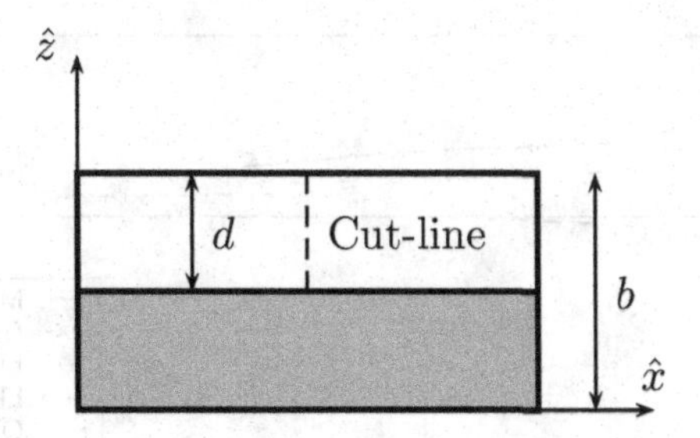

Figure 11.10 The capacitive iris. (After [11], ©2003 IEEE, reprinted with permission.)

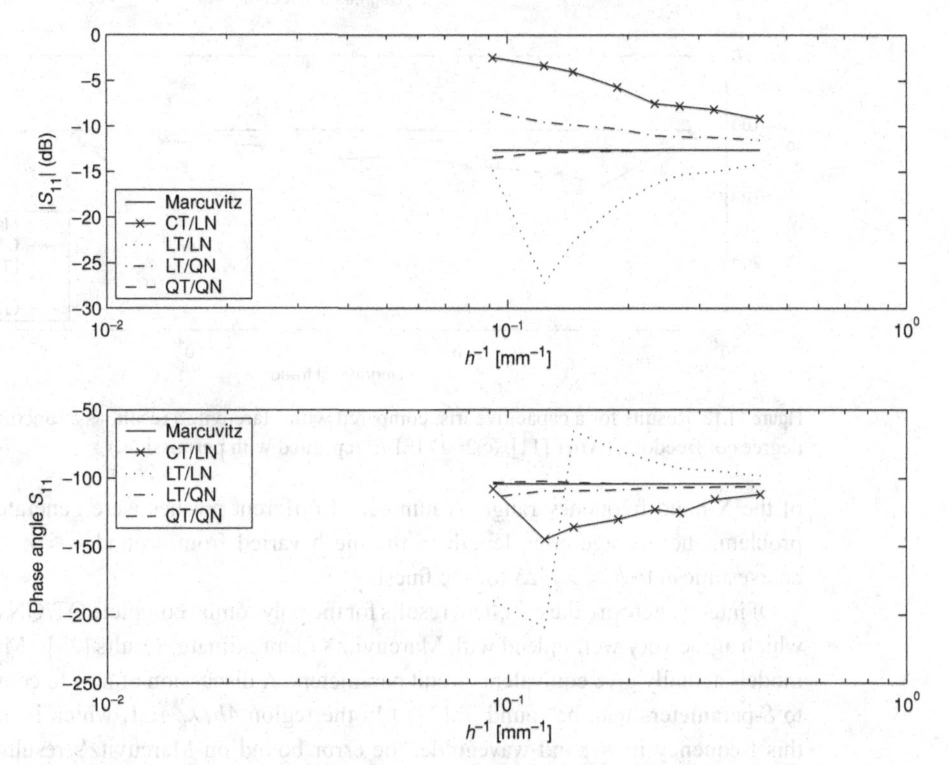

Figure 11.11 Results for a capacitive iris, compared with Marcuvitz's result, as a function of (inverse) mesh size. (After [11], ©2003 IEEE, reprinted with permission.)

quasi-static electric field behavior. Hence it can be expected that full-order elements should be useful for such problems.

Results

Here, a capacitive iris is considered.[8] The metallic iris, shown in Fig. 11.10, is half the height of the waveguide, and again, the analysis is performed at X-band. The results shown in Figs. 11.11 and 11.12 were computed at 8.25 GHz, towards the bottom end

[8] This example was first published as [11].

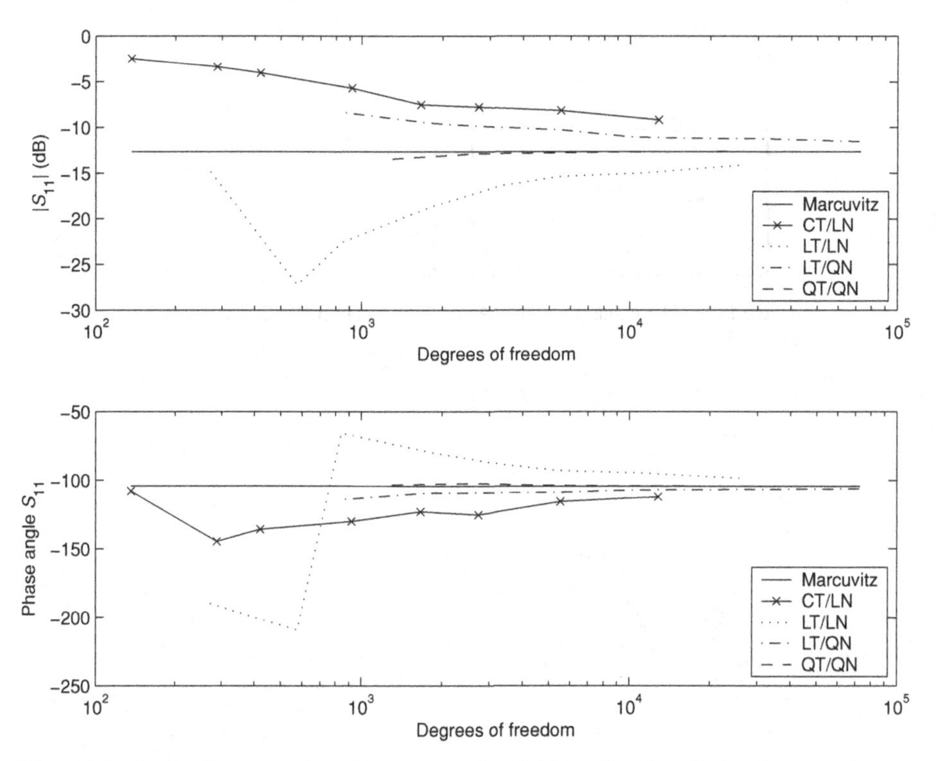

Figure 11.12 Results for a capacitive iris, compared with Marcuvitz's result, as a function of degrees of freedom. (After [11], ©2003 IEEE, reprinted with permission.)

of the X-band frequency range. A number of different meshes were generated for the problem; the average edge length in the mesh varied from around $h \approx \lambda_g/6$ for the coarsest mesh to $h \approx \lambda_g/25$ for the finest.

Of interest here are the excellent results for the polynomial complete QT/QN elements, which agree very well indeed with Marcuvitz's (approximate) results [25]. (Marcuvitz's models actually give equivalent circuit parameters. A discussion of how to convert these to S-parameters may be found in [27].) In the region $4b/\lambda_g < 1$, which is the case at this frequency in X-band waveguide, the error bound on Marcuvitz's results is given as within 1%, a result verified by this QT/QN FEM solution. It is clear that LT/QN elements converge very slowly to the correct solution for this problem. A commercial FEM code using conventional mixed-order elements also produced unconverged results for this problem, despite incorporating adaptive mesh refinement techniques.

In retrospect, it is clear that this is an especially difficult problem for general-purpose finite element solvers. Even with quite a fine mesh overall, it is likely that the mesh above the iris may only be two or three elements "thick" (this could be improved by manual "seeding"). This of course is precisely the direction in which the field is changing most rapidly, and furthermore, the electric field is strongly dominated by the quasi-static field with singular behavior. To describe this field adequately, one would expect to need full-order elements, thus also approximating the gradient space as accurately as possible (for the given maximum element order available).

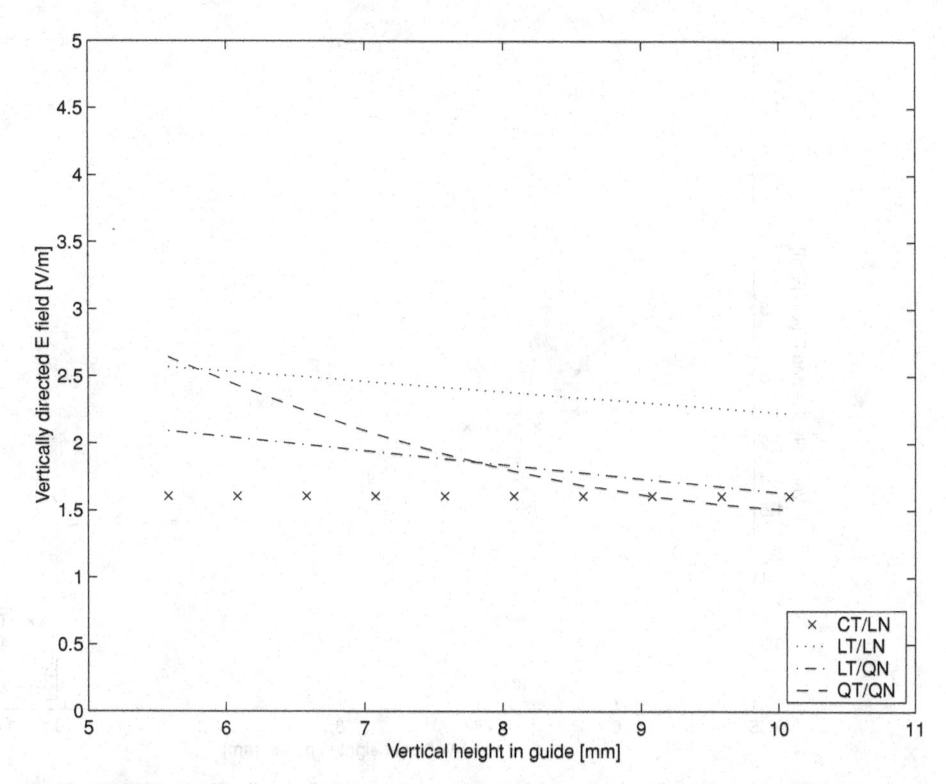

Figure 11.13 Vertically directed electric field along a line in the center of the guide, directly above the iris. Coarse mesh, $h \approx \lambda_g/6$. (After [11], ©2003 IEEE, reprinted with permission.)

It is also of interest to note that the relative improvement of the QT/QN elements compared to the LT/QN ones appears more marked than the improvement of LT/LN over CT/LN. It is quite possible that the mesh in the vicinity of the iris (as discussed above) is limiting the performance of the linear elements – indeed, only the finest mesh in the above results had three elements' "thickness" above the iris.

The above results, using S-parameters, concentrate on what are essentially integrated field quantities (also known as observables). It is also of interest to examine the actual field behavior in the vicinity of the iris. In Figs. 11.13 and 11.14, the vertically directed electric field (the field component aligned with the TE_{10} mode electric field) on a vertical line directly above the iris is plotted. (The modal excitation E_0, see Eq. (11.38), at the port was 1 V/m, in this and subsequent plots.) The cut-line is located in the center of the width of the waveguide, shown in Fig. 11.10 by the dashed line. The half-height iris runs from 0 to 5.08 mm; the figures show the field from 5.08 to 10.16 mm, the roof of the guide. The superior performance of the QT/QN elements is clear in these figures; even in the fine mesh case, Fig. 11.14, the CT/LN results are poor, and evidence a considerable (and non-physical) discontinuity at around 7.5 mm. The LT/LN results are close to the LT/QN results, and the discontinuity evidenced by the CT/LN results has gone. The QT/QN results in both cases give the largest field value at the iris, indicating superior modelling of the field in this case.

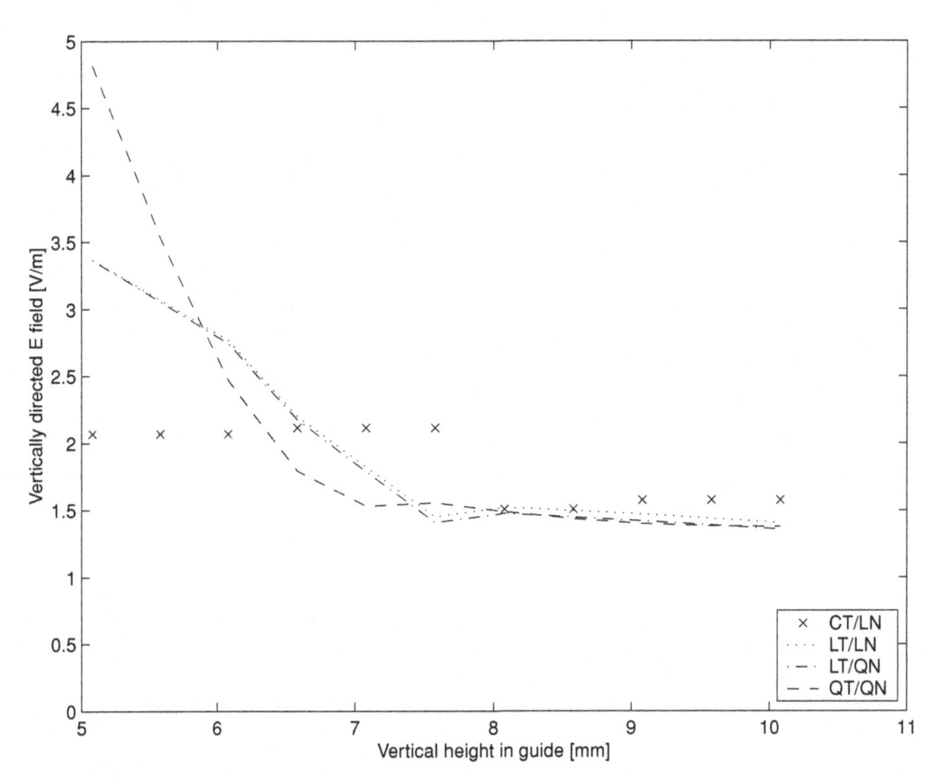

Figure 11.14 Vertically directed electric field along a line in the center of the guide, directly above the iris. Fine mesh, $h \approx \lambda_g/20$. (After [11], ©2003 IEEE, reprinted with permission.)

Some further comments here, especially on the CT/LN results, are called for. In the coarse mesh case, the mesh generator produced only one row of elements above the iris; for the finer mesh, it produced two. In Fig. 11.13, the cut-line ran on the boundary of an element, hence the uniform CT/LN result is to be expected. In Fig. 11.14, the cut-line went through four elements (the mesh was not symmetrical about the center-line), hence the four distinct and different values on the plot. In both these figures the CT/LN results are plotted *only* at the points where the field was computed, to avoid an incorrect linear interpolation being imposed by the plotting program.

It might be argued that this comparison is unfair, since obviously the QT/QN solution involves many more degrees of freedom than, for example, the CT/LN solution on the same mesh. This is not so in this case. The CT/LN results shown in Fig. 11.14 used 5523 degrees of freedom; the QT/QN solution in Fig. 11.13 used 1302, and the solution quality of the latter is clearly far better than that of the former, for fewer degrees of freedom. (The issue of the potentially slower convergence of the higher-order elements will not be considered, since appropriate preconditioners can rectify this problem [16].)

Discussion

This capacitive iris problem has clearly highlighted the utility of full-order elements for problems where quasi-static electric fields dominate the solution. Furthermore, electric field results for this problem have demonstrated that full-order elements can provide enhanced field modelling for a similar (or sometimes even smaller) computational effort in situations where the field itself, rather than an integrated quantity such as the transmission or reflection coefficient, is of primary concern. An interesting idea is to consider how finite element solvers might automatically identify the appropriate element type in different regions; some preliminary results show promise [24, Chapter 6; 17]. Work has also recently appeared on independently controlling the gradient and rotational polynomial orders [28].

11.6 Open-region finite element method formulations: absorbing boundary conditions (ABCs)

As we have seen, finite element formulations offer powerful methods for the numerical solution of electromagnetic fields in inhomogeneous media. However, when simulating open (or unbounded) regions, some method to terminate the finite element mesh at a finite distance is required. Such methods are generally known as radiation boundary conditions (RBCs). Various RBC schemes have been proposed and implemented, including mathematical absorbing boundary conditions – requiring special treatment of "boundary" elements – and more recently, perfectly matched layers. In Chapter 3, we studied the application of both these methods within the context of the FDTD, and these methods have also been used in FEM approaches. A very simple (although not very accurate) first-order scheme will now be discussed, followed by a brief review of some other approaches. In the next chapter, a rigorous scheme for mesh termination, viz. the hybrid FEM/MoM, will be presented.

RBCs based on the physical properties of scattered/radiated fields are amongst the most established types of mesh termination schemes, and are widely known simply as absorbing boundary conditions (ABCs). These are based on constructing a surface differential operator which annihilates outgoing waves of some given format. In 1980, Bayliss and Turkel proposed a scheme for spherical boundaries, where an mth-order differential operator is constructed which annihilates the first m terms in a Wilcox expansion of the outgoing field (a convergent series in inverse powers of r) [29]. This approach has been applied to the electromagnetic vector wave equation in [30] and [31]. In the following, the approach and notation of [26, Chapter 9] is largely followed.

The widely used first-order accurate ABC is given by

$$\hat{r} \times (\nabla \times \boldsymbol{E}^{\mathrm{sc}}) \approx -jk_0 \hat{r} \times (\hat{r} \times \boldsymbol{E}^{\mathrm{sc}}) \tag{11.46}$$

in a free-space region. (It can be generalized to a uniform region.) It is clear that this is a Neumann boundary condition. Reference to Eq. (11.30) shows that on a spherical bounding surface with $\hat{r} = \hat{n}$, $N = -jk_0 \hat{r} \times (\hat{r} \times \boldsymbol{E}^{\mathrm{sc}})$. This can be written as

$$\hat{r} \times (\nabla \times \boldsymbol{E}^{\mathrm{sc}}) + P(\boldsymbol{E}^{\mathrm{sc}}) = 0 \tag{11.47}$$

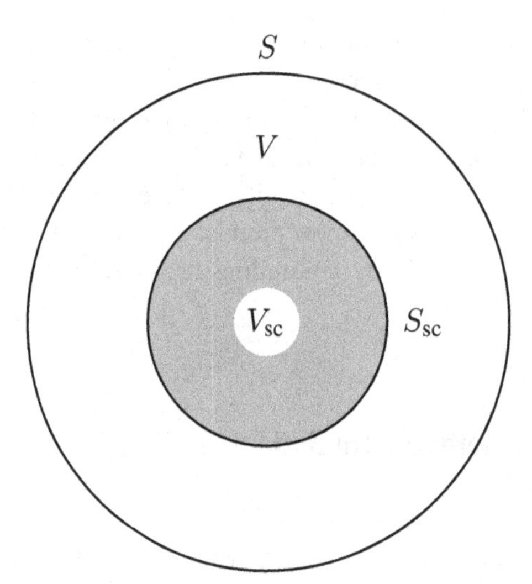

Figure 11.15 The scattered field formulation: definition of volumes and surfaces. V (and associated closure S) is the entire FEM domain, including V_{sc}; V_{sc} (and associated closure S_{sc}) is the region containing the scatterer.

with

$$P(\boldsymbol{E}^{sc}) = jk_0\hat{r} \times (\hat{r} \times \boldsymbol{E}^{sc}) \tag{11.48}$$

11.6.1 Formulation in terms of the scattered field

It will be noted that the ABC operates on the *scattered* field. By decomposing the total field into the scattered field and incident field (as discussed at length in Section 3.2.3), and since the incident field also of course satisfies the wave equation, a formulation may be derived entirely in terms of the scattered fields in V. This can substituted directly in Eq. (11.35). It may be shown that the variational functional, Eq. (11.37), in terms of the scattered field, is given by [26, Eq. (9.75)]:

$$\begin{aligned} F(\boldsymbol{E}^{sc}) = &\frac{1}{2}\iiint_V \left[\frac{1}{\mu_r}(\nabla \times \boldsymbol{E}^{sc})\cdot(\nabla \times \boldsymbol{E}^{sc}) - k_0^2\varepsilon_r \boldsymbol{E}^{sc}\cdot\boldsymbol{E}^{sc}\right] dV \\ &+ \iiint_{V_{sc}} \left[\frac{1}{\mu_r}(\nabla \times \boldsymbol{E}^{sc})\cdot(\nabla \times \boldsymbol{E}^{inc}) - k_0^2\varepsilon_r \boldsymbol{E}^{sc}\cdot\boldsymbol{E}^{inc}\right] dV \\ &- \frac{1}{2}\iint_S \boldsymbol{E}^{sc}\cdot P(\boldsymbol{E}^{sc})dS + \iint_{S_{sc}} \boldsymbol{E}^{sc}\cdot(\hat{n}\times\nabla\times\boldsymbol{E}^{inc})dS \end{aligned} \tag{11.49}$$

In the above, V is the entire FE volume, with bounding surface S, and V_{sc} the volume of the scatterer, and S_{sc} the surface of the scatterer. Internal sources have been excluded for simplicity (see Fig. 11.15). Obviously, V_{sc} must be entirely contained within V, and similarly S_{sc} within S. Using the vector identity $\boldsymbol{a}\cdot(\boldsymbol{b}\times\boldsymbol{c}) = \boldsymbol{c}\cdot(\boldsymbol{a}\times\boldsymbol{b})$, this can also be written as

$$F(\boldsymbol{E}) = \frac{1}{2}\iiint_V \left[\frac{1}{\mu_r}(\nabla\times\boldsymbol{E}^{\text{sc}})\cdot(\nabla\times\boldsymbol{E}^{\text{sc}}) - k_0^2\varepsilon_r\boldsymbol{E}^{\text{sc}}\cdot\boldsymbol{E}^{\text{sc}}\right]dV$$
$$+\iiint_{V_{\text{sc}}}\left[\frac{1}{\mu_r}(\nabla\times\boldsymbol{E}^{\text{sc}})\cdot(\nabla\times\boldsymbol{E}^{\text{inc}}) - k_0^2\varepsilon_r\boldsymbol{E}^{\text{sc}}\cdot\boldsymbol{E}^{\text{inc}}\right]dV$$
$$+\frac{1}{2}\iint_S jk_0(\hat{r}\times\boldsymbol{E}^{\text{sc}})\cdot(\hat{r}\times\boldsymbol{E}^{\text{sc}})dS + \iint_{S_{\text{sc}}}\boldsymbol{E}^{\text{sc}}\cdot(\hat{n}\times\nabla\times\boldsymbol{E}^{\text{inc}})dS \tag{11.50}$$

For a PEC scatterer, the volume integral over V_{sc} is zero, and this simplifies appreciably to

$$F(\boldsymbol{E}) = \frac{1}{2}\iiint_V [(\nabla\times\boldsymbol{E}^{\text{sc}})\cdot(\nabla\times\boldsymbol{E}^{\text{sc}}) - k_0^2\boldsymbol{E}^{\text{sc}}\cdot\boldsymbol{E}^{\text{sc}}]\,dV$$
$$-\frac{1}{2}\iint_S \boldsymbol{E}^{\text{sc}}\cdot P(\boldsymbol{E}^{\text{sc}})dS \tag{11.51}$$

For a PEC scatterer, the integral over S_{sc} in Eq. (11.50) involves only prescribed values, and vanishes when the functional is rendering stationary, making no contribution to the system matrix. The incident field enters *only* as a prescribed (Dirichlet) boundary condition on the surface S_{sc} of the scatterer, via the boundary condition for a PEC, $\hat{n}\times\boldsymbol{E}^{\text{sc}} = \hat{n}\times\boldsymbol{E}^{\text{inc}}$.

11.6.2 Formulation in terms of the total field

It is frequently convenient to have a formulation in terms of the total field as a working variable. Substituting $\boldsymbol{E}^{\text{sc}} = \boldsymbol{E} - \boldsymbol{E}^{\text{sc}}$ into Eq. (11.46), an expression in terms of the total field is obtained:

$$\hat{r}\times(\nabla\times\boldsymbol{E}) + P(\boldsymbol{E}) = \hat{r}\times(\nabla\times\boldsymbol{E}^{\text{inc}}) + P(\boldsymbol{E}^{\text{inc}}) \tag{11.52}$$

This is a Cauchy boundary condition. The appropriate variational functional is [26, Eq. (9.61)]

$$F(\boldsymbol{E}) = \frac{1}{2}\iiint_V \left[\frac{1}{\mu_r}(\nabla\times\boldsymbol{E})\cdot(\nabla\times\boldsymbol{E}) - k_0^2\varepsilon_r\boldsymbol{E}\cdot\boldsymbol{E}\right]dV$$
$$+\frac{1}{2}\iint_S [jk(\hat{r}\times\boldsymbol{E})\cdot(\hat{r}\times\boldsymbol{E}) + \boldsymbol{E}\cdot\boldsymbol{U}^{\text{inc}}]\,dS \tag{11.53}$$

with

$$\boldsymbol{U}^{\text{inc}} = \hat{r}\times(\nabla\times\boldsymbol{E}^{\text{inc}}) + P(\boldsymbol{E}^{\text{inc}}) \tag{11.54}$$

When implementing Eq. (11.54), one can use one of Maxwell's curl equations in free-space, $\nabla\times\boldsymbol{E}^{\text{inc}} = -j\omega\mu_0\boldsymbol{H}^{\text{inc}}$, to avoid having to take the curl of the incident electric field, using instead the corresponding magnetic field (assuming this is known).

11.6.3 Discussion

A few points should be made about this boundary condition. Firstly, in the first-order form, the radius at which the ABC is applied does not appear in the formulation; it does, however, in higher-order forms. Secondly, this boundary condition is strictly speaking for a spherical outer boundary, with $\hat{n} = \hat{r}$, although it can also be used (approximately) for other boundaries. This is discussed in more detail in [4, Chapter 6]. Thirdly, the formulation as given assumes that the ABC is applied in a free-space region, but it can be generalized to any uniform region.

Regarding the choice of total field vs. scattered field formulation, both methods have advantages and disadvantages. The incident field enters only via the surface integral on the closure S when using the total field formulation, whereas in the scattered field formulation it must be integrated throughout the region containing the scatterer (assuming a penetrable scatterer). On the other hand, in the former case, the incident field has to numerically propagate to the scatterer, undergoing numerical dispersion, whereas in the latter case, the field is analytically propagated to the scatterer. (If the scatterer is highly conducting, then the incident field in the scattered field formulation enters as a Dirichlet boundary condition on the surface of the scatterer. In this case, using either formulation, the region internal to the scatterer need not be discretized.) In practice, the present author has found the former slightly easier to implement, but both merit investigation.

Note that if an internal source $\boldsymbol{J}$ is also present, this relevant term must be added to the formulation; this would be the last term in Eq. (11.35). If *only* an internal source is present, then the scattered and total field formulations reduce to the same result.

The second-order version of this ABC appears attractive, and the basic formulation may be readily found in the literature, e.g. [26], but its implementation is problematic from a theoretical viewpoint when using the usual curl-conforming elements, a point frequently not clearly made. Different approaches for tackling this issue were proposed independently in [32] and [33]. The former uses a vector auxiliary variable combined with Lagrange multipliers; the latter, a scalar auxiliary variable combined with a Galerkin procedure to evaluate the variable.

On the subject of mesh termination, spherical PMLs can be attractive for terminating FEM meshes. The basic theory may be found in [34, 35, 36]; the implementation and evaluation of such an absorber is discussed in [37].

11.7 Further reading

Most of the references cited in Section 10.10 are also of course relevant here. Jin's second edition [26] is probably the best single-volume reference regarding further coverage of the material addressed in this chapter.

In the context of higher-order vector elements, it should be noted that there is another school of thought regarding the construction of higher-order basis functions, which might be described as the *degree of freedom centered* approach (as opposed to that given in this chapter, which could be described as the *basis function centered* approach).

Salazar-Palma *et al.* [38] use elements from the Nedelec polynomial space and enforce Lagrangian interpolatory properties on the degrees of freedom. This produces *interpolatory* elements with well-defined degrees of freedom at *points*, but, at the time of writing, no-one had yet succeeded in doing this in general with higher-order *hierarchal* elements. Yioultsis and Tsiboukis take a similar degree of freedom centered approach, but working with simplex instead of Cartesian coordinates [39].

The work of Hiptmair should also be mentioned; he has also published a general scheme for the construction of higher-order elements, but from a far more mathematical viewpoint, and couched in the language of differential forms [40].

Although an obvious application of the FEM, discontinuities in rectangular waveguides have not been as widely addressed in the literature as one might expect. Ise *et al.* [41] used "brick" elements of "first" order (CT/LN) to analyze both a dielectric post and a concentric step discontinuity in a rectangular waveguide; Jin presented a detailed formulation in [26, Chapter 8], also using CT/LN elements; Webb's review paper discussed a number of related issues [42]; and Pekel and Lee addressed theoretical aspects of mesh refinement using an empty piece of waveguides [43]. Scott addressed rotationally symmetric waveguides and obtained very good results using special-purpose higher-order elements [44]. The present author studied LT/QN elements in [27], and then considered the use of both mixed- and full-order elements in [11].

Ferrari published a new formulation for the analysis of scattering discontinuities in waveguides, using an extended Huygens' principle [45]. The scatterer is discretized using finite elements, and the waveguide Green functions are used on the boundary of the scatterer, so this is a type of FEM/MoM hybrid. Geschke *et al.* reported the first successful implementation of the formulation in [46]; further details were were published in [47].

11.8 Conclusions

This chapter brings to a conclusion our study of the FEM in isolation. In the next chapter, we will primarily discuss hybrid FEM schemes. We start by studying the hybrid FEM/MoM. Following this, time-domain FEM is discussed. Time-domain finite elements (FETD) turn out to be closely related to the FDTD method, and indeed it is possible to hybridize them. Hence we come full circle, with the major techniques in CEM interacting efficiently with one another, each operating in its own region of excellence. Some other advanced topics on the FEM will also be considered.

References

[1] J. S. Savage and A. F. Peterson, "Higher-order vector finite elements for tetrahedral cells," *IEEE Trans. Microwave Theory Tech.*, **44**, 874–879, June 1996.

[2] J.-F. Lee and R. Mittra, "A note on the application of edge-elements for modeling three-dimensional inhomogeneously-filled cavities," *IEEE Trans. Microwave Theory Tech.*, **40**, 1767–1773, September 1992.

[3] P. P. Silvester and R. L. Ferrari, *Finite Elements for Electrical Engineers*. Cambridge: Cambridge University Press, 2nd edn., 1990.

[4] J. Volakis, A. Chatterjee and L. Kempel, *Finite Element Method for electromagnetics: Antennas, Microwave Cicuits and Scattering Applications*. Oxford & New York: Oxford University Press and IEEE Press, 1998.

[5] A. Awadhiya, P. Barba and L. Kempel, "Finite element method programming made easy???," *IEEE Antennas Propag. Soc. Mag.*, **45**, 73–79, August 2003.

[6] S. Ramo, J. R. Whinnery and T. van Duzer, *Fields and Waves in Communication Electronics*. Chichester: Wiley, 3rd edn., 1994.

[7] D. M. Pozar, *Microwave Engineering*. New York: Wiley, 2nd edn., 1998.

[8] P. P. Silvester and R. L. Ferrari, *Finite Elements for Electrical Engineers*. Cambridge: Cambridge University Press, 3rd edn., 1996.

[9] J. P. Webb, "Edge elements and what they can do for you," *IEEE Trans. Magn.*, **29**, 1460–1465, March 1993.

[10] J. P. Webb, "Hierarchal vector basis functions of arbitrary order for triangular and tetrahedral finite elements," *IEEE Antennas Propagat.*, **47**, 1244–1253, August 1999.

[11] D. B. Davidson, "An evaluation of mixed-order versus full-order vector finite elements," *IEEE Trans. Antennas Propagat.*, **51**, 2430–2441, September 2003.

[12] R. D. Graglia, D. R. Wilton and A. F. Peterson, "Higher order interpolatory vector bases for computational electromagnetics," *IEEE Trans. Antennas Propagat.*, **45**, 329–342, March 1997.

[13] J. S. Savage, "Comparing high order vector basis functions," in *Proceedings of the 14th Annual Review of Progress in Applied Computational Electromagnetics*, Monterey, CA, pp. 742–749, March 1998.

[14] J. P. Webb and B. Forghani, "Hierarchal scalar and vector tetrahedra," *IEEE Trans. Magn.*, **29**, 1495–1498, March 1993.

[15] L. S. Andersen and J. L. Volakis, "Hierarchical tangential vector finite elements for tetrahedra," *IEEE Microwave Guided Wave Lett.*, **8**, 127–129, March 1998.

[16] D. Sun, J. Lee and Z. Cendes, "Construction of nearly orthogonal Nedelec bases for rapid convergence with multilevel preconditioned solvers," *Siam. J. Sci. Comput.*, **23**(4), 1053–1076, 2001.

[17] P. Ingelström, "A new set of H(curl)-conforming hierarchical basis functions for tetrahedral meshes," *IEEE Trans. Microwave Theory Tech.*, **54**, 106–114, January 2006.

[18] M. M. Botha, "Fully hierarchical divergence-conforming basis functions on tetrahedral cells, with applications," *Int. J. Numer. Meth. Eng.*, **71**, 127–148, 2007.

[19] A. F. Peterson and R. D. Graglia, "Scale factors and matrix conditioning associated with triangular-cell hierarchical vector basis functions," *IEEE Antennas Wireless Propag. Lett.*, **9**, 40–43, 2010.

[20] J. C. Nedelec, "Mixed finite elements in $\Re^3$," *Numerische Mathematik*, **35**, 315–341, 1980.

[21] P. Keast, "Moderate-degree tetrahedral quadrature formulas," *Computer Meth. Appl. Mechan. Eng.*, **55**(3), 339–348, 1986.

[22] J. P. Webb, "Matching a given field using hierarchal vector basis functions," *Electromagnetics*, **24**, 113–122, January–March 2004.

[23] L. E. R. Petersson and J. Jin, "An efficient procedure for the projection of a given field onto hierarchical vector basis functions of arbitrary order," *Electromagnetics*, **25**, 81–91, 2005.

[24] M. M. Botha, Efficient finite element electromagnetic analysis of antennas and microwave devices: the FE-BI-FMM formulation and a posteriori error estimation for p adaptive

analysis. PhD thesis, Dept. Electrical & Electronic Engineering, University of Stellenbosch, 2002.

[25] N. Marcuvitz, *Waveguide Handbook*. London: Peter Peregrinus, on behalf of IEE, 1986. Originally published 1951.

[26] J.-M. Jin, *The Finite Element Method in Electromagnetics*. New York: Wiley, 2nd edn., 2002.

[27] D. B. Davidson, "Higher-order (LT/QN) vector finite elements for waveguide analysis," Special Issue on Approaches to Better Accuracy/Resolution in Computational Electromagnetics, *Appl. Comput. Electromagn. Soc. J.*, **17**, 1–10, March 2002.

[28] J. P. Webb, "P-adaptive methods for electromagnetic wave problems using hierarchal tetrahedral edge elements," *Electromagnetics*, **22**, 443–451, July 2002.

[29] A. Bayliss and E. Turkel, "Radiation boundary conditions for wave-like equations," *Commun. Pure Appl. Mathem.*, **33**, 707–725, 1980.

[30] A. F. Peterson, "Absorbing boundary conditions for the vector wave equation," *Microwave Optical Technol. Lett.*, **1**, 62–64, April 1988.

[31] J. P. Webb and V. N. Kanellopoulos, "Absorbing boundary conditions for the finite element solution of the vector wave equation," *Microwave Optical Technol. Lett.*, **2**, 370–372, October 1989.

[32] B. Stupfel and M. Mognot, "Implementation and derivation of conformal absorbing boundary conditions for the vector wave equation," *J. Electromagn. Waves Appl.*, **12**, 1653–1677, 1998.

[33] M. M. Botha and D. B. Davidson, "Rigorous, auxiliary variable-based implementation of a second-order ABC for the vector FEM," *IEEE Trans. Antennas Propagat.*, **54**, 3499–3504, November 2006.

[34] F. L. Teixeira and W. C. Chew, "PML-FDTD in cylindrical and spherical grids," *IEEE Microwave Guided Wave Lett.*, **7**(9), 285–287, 1997.

[35] C. W. Teixeira, "Systematic derivation of anisotropic PML absorbing media in cylindrical and spherical coordinates," *IEEE Microwave Guided Wave Lett.*, **7**(11), 371–373, 1997.

[36] P. G. Petropoulos, "Reflectionless sponge layers as absorbing boundary conditions for the numerical solution of Maxwell equations in rectangular, cylindrical, and spherical coordinates," *SIAM J. Appl. Mathem.*, **60**(3), 1037–1058, 2000.

[37] D. B. Davidson and M. M. Botha, "Evaluation of a spherical PML for vector FEM applications," *IEEE Trans. Antennas Propagat.*, **55**, 494–498, Feb 2007.

[38] M. Salazar-Palma, T. K. Sarkar, L. E. García-Castillo, T. Roy and A. Djordjević, *Iterative and Self-Adaptive Finite-Elements in Electromagnetic Modelling*. Boston, MA: Artech House, 1998.

[39] T. V. Yioultsis and T. D. Tsiboukis, "Development and implementation of second and third order vector finite elements in various 3-D electromagnetic field problems," *IEEE Trans. Magn.*, **33**, 1812–1815, March 1997.

[40] R. Hiptmair, "Canonical construction of finite elements," *Mathem. Comput.*, **68**, 1325–1346, May 1999.

[41] K. Ise, K. Inoue and M. Koshiba, "Three-dimensional finite-element method with edge elements for electromagnetic waveguide discontinuities," *IEEE Trans. Microwave Theory Tech.*, **39**, 1289–1295, August 1991.

[42] J. P. Webb, "Finite element methods for junctions of microwave and optical waveguides," *IEEE Trans. Magn.*, **26**, 1754–1758, September 1990.

[43] Ü. Pekel and R. Lee, "An a posteriori error reduction scheme for the three-dimensional finite-element solution of Maxwell's equations," *IEEE Trans. Microwave Theory Tech.*, **43**, 421–427, February 1995.

[44] W. R. Scott, "Accurate modelling of axisymmetric two-port junctions in coaxial lines using the finite element method," *IEEE Trans. Microwave Theory Tech.*, **40**, 1712–1716, August 1992.
[45] R. L. Ferrari, "An extended Huygens' principle for modeling scattering from general discontinuities within hollow waveguides," *Int. J. Numer. Modelling: Electronic Networks, Devices Fields*, **14**(5), 411–422, 2002.
[46] R. H. Geschke, R. L. Ferrari, D. B. Davidson and P. Meyer, "Application of extended Huygens' principle to scattering discontinuities in waveguide," in *IEEE AFRICON-02 Proceedings*, pp. 555–558, October 2002.
[47] R. H. Geschke, R. L. Ferrari, D. Davidson and P. Meyer, "The solution of waveguide scattering problems by application of an extended Huygens formulation," *IEEE Trans. Microwave Theory Tech.*, **54**(10), 3698–3705, 2006.
[48] A. F. Peterson and D. R. Wilton, "Curl-conforming mixed-order edge elements for discretizing the 2D and 3D vector Helmholtz equation," in *Finite Element Software for Microwave Engineering* (T. Itoh, G. Pelosi and P. P. Silvester, eds.), Chapter 5. New York: Wiley, 1996.

Problems and assignments

Problems

These problems focus on the properties of vector-based functions.

P11.1 A complete first-order function on a triangle can be written in the following form, with six degrees of freedom [48]:

$$\boldsymbol{B}(x, y) = (A + Bx + Cy)\hat{x} + (D + Ex + Fy)\hat{y} \tag{11.55}$$

This function can be divided into two functions:

$$\boldsymbol{B}_{\text{rot}}(x, y) = \left(A + \frac{C - E}{2}y\right)\hat{x} + \left(D + \frac{E - C}{2}x\right)\hat{y} \tag{11.56}$$

$$\boldsymbol{B}_{\text{grad}}(x, y) = \left(Bx + \frac{C + E}{2}y\right)\hat{x} + \left(\frac{E + C}{2}x + Fy\right)\hat{y} \tag{11.57}$$

In Eq. (11.56) there are effectively three degrees of freedom; the Whitney functions are a convenient way of expressing Eq. (11.56) in simplex coordinates.
(a) Verify that $\boldsymbol{B}(x, y) = \boldsymbol{B}_{\text{rot}}(x, y) + \boldsymbol{B}_{\text{grad}}(x, y)$.

(b) Show that

$$\boldsymbol{B}_{\text{grad}}(x, y) = \nabla\left[\frac{B}{2}x^2 + \frac{E + C}{2}xy + \frac{F}{2}y^2\right]$$

and hence $\nabla \times \boldsymbol{B}_{\text{grad}} = 0$.

(c) Show that $\nabla \times \boldsymbol{B}_{\text{rot}} = \nabla \times \boldsymbol{B} = (E - C)\hat{z}$.

P11.2 On a triangular element, the first-order Whitney space is spanned by the following three functions:

$$\begin{aligned} \boldsymbol{w}_{12} &= \lambda_1 \nabla \lambda_2 - \lambda_2 \nabla \lambda_1 \\ \boldsymbol{w}_{13} &= \lambda_1 \nabla \lambda_3 - \lambda_3 \nabla \lambda_1 \\ \boldsymbol{w}_{23} &= \lambda_2 \nabla \lambda_3 - \lambda_3 \nabla \lambda_2 \end{aligned} \tag{11.58}$$

(It is convenient to use the notation of Chapter 10 for these individual elements of the Whitney space in Table 11.3, $W_1^{(e)}$.) Using the identity $\lambda_3 = 1 - \lambda_1 - \lambda_2$, show that these three functions can be written as one rotational function $\boldsymbol{w}_{12}$ and two constant vector functions (which can be interpreted as the gradient of a first-order scalar function):

$$\begin{aligned} \boldsymbol{w}_{13} &= -\boldsymbol{w}_{12} - \nabla \lambda_1 \\ \boldsymbol{w}_{23} &= +\boldsymbol{w}_{12} - \nabla \lambda_2 \end{aligned} \tag{11.59}$$

P11.3 On a tetrahedral element, the first-order Whitney space is spanned by six functions; three as in Eq. (11.58) and three additional ones as follows:

$$\begin{aligned} \boldsymbol{w}_{14} &= \lambda_1 \nabla \lambda_4 - \lambda_4 \nabla \lambda_1 \\ \boldsymbol{w}_{24} &= \lambda_2 \nabla \lambda_4 - \lambda_4 \nabla \lambda_2 \\ \boldsymbol{w}_{34} &= \lambda_3 \nabla \lambda_4 - \lambda_4 \nabla \lambda_3 \end{aligned} \tag{11.60}$$

As for the two-dimensional case, using the appropriate identity $\lambda_4 = 1 - \lambda_1 - \lambda_2 - \lambda_3$, show that these six functions can be written as three rotational functions, viz. $\boldsymbol{w}_{12}$, $\boldsymbol{w}_{13}$, $\boldsymbol{w}_{23}$, and three constant vector functions:

$$\begin{aligned} \boldsymbol{w}_{14} &= -\boldsymbol{w}_{12} - \boldsymbol{w}_{13} - \nabla \lambda_1 \\ \boldsymbol{w}_{24} &= +\boldsymbol{w}_{12} - \boldsymbol{w}_{23} - \nabla \lambda_2 \\ \boldsymbol{w}_{34} &= +\boldsymbol{w}_{13} + \boldsymbol{w}_{23} - \nabla \lambda_3 \end{aligned} \tag{11.61}$$

P11.4 Although the divergence of curl-conforming elements is usually not of great concern, on occasions knowledge of this is required. From Eq. (11.56), show that within the element, $\nabla \cdot \boldsymbol{B}_{\mathrm{rot}}(x, y) = 0$.

P11.5 It should be appreciated that in general, only the lowest-order curl-conforming elements – as in the previous problem – will have zero divergence. This can be demonstrated as follows. Any properly constructed set of mixed-order hierarchal functions of order exceeding one will contain functions of the form of Eq. (11.57). Show that $\nabla \cdot \boldsymbol{B}_{\mathrm{grad}}(x, y) = B + F$, and, hence, that the divergence of such higher-order elements is non-zero.

P11.6 This problem continues the theme of the divergence of curl-conforming elements, by explicitly evaluating the divergence of the widely used set due to Webb:

(a) For a function of the form $\nabla(\lambda_i\lambda_j)$ (i.e. $\boldsymbol{G}_1^{(e)}$ in Table 11.3), show that its divergence is given by $2\nabla\lambda_i \cdot \nabla\lambda_j$.
(Hint: make use of the vector identity $\nabla \cdot (a\boldsymbol{b}) = a\nabla \cdot \boldsymbol{b} + \boldsymbol{b} \cdot \nabla a$.)

(b) Similarly, for a function of the form $\lambda_j\lambda_k\nabla\lambda_i + \lambda_i\lambda_k\nabla\lambda_j - 2\lambda_i\lambda_j\nabla\lambda_k$ (i.e. $\boldsymbol{R}_{20}^{(f)}$ in Table 11.3), show that its divergence is given by $\nabla\lambda_j \cdot \boldsymbol{w}_{ik} - \nabla\lambda_i \cdot \boldsymbol{w}_{jk}$.
(Hint: note that the gradient of a simplex coordinate is constant, so the divergence thereof is zero.)

P11.6 Refer to Tables 11.3 and 11.4. Show that the LT/QN elements proposed by Webb can be written as linear combinations of those proposed by Andersen and Volakis (specifically, as the difference and negated sum of the latter functions).

Assignments

A11.1 Develop a code to compute the eigenvalues of a rectangular PEC cavity, using Whitney elements (as in Section 11.1).

A11.2 Extend this code to use LT/QN elements, as in Section 11.2.

12 A selection of more advanced topics in full-wave computational electromagnetics

In this final chapter, we conclude the main part of this book with a selection of more advanced topics. Although primarily relating to the finite element method, hybridization with both the MoM and FDTD will be discussed, so as a final chapter it appropriately draws together these three apparently quite different methods.

The previous chapter concluded with one method for terminating open-region problems with radiation boundary conditions, specifically the use of an absorbing boundary condition. An alternative to the application of an approximate local ABC is to use an exact global RBC, the MoM; this leads to the hybrid FEM/MoM formulation. This approach has proven very powerful for specialized applications, and an application to radiation exposure assessment near a base-station antenna will be presented.

To this point in this book, all the finite element work has proceeded in the frequency domain; in this chapter, time domain finite element analysis (FETD) is discussed, and a connection made with the FDTD, which is explored in detail, revisiting the one-dimensional wave analysis problem introduced in Chapter 9 in the time domain. This connection permits the consideration of hybrid FETD/FDTD schemes.

We conclude the chapter with a discussion on two issues which impact on efficiency. Firstly, sparse matrix storage schemes are briefly outlined, and secondly, error estimation and the use of mesh adaptation based on this is discussed.

The coverage in this chapter is at a higher level than in much of the rest of this book. Generally, the topics discussed are too complex to permit a simple implementation, and the intention of this material is rather to sensitize the reader to current topics of interest in the field. Nonetheless, with the exception of time domain FEM, aspects of all the topics discussed have been incorporated in commercial codes.

12.1 Hybrid finite element/method of moments formulations

12.1.1 Introduction

In the preceding chapter, the use of absorbing boundary conditions to terminate the finite element mesh at a finite distance was discussed. In this section, we will instead consider an "exact" termination scheme, which effectively uses the method of moments applied on the open boundary to terminate the FEM region, producing the FEM/MoM hybrid method. (This method is also sometimes called the boundary element/finite element method, or boundary integral/finite element method.)

12.1.2 Theoretical background

Before addressing the theory of the FEM/MoM hybrid method, a connection between the Rao–Wilton–Glisson (RWG) element [1], widely used in MoM formulations, and the Whitney (CT/LN) element which we have discussed extensively here needs to be highlighted. Much earlier, in Chapter 6, it was commented that the RWG element [1] and the Whitney element are intimately connected. The relationship is the following: by simply taking the normal crossed with the Whitney element, the RWG element is obtained. It will be recalled that the Whitney element is also sometimes called "curl conforming"; the RWG element is an example of a "divergence conforming" element. (Nedelec's original work also considered such elements, although the RWG element was derived independently.) This close relationship is fortunate and not by any means serendipitous: the underlying requirements of field continuity are the reason for the close relationship. This is an important practical point, because it implies that edge-element FEM codes, with volumetric fields as unknowns, and RWG-based MoM codes, with surface currents as unknowns,[1] can at least potentially conform on a boundary.

With this background, we can now consider the FEM/MoM formulation. The following is based on the presentation in [2, Chapter 9]. Within a region Ω, with closure (bounding surface) S, and free space in the exterior region, a finite element discretization of Maxwell's equations, via the stationary functional as in Section 11.3, results in the following matrix equation:

$$[A]^E\{e\} + [B]^E\{h\}_S = \{c\}^E \tag{12.1}$$

In this equation, the superscript E indicates that the $\boldsymbol{E}$ field has been chosen as the main working variable. Matrix $[A]$ is the usual FEM matrix obtained from the bilinear functional applied throughout the volume; vector $\{e\}$ is the vector of unknown coefficients of the electric field in the volume; matrix $[B]$ represents the Neumann boundary condition applied on the surface;[2] and vector $\{h\}$ is the vector of unknown coefficients of the magnetic field on the closure. Finally, vector $\{c\}$ accounts for current sources internal to the volume. Specifically, the elements of each are

[1] Recall that an equivalent surface current is obtained from the normal crossed with the appropriate tangential field component.

[2] Recall Eq. (11.31) and the connection with the tangential magnetic fields.

given by

$$A_{ij}^E = \int_\Omega \{\mu_r^{-1}(\nabla \times \boldsymbol{N}_i)\cdot(\nabla \times \boldsymbol{N}_j) - k^2\epsilon_r \boldsymbol{N}_i \cdot \boldsymbol{N}_j\}\, d\Omega,$$
$$\forall\, i \text{ and } j = 1, \ldots, N \tag{12.2}$$

$$B_{ij}^E = jk\eta \oint_S \boldsymbol{N}_i \cdot (\boldsymbol{N}_j \times \hat{n})\, dS,$$
$$\forall\, i = 1, \ldots, N, \;\; j = 1, \ldots, N_S \tag{12.3}$$

$$c^E = -\int_\Omega \boldsymbol{N}_i \cdot \{jk\eta \boldsymbol{J}^{\text{int}} + \nabla \times (\mu_r^{-1}\boldsymbol{K}^{\text{int}})\}\, d\Omega,$$
$$\forall\, i = 1, \ldots, N \tag{12.4}$$

In the above expressions, $\boldsymbol{N}_i$ and $\boldsymbol{N}_j$ are the element shape functions. The elements of $[A]$ are immediately recognized as the $[S]$ and $[T]$ matrix elements discussed in Section 11.1, for which closed-form expressions are available. $\boldsymbol{J}^{\text{int}}$ and $\boldsymbol{K}^{\text{int}}$ represent sources internal to Ω.

The problem is clear: there are $N + N_S$ degrees of freedom (N unknowns in $\{e\}$ and a further N_S unknowns in $\{h\}_S$, the latter is the $\boldsymbol{H}$ field on the boundary surface S). An additional constraint is required to connect the surface magnetic fields with the volumetric electric fields (which also of course exist on the closing surface). In the waveguide formulation, knowledge of the modal structure of the field was sufficient, but now a further matrix equation must be derived in terms of the surface fields.

Deriving essentially the EFIE and MFIE, one can obtain the following, suitable for an MoM representation on the boundary S:[3]

$$\boldsymbol{E}(\boldsymbol{r}) = \boldsymbol{E}^{\text{inc}}(\boldsymbol{r}) + \oint_S \Big(\nabla \times \bar{\bar{G}}(\boldsymbol{r}, \boldsymbol{r}')\cdot\{\hat{n}' \times \boldsymbol{E}_S(\boldsymbol{r}')\}$$
$$- jk\eta \bar{\bar{G}}(\boldsymbol{r}, \boldsymbol{r}')\cdot\{\hat{n}' \times \boldsymbol{H}_S(\boldsymbol{r}')\}\Big)\, dS' \tag{12.5}$$

and

$$\boldsymbol{H}(\boldsymbol{r}) = \boldsymbol{H}^{\text{inc}}(\boldsymbol{r}) + \oint_S \Bigg(\nabla \times \bar{\bar{G}}(\boldsymbol{r}, \boldsymbol{r}')\cdot\{\hat{n}' \times \boldsymbol{H}_S(\boldsymbol{r}')\}$$
$$+ \frac{jk}{\eta}\bar{\bar{G}}(\boldsymbol{r}, \boldsymbol{r}')\cdot\{\hat{n}' \times \boldsymbol{E}_S(\boldsymbol{r}')\}\Bigg)\, dS' \tag{12.6}$$

Note that the S subscript refers to quantities on surface S, *not* the scattered field. $\bar{\bar{G}}$ is the dyadic free-space Green function, and $\hat{n}'$ is the outward directed normal.

Writing these in a more compact notation, one obtains

$$-\boldsymbol{E} + L_{e1}^S(\boldsymbol{E}_s \times \hat{n}') + L_{e2}^S(\boldsymbol{H}_s \times \hat{n}') + \boldsymbol{E}^{\text{inc}}(\boldsymbol{r}) = 0 \tag{12.7}$$

Using a Galerkin procedure, this may be discretized as

$$[B]^M\{e\}_S + [P]^E\{e\}_S + [Q]^E\{h\}_S + \{y\}^E = 0 \tag{12.8}$$

[3] Also known as a *Huygen's* integral representation.

The matrix $[B]^M$ in the above is of the same form as $[B]^E$ in Eq. (12.3): the only difference is that the constant term is $-jk/\eta$ instead of $jk\eta$ [2, pp. 408–409]. The other matrices are given by

$$P_{ij}^E = j\frac{k}{\eta}\oint_S \boldsymbol{N}_i \cdot \{L_{e1}^S(\boldsymbol{N}_j \times \hat{n}) \times \hat{n}\}\, dS \tag{12.9}$$

$$Q_{ij}^E = j\frac{k}{\eta}\oint_S \boldsymbol{N}_i \cdot \{L_{e2}^S(\boldsymbol{N}_j \times \hat{n}) \times \hat{n}\}\, dS \tag{12.10}$$

$$y_i^E = j\frac{k}{\eta}\oint_S \boldsymbol{N}_i \cdot (\boldsymbol{E}^{\text{inc}} \times \hat{n})\, dS \tag{12.11}$$

The matrix size of $[B]^M$ is $N \times N_S$, but for the boundary element terms in Eq. (12.8), only the relevant $N_S \times N_S$ submatrix is retained, so that the above matrix equation (12.8) is of dimension N_S. Similarly, Eq. (12.6) can be discretized to yield

$$[B]^E\{h\}_S + [P]^M\{h\}_S + [Q]^M\{e\}_S + \{y\}^M = 0 \tag{12.12}$$

Either Eq. (12.8) or (12.12) is sufficient to eliminate $\{h\}_S$ in terms of $\{e\}_S$, which is then substituted into Eq. (12.1). (Note that $\{e\}_S \subset \{e\}$, since these are just the components of electric field on the surface.)

The $[P]$ and $[Q]$ matrices are not straightforward to compute, since they involve integrals of Green's functions, containing integrable singularities, acting on the basis functions; see [2, p. 413]. (As we saw in Chapter 6, this is standard in MoM formulations involving a rigorous surface current treatment.) The case of a cavity in a conducting half-space has been worked further by Jin [3, Chapter 10]; his results are also summarized in [2, Chapter 9]. For more general problems, see [3, Chapter 10; 28, Chapter 11; 4, Chapter 7].

A computational problem which emerges is that the resulting system of linear equations is overwhelmingly sparse, but contains a dense submatrix representing the MoM (BEM) interactions. The overall matrix is also not, in general, symmetric.

An example of the use of the FEM/MoM hybrid in FEKO, using the above scheme, was shown in Fig. 6.10.

12.2 An application of the FEM/MoM hybrid – GSM base stations

12.2.1 Applications of FEM/MoM hybrid formulations

The hybrid FEM/MoM formulation outlined above is applicable to many problems. In general antenna analysis, the FEM is not the method of choice for wire antennas, where the standard MoM formulation provides a straightforward and robust solution. However, when such antennas are radiating in the presence of electromagnetically penetrable bodies, the FEM/MoM hybrid comes into its own. Modelling the interaction of operators and personal communications systems, in particular cellular phones, has emerged as an important field of application of this formulation, and the example presented here is a variant on this theme. However, there are a number of other applications, which will now be outlined.

Cavity-backed antennas were one of the first applications of the FEM/MoM (BEM) hybrid formulations, see [3], and they continue to attract interest [4]. Although the original formulation assumed that the cavity was recessed into an infinite ground plane, recently work has extended this to cavities on elliptical shapes, permitting analysis of conformal airborne antennas. Microstrip antennas have also been efficiently analyzed using this approach; since the substrate, which is discretized with the FEM, need not be uniform in this approach, some interesting work has been done on the use of perforated substrates (a type of electromagnetic band-gap material) to reduce mutual coupling [5]. An important class of cavity-backed antenna is the spiral, both Archimedes and logarithmic. Again, stratified media MoM codes assume infinite planar media, whereas an FEM/BEM formulation need not.

General FEM/MoM hybrids also permit the analysis of microstrip antennas, removing the assumptions of infinite substrate and permitting the effect of edge diffraction to be studied. However, this is computationally quite expensive.

The use of CEM tools in what are often EMC problems can be problematic, due to the great complexity of the systems. Work by Hubing's group has proposed the use of the FEM for regions of geometric and material complexity, combined with a MoM treatment of the interconnects [6]. Work has also been done on the coupling of energy through deep slots using FEM/MoM hybrids.

Inhomogeneous objects buried in stratified media are another interesting application; perhaps the most obvious candidates here are landmines and other unexploded ordnance. The formulation required becomes extremely complex, since the "exterior" Green functions involve the Sommerfeld potentials. Eibert and Hansen present the necessary formulation in [7].

12.2.2 Human exposure assessment near GSM base stations

The widespread adoption of personal communication devices, in particular mobile (cellular) telephones, during the 1990s and the continuing growth in the present decade presented significant challenges for CEM analysis. When first introduced, there were widespread concerns over safety issues associated with the widespread and prolonged use of mobile handsets, perhaps triggered by the term "radiation."[4] After much research, it would appear that these concerns were fortunately unfounded, due primarily to the low power levels of the handsets. However, a case where there are indeed valid concerns for

[4] It must say something of human nature that a number of users who express such concerns are prepared to operate their mobile phones while driving, a well-known and much documented hazard, and illegal in many countries for precisely this reason!

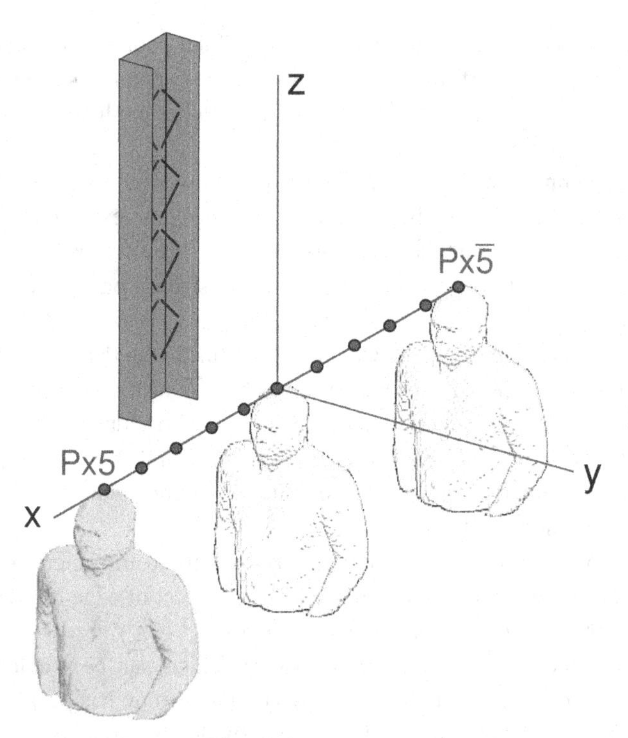

Figure 12.1 Near-field base-station exposure setup. (After [8], ©2003 IEEE, reprinted with permission.)

health issues is that of *base stations*, due to the much higher power levels encountered there (60 W is typical) and the requirement for maintenance workers to operate close to the antennas.

The FDTD, FEM with ABC, FEM/MoM and also volume equivalence principle MoM formulations have all been used successfully for the analysis of human exposure assessment of radiation from handsets. However, for base stations, one has the problem both of complex wire antennas, typically mounted on a mast, and considerable distance between the human phantom and the antenna. Figure 12.1 shows an example of a typical setup. (Handsets are usually analyzed in very close proximity to the head, which has been the major health concern.) Although this was not discussed in our theoretical development, a very powerful feature of the FEM/MoM formulation is that the exterior region *may also contain scatterers/radiators*; i.e. it need not be purely free space, as shown in Fig. 12.2. These scatterers and radiators are treated with the MoM in a self-consistent manner.

Meyer has implemented an outward-looking FEM/MoM hybrid (using a number of FEKO routines, as well as elements of the FEM code discussed in Section 11.5), and in [8] results are shown for base-station exposure assessment in terms of ICNIRP[5] guidelines.

[5] International Commission on Non-Ionizing Radiation Protection.

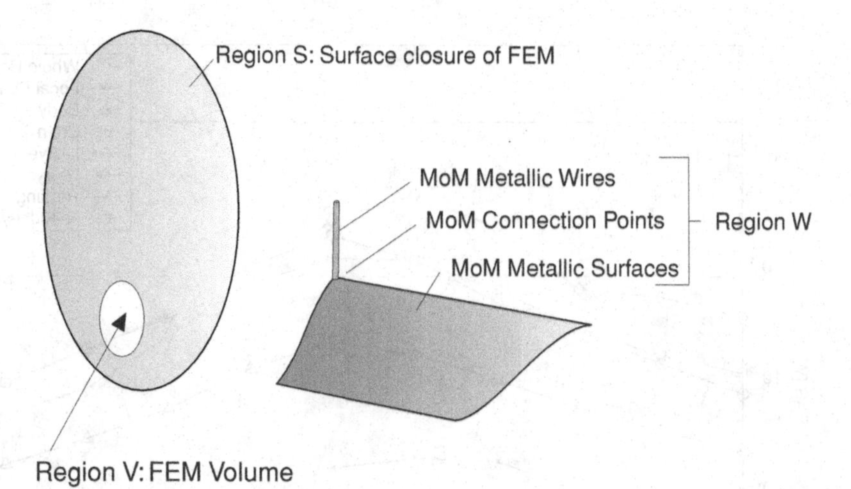

Figure 12.2 Hybrid FEM/MoM problem setup. (After [8], ©2003 IEEE, reprinted with permission.)

In that paper, results are also shown for careful validation of some smaller problems, using both an FDTD code and FEKO; readers are referred to the paper as a good example of this process for complex problems. Here, results for only the FEM/MoM hybrid will be shown. Of particular interest are exposure results for particular organs, shown in Figs. 12.3 and 12.4. As discussed in [8], this particular problem could not be analyzed in any way other than the FEM/MoM, since it was electrically too large for both the FDTD and the MoM volume equivalence principle.

12.3 Time domain FEM

Time domain finite elements are widely used in other fields of engineering, but have not seen especially widespread use in CEM. This probably reflects both the technological driving forces behind the development of CEM, which until the 1980s emphasized the development of frequency domain formulations (since most RF communication and radar systems were inherently narrowband), as well as the competing algorithm in the time domain, the FDTD, which is so firmly established in CEM and has produced so many excellent results that it is difficult to "sell" another time domain formulation. Nonetheless, the finite element time domain (FETD) method has seen a considerable amount of work and development in CEM over the last decade. In particular for devices with fine geometrical detail, it can be expected to emerge as a competitor for specialized applications. Perhaps the most interesting use is as a hybrid form with the FDTD, which exploits the superior geometrical modelling ability of finite elements with the robustness and speed of the FDTD; no commercial implementation is presently available, nor is

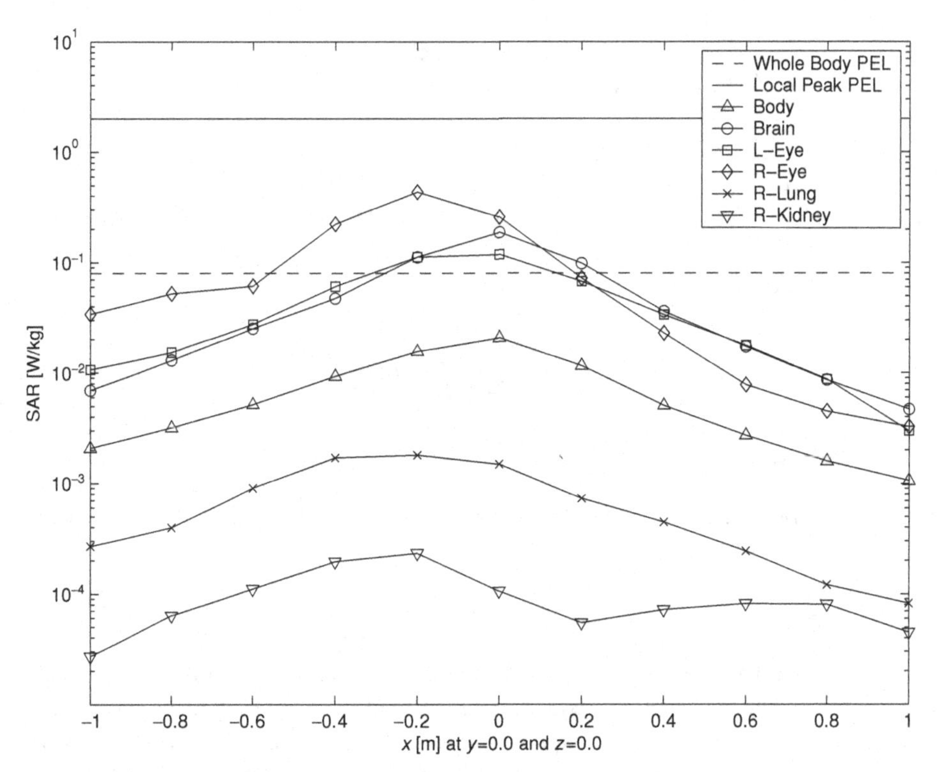

Figure 12.3 Average specific absorption rate (SAR) in different body organs compared with whole-body (0.08 W/kg) and spatial-peak (2 W/kg) ICNIRP basic restriction, x-direction (transverse across antenna), for the base-station/half-body problem shown in Fig. 12.1. $P_{\text{rad}} = 60$ W. Front of base-station antenna at $y = -0.428$ m and top-center of phantom head at $y = 0$ m. (Adapted from [8], ©2003 IEEE, reprinted with permission.)

likely to be for some time, but recent research has produced good results [9, 10, 11]. In a book which is otherwise devoted to methods already implemented in widespread public domain and commercial codes, coverage of this method may seem slightly anomalous, but at least one interesting point which emerges is a more general view of the FDTD, which can be viewed as a special case of the general FETD formulation, and furthermore, this is a method which we can expect to see more of in the future.

12.3.1 Basic formulation and implementation

Basic finite element formulation

The following formulation is based on the second-order (curl-curl) wave equation approach, presented in [3, Section 12.1]:

$$\nabla \times \left[\frac{1}{\mu}\nabla \times \boldsymbol{E}(\boldsymbol{r},t)\right] + \varepsilon\frac{\partial^2}{\partial t^2}\boldsymbol{E}(\boldsymbol{r},t) + \sigma\frac{\partial}{\partial t}\boldsymbol{E}(\boldsymbol{r},t) = -\frac{\partial}{\partial t}\boldsymbol{J}_i(\boldsymbol{r},t), \qquad \boldsymbol{r} \in V \tag{12.13}$$

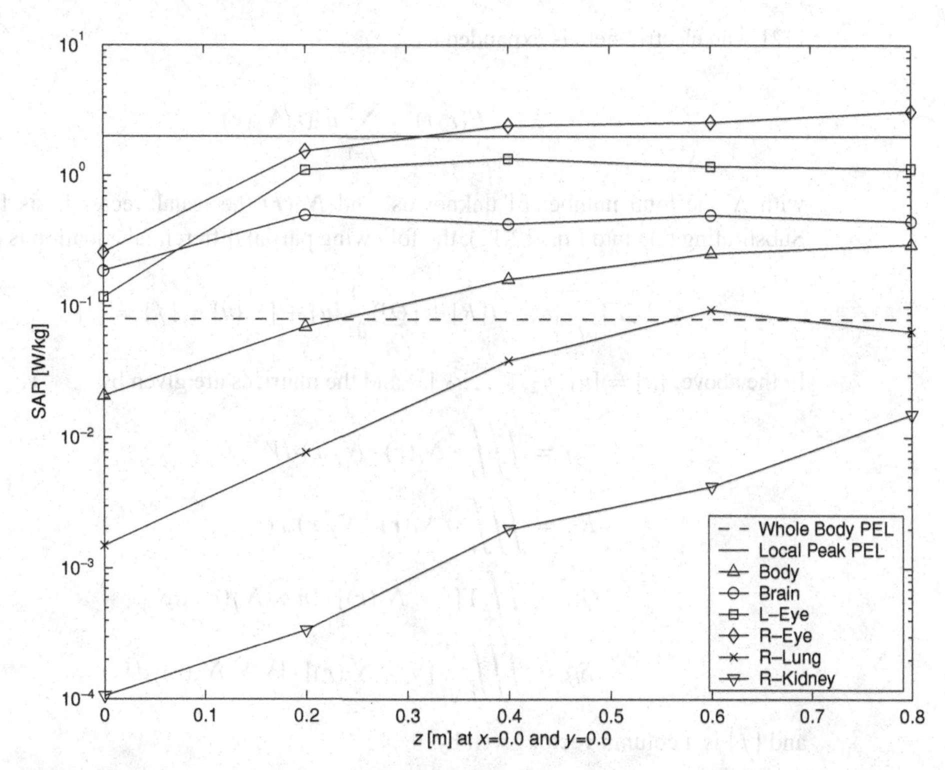

Figure 12.4 Average SAR specific absorption rate in different body organs for the z-direction (along antenna length), for the base-station/half-body problem. Details as in Fig. 12.3. (Adapted from [8], ©2003 IEEE, reprinted with permission.)

The boundary condition is

$$\hat{n} \times \left[\frac{1}{\mu}\nabla \times \boldsymbol{E}(\boldsymbol{r}, t)\right] + Y\frac{\partial}{\partial t}[\hat{n} \times \hat{n} \times \boldsymbol{E}(\boldsymbol{r}, t)] = \boldsymbol{U}(\boldsymbol{r}, t), \qquad \boldsymbol{r} \in S \tag{12.14}$$

Y is the surface admittance of the boundary, $\hat{n}$ is the outward unit normal to S and $\boldsymbol{U}$ is a known quantity representing the boundary source (if present).

The corresponding weak-form solution of the boundary value problem is given by

$$\begin{aligned}
&\iiint_V \left\{\frac{1}{\mu}[\nabla \times \boldsymbol{N}_i(\boldsymbol{r}) \cdot \nabla \times \boldsymbol{E}(\boldsymbol{r}, t)] + \varepsilon N_i(\boldsymbol{r}) \cdot \frac{\partial^2}{\partial t^2}\boldsymbol{E}(\boldsymbol{r}, t)\right. \\
&\quad \left. + \sigma N_i(\boldsymbol{r}) \cdot \frac{\partial}{\partial t}\boldsymbol{E}(\boldsymbol{r}, t) + N_i(\boldsymbol{r}) \cdot \frac{\partial}{\partial t}\boldsymbol{J}_i(\boldsymbol{r}, t)\right\} dV \\
&\quad + \iint_S \left\{Y\,[\hat{n} \times N_i(\boldsymbol{r})] \cdot \left[\frac{\partial}{\partial t}\hat{n} \times \boldsymbol{E}(\boldsymbol{r}, t)\right] + \boldsymbol{N}_i(\boldsymbol{r}) \cdot \boldsymbol{U}(\boldsymbol{r})\right\} dS = 0
\end{aligned} \tag{12.15}$$

At this stage, the problem is still posed in continuous form. It will now be discretized in space, a process which has been rather descriptively named "semi-discretization"

[12]. The electric field is expanded as

$$\boldsymbol{E}(\boldsymbol{r}, t) = \sum_{j=1}^{N} u_j(t)\boldsymbol{N}_j(\boldsymbol{r}) \tag{12.16}$$

with N the total number of unknowns, and $\boldsymbol{N}_j(\boldsymbol{r})$ the usual vector basis functions. Substituting this into Eq. (12.15), the following partial differential equation is obtained:

$$[T]\frac{\partial^2}{\partial t^2}\{u\} + ([R] + [Q])\frac{\partial}{\partial t}\{u\} + [S]\{u\} + \{f\} = \{0\} \tag{12.17}$$

In the above, $\{u\} = [u_1, u_2, \ldots, u_N]^T$; and the matrices are given by

$$T_{ij} = \iiint_V \varepsilon \boldsymbol{N}_i(\boldsymbol{r}) \cdot \boldsymbol{N}_j(\boldsymbol{r})\, dV \tag{12.18}$$

$$R_{ij} = \iiint_V \sigma \boldsymbol{N}_i(\boldsymbol{r}) \cdot \boldsymbol{N}_j(\boldsymbol{r})\, dV \tag{12.19}$$

$$Q_{ij} = \iint_S Y[\hat{n} \times \boldsymbol{N}_i(\boldsymbol{r})] \cdot [\hat{n} \times \boldsymbol{N}_j(\boldsymbol{r})]\, dS \tag{12.20}$$

$$S_{ij} = \iiint_V \frac{1}{\mu}[\nabla \times \boldsymbol{N}_i(\boldsymbol{r})] \cdot [\nabla \times \boldsymbol{N}_j(\boldsymbol{r})]\, dV \tag{12.21}$$

and $\{f\}$ is a column vector given by

$$f_i = \iiint_V \boldsymbol{N}_i(\boldsymbol{r}) \cdot \frac{\partial}{\partial t}\boldsymbol{J}_i(\boldsymbol{r}, t)\, dV + \iint_S \boldsymbol{N}_i(\boldsymbol{r}) \cdot \boldsymbol{U}(\boldsymbol{r}, t)\, dS \tag{12.22}$$

Equation (12.17) is an ordinary differential equation in the time domain and can be solved using a direct integration or finite difference method.

Before departing from this, it should be commented that these semi-discrete equations are essentially identical in form to those arising in standard frequency domain formulations, and the matrices already computed within a typical finite element frequency domain code can be largely re-used.

Time integration

For the time domain discretization, the Newmark-β method is used. (An outline of the derivation of the method is given in Appendix B.) The equation to be solved at each timestep is the following:

$$\begin{aligned}
&\left\{\frac{1}{(\Delta t)^2}[T] + \frac{1}{2\Delta t}[T_\sigma] + \beta[S]\right\}\{u\}^{n+1} \\
&= \left\{\frac{2}{(\Delta t)^2}[T] - (1 - 2\beta)[S]\right\}\{u\}^n \\
&\quad - \left\{\frac{1}{(\Delta t)^2}[T] - \frac{1}{2\Delta t}[T_\sigma] + \beta[S]\right\}\{u\}^{n-1} \\
&\quad - \left[\beta\{f\}^{n+1} + (1 - 2\beta)\{f\}^n + \beta\{f\}^{n-1}\right]
\end{aligned} \tag{12.23}$$

with

$$[T_\sigma] = [R] + [Q] \tag{12.24}$$

This can be more conveniently written as

$$[A]\{u\}^{n+1} = [B]\{u\}^n + [C]\{u\}^{n-1} - \left[\beta\{f\}^{n+1} + (1-2\beta)\{f\}^n + \beta\{f\}^{n-1}\right] \tag{12.25}$$

Clearly, the solution of this is

$$\{u\}^{n+1} = [A]^{-1}\left([B]\{u\}^n + [C]\{u\}^{n-1} - \left[\beta\{f\}^{n+1} + (1-2\beta)\{f\}^n + \beta\{f\}^{n-1}\right]\right) \tag{12.26}$$

The matrix $[A]$ is time invariant, and may be factored once, each timestep requiring then just a backward and forward substitution to establish the next solution vector $\{u\}^{n+1}$. With $\beta \geq 0.25$, the method is unconditionally stable, i.e. there is no upper bound on the timestep.

Analysis of the stability of this scheme follows the usual von Neumann approach already encountered in the FDTD context, using an amplification matrix whose eigenvalues must have magnitude less than or equal to unity for stability. (There are other equivalent methods of deriving this; see, for instance, [3, Section 12.3].) It should be noted that this type of stability proof only guarantees the absence of *exponentially* growing instabilities; it is a necessary but not sufficient condition for stability. (However, other forms of stability are usually too difficult to show theoretically, hence the reliance on von Neumann analysis, as mentioned in Chapter 2.) Stability of the Newmark-β system was analysed in detail in by Chilton and Lee in [13], and their contribution is briefly summarized here. They demonstrated the existence of linear growth modes, which are, however, non-physical. If only stable, physical modes are initially excited, the unstable modes should never manifest themselves, since multiplication by the amplification matrix is closed for the stable electrodynamic and electrostatic spaces. Assuming exact arithmetic, a properly set-up problem should therefore, in theory, never suffer from linear growth. However, numerical solutions suffer from errors due to numeric precision and also residual error if iterative matrix solution is used. This inevitably leads to the eventual excitation of unstable modes. A method for identifying and explicitly removing unstable modes from the numerical solution at intervals was also presented in their paper.

12.3.2 Preliminary results

To test the time domain formulation, propagating a plane wave through the mesh is usually a good initial test, since one has a simple analytical solution to compare with the results. A differentiated Gaussian pulse with $\sigma = 1 \times 10^{-10}$ was used; as in Chapter 2, $m = 4\sigma$ was used. This produces a wideband pulse with significant spectral content to around 3 GHz. A cuboidal free-space volume $0.1 \times 0.1 \times 0.2$ m^3 was meshed using tetrahedral elements; some 804 elements produced a mesh with an average edge length of about 0.0285 m. The plane wave was injected travelling in the $-\hat{z}$-direction. The result in Fig. 12.5 shows the plane wave at three points in the mesh. Firstly, $z = 0.19$ m is

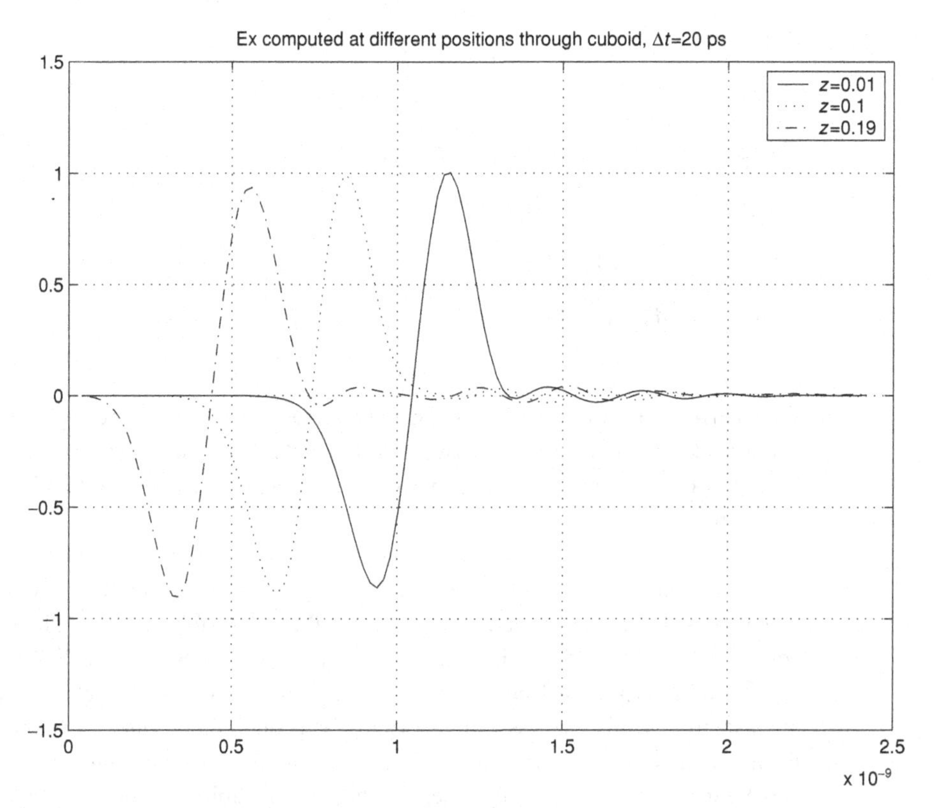

Figure 12.5 The differentiated Gaussian having propagated through a free-space "box" meshed using tetrahedral Whitney finite elements. $\Delta t = 20$ ps.

illuminated, then $z = 0.1$ m and finally $z = 0.01$ m. Measuring the distance between the first and last peaks shows a delay of 0.60 ps (within the accuracy with which the graph can be read); to cover a distance of 0.18 m at the speed of light takes 0.6 ps, so this is very accurate. In particular, considering the coarse mesh, the result is really surprisingly good; at 3 GHz the mesh density is less than four unknowns per wavelength. (At the center frequency of the signal, around 1.5 GHz, there are around seven, somewhat better but still a very coarse discretization using Whitney (CT/LN) elements.) Incidentally, the pulse may appear to have undergone a 180° phase reversal, but this was simply due to a coding convention. The late time signal is very likely a reflection from the absorbing boundary condition; the value is around 1/20 of the incident signal, or −26 dB, not by any means excellent absorber performance, but not out of line with what is expected from a first-order ABC.

Figures 12.6 and 12.7 show the results for $\Delta t = 50$ ps and 100 ps respectively. Clearly, the result in Fig. 12.7 is very poor, but it is still stable, and what is significant is the size of Δt. For a similar FDTD mesh with spatial step size 0.0285 m, the Courant limit would require $\Delta t < 54.8$ ps. The FETD code has remained stable at almost twice this limit. (Theoretically of course there *is* no limit for the Newmark-β scheme, but it is gratifying to have this confirmed by numerical experimentation.)

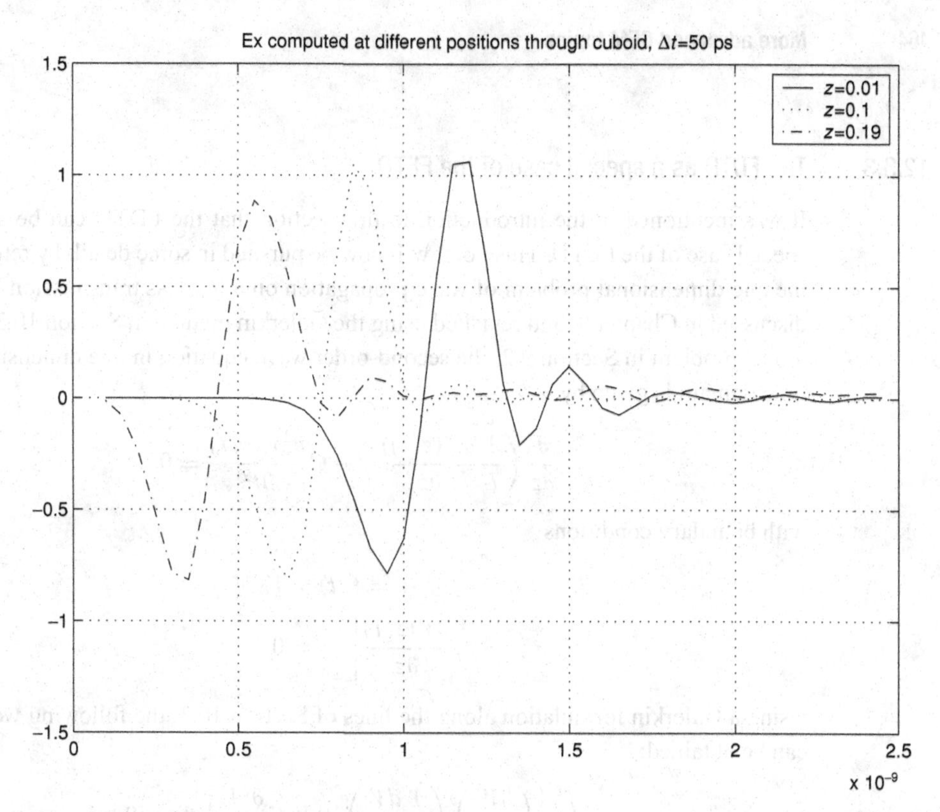

Figure 12.6 As for Fig. 12.5, but with $\Delta t = 50\,\text{ps}$.

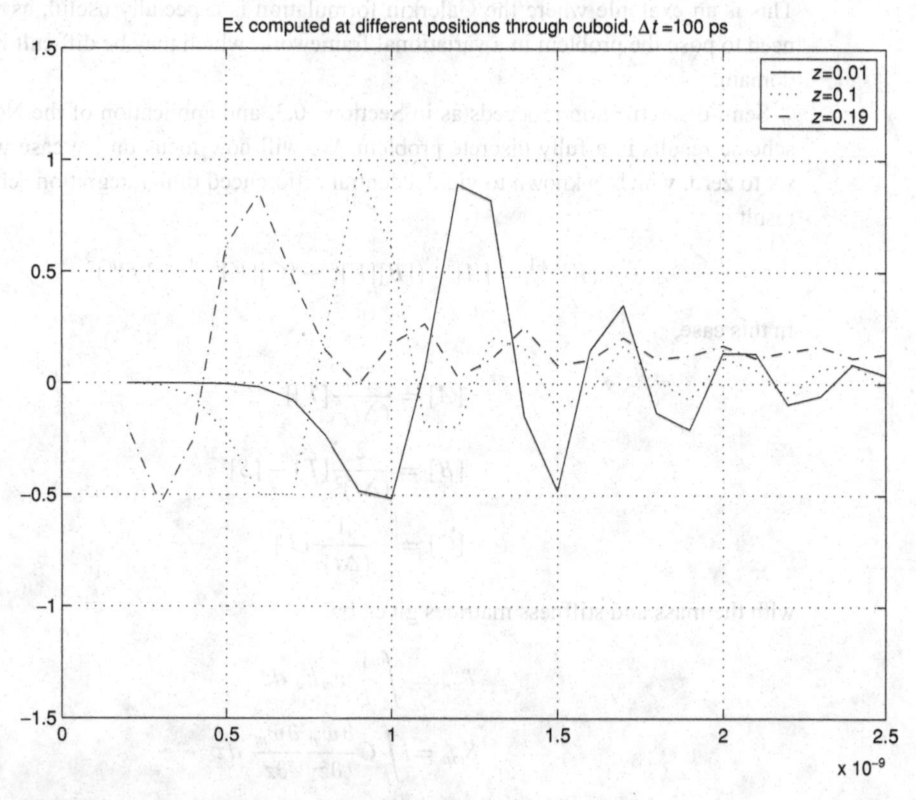

Figure 12.7 As for Fig. 12.5, but with $\Delta t = 100\,\text{ps}$.

12.3.3 The FDTD as a special case of the FETD

It was mentioned in the introduction to this section that the FDTD can be seen as a special case of the FETD. This point will now be pursued in some detail, by returning to the one-dimensional problem of wave propagation on a lossless transmission line, first discussed in Chapter 9, and revisited using the Galerkin method in Section 10.3. For the model problem in Section 9.2, the second-order wave equation in one dimension in the time domain is given by

$$\frac{\partial}{\partial z}\left(\frac{1}{L}\frac{\partial V(z,t))}{\partial z}\right) - C\frac{\partial^2 V(z,t)}{\partial t^2} = 0 \tag{12.27}$$

with boundary conditions

$$\begin{aligned} V(z=\ell,t) &= V_0 \\ \left.\frac{\partial V(z,t)}{\partial z}\right|_{z=0} &= 0 \end{aligned} \tag{12.28}$$

Using a Galerkin formulation along the lines of Section 10.3, the following weak form can be obtained:

$$\int_0^{\ell}\left[\left(\frac{\partial W}{\partial z}\right)\left(\frac{1}{L}\frac{dV}{dz}\right) + WC\frac{\partial^2 V}{\partial t^2}\right] dz = 0 \tag{12.29}$$

This is an example where the Galerkin formulation is especially useful, as we do not need to pose the problem in a variational framework, which may be difficult in the time domain.

Semi-discretization proceeds as in Section 10.3, and application of the Newmark-β scheme results in a fully discrete problem. We will now focus on the case where β is set to zero, which is known to yield a central differenced time integration scheme. The result is

$$\{V\}^{n+1} = [A]^{-1}\left([B]\{V\}^n + [C]\{V\}^{n-1} - \{f\}^n\right) \tag{12.30}$$

In this case,

$$[A] = \frac{1}{(\Delta t)^2}[T]] \tag{12.31}$$

$$[B] = \frac{2}{(\Delta t)^2}[T] - [S] \tag{12.32}$$

$$[C] = -\frac{1}{(\Delta t)^2}[T] \tag{12.33}$$

with the mass and stiffness matrices given by

$$T_{mn} = \int_{\ell}\frac{1}{L} w_m h_n \, dz \tag{12.34}$$

$$S_{mn} = \int_{\ell} C\frac{\partial w_m}{\partial z}\frac{\partial w_m}{\partial z}\, dz \tag{12.35}$$

with the first-order basis and testing functions given by Eqs. (10.64), (10.65), and (10.66) and (10.67). Although $\beta = 0$ produces a central differenced time integration scheme, which is temporally explicit, we note that in Eq. (12.30) the inverse of the mass matrix $[T]$ is required; as this matrix is not diagonal, the scheme is not operationally explicit – i.e. a matrix equation still needs to be solved at each time step. Clearly, making the matrix diagonal in some way is necessary to obtain a fully explicit scheme.

Methods to render $[T]$ diagonal originated in structural mechanics, and are often referred to as "mass lumping" methods. The simplest approach has been to add the off-diagonal elements to the corresponding diagonal elements, hence the term "lumping." Unsurprisingly, such simple methods often introduce significant errors, and possible instability [3, p. 542]. A good summary of mass lumping methods is [12, §7.3.2].

A common strategy used to accomplish this is reduced-order integration, which sounds counter-intuitive initially. The idea is to use an interpolatory basis together with an inner product based on an approximate numerical quadrature rule, where the quadrature points coincide with the interpolation points. The interpolation points should then be chosen to optimise the accuracy of the quadrature rule. The quadrature rules used previously in this book are all standard Gauss–Legendre, which contain quadrature nodes internal to interval $[a, b]$. There is another class of quadrature rules, known as Gauss–Lobatto.

The Gauss–Lobatto and Gauss–Legendre interpolation points both arise as the sampling points of optimal numerical integration rules. The Gauss–Legendre points form the basis of standard Gaussian quadrature. For a given number of quadrature points n over an interval (a, b), the Gauss–Legendre points together with suitable weights can integrate a polynomial function of order $(2n - 1)$ exactly. This order of integration is optimal (in that no other set of n points could integrate a higher-degree polynomial exactly); however, all the points are always interior to the domain, that is they never coincide with the ends of the integration domain a or b. The Gauss–Lobatto points, on the other hand, provide integration of optimal order given the constraint that a and b must be included, and can exactly integrate polynomials up to order $(2n - 3)$. The simplest Gauss–Lobatto rule for one-dimensional quadrature on an interval $[0, 1]$ is

$$\int f(x) d\Omega = \sum_{i=1}^{2} w_i f(x_i) \tag{12.36}$$

with weights $x_i = 1/2$, and nodes $x_1 = 0$, $x_2 = 1$.

Applied now to the mass matrix in the present example, this produces a diagonal matrix. It is easiest to do this for one element.[6] This is illustrated in Fig. 12.8. In the lower plot, it is clear that the product $\alpha_l \alpha_r$ is zero at both end points, whereas the product $\alpha_r \alpha_r$ is non-zero at the right-hand node. (Obviously, similar arguments apply to $\alpha_r \alpha_l$

[6] Note that the equivalence between the Galerkin and variational approaches is being exploited here, to permit one to now work at the elemental level.

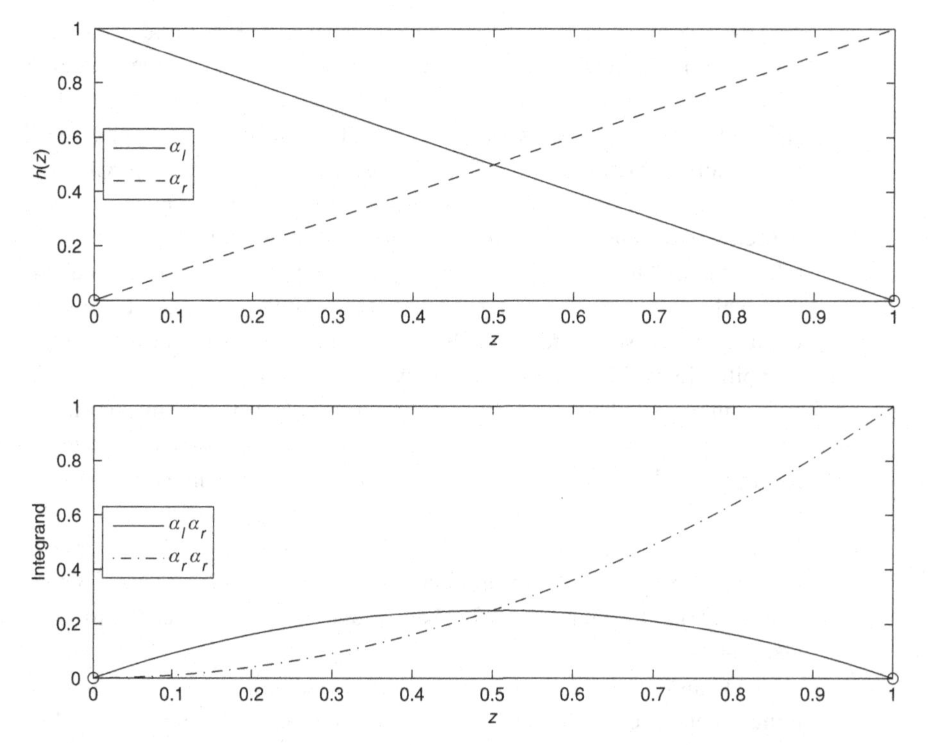

Figure 12.8 Upper plot: the first-order basis functions α_l and α_r. Lower plot: the products $\alpha_l\alpha_r$ and $\alpha_r\alpha_r$, which form the integrand of the relevant entries of the mass matrix. The $\circ$ markers at $z = 0$ and $z = 1$ indicate the location of the 2-point Gauss–Lobatto quadrature nodes.

and $\alpha_l\alpha_l$.) For an element, the normalized mass matrix is

$$[T] = \frac{1}{2}\begin{bmatrix} 1 & 0 \\ 0 & 1 \end{bmatrix} \tag{12.37}$$

The exact result is Eq. (9.30), and it is noted that not only are the off-diagonal terms in the mass lumped system zero, but the diagonal terms are also different. Nonetheless, this procedure produces good results.

Continuing with the variational approach, when assembled, one obtains

$$[T] = hC\begin{bmatrix} \frac{1}{2} & & & & \\ & 1 & & & \\ & & 1 & & \\ & & & \cdots & \\ & & & & \frac{1}{2} \end{bmatrix} \tag{12.38}$$

The stiffness matrix entries (each entry being a product of two constant functions for first-order elements, which is in turn a constant) is integrated exactly by this 2-point Gauss–Lobatto rule, and the off-diagonal entries are retained. (Where the stiffness matrix also to be diagonalized, the familiar finite difference stencil would not emerge.)

When assembled, it has the following form:

$$[S] = \frac{1}{hL}\begin{bmatrix} 1 & & & & \\ -1 & 2 & -1 & & \\ & -1 & 2 & -1 & \\ & & & \dots & \\ & & & & 1 \end{bmatrix} \tag{12.39}$$

Our interest is in the interior nodes; since the forcing function only enters at the RHS node in our model problem, it can be omitted in the following. Given that $[T]$, and hence $[A]$, is now diagonal, the system is now fully explicit, and for an interior node m, the mth diagonal element of $[A]^{-1}$ is simply $(\Delta t)^2/hC$. From Eq. (12.30) the update equation for one interior node can be written as

$$V_m^{n+1} = \frac{(\Delta t)^2}{hC}\left(\left[hC\frac{2}{(\Delta t)^2} - \frac{2}{hL}\right]V_m^n + \frac{1}{hL}V_{m+1}^n + \frac{1}{hL}V_{m-1}^n - \frac{hC}{(\Delta t)^2}V_m^{n-1}\right) \tag{12.40}$$

In FDTD notation, $h = \Delta z$. Noting further that $v = 1/\sqrt{LC}$, this can be re-arranged to obtain

$$V_m^{n+1} = 2V_m^n - V_m^{n-1} + \left(\frac{v\Delta t}{\Delta z}\right)^2\left[V_{m+1}^n + -2V_m^n + V_{m-1}^n\right] \tag{12.41}$$

which is equivalent to Eq. (2.115). Since the second-order FD scheme can be obtained from the coupled first order FDTD scheme (by eliminating I between the two first-order update equations, as discussed in Section 2.6), this shows the equivalence of the methods, at least in one dimension.

To show the equivalence in 2D and 3D, mixed first-order edge elements (for the 2D case, as in Section 10.7.2) are used as basis (and weighting) functions. Almost all sources in the literature indicate that the mass matrix is diagonalized by using trapezoidal integration. This essentially entails sampling the basis function values at the nodes of the hexahedral elements. While this approach generates the correct equivalence, it seems somewhat unintuitive, since the fields are not sampled at the same positions as they would have been in the FDTD.

Using instead the mixed Gauss–Lobatto reduced integration described in [14], a more satisfying result is obtained [11]. As previously noted, the trapezoidal rule is the same as the lowest-order Lobatto rule, and applying the mixed rule or the all-Lobatto rule yields the same result. If the mixed Gauss–Lobatto rule is used, the numerical integration points fall on the same place where the Yee FDTD cell has its degrees of freedom. The Gauss–Lobatto points of the electric degrees of freedom fall in the middle of the element's edges, and the points of the magnetic degrees of freedom fall in the element's face centers. In 2D, a second-order FDTD scheme in the relevant z-directed field component can be derived to complete the equivalence, and it is possible in 3D to eliminate three of the six field vectors, producing a second-order scheme in $\boldsymbol{E}$ (or $\boldsymbol{H}$) [15]. One can also show the equivalence between a coupled first-order FETD scheme and the Newmark-β second-order one. An analysis of several coupled first-order FETD schemes was presented in [16]. Chilton and Lee independently approached high-order FDTD using the Gauss–Lobatto scheme; see [17].

12.3.4 Hybrid FDTD/FETD schemes

The equivalence shown in the preceding section between the FDTD and FETD is not merely of academic interest. Both the FDTD and FETD are powerful methods, but deployed individually, both have issues requiring attention. At heart, the main claim of the FDTD is great computational efficiency, achieved due to the regular grid and explicit time stepping scheme; whereas the main claim of the FETD[7] is high-fidelity geometric modelling, using unstructured tetrahedral meshes. However, the FDTD pays the price of inefficient modelling of non-rectilinear geometries, whereas the FETD is computationally an expensive method. A hybrid scheme would certainly be attractive; using the unconditional stable FETD in regions where fine geometrical modelling is required, but the FDTD in the rest of the computational domain, should offer significant benefits. Expanding further on this point, the FETD is well suited to modelling complex geometrical components of the overall domain, since the unconditional stability means that small elements do not limit the overall Δt (as would be the case with an explicit scheme), whereas the speed of the FDTD can be exploited to the full in the rest of the region, and well-established mesh termination schemes such as PMLs can be used. Another point of contrast between the methods is that the FETD generalizes to higher order in a straighforward fashion (as seen repeatedly seen in this text). However, the FDTD has proven difficult to generalize to higher order, not least because the larger support of the finite differencc stencil for higher-order approximations has proven very difficult to reconcile with a precise location of material interfaces. (One such successful application [18, 19] used high-order FDTD in uniform regions, hybridized with conventional FDTD near boundaries and scatterers.)

The potential benefits of such a scheme were noted many years ago in the community, but it was the work of Rylander and Bondeson [9] that first provided a provably stable FETD/FDTD hybrid. For the first time, the methods were brought together without the weak instability which had plagued previous attempts (see [9] for these). In [10], an improved scheme was presented. Marais and the present author generalized the Rylander–Bondeson scheme to higher order in [11]; a method for directly connecting high-order hexahedral ("brick") and tetrahedral meshes was also described by us in [16], which simplified construction of the mesh.

12.4 Sparse matrix solvers

The development of an FEM code often goes through two major stages: the first concentrates on getting the code to work; the second concentrates on optimizing the code with regard to both memory usage and runtime. In Chapters 9–11, for instance, we focussed

[7] In this discussion, FETD implies the Newmark-β scheme, with $\beta = 1/4$ ensuring unconditional stability.

exclusively on the former. This process is frequently iterative, since new theoretical extensions must again be validated first, and then optimized. Furthermore, certain validation can only be undertaken once some optimization is already in place. Since the finite element system matrices are usually highly sparse (i.e. have a very large number of zero entries), the efficiency of the sparse solver(s) is probably the single most important factor in determining the overall efficiency on an FEM code, since the matrix solution time usually dominates all other contributors to the total runtime, and FEM codes cannot work efficiently unless the sparsity of the finite element system matrix is properly exploited. There are two choices to make when exploiting sparsity:

Iterative solvers
Iterative matrix solvers have the major advantage of requiring no additional memory beyond that required to store the coefficient matrices. They have the major disadvantage that each new solution of the system requires the iterative process to be repeated from scratch.

Direct solvers
These are usually variations on the LU decomposition theme, and factor the matrix into a lower (and an upper, if the matrix is not symmetric) triangular matrix which permits very rapid subsequent solution of the system. However, they have the major disadvantage that the factorization process generates a number of non-zero entries in the matrix; this is known as "fill-in." Various methods are used to handle this; here, a method called "skyline storage" will be used.

Which choice is best is in general problem dependent; surprisingly, even in the case of a finite element time domain solver, where the same matrix is involved at each timestep, a direct solver is not necessarily the best solution. (In that specific case, the real valued system generated appears to be well conditioned, resulting in very rapid convergence of the iterative process. Dibben and Metaxas reported this in some of the earlier work in the field [21].) The memory overhead of the profiled storage scheme can also be prohibitive. For frequency domain solvers, the complex valued matrix can become very ill-conditioned, and generally some form of preconditioning is required if an iterative solver is used.

First, two methods for storing a sparse matrix will be discussed.

12.4.1 Profile-in skyline storage

Consider the symmetrical matrix $[A]$:

$$[A] = \begin{bmatrix} a_{11} & a_{12} & & & \\ a_{21} & a_{22} & & a_{24} & a_{25} \\ & & a_{33} & 0 & 0 \\ & a_{42} & 0 & a_{44} & 0 \\ & a_{52} & 0 & 0 & a_{55} \end{bmatrix} \tag{12.42}$$

with $a_{12} = a_{21}$, etc. Here, an observation will be made. If this matrix is factored, without pivoting, then *possible*[8] fill-ins will occur in $[L]$ to the right of the first non-zero entry in a row across to the diagonal (and similarly, in $[U]$ under the first non-zero entry in a column down to the diagonal). Hence, if all the zeros indicated above are stored, the factored matrix is guaranteed to fit into the data structure. This type of data structure is called a "skyline" matrix. There are several methods for storing the data: the one adopted here is called "profile-in," and what is stored is the elements in each row (column) from the first non-zero element to the diagonal (hence "in," since one moves inwards to the diagonal). Additionally, the index of the diagonal element is stored. For this matrix, the profile-in storage looks as follows:

$$AL = [a_{11}, a_{21}, a_{22}, a_{33}, a_{42}, 0, a_{44}, a_{52}, 0, 0, a_{55}]$$
$$IALDIAG = [1, 3, 4, 7, 11] \tag{12.43}$$

Since the matrix is symmetric, these structures could equally have been AU and $IAUDIAG$. The dimension of $IALDIAG$ is n. The dimension of AL is *at least* nz_s, the number of non-zeros in the lower (or upper) triangular half. Unfortunately, it is frequently many times this number.

12.4.2 Compressed row storage

Skyline storage is convenient when factoring a matrix but has a very high overhead, which only becomes clear when much larger finite element matrices are considered. The percentage of non-zero elements rapidly drops under one percent, but the profiled storage results in a very large number of zeros being stored, frequently an appreciable fraction of the original matrix. For iterative solvers, which require only a matrix-vector product, a much more efficient scheme is compressed row storage (CRS). Here, absolutely *only* the non-zero elements are stored. Since the storage requirements of a CRS matrix are so small, it is convenient to store each row completely, even if the matrix is symmetrical – this makes the sparse matrix-vector product far easier to write. In addition to an array storing the non-zero matrix elements, two other pointer arrays are needed. One stores the starting index of each row, the other stores the column indices. For the above matrix, the CRS equivalent is

$$A\_CRS = [a_{11}, a_{12}, a_{21}, a_{22}, a_{24}, a_{25}, a_{33}, a_{42}, a_{44}, a_{52}, a_{55}]$$
$$JA = [1, 2, 1, 2, 4, 5, 3, 2, 4, 2, 5] \tag{12.44}$$
$$IA = [1, 3, 7, 8, 10, 12] \tag{12.45}$$

The $n + 1$-element of IA is $nz + 1$, where nz is the number of non-zeros. The dimension of IA is $n + 1$, and the dimensions of both A_CRS and JA are nnz. These are known a priori, as soon as the matrix entries are known.

This storage scheme is also known as "general storage by rows."

[8] Not all these positions will indeed be filled. More sophisticated methods do a better job of this process.

A very similar storage scheme (and the one implemented in MATLAB) is compressed column storage; the procedure simply interchanges the storage direction. Since finite element matrices are generally symmetrical (unless one is dealing with non-reciprocal materials) the schemes are in practice essentially identical for finite element applications.

12.4.3 Implementation of matrix solution using these storage schemes

Sparse matrices are important for two reasons: firstly, to save memory, and secondly, to reduce runtime. Unfortunately, at the time of writing there is no analogy of the excellent public domain LAPACK routines for sparse matrices. If working with languages such as FORTRAN 90, sparse libraries may be available, either bundled with the compiler or for purchase separately. However, actually *storing* the matrix in sparse form is a complex book-keeping task; one has firstly to establish the connections between all the degrees of freedom present (and this becomes increasingly more complex as higher-order elements are added) to determine the number of non-zero entries, following which the compressed matrices may then be filled as the matrix is assembled. Alternatively, and rather more easily, a full matrix may be generated first, and a sparse matrix then generated from this – the MATLAB function `sparse` does precisely this. However, the requirement to store the full matrix first wastes large amounts of memory, and is not practical for FEM codes designed for electromagnetically large problems.

It should be mentioned that especially higher-order elements appear to generate ill-conditioned matrices. When using iterative methods, such as conjugate gradient schemes (CG, Bi-CG), QMR and GMRES, convergence tends to be erratic. (For a description and discussion of these algorithms, see [3].) Some recent approaches have focussed on the use of more sophisticated preconditioners. Incomplete LU preconditioning is one possibility; another is the use of a direct solution of the CT/LN solution (which can generally be computed quite cheaply) as a preconditioner for the LT/QN matrix. This has been extended to higher-order schemes by [22]. Most of the more sophisticated preconditioner schemes trade off quicker convergence for increased matrix storage requirements.

Direct solvers have a place; generally, ill-conditioning is far less problematic, but the fill-ins can result in very large matrices indeed. Renumbering schemes can ameliorate this, but unfortunately 3D finite element meshes tend to generate meshes with significant "bandwidth."

12.4.4 Results for sparse storage schemes

Some results illustrating the impact of sparse matrix solvers on an FEM code – in this case the FETD implementation by the author discussed in Section 12.3 – are shown in Figs. 12.9 and 12.10. The times shown in Fig. 12.9 compare the time using the sparse skyline or iterative CG solver (using CRS) with those of a full matrix solver (the latter not exploiting symmetry, i.e. worst case). Similarly, the memory percentages

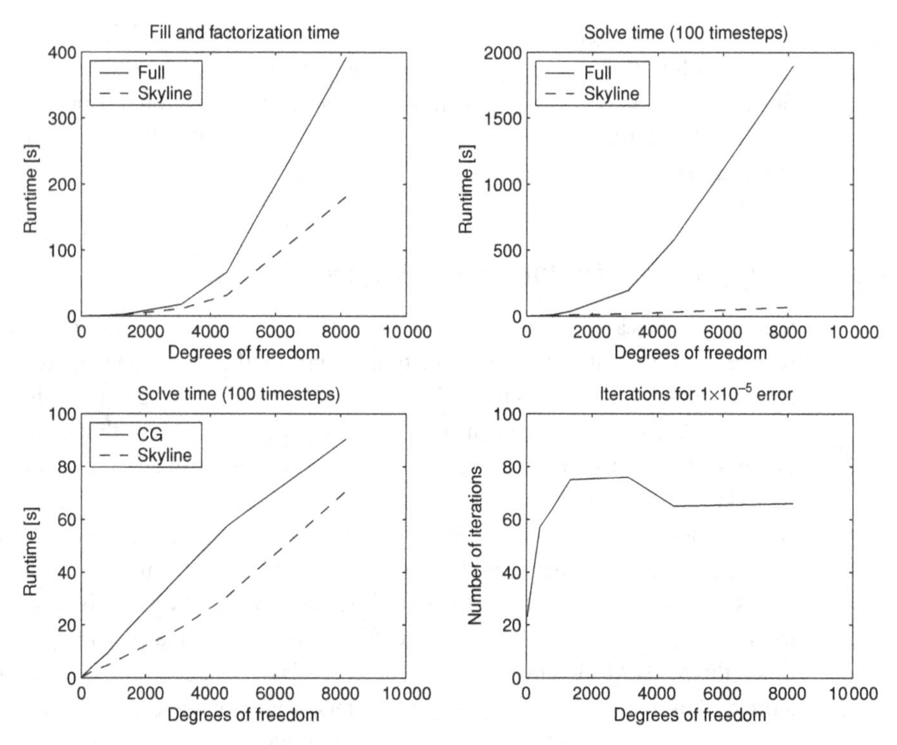

Figure 12.9 Solver times for the cubic example given in Section 12.3.2. Solve times are for 100 timesteps. The CG solver normalized residual target was 1×10^{-5}.

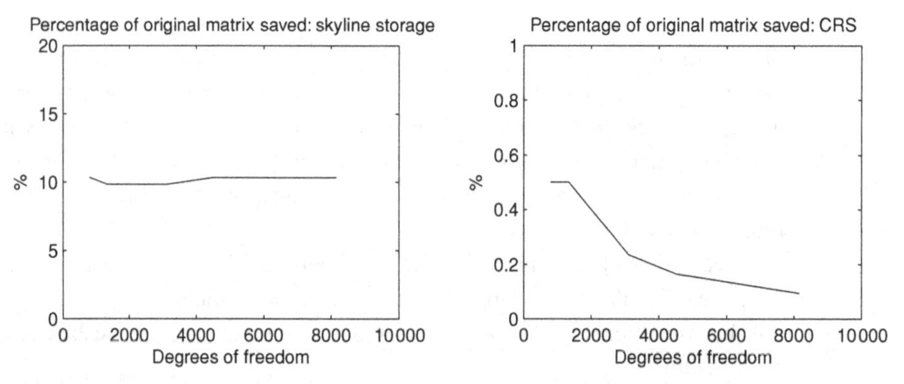

Figure 12.10 Memory usage for the cubic example given in Section 12.3.2. This is expressed as a percentage of the memory required by a full matrix scheme making no use of symmetry.

shown in Fig. 12.10 compare the relevant storage scheme with a full matrix scheme not using symmetry. The skyline storage is actually considerably less efficient than might be inferred from Fig. 12.10. Because with either of the sparse schemes, the $[B]$ and $[C]$ matrices in Eq. (12.25) can be (and are) stored in the much more efficient CRS form, whereas in a full matrix scheme they are of course stored as full matrices, there is already a saving by a factor of very close to three which is reflected in this figure. (CRS stored matrices require negligible storage compared to even skyline schemes, hence the factor of approximately three.)

These results are significant for code developers. Firstly, the timing results in Fig. 12.9 indicate that *any* sparse scheme is significantly better than none, as would be expected. Another interesting result is the comparison of the runtime of the CG solver with the skyline solver (lower left-hand plot in Fig. 12.9). The reason is that, for this problem at least, the number of iterations is almost constant, irrespective of problem size. This is shown in the lower right-hand graph in Fig. 12.9. (Although not shown on the figures, the number of iterations required also did not change from timestep to timestep.) One must be cautious of extrapolating this result to electromagnetically larger and more complex problems. These results were generated by increasingly refining the *same* problem. It is well known that the convergence rate of iterative solvers is a function of the ratio of maximum to minimum eigenvalues; furthermore, for any given electromagnetic problem, discretization beyond a certain point does not yield more significant eigenvalues. Electromagnetically larger problems may of course contain a wider eigenvalue spectrum. This note of caution notwithstanding, the results for the iterative solvers are highly encouraging, since no effort was made with these results to increase the rate of convergence, and an entire class of methods using various preconditioners exists which can still be applied. The memory savings of the iterative solver are of course very impressive (the right-hand graph in Fig. 12.10) and imply that the limit on large problems is more likely to be runtime than memory.

A final comment on these results. The graphs comparing memory savings are actually in terms of memory *locations* required, rather than actual Mbytes of RAM used. The libraries used were double precision, so in RAM the percentages are twice that shown in the graphs. (Double precision was presumably used since the sparse factorizer does not apply pivoting. All the results were tested against a full matrix direct solver, and results were generally identical within working precision, around 4–5 significant figures after 100 timesteps. The CG solver normalized residual was set as 1×10^{-5} to ensure that inaccuracies did not accumulate during timestepping.) This is a peculiarity of the particular implementation rather than the method per se. It also means that the computation times using the sparse schemes are slightly longer than would be the case if single precision were used.

12.5 A posteriori error estimation and adaptive meshing

As a final topic, some interesting recent work by Botha on the problem of estimating errors in the finite element solution will be outlined [23, 24, 25]. One of the main advantages of the FEM over the FDTD is that, theoretically at least, it is easy to *refine* a finite element mesh selectively. This can either be done by increasing the element order (p adaptation), decreasing the element size (h adaptation), or doing both ($h-p$ adaptation). In practice of course, mesh refinement does bring some complexity.

However, before one can undertake any form of mesh refinement, one needs an idea of in which part of the mesh the greatest benefits will be obtained. (Simply refining the entire mesh is of course a valid process, but computationally expensive. This is sometimes known as *uniform* mesh refinement.) It is here that the complex topic of *error estimation* comes to the fore. Here, one needs firstly to distinguish between a priori and a posteriori error estimates. The former are derived theoretically, and do not use the specific geometrical data represented by the mesh; examples are the analysis of dispersion error in a finite element or finite difference mesh. The latter are derived from the approximate solution, and it is these that will be considered here.

A posteriori error estimates can themselves be categorized as follows:

Explicit, residual-based
These estimators are usually rigorously derived in the sense that the sum of the errors in each element is an upper bound on the error. (Here, we assume some suitable norm is available; often, the energy norm, discussed subsequently, is used.) Typically, field discontinuities at element edges and faces are evaluated.

Implicit, residual-based
These estimators are based on the solution of local variational boundary value problems, usually on an element-wise basis. Usually, an estimate of the error is made using additional basis functions of higher order than the initial solution. Since this is done on an element-by-element basis, this is not prohibitively expensive computationally – certainly not when compared to uniform refinement.

Estimation through post-processing
These methods estimate the error in a derivative of the solution field, by comparing it with an improved version. Although this may seem counter-intuitive, some methods are available for computing improved versions of the solution field and its derivatives.

Targeted quantities
These are also known as *goal-oriented* or *targeted* error estimation. They attempt to bound the error of a quantity based on some functional output of the solution field. An example is the *S*-parameters discussed in the context of the waveguide formulation.

Botha's work focussed on explicit and implicit residual-based methods; the best results in general were obtained with the former, and a very brief summary of the method will now be presented.

12.5.1 Explicit, residual-based error estimators

Firstly, one must define the error in the solution as

$$\boldsymbol{e}_h = \boldsymbol{E} - \boldsymbol{E}_h \tag{12.46}$$

where $\boldsymbol{E}$ is the (usually unknown) exact solution of the problem, and $\boldsymbol{E}_h$ is the approximate, finite element computed solution. Botha showed that an estimate of the error in the CT/LN solution may be obtained as

$$||\boldsymbol{e}_h||^2_{E^a(V,\tau,1)} \leq C \sum_{i=1}^{N} \left(h_i^2 ||\boldsymbol{R}_v||^2_{L^2(K_i)} + 0.5 \sum_{f \subset \partial K_i} h_f ||\boldsymbol{R}_f||^2_{L^2(K_i)} \right) \tag{12.47}$$

N is the number of elements in the mesh; τ refers to the current discretization and solution, which will be used to compute the error indicators. The constant C is in general unknown, but is independent of solution field and source terms; error estimates usually contain such constants.

The term $||\boldsymbol{R}_v||^2_{L^2(K_i)}$ is the *volume* residual on element i, with volume K_i, measured in the L^2 norm – the space of square integrable functions. The volume residual in element i is computed from

$$\boldsymbol{R}_v = -\nabla \times \frac{1}{\mu_r} \nabla \times \boldsymbol{E}_h + k_0^2 \varepsilon_r \boldsymbol{E}_h - jk_0 Z_o \boldsymbol{J} \text{ in } K_i \tag{12.48}$$

In other words, this is the difference between the finite element computed solution, and the specified impressed current – in short, the residual of the vector wave equation. (If the latter is zero, then this term should of course be zero.) Were the solution exact, then this residual would be zero throughout the finite element volume.

The face residual on the surfaces of element i is computed from

$$\boldsymbol{R}_f = \hat{n}_{(12)} \times \left[\frac{1}{\mu_r^{(1)}} \nabla \times \boldsymbol{E}_h^{(1)} - \frac{1}{\mu_r^{(2)}} \nabla \times \boldsymbol{E}_h^{(2)} \right] \text{ on } f_m, \quad m = 1, 4 \tag{12.49}$$

f_m is a specific face of the element, and the superscripts (1) and (2) indicate the elements shared by a particular face. In other words, this is the discontinuity in tangential magnetic field on the inter-element boundaries; again, were the solution exact, then this residual would be zero at all inter-element boundaries. Note that a special treatment, not shown here, is required at the Neumann boundary.

Whilst it may seem obvious that such residuals provide an indication of the error in the solution, some subtle mathematical arguments are required to show that the sum of residuals in Eq. (12.47) does indeed produce a bounded estimate of the overall error; the details may be found in [23, Chapter 5].

It should also be commented that the "norm" on $\boldsymbol{e}_h$ on the left-hand side of Eq. (12.47) is not a proper norm of the error field, but rather an approximate *energy norm.* (The reason that this does not conform to the usual definition is that this energy norm can be zero, without the field being zero. However, the converse is indeed true, i.e. the energy norm of a zero-valued field is zero.) The reason that this needs to be introduced is rooted in the complex valued nature of the functional. The approximate energy norm for space of maximum (but not necessarily complete) polynomial order p is defined as

$$||\boldsymbol{v}||_{E^a(V,\tau,p)} \equiv \frac{|\iiint_V \frac{1}{\mu_r} \nabla \times \boldsymbol{v} \cdot \nabla \times \boldsymbol{v} - k_0^2 \varepsilon_r \boldsymbol{v} \cdot \boldsymbol{v} \, dV|}{|\sum_{m=1}^{N} |\boldsymbol{v}|^2_{(H^p(K_m))^3}|^{1/2}} \tag{12.50}$$

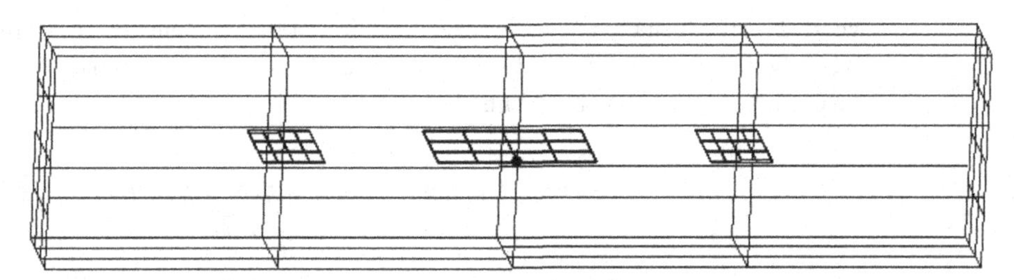

Figure 12.11 The waveguide filter geometry. (After [26], ©2002 IEEE, reprinted with permission.)

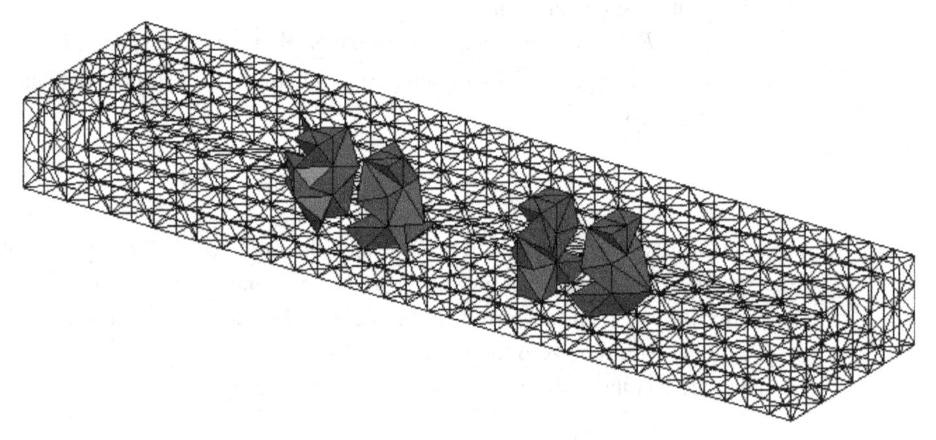

Figure 12.12 The waveguide filter: 2.5% elements with highest indicated error. (After [26], ©2002 IEEE, reprinted with permission.)

The term in the denominator, $|v|^2_{(H^p(K_m))^3}$, represents the vector Sobolev semi-norm of derivative order p on domain K_m. Details of its evaluation may be found in [23].

12.5.2 An example of the application of an error estimator

An insightful example of the application of an error estimator may be found in [26]. The problem is an X-band waveguide filter (Fig. 12.11), with three metal septa along its center, normal to the broad walls of the waveguide (we have already encountered this problem in Chapter 3). The explicit residual-based error indicator was applied, and results were obtained indicating where the computed errors were the highest. These are shown in Figs. 12.12–12.15. As expected, the errors cluster around the edges of the septa.

Once one has an indication of where the errors are most serious, one has various options to improve the solution. In this case, the results were used to drive a p-adaptive scheme, using the hierarchal elements of both mixed and complete order discussed earlier in this chapter. This permits a variety of possible schemes. The original solution was obtained with CT/LN elements; one possibility is to upgrade all the elements with the highest indicated error to QT/QN (which was the highest order available within

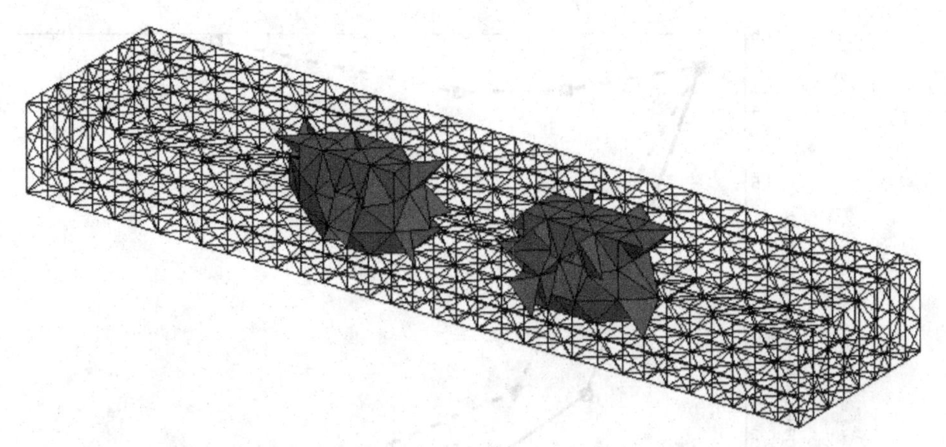

Figure 12.13 The waveguide filter: 5% elements with highest indicated error. (After [26], ©2002 IEEE, reprinted with permission.)

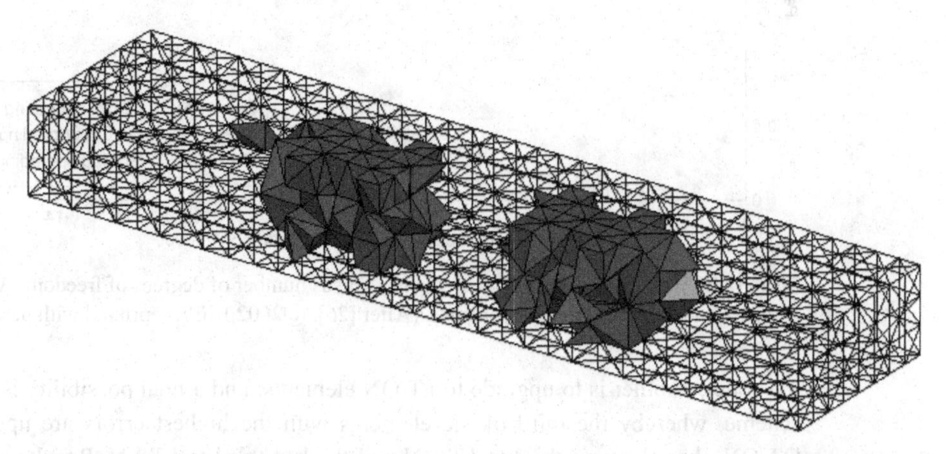

Figure 12.14 The waveguide filter: 10% elements with highest indicated error. (After [26], ©2002 IEEE, reprinted with permission.)

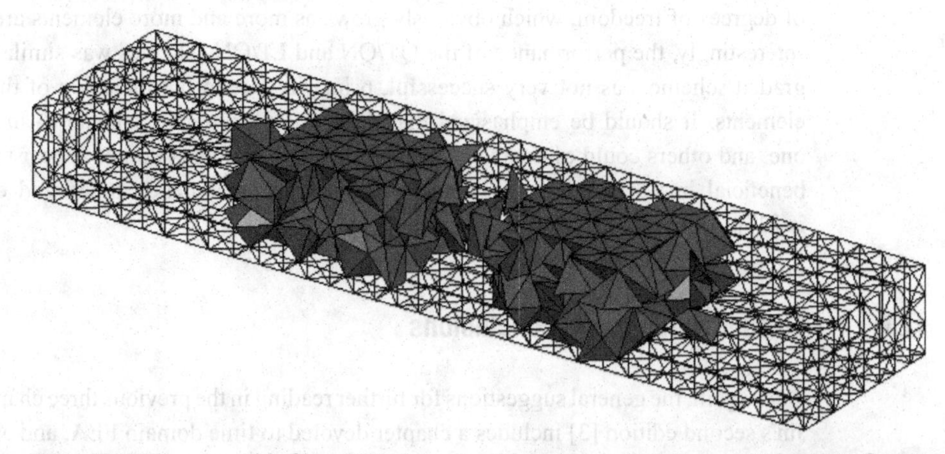

Figure 12.15 The waveguide filter: 20% elements with highest indicated error. (After [26], ©2002 IEEE, reprinted with permission.)

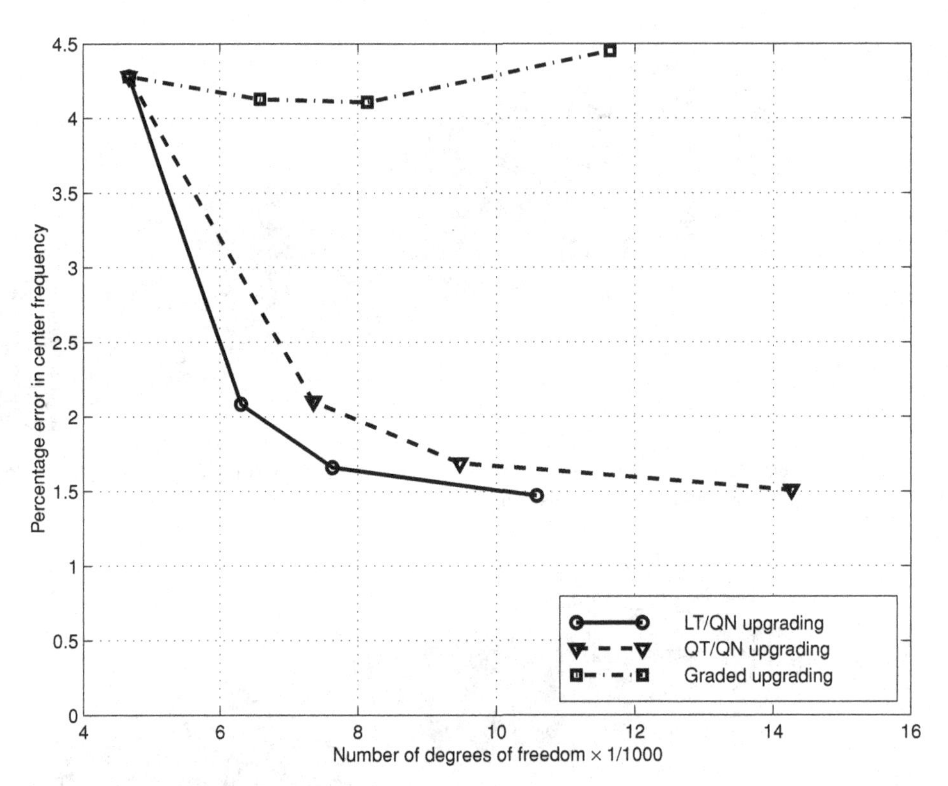

Figure 12.16 Waveguide filter center frequency versus number of degrees of freedom. A comparison of three upgrading schemes. (After [26], ©2002 IEEE, reprinted with permission.)

the code); another is to upgrade to LT/QN elements; and a final possibility is a *graded* scheme, whereby the third of the elements with the highest errors are upgraded to QT/QN, then the next third to LT/QN and the last third to LT/LN. Results are shown in Fig. 12.16. The percentage error in center frequency is plotted against the number of degrees of freedom, which obviously grows as more and more elements are refined. Interestingly, the performance of the QT/QN and LT/QN schemes was similar, but the graded scheme was not very successful, primarily due to the inclusion of the LT/LN elements. It should be emphasized that this particular graded scheme is an heuristic one, and others could of course be proposed. These elements do not appear to be very beneficial in waveguide finite element analysis, a phenomenon noted and discussed in [27].

12.6 Further reading and conclusions

Once again, the general suggestions for further reading in the previous three chapter refer. Jin's second edition [3] includes a chapter devoted to time domain FEA, and another to matrix solution. The text by Bondeson *et al.* [15] can once again be recommended for its combination of brevity and completeness; time domain FEA is covered there, and the book also discusses applications of the FEM to magneto-quasistatics, which brings with it some additional complications in terms of gauging the magnetic vector potential which have not been considered here. (Ironically, the quasistatic case is not necessarily easier to work than the dynamic case considered throughout this book.) In terms of the specific material in this chapter:

Regarding the FEM/MoM hybrid formulation, Peterson *et al.* [28] proposed that FEM/MoM hybrids be classified as either *outward-looking* or *inward-looking*. In the former case, the surface integral formulation represented by the MoM is used to augment the variational functional form of the vector wave equation, and this was the approach used in this chapter. It is also the most commonly encountered in the literature and in practice, since the effect is to increase the size of the FEM matrix somewhat. Furthermore, this outward-looking approach is also readily amenable to the introduction of approximate radiation boundary conditions, such as absorbing boundary conditions, rather than the rigorous Green function approach implicit in the MoM. Inward-looking formulations use the interior problem to constrain equivalent sources on the bounding surface. In this case, a large FEM matrix must first be solved before a smaller dense matrix can be constructed. Examples of the latter approach are the unimoment method, first suggested by Mei in 1974. More details on this topic may be found in [28, Chapter 3]. One problem with the outward-looking approach outlined here is that the matrix symmetry is generally destroyed. Botha and Jin proposed a formulation which hybridizes the FEM and MoM on the formulation level (rather than on the matrix level) and which preserves the matrix symmetry [29]. It also offers some alternative approaches, including one which uses both $\boldsymbol{E}$ and $\boldsymbol{H}$ as working variables, permitting both fields to be computed to the same level of accuracy. The FEM/MoM formulation given in this chapter can suffer from internal resonances; the problem comes from the MoM treatment, and has already been mentioned in Section 6.10. The usual solution is to combine the EFIE and MFIE on the boundary. The formulation of Botha and Jin apparently also solves the problem, and computed results support the claim [29].

In terms of time domain formulations and applications, basic reference material can usefully be found in the structural mechanics textbook by Hughes [12], as already noted. The paper by Gedney and Navsariwala [30] is one of the earlier in the field of CEM to discuss the FETD. It discusses an unconditionally stable formulation using the Newmark-β method. Although brief, it discusses most of the important topics and provides a stability analysis. The formulation is similar to that presented recently in [3]. The paper by Dibben and Metaxas [21] is also one of the earlier publications, and also uses the Newmark method. Together, these two represent well some of the earlier work on FETD formulations and implementations within CEM. The review paper by Lee *et al.* [31] appeared in a special issue of the *IEEE Transactions on Antennas and Propagation* on numerical methods some years back; it presents a good overview of the state of the art at that time, discussing an elegant theoretical framework for the general class of FETD methods, and it is more general than the approach presented in [3]; the latter focusses on the conventional electromagtnetic curl-curl functional formulation for the semi-discretization. It should be noted that the coupled first-order four coupled scheme proposed in [31] was later shown to have significant deficiencies when implemented [16]; that paper compares several schemes and in a number of respects updates [31]. Since the first edition of this book appeared, there has been significant advances in the understanding of the FETD; the properties of the primal and dual meshes used in coupled first order schemes were analyzed by He and Teixeira [32]. Chilton and Lee's work on the stability of the FETD [13] has already been mentioned.

One very troublesome problem with FETD methods has been the development of efficient ABC-based boundary conditions; the paper by Jiao *et al.* [33] appears to have been the first to report a rigorous PML-type implementation for the FETD, although several workers, including the present author, have encountered problems with this implementation, in particular regarding stability. Reference should also be made to later work in this regard, e.g. [34]. The problem arises from the second order time derivative in the curl-curl formulation. Use of the hybrid FETD/FDTD scheme discussed in Section 12.3.4 is attractive in that it permits the use of the standard PML approach in the FDTD region.

Error estimation and mesh adaptation has a rather small bibliography in the engineering electromagnetics literature. Earlier work on this was done by Meyer, in the context of scalar, two-dimensional electromagnetic scattering and radiation problems, and results may be found in [35] and [36]. His results remain one of the most complete investigations of that specific problem. Some of the earlier work on the three-dimensional vector problem was done by Pekel and Lee [37].

In a field as large as finite elements, it is inevitable that there will be some important topics which we have not discussed at all. One of these is curvilinear elements. Whilst higher-order elements can do an excellent job of representing the fields very accurately, the limitations imposed by straight-sided triangular or tetrahedral elements in terms of accurate modelling of curved geometries can be very significant for many practical problems. There are in essence two questions to answer here: firstly, given a geometrical transformation, how does one implement this as a curvilinear element, and secondly, what transformation should be used? The former is the more theoretical issue, the latter a more practical one. Although curvilinear elements have been used in CEM, the literature on this is rather incomplete from a practical perspective, in particular in the context of vector elements. The following references either deal with the issue in passing, touch on the issue, or summarize some aspects thereof: [38, 39, 40, 41]. In the context of nodal elements, the discussion in [2, Chapter 7] is also useful. Two recent references by the present author and his students address some practical aspects in detail: [42] considers two-dimensional problems, whilst [43] provides a comprehensive analysis of a second-order curvilinear element using the Webb vector basis functions.

Another important topic which we have omitted mention of, and which has produced important and interesting results, is the analysis of dispersion error in finite element meshes (this is also sometimes called pollution error). The work of Cangellaris and Lee is an important reference here [44]; an overview of more recent work may be found in [3].

Modelling microwave ovens for commercial electro-heat applications has been a significant radio-frequency application of the FEM, using both eigenvalue and driven problem analysis. Details may be found in the books by Metaxas [45] and Chan and Reader [46].

Finally, serious students of the FEM who would like to read the large applied mechanics and applied mathematics literature will find that much of it uses the language of functional analysis. A very readable introduction is the text by Reddy [47], not least since it focusses on FEM formulations, unlike many of the more general texts on functional analysis.

References

[1] S. M. Rao, D. R. Wilton and A. W. Glisson, "Electromagnetic scattering by surfaces of arbitrary shape," *IEEE Trans. Antennas Propagation*, **30**, 409–418, May 1982.

[2] P. P. Silvester and R. L. Ferrari, *Finite Elements for Electrical Engineers*. Cambridge: Cambridge University Press, 3rd edn., 1996.

[3] J.-M. Jin, *The Finite Element Method in Electromagnetics*. New York: Wiley, 2nd edn., 2002.

[4] J. Volakis, A. Chatterjee and L. Kempel, *Finite Element Method for electromagnetics: Antennas, Microwave Cicuits and Scattering Applications*. Oxford & New York: Oxford University Press and IEEE Press, 1998.

[5] M. M. Botha and D. B. Davidson, "Analyzing cavity backed, perforated substrate, microstrip patch antennas with a FMM, FE-BI hybrid formulation," in *Proceedings of the 2001 URSI International Symposium on Electromagnetic Theory*, pp. 627–629, May 2001.

[6] M. W. Ali, T. H. Hubing and V. M. Drewniak, "A hybrid FEM/MOM technique for electromagnetic scatttering and radiation from dielectric objects with attached wires," *IEEE Trans. Electromag. Compat.*, **39**, 304–314, November 1997.

[7] T. F. Eibert and V. Hansen, "3-D FEM/BEM-hybrid approach based on a general formulation of Huygens' principle for planar layered media," *IEEE Trans. Microwave Theory Tech.*, **45**, 1105–1112, July 1997.

[8] F. J. C. Meyer, D. B. Davidson, U. Jakobus and M. A. Stuchly, "Human exposure assessment in the near field of GSM base-station antennas using the hybrid finite element/method of momemnts technique," *IEEE Trans. Biomedical Engineering*, **50**(2), 224–233, 2003.

[9] T. Rylander and A. Bondeson, "Stable FEM-FDTD hybrid method for Maxwell's equations," *Comput. Phys. Commun.*, **125**, 75–82, 2000.

[10] T. Rylander and A. Bondeson, "Stability of explicit-implicit hybrid time-stepping schemes for Maxwell's equations," *J. Comput. Phys.*, **179**, 426–438, July 2002.

[11] N. Marais and D. B. Davidson, "Efficient high-order time domain hybrid implicit/explicit FEM methods for microwave electromagnetics," *Electromagnetics*, **30**(1–2), 127–148, January 2010.

[12] T. J. R. Hughes, *The Finite Element Method: Linear Static and Dynamic Finite Element Analysis*. Englewood Cliffs, NJ: Prentice-Hall, 1987. Dover re-print, 2000.

[13] R. A. Chilton and R. Lee, "The discrete origin of FETD-Newmark late time instability, and a correction scheme," *J. Comput. Phys.*, **224**, 1293–1306, June 2007.

[14] G. Cohen and P. Monk, "Gauss point mass lumping schemes for Maxwell's equations," *Numer. Meth. Partial Differential Equations*, **14**, 63–88, 1998.

[15] A. Bondeson, T. Rylander and P. Ingelström, *Computational Electromagnetics*. New York: Springer Science, 2005.

[16] N. Marais and D. B. Davidson, "Conforming arbitrary order hexahedral/tetrahedral hybrid discretisation," *Electronics Lett.*, **44**, 1384–1385, November 2008.

[17] R. A. Chilton and R. Lee, "The Lobatto cell: robust, explicit, higher order FDTD that handles inhomogeneous media," *IEEE Trans. Antennas Propagat.*, **56**, 2167–77, August 2008.

[18] S. V. Georgakopoulos, C. R. Birtcher, C. A. Balanis and R. A. Renaut, "Higher-order finite-difference schemes for electromagnetic radiation, scattering, and penetration. Part 1: theory," *IEEE Antennas Propag. Soc. Mag.*, **44**, 134–142, February 2002.

[19] S. V. Georgakopoulos, C. R. Birtcher, C. A. Balanis and R. A. Renaut, "Higher-order finite-difference schemes for electromagnetic radiation, scattering, and penetration. Part 2: applications," *IEEE Antennas Propag. Soc. Mag.*, **44**, 92–101, April 2002.

[20] N. Marais and D. B. Davidson, "Numerical evaluation of high order finite element time domain formulations in electromagnetics," *IEEE Trans. Antennas Propagat.*, **56**, 3743–3751, December 2008.

[21] D. C. Dibben and R. Metaxas, "Time domain finite element analyis of multimode microwave applicators," *IEEE Trans. Magn.*, **32**, 942–945, May 1996.

[22] D. Sun, J. Lee and Z. Cendes, "Construction of nearly orthogonal Nedelec bases for rapid convergence with multilevel preconditioned solvers," *Siam. J. Sci. Comput.*, **23**(4), 1053–1076, 2001.

[23] M. M. Botha, Efficient finite element electromagnetic analysis of antennas and microwave devices: the FE-BI-FMM formulation and a posteriori error estimation for p adaptive analysis. PhD thesis, Dept. Electrical & Electronic Engineering, University of Stellenbosch, 2002.

[24] M. M. Botha and D. B. Davidson, "An explicit a posteriori error indicator for electromagnetic, finite element analysis in 3D," *IEEE Trans. Antennas Propagat.*, **53**, 3717–3725, November 2005.

[25] M. M. Botha and D. B. Davidson, "The implicit, element residual method for a posteriori error estimation in FE-BI analysis," *IEEE Trans. Antennas Propagat.*, **54**, 255–258, January 2006.

[26] M. M. Botha and D. B. Davidson, "A posteriori error estimates for the FEM analysis of a waveguide filter," in *IEEE AFRICON-02 Proceedings*, pp. 541–544, October 2002.

[27] D. B. Davidson, "An evaluation of mixed-order versus full-order vector finite elements," *IEEE Trans. Antennas Propagat.*, **51**, 2430–2441, September 2003.

[28] A. F. Peterson, S. L. Ray and R. Mittra, *Computational Methods for Electromagnetics*. Oxford & New York: Oxford University Press and IEEE Press, 1998.

[29] M. M. Botha and J. Jin, "On the variational formulation of hybrid finite element – boundary integral techniques for time-harmonic electromagnetic analysis in 3D," *IEEE Trans. Antennas Propagat.*, **52**, 3037–3047, November 2004.

[30] S. D. Gedney and U. Navsariwala, "An unconditionally stable finite element time-domain solution of the vector wave equation," *IEEE Microwave Guided Wave Lett.*, **5**, 332–334, October 1995.

[31] J.-F. Lee, R. Lee and A. Cangellaris, "Time-domain finite-element methods," *IEEE Trans. Antennas Propagat.*, **45**, 430–442, March 1997.

[32] B. He and F. Teixeira, "Geometric finite element discretization of Maxwell equations in primal and dual spaces," *Phys. Lett. A*, **349**, 1–14, 2006.

[33] D. Jiao, J. Jin, E. Michielssen and D. J. Riley, "Time-domain finite-element simulation of three-dimensional scattering and radiation problems using perfectly matched layers," *IEEE Trans. Antennas Propagat.*, **51**, 296–305, February 2003.

[34] T. Rylander and J. Jin, "Perfectly matched layer in three dimensions for the time-domain finite-element method applied to radiation problems," *IEEE Trans. Antennas Propagat.*, **53**, 1489–1499, April 2005.

[35] F. J. C. Meyer and D. B. Davidson, "A posteriori error estimates for the two-dimensional finite element/boundary element solution of electromagnetic scattering and radiation problems," *Appl. Comput. Electromagn. Soc. J.*, **11**, 40–54, July 1996.

[36] F. J. C. Meyer and D. B. Davidson, "Adaptive-mesh refinement of finite-element solutions for two-dimensional electromagnetic problems," *IEEE Antennas Propag. Soc. Mag.*, **37**, 77–83, October 1996.

[37] Ü. Pekel and R. Lee, "An a posteriori error reduction scheme for the three-dimensional finite-element solution of Maxwell's equations," *IEEE Trans. Microwave Theory Tech.*, **43**, 421–427, February 1995.

[38] R. D. Graglia, D. R. Wilton and A. F. Peterson, "Higher order interpolatory vector bases for computational electromagnetics," *IEEE Trans. Antennas Propagat.*, **45**, 329–342, March 1997.

[39] J. Wang and J. Webb, "Hierarchal vector boundary elements and p-adaption for 3-D electromagnetic scattering," *IEEE Trans. Antennas Propagat.*, **45**, 1869–1879, December 1997.

[40] C. W. Crowley, P. P. Silvester and H. Hurwitz, "Covariant projection elements for 3D vector field problems," *IEEE Trans. Magn.*, **24**, 397–400, 1988.

[41] A. F. Peterson and D. R. Wilton, "Curl-conforming mixed-order edge elements for discretizing the 2D and 3D vector Helmholtz equation," in *Finite Element Software for Microwave Engineering* (T. Itoh, G. Pelosi and P. P. Silvester, eds.), Chapter 5. New York: Wiley, 1996.

[42] N. Marais and D. B. Davidson, "Numerical evaluation of hierarchical vector finite elements on curvilinear domains in 2-D," *IEEE Trans. Antennas Propagat.*, **54**, 734–738, February 2006.

[43] J. P. Swartz and D. B. Davidson, "Curvilinear vector finite elements using a set of hierarchical basis functions," *IEEE Trans. Antennas Propagat.*, **55**, 440–446, February 2007.

[44] A. C. Cangellaris and R. Lee, "On the accuracy of numerical wave simulations based on finite methods," *J. Electromagn. Waves Appl.*, **6**(12), 1635–1653, 1992.

[45] A. C. Metaxas, *Foundations of Electroheat: a Unified Approach*. Chichester: Wiley, 1996.

[46] T. V. C. T. Chan and H. C. Reader, *Understanding Microwave Heating Cavities*. Boston, MA: Artech House, 2000.

[47] B. D. Reddy, *Introductory Functional Analysis with Applications to Boundary Value Problems and Finite Elements*. New York: Springer-Verlag, 1998.

Appendix A The Whitney element

The Whitney form $\lambda_i \nabla \lambda_j - \lambda_j \nabla \lambda_i$ is so widely used in vector elements that it is worth discussing in more detail. The development here is for two-dimensional elements, which has the benefit of simplicity; however, the essential argument is the same for the three-dimensional case.

Firstly, note the following very important property of the gradient of a simplex coordinate: it is constant, and is directed perpendicular to an edge. As an example, for the triangle shown in Fig. A1, $\nabla \lambda_1$ is perpendicular to edge 1, opposite vertex (node) 1. The formula is

$$\nabla \lambda_i = \frac{l_i}{2A} \hat{n}_i \tag{A.1}$$

with A the area of the triangle and $\hat{n}_i$ the normal on edge i.

We now investigate the properties of an approximation using the Whitney basis functions

$$\boldsymbol{E} \approx E_3(\lambda_1 \nabla \lambda_2 - \lambda_2 \nabla \lambda_1) + E_2(\lambda_1 \nabla \lambda_3 - \lambda_3 \nabla \lambda_1) + E_1(\lambda_2 \nabla \lambda_3 - \lambda_3 \nabla \lambda_2) \tag{A.2}$$

where E_1, E_2 and E_3 are constants whose physical meaning will shortly become clear.

We consider edge 3; anywhere on edge 3, $\lambda_3 \equiv 0$, and $\nabla \lambda_3$ is perpendicular to it. Finding the tangential component of the field on edge 3, we obtain:

$$\begin{aligned} \hat{e}_3 \cdot \boldsymbol{E} &= E_3(\lambda_1 \nabla \lambda_2 - \lambda_2 \nabla \lambda_1) + E_2 \cdot 0 + E_1 \cdot 0 \\ &= \boldsymbol{E}_{\tan}\Big|_{\text{edge}_3} \end{aligned} \tag{A.3}$$

where the second and third terms are zero due either to $\lambda_3 = 0$ on edge 3 or $\nabla \lambda_3$ being perpendicular to this edge.

Using the sine rule for triangles (that the ratios of edge lengths and sines of opposite angles are equal) and Eq. (A.1), and the geometrical meaning of the dot product, we find

$$\begin{aligned} \boldsymbol{E}_{\tan}\Big|_{\text{edge}_3} &= E_3 \left(\lambda_1 \frac{\ell_2}{2A} \hat{n}_2 \cdot \hat{e}_3 - \lambda_2 \frac{\ell_1}{2A} \hat{n}_1 \cdot \hat{e}_3 \right) \\ &= E_3 \frac{1}{2A} (\lambda_1 \ell_2 \sin \theta_1 + \lambda_2 \ell_1 \sin \theta_2) \\ &= E_3 \frac{1}{2A} \ell_2 \sin \theta_1 (\lambda_1 + \lambda_2) \\ &= E_3 \frac{1}{2A} \ell_2 \sin \theta_1 \end{aligned} \tag{A.4}$$

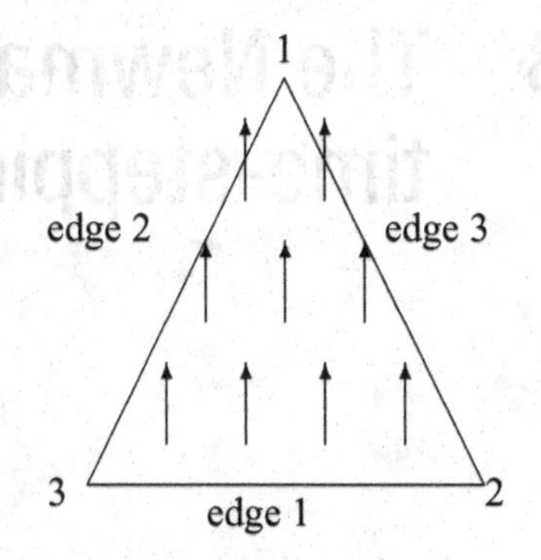

Figure A1 $\nabla\lambda_1$.

where the identity $\lambda_1 + \lambda_2 + \lambda_3 \equiv 1$ has been used in the last line (noting that $\lambda_3 = 0$ on this edge). $\sin\theta_i$ is the included angle at vertex i. Clearly, this is *constant*; now it is clear that E_3 is the tangential field on edge 3. Similar results follow for E_1 and edge 1, and E_2 and edge 2.

It can be simplified further by noting (from Fig. A1) that $\ell_2 \sin\theta_1$ is just the height of the triangle above edge 3. Since the area of the triangle is $A = (1/2)\ell h = (1/2)\ell_3\ell_2 \sin\theta_1$, it follows that

$$\boldsymbol{E}_{\tan}\Big|_{\text{edge}_3} = \frac{E_3}{\ell_3} \tag{A.5}$$

This is a well-known result, derived independently here. The general form for edge i is

$$\boldsymbol{E}_{\tan}\Big|_{\text{edge}_i} = \frac{E_i}{\ell_i} \tag{A.6}$$

If the vector basis function includes the edge length, as some published versions have, then the result is

$$\boldsymbol{E}_{\tan}\Big|_{\text{edge}_i} = E_i \tag{A.7}$$

Appendix B The Newmark-β time-stepping algorithm

The Newmark-β algorithm is rather more challenging to derive than is generally indicated in the literature, and its derivation is worth outlining. Most references cite the original paper by Newmark [1], which perhaps surprisingly does *not* derive the recurrence relationship, Eq. (12.23), which is generally associated with the name. This recurrence relationship was first given in a much later and very important paper by Zienkiewicz [2] published almost 20 years after the original method appeared. It is worth outlining the formulation, since it underlies the timestepping approach implemented and does not appear to be available anywhere apart from Zienkiewicz's paper, which can be difficult to obtain.

The method is *only* relevant to the following differential equation representing a general second-order system with damping:

$$M\ddot{x} + C\dot{x} + Kx + f = 0 \tag{B.1}$$

It was derived for structural mechanics, where x is the displacement,[1] and $\dot{x}$ and $\ddot{x}$ the velocity and acceleration respectively. It is also based on a Taylor series expansion. For discrete samples at $t = n\Delta t$ and $t = (n+1)\Delta t$, the Taylor series expansion of the first derivative is

$$\dot{x}_{n+1} = \dot{x}_n + \ddot{x}_n \Delta t + \dddot{x}_n \frac{\Delta t^2}{2} + \cdots \tag{B.2}$$

Newmark proposed that for sufficiently smooth functions this can be evaluated as

$$\dot{x}_{n+1} = \dot{x}_n + \widehat{\ddot{x}} \Delta t \tag{B.3}$$

where $\widehat{\ddot{x}}$ represents some value of $\ddot{x}$ (in structural dynamics, the acceleration) intermediate between $\ddot{x}_n$ and $\ddot{x}_{n+1}$. This is where the parameter γ in Newmark's scheme is introduced:

$$\dot{x}_{n+1} = \dot{x}_n + (1-\gamma)\ddot{x}_n \Delta t + \gamma \ddot{x}_{n+1} \Delta t \tag{B.4}$$

Clearly, this is a second-order accurate scheme (for sufficiently smooth functions). The Newmark-β scheme uses $\gamma = 1/2$, hence the approximation of the second differential places equal weight on the values at n and $n+1$. The function itself (in structural dynamics, the displacement) is approximated in a similar fashion, although in this case

[1] The extension to two and three dimensions is straightforward, x is replaced by $\boldsymbol{x}$.

retaining an additional term in the Taylor series:

$$x_{n+1} = x_n + \dot{x}_n \Delta t + (1 - 2\beta)\ddot{x}_n \Delta t^2/2 + 2\beta \ddot{x}_{n+1} \Delta t^2/2 \tag{B.5}$$

Note that this is *not* the time integral of the approximate velocity, but rather the expansion of the displacement.

Most textbooks which discuss the technique indicate that by writing Eq. (B.1) at timestep $n + 1$

$$M\ddot{x}_{n+1} + C\dot{x}_{n+1} + Kx_{n+1} + f_{n+1} = 0 \tag{B.6}$$

and by also using Eqs. (B.4) and (B.5), one obtains values for x_{n+1}, $\dot{x}_{n+1}$ and $\ddot{x}_{n+1}$ in terms of x_n, $\dot{x}_n$ and $\ddot{x}_n$ and this is what Newmark implied in his original paper. This, however, is *not* the desired recurrence relation, Eq. (12.23). Zienkiewicz indicates the process required to obtain this. One writes the governing equation, Eq. (B.1), additionally at the timesteps n and $n - 1$; further, the integration formulas, Eqs. (B.4) and (B.5), are written at timestep $n - 1, n$. This provides seven equations in nine unknowns (three displacements, three velocities and three accelerations) from which all the velocities and accelerations can be eliminated to produce the conventional recurrence scheme:

$$\begin{aligned}
&\left[M + \gamma \Delta t C + \beta \Delta t^2 K\right] x_{n+1} \\
&\quad + \left[-2M + (1 - 2\gamma)\Delta t C + \left(\frac{1}{2} + \gamma - 2\beta\right) \Delta t^2 K\right] x_n \\
&\quad + \left[M + (-1 + \gamma)\Delta t C + \left(\frac{1}{2} - \gamma + 2\beta\right) \Delta t^2 K\right] x_{n-1} \\
&\quad + (\beta \Delta t^2) f_{n+1} \\
&\quad + \left(\frac{1}{2} + \gamma - 2\beta\right) f_n \Delta t^2 + \left(\frac{1}{2} - \gamma + 2\beta\right) f_{n-1} \Delta t^2 = 0
\end{aligned} \tag{B.7}$$

The derivation as outlined above does not appear ever to have been published, only the results.

Importantly, Zienkiewicz then proposed that this recurrence relation can alternatively be derived by applying a weighted residual process to Eq. (B.1). In addition to providing an independent check of Eq. (B.7), this procedure permits a far more general approach to the problem, and proceeds as follows. Firstly, x is approximated by the three-term expansion:

$$x \approx \sum_i N_i x_i, \qquad i = n - 1, n, n + 1 \tag{B.8}$$

Obviously, this will support a second-order expansion in time, as required by the second-order derivative in Eq. (B.1). The shape functions N_i (which represent the temporal expansion functions) are the usual node-based quadratic functions and are given in detail in [2, Eq. (10)]. It is further assumed that x_n and x_{n-1} are known, and that the only unknown is x_{n+1}. Hence only one weighting function is required. Replacing the interval $[-\Delta t; \Delta t]$ with the normalized variable $-1 \leq \xi = t/\Delta t \leq 1$, Zienkiewicz shows that if we identify

$$\gamma = \left[\int_{-1}^{1} W\xi \, d\xi \Big/ \int_{-1}^{1} W \, d\xi\right] + \frac{1}{2}$$

$$\beta = \frac{1}{2}\int_{-1}^{1} W\xi(1+\xi)\,d\xi \Big/ \int_{-1}^{1} W\,d\xi = \frac{1}{2}\left(\gamma - \frac{1}{2}\right) + \frac{1}{2}\int_{-1}^{1} W\xi^2\,d\xi \Big/ \int_{-1}^{1} W\,d\xi \tag{B.9}$$

then we obtain Eq. (B.7). This is a very useful result, since it makes the approximations involved far clearer. It also permits us to extend the Newmark scheme if necessary. Zienkiewicz used the result to show how a variety of weighting functions yield different three "timestations" timestepping schemes, of which the Newmark scheme is the most general. For instance, with $\gamma = 1/2$ and $\beta = 0$, the weighting function is a Dirac delta at $t = n$, and the central difference scheme results. The Newmark-β scheme, on the limit of stability with $\gamma = 1/2$ and $\beta = 1/4$, corresponds to the "average acceleration" scheme and the weighting function is the linear function $|\xi|$, zero at the center of the interval ($t = n$) and unity at the ends of the interval ($n - 1$ and $n + 1$) [2, Fig. 1]. It is also possible to produce higher-order schemes. Using cubic functions, for instance, a third-order scheme can be derived with four time-stations and Zienkiewicz also outlines this.

References

[1] N. M. Newmark, "A method of computation for structural dynamics," *J. Eng. Mech. Div., Am. Soc. Civil Eng.*, **85**, 67–94, July 1959.

[2] O. C. Zienkiewicz, "A new look at the Newmark, Houboult and other time stepping formulas. a weighted residual approach," *Earthquake Eng. Struct. Dynamics*, **5**, 413–418, 1977.

Appendix C On the convergence of the MoM

Throughout this book, checking convergence numerically has been continually emphasized. However, we have not discussed the more theoretical issues of whether the underlying numerical formulations are indeed convergent, in the sense that the approximate numerical solution f^N of the continuous operator equation $Lf = g$ has the property $f^N \to f$ as $N \to \infty$. The aim of this appendix is to give a brief summary of the current status of this – which readers may be surprised to learn is far from a closed subject.

With the FDTD, the Lax equivalence theorem (discussed in Chapter 2) provides us with confidence that refining the FDTD mesh will indeed result in a convergent solution. With the FEM, work in applied mechanics has provided a rich set of convergence results – although we should note that convergence for high-frequency electromagnetics problems is often in terms of the energy norm, as discussed in Chapter 12. This is a slightly weaker statement of convergence, since the energy norm does not satisfy all the properties of the norm. Also, these proofs are usually in terms of interpolation error; as has been noted, dispersion (or pollution) error is a different problem specific to the differential equation based solvers, but can usually be controlled by adequate meshing. (Integral equation formulations using exact Green functions do not suffer from this problem of cumulative error resulting from dispersion error [1, p. 200].)

However, with the MoM, the problem has been studied somewhat less, presumably since the Green function is specific to electromagnetics. Rather surprisingly, only one form of operator has been rigorously shown to be convergent. (A summary may be found in [1, Chapter 5], which we summarize very briefly here.) This is the "identity plus compact" operator, of which the (two-dimensional) TE MFIE is an example. Proofs follow either via Galerkin's method, or via degenerate kernel analysis. Other types of operators are "compact" (the TM EFIE) and "unbounded" (the TE EFIE) – for neither of these do rigorous convergence proofs currently exist. (Incidentally, this nomenclature derives from the behavior of the eigenvalues of the operators.)

On the one hand, this is a somewhat disturbing situation, since important engineering designs are based on a field of mathematics which it transpires is far from complete. On the other hand, over forty years of development of the MoM has produced methods which have solved an enormous number of practical engineering problems with great accuracy, so it would appear most likely that what is missing is a convergence proof, rather than a fundamental problem. It would be satisfying were such proofs to be provided – or if they exist, popularized in the engineering literature. Here, we can but quote Peterson *et al.* [1, p. 224]:

Our previous experience with integral equation formulations supports the notion that, if constructed with sufficient care, numerical solutions appear to converge under much more general conditions. Despite this observation, the authors are not aware of more general convergence proofs applicable to the specific integral operators arising in electromagnetic scattering.

On the subject of convergence, another topic which has aroused controversy is whether the Galerkin formulation is superior to other forms of testing procedure. The controversy arose because the far-zone characteristics of the antenna or scatterer can be expressed as quadratic functionals of the surface current, which can sometimes be defined in such a way that they have a stationary point at the true solution. The work of Peterson *et al.* [1, Section 5.12] has shed new light on this matter: they have shown that provided the testing functions have similar accuracy properties as the basis functions, the overall error from either a true stationary functional (as can be obtained using a Galerkin procedure) or a general continuous functional form is of similar magnitude. They took this further, by numerical tests using high-order spline basis and testing functions; their results support the contention that the error is actually a function of the combined order of basis (P) and testing function (Q), and that a Galerkin solution with $P = Q$ is no more accurate than a non-Galerkin solution with the same total $P + Q$.

Reference

[1] A. F. Peterson, S. L. Ray and R. Mittra, *Computational Methods for Electromagnetics*. Oxford & New York: Oxford University Press and IEEE Press, 1998.

Appendix D Useful formulas for simplex coordinates

Basic properties

On a triangle:

$$\lambda_1 + \lambda_2 + \lambda_3 = 1 \tag{D.1}$$

On a tetrahedron:

$$\lambda_1 + \lambda_2 + \lambda_3 + \lambda_4 = 1 \tag{D.2}$$

Integration

Integration over a triangle:

$$\iint_S \lambda_1^i \lambda_2^j \lambda_3^k \, dS = \frac{2!\, i!\, j!\, k!}{(2+i+j+k)!} A \tag{D.3}$$

where A is the area of the triangle.

Integration over a tetrahedron:

$$\iiint_V \lambda_1^i \lambda_2^j \lambda_3^k \lambda_4^l \, dS = \frac{3!\, i!\, j!\, k!\, l!}{(3+i+j+k+l)!} V \tag{D.4}$$

where V is the volume of the tetrahedron.

Gradient

Gradient on a triangle:

$$\nabla \lambda_i = \frac{l_i}{2A} \hat{n}_i \tag{D.5}$$

with A the area of the triangle, l_i the length of edge i and $\hat{n}_i$ the normal on edge i, pointing into the triangle.

Gradient on a tetrahedron:

$$\nabla\lambda_i = \frac{\boldsymbol{A}_i}{3V} \tag{D.6}$$

with V the volume of the tetrahedron and $\boldsymbol{A}_i$ the area of face $\{j, k, l\}$, with normal pointing into the tetrahedron.

Appendix E Web resources

These sites, which include a number of commercial companies, were correct as of the date of publication – websites do change from time to time. This list is far from exhaustive, but gives a flavor of the variety of CEM products on offer, as well as the international technology base in this regard.

Ansoft Corporation
A Pittsburgh, USA-based company specializing in commercial FEM code suites, recently bought out by ANSYS (who market FEM codes for many fields of engineering application).
URL: `http://www.ansoft.com/`

Antenna Magus
An antenna expert design system, interfacing seamlessly with FEKO and MWS.
URL: `http://www.antennamagus.com/`

Applied Computational Electromagnetics Society
An organization supporting the development, validation, and distribution of numerical EM modelling codes. Presently hosted by the University of Mississippi. Contains a number of very useful CEM links, including links to the public domain code NEC2.
URL: `http://aces.ee.olemiss.edu/`

Computer Simulation Technology
Based in Darmstadt, Germany, this company specializes in commercial Finite Integration Technique (largely FDTD) code suites, in particular MWS.
URL: `http://www.cst.de/` or `http://www.cst-world.com/`

COMSOL
A Swedish company, their main product is Multiphysics. The name is self-explanatory: a multi-physics FEM solver.
URL: `http://www.comsol.se/`

EMSS (Electromagnetic Software and Systems)
Originally based in Stellenbosch, South Africa, this company now also has a German branch and US offices. Their main product is FEKO.
URL: `http://www.emss.co.za/` or `http://www.feko.info/`

MININEC website
EM Scientific, Inc market a professional version of this code.
URL: `http://www.emsci.com/`

NEC2 homepage
An unofficial homepage with a number of a number of links, as well as much of the NEC2 documentation.
URL: `http://www.nec2.org/`.

Netlib
The premium international collection of mathematical software (including the industry-standard LAPACK), papers, and databases.
URL: `http://www.netlib.org/`

REMCON
A US company, offering XFDTD, an FDTD-based package.
URL: `http://www.remcom.com/`

Poynting Software
Another South African company, based in Johannesburg, offering SuperNEC.
URL: `http://www.supernec.com/`

SEMCAD
Based in Switzerland, this company offers a 3D FDTD solver, with supporting GUI.
URL: `http://www.semcad.com/`

Top 500
The Top 500 supercomputer sites in the world, updated biennially.
URL: `http://www.top500.org/`

Vector Fields Software (now trading as Cobham Technical Services)
A UK-based company. Their main offering in this field is CONCERTO, which includes an FDTD 3D solver, a MoM frequency domain module and a 3D geometric modeller.
URL: `http://www.vectorfields.com/`

WIPL-D
This company, orignally based in Serbia, but with strong US links, offers the eponymous WIPL-D code. This is a MoM-based tool, incorporating a treatment for both metallic and dielectric structures.
URL: `http://www.wipl-d.com/`

Zeland Software
Based in California, their best-known product is probably IE3D, a planar and 3D MoM simulation package. It is widely used for microstrip structure simulation.
URL: `http://www.zeland.com/`

Appendix F MATLAB files supporting this text

A substantial number of MATLAB files are available on the publisher's website www.cambridge.org/Davidson, supporting the theoretical material in this book. This appendix briefly describes the functionality of each main script. (Most have a number of functions associated with them.)

FDTD codes

- `fdtd_1D_demo` This implements the 1D FDTD theory described in Section 2.4, for a transmission line with a sinusoidal excitation.
- `fdtd_1D_WB_demo` This implements the 1D FDTD theory described in Section 2.5, for a transmission line with a wideband excitation.
- `fdtd_2D_demo` This implements the 2D FDTD theory described in Section 3.2, for TE scattering from a PEC cylinder.
- `fdtd_2D_pml_demo` This implements the PML described in Section 3.3.
- `fdtd_3D_demo` This implements the 3D FDTD theory described in Section 3.4, for eigenanalysis of a PEC cavity.

MoM codes

- `static_mom` This implements the 1D MoM theory described in Section 4.2, for a charged thin wire.
- `thin_dipole` This implements the 1D MoM theory described in Section 4.3, for radiation from a center-fed thin dipole.
- `MoM_2D_TM` This implements the MoM theory described in Section 4.6, for TM scattering from a PEC cylinder.
- `MoM3D_demo` This implements the MoM theory described in Section 6.3, for scattering from a flat plate using the mixed potential EFIE using RWG basis functions.
- `MoM_Som` This computes the input impedance of a thin printed dipole using the Sommerfeld MPIE MoM formulation of Chapter 7.

FEM codes

- `FEM_1D_1st` This implements the 1D FEM theory described in Section 9.2, for a transmission line, using first-order nodal elements.
- `Static2D` This implements the 2D FEM theory described in Section 10.2, for the quasi-static analysis of microstrip using first-order nodal triangular elements.
- `Eigen2D` This implements the FEM theory described in Section 10.8, for waveguide eigenanalysis, using triangular CT/LN vector elements.
- `Eigen2D_LTQN` This implements the above using triangular LT/QN vector elements, as in Section 10.8.
- `Eigen3D_CTLN` This implements the FEM theory described in Section 11.1, for cavity eigenanalysis, using tetrahedral CT/LN vector elements.
- `Eigen3D_LTQN` This implements the above using tetrahedral LT/QN vector elements, as described in Section 11.2.

Index

Printed in the United States
By Bookmasters

Printed in the United States
By Bookmasters